Handbook of Fisheries

Handbook of Fisheries

Edited by Halsey Parkinson

Syrawood
PUBLISHING HOUSE

New York

Published by Syrawood Publishing House,
750 Third Avenue, 9th Floor,
New York, NY 10017, USA
www.syrawoodpublishinghouse.com

Handbook of Fisheries
Edited by Halsey Parkinson

International Standard Book Number: 978-1-68286-724-2 (Hardback)

Cataloging-in-Publication Data

Handbook of fisheries / edited by Halsey Parkinson.
    p. cm.
Includes bibliographical references and index.
ISBN 978-1-68286-724-2
1. Fisheries. 2. Fishery sciences. I. Parkinson, Halsey.
SH331 .H36 2019
639.2--dc23

TABLE OF CONTENTS

Preface IX

Chapter 1   Expression of phospholipase C β1 in olive flounder (*Paralichthys olivaceus*)
              following external stress stimulation 1
              Soo Ji Woo, Hee Young Jang, Hyung Ho Lee and Joon Ki Chung

Chapter 2   Re-evaluation of the optimum dietary protein level for maximum growth
              of juvenile barred knifejaw *Oplegnathus fasciatus* reared in cages 11
              Kang-Woong Kim, Mohammad Moniruzzaman, Kyoung-Duck Kim,
              Hyon Sob Han, Hyeonho Yun, Seunghan Lee and Sungchul C. Bai

Chapter 3   Genetic identification of anisakid nematodes isolated from largehead hairtail
              (*Trichiurus japonicus*) in Korea 17
              Jeong-Ho Kim, Woo-Hwa Nam and Chan-Hyeok Jeon

Chapter 4   Apparent digestibility coefficients of the extruded pellet diets containing
              various fish meals for olive flounder, *Paralichthys olivaceus* 25
              Md Mostafizur Rahman, Hyon-Sob Han, Kang-Woong Kim, Kyoung-Duck Kim,
              Bong-Joo Lee and Sang-Min Lee

Chapter 5   Antioxidant and angiotensin I-converting enzyme inhibitory activities
              of northern shrimp (*Pandalus borealis*) by-products hydrolysate
              by enzymatic hydrolysis... 33
              Sang-Bo Kim, Na Young Yoon, Kil-Bo Shim and Chi-Won Lim

Chapter 6   Photoinactivation of major bacterial pathogens in aquaculture 39
              Heyong Jin Roh, Ahran Kim, Gyoung Sik Kang and Do-Hyung Kim

Chapter 7   Production optimization of flying fish roe analogs using calcium alginate
              hydrogel beads 46
              Bom-Bi Ha, Eun-Hee Jo, Suengmok Cho and Seon-Bong Kim

Chapter 8   Effects of extruded pellet and moist pellet on growth performance,
              body composition, and hematology of juvenile olive flounder,
              Paralichthys olivaceus 53
              Seunghan Lee, Mohammad Moniruzzaman, Jinho Bae, Minji Seong, Yu-jin Song,
              Bakshish Dosanjh and Sungchul C. Bai

Chapter 9   Cloning, expression, and activity of type IV antifreeze protein from cultured
              subtropical olive flounder (*Paralichthys olivaceus*) 59
              Jong Kyu Lee and Hak Jun Kim

Chapter 10   Galatheoid squat lobsters (Crustacea: Decapoda: Anomura) from Korean waters 66
              Jung Nyun Kim, Mi Hyang Kim, Jung Hwa Choi and Yang Jae Im

Chapter 11   **A new record of the Axiid shrimp** *Balssaxius habereri* **(Balss, 1913)**
(Crustacea: Decapoda: Axiidea) in Korean waters 78
Jung Nyun Kim, Jung Hwa Choi, Yang Jae Im and Hyun-su Jo

Chapter 12   *Miuraea migitae*, **a new record of the order Bangiales**
(Bangiophyceae, Rhodophyta) from Korea 81
Young Ho Koh, Hyung Woo Lee and Myung Sook Kim

Chapter 13   **Appropriate feeding for early juvenile stages of eunicid polychaete**
Marphysa sanguinea 86
Kyeong Hun Kim, Byoung Kwon Kim, Sung Kyun Kim, War War Phoo,
B. A. Venmathi Maran and Chang-Hoon Kim

Chapter 14   **Taurine supplementation in diet for olive flounder at low water temperature** 95
Joo-Min Kim, G. H. T. Malintha, G. L. B. E. Gunathilaka, Chorong Lee,
Min-Gi Kim, Bong-Joo Lee, Jeong-Dae Kim and Kyeong-Jun Lee

Chapter 15   **Evaluation of visible fluorescent elastomer tags implanted in marine medaka,**
Oryzias dancena 103
Jae Hyun Im, Hyun Woo Gil, In-Seok Park, Cheol Young Choi, Tae Ho Lee,
Kwang Yeol Yoo, Chi Hong Kim and Bong Seok Kim

Chapter 16   **Protective effects of alginate-free residue of sea tangle against**
hyperlipidemic and oxidant activities in rats 113
Mi-Jin Yim, Grace Choi, Jeong Min Lee, Soon-Yeong Cho and Dae-Sung Lee

Chapter 17   **Characterization and expression profiles of aquaporins (AQPs) 1a and 3a in mud**
loach *Misgurnus mizolepis* **after experimental challenges** 119
Sang Yoon Lee, Yoon Kwon Nam and Yi Kyung Kim

Chapter 18   **Effects of different algae in diet on growth and interleukin (IL)-10 production**
of juvenile sea cucumber *Apostichopus japonicus* 128
Md Anisuzzaman, Jeong U-Cheol, Jin Feng, Choi Jong-Kuk, Kabery Kamrunnahar,
Lee Da-In, Yu Hak Sun and Kang Seok-Joong

Chapter 19   **New record of an economic marine alga,** *Ahnfeltiopsis concinna*, **in Korea** 136
Pil Joon Kang and Ki Wan Nam

Chapter 20   **Comparison of different ploidy detection methods in** *Oncorhynchus mykiss*,
the rainbow trout 141
Hong Seab Kim, Ki-Hwa Chung and Jung-Ho Son

Chapter 21   **Effects of water physico-chemical parameters on tilapia (***Oreochromis niloticus***)**
growth in earthen ponds in Teso North Sub-County, Busia County 148
Agano J. Makori, Paul O. Abuom, Raphael Kapiyo, Douglas N. Anyona and
Gabriel O. Dida

Chapter 22   **Gene structure and expression characteristics of liver-expressed antimicrobial peptide-2 isoforms in mud loach (*Misgurnus mizolepis*, Cypriniformes)** 158
Sang Yoon Lee and Yoon Kwon Nam

Chapter 23   **The effect of feeding frequency, water temperature, and stocking density on the growth of river puffer *Takifugu obscurus* reared in a zero-exchange water system** 169
Gwang-Yeol Yoo and Jeong-Yeol Lee

Chapter 24   **Osteological development of wild-captured larvae and a juvenile *Sebastes koreanus* (Pisces, Scorpaenoidei) from the Yellow Sea** 176
Hyo Jae Yu and Jin-Koo Kim

Chapter 25   **Bioaccumulation, alterations of metallothionein, and antioxidant enzymes in the mullet *Mugil cephalus* exposed to hexavalent chromium** 188
Eun Young Min, Tae Young Ahn and Ju-Chan Kang

Chapter 26   **Evaluation of reference genes for RT-qPCR study in abalone *Haliotis discus hannai* during heavy metal overload stress** 195
Sang Yoon Lee and Yoon Kwon Nam

Permissions

List of Contributors

Index

PREFACE

The main aim of this book is to educate learners and enhance their research focus by presenting diverse topics covering this vast field. This is an advanced book which compiles significant studies by distinguished experts in the area of analysis. This book addresses successive solutions to the challenges arising in the area of application, along with it; the book provides scope for future developments.

Fisheries science is the scientific study of fisheries. It integrates the principles of biology, ecology, limnology and marine science to develop a holistic understanding of fisheries. It is concerned with capturing wild fish as well as raising fish through fish farming. The primary sources of fisheries are oceans and seas. Fisheries can be classified as industrial scale, recreational and artisanal fisheries. The management of fisheries is concerned with the sustainable exploitation of resources. This book provides comprehensive insights into this area of study. It unfolds the innovative aspects of fisheries science which will be crucial for the progress of this field in the future. This book is appropriate for students seeking detailed information in this discipline as well as for experts.

It was a great honour to edit this book, though there were challenges, as it involved a lot of communication and networking between me and the editorial team. However, the end result was this all-inclusive book covering diverse themes in the field.

Finally, it is important to acknowledge the efforts of the contributors for their excellent chapters, through which a wide variety of issues have been addressed. I would also like to thank my colleagues for their valuable feedback during the making of this book.

Editor

Expression of phospholipase C β1 in olive flounder (*Paralichthys olivaceus*) following external stress stimulation

Soo Ji Woo[1†], Hee Young Jang[1†], Hyung Ho Lee[2] and Joon Ki Chung[1*]

Abstract

In this study, to clarify the function of PoPLC-β1, in response to stress challenge, we examined the PoPLC-β1 expression pattern in response to external stress (pathogen-associated molecular pathogen challenge and environmental challenge including temperature and salinity). PoPLC-β1 expression analysis of tissue from olive flounder showed that the messenger RNA (mRNA) was predominantly expressed in the brain, heart, eye, liver, spleen, and stomach. We also tested the mRNA expression of the PoPLC-β1 in the spleen and kidney of olive flounder by RT-PCR and real-time PCR following stimulation with lipopolysaccharide (LPS), concanavalin A (ConA), or polyinosinic:polycytidylic acid (PolyI:C) and compared with the inflammatory cytokines IL-1b and IL-6 in the stimulated flounder tissues. Each of the spleen and kidney and mRNA transcripts of PoPLC-β1 were increased 30- and 10-fold than normal tissue at 1–6 h post injection (HPI) with PolyI:C when the expression of PoPLC-β1 transcript was similar to LPS and ConA. We also tested the expression of PoPLC-β1 in response to temperature and salinity stress. The expression of PoPLC-β1 also was affected by temperature and salinity stress. Our results provide clear evidence that the olive flounder PLC-β1 signal pathways may play a critical role in immune function at the cellular level and in inflammation reactions. In addition, PLC-β1 appears to act as an oxidative-stress suppressor to prevent cell damage in fish.

Keywords: Phospholipase C beta 1 (PLC-β1), Olive flounder, *Paralichthys olivaceus*, Stress response, mRNA expression

Background

In marine organisms, environmental changes in the water cause a variety of physiological stress responses, such as changes in oxidative stress, decline in immunity, and changes in plasma components. Among these responses, changes in the immune system are particularly important in the context of fish disease outbreaks in the face of environmental change. Physical factors, such as extreme salinity and temperature changes, affect growth, metabolism, osmoregulation reproductive function, and immune function (Ackerman et al. 2000; Borski et al. 1994; Sampaio and Bianchini 2002; Bly and Clem 1992; Bowden 2008; Schreck et al. 1989; Britoa et al. 2000). Furthermore, stress induced by changes in salinity and

temperature has been associated with enhanced generation of reactive oxygen species (ROS) in fish, which may seriously affect their immune function and antioxidant defense abilities (Paital and Chainy 2010; An et al. 2010; Kim and Kang 2015).

Phosphoinositide-specific phospholipase C (PLC) enzymes have been shown to play a role in immune responses in mammals; however, little is known about their function in marine organisms in response to stress. PLC is a family of enzymes that hydrolyzes the inner membrane element phosphatidylinositol-4,5-bisphosphate (PIP_2) to produce inositol 1,4,5-trisphosphate (IP_3) and diacylglycerol (DAG) (Rhee 2001). IP_3 binds to IP_3-specific receptors on the membrane of the endoplasmic reticulum (ER), inducing the release of Ca^{2+} level from the intracellular cytoplasm. While DAG initiates the activation of Ras proteins along with this Ca^{2+} release, which in turn activates protein kinase C

* Correspondence: jkchung@pknu.ac.kr
†Equal contributors
[1]Department of Aquatic Life Medicine, Pukyong National University, Busan 608-737, South Korea
Full list of author information is available at the end of the article

(PKC) (Bunney and Katan 2011). Thirteen PLC isozymes have been cloned from mammalian species to date and have been classified into six classes according to their structures and mechanism of actions: PLC-β (1–4), PLC-γ (1 and 2), PLC-δ (1, 3, and 4), PLC-ε, PLC-ζ, and the recently identified PLC-η (1 and 2) (Rebecchi and Pentyala 2000). The PLC-β subfamily (1–4) has a significantly conserved structure of an N-terminal pleckstrin homology (PH) domain, four EF hand repeats, a triose phosphate isomerase (TIM)-like barrel domain split into an X and Y catalytic domain, a C2 domain, and a C-terminal (CT) domain (Otaegui et al. 2010). The PH domain plays a minor role in association to the membrane (Tall et al. 1997), whereas its most important role is in mediating protein-protein interactions. The EF hand repeats support the loop related to stimulation of GTP hydrolysis. The catalytic X and Y domains are necessary for the catalysis of PIP_2 to IP_3 and DAG (Song et al. 2001). The C2 domain serves as an intra- and intermolecular regulatory binding position, and the CT domain, the most unique feature, interacts with the GTP-bound α subunit of heterotrimeric G proteins to stimulate PLC-β (Singer et al. 2002).

The identified signaling molecules of PLCs have also been suggested to perform a role in the regulation of chemokine-mediated cell migration. There is substantial evidence suggesting a role of PLC-β in the immune response. PLC-β activity is started by G protein coupled receptor (GPCR)-mediated signal pathway (Kim et al. 2011). PLC-β plays an essential role in T cell chemotaxis (Bach et al. 2007) and has previously been shown to participate in several aspects of lymphocyte function, including cell proliferation, rescue from apoptosis, and $CD4^+$ and $CD8^+$ T cell differentiation (Sasaki et al. 2000; Ward and Cantrell 2001).

The stimulatory effects of lipopolysaccharide (LPS), concanavalin A (ConA), and polyinosinic:polycytidylic acid (PolyI:C) induced the activity of PKC and other mitogens. Previous studies have shown that LPS, ConA, and PolyI:C stimulation induced phosphoinositide-specific PLC (PI-PLC) and phosphatidylcholine-specific PLC (PC-PLC) expression of the kidney and spleen. (Chen et al. 1998; Wang et al. 1998; Schütze et al. 1992). Thus, the stimulatory effect contributes to both the innate and adaptive immune responses. However, unlike the signaling pathways of mammalian PLC isoforms, the intra and intercellular signaling pathways of marine organisms remain relatively elusive.

In particular, at the moment, only a few studies have examined the function of PLC-β1 in response to stress challenge. Here, therefore, we evaluated changes in PLC-β1 expression in response to external stress, elicited by pathogen-associated molecular pathogen (PAMP) challenge and environmental challenge (temperature and salinity) in the olive flounder (*Paralichthys olivaceus*).

Methods

Stimulation with LPS, ConA, and PolyI:C in the kidney and spleen

LPS is a major component of the gram-negative cell wall. ConA triggers a release mechanism similar to that elicited by the specific worm allergen (Keller 1973), and PolyI:C is structurally similar to a double-stranded RNA virus and potent inducer of interferon (IFN) (Akiyama et al. 1985; Manetti et al. 1995). To study the immune response of PLC-β1 in the olive flounder (*P. olivaceus*), we performed immune challenge experiments by using commercially available LPS (Sigma), ConA (Sigma), and PolyI:C (Sigma). Healthy juvenile *P. olivaceus* (40 ± 12 g) were anesthetized with MS-222 (3-aminobenzoic acid ethyl ester; Sigma, USA) and were intraperitoneally injected with 500 μl of LPS (100 μg/ml), ConA (100 μg/ml), PolyI:C (100 μg/ml), and PBS (500 μl; as a control), as per the body mass of the fish. Target tissues were removed from *P. olivaceus* at 0, 1, 3, 6, and 24 h after injection. All procedures were performed according to the American Veterinary Medical Association guidelines on euthanasia.

Changes in temperature and salinity

P. olivaceus with a mean weight of 32 ± 7 g were obtained from a commercial fish farm. The fish were conditioned in aerated 200-l tanks at 20 °C for a week prior to experimentation. During the acclimatization, the fish were fed with a commercial pelleted diet twice a day (at 10:00 and 17:00). Half of the water (30‰ ± 0.2) in the tank was changed every day. Salinity was decreased at 5‰ per hour by adding fresh water (Bio Safe 50 ml/ 200 l water) and was regulated from 30‰ to zero. The water temperature was rapidly increased from 20 to 30 °C at a rate of 2.5 °C/h. The brain, gill, heart, liver, spleen, stomach, intestine, kidney, and muscle tissue were analyzed after 24 h. No mortality was observed following injection and environmental stress.

mRNA isolation and cDNA synthesis from *P. olivaceus*

The total RNA was isolated using the TRIzol® kit (Invitrogen), according to the method described by the manufacturer. Complementary DNA (cDNA) was synthesized from this isolated mRNA by using the Transcriptor First Strand cDNA Synthesis Kit (Roche) and then was used as the template for amplification. Purified RNA was quantified on the basis of its optical density at 260 nm by using a UV spectrophotometer (Ultrospec 6300 pro, Amersham Biosciences). Two micrograms of total RNA was reverse-transcribed with an oligo (dT)18 and random hexamer primers and Superscript™ III reverse transcriptase (Invitrogen, USA), as per the manufacturer's instruction. Reverse transcription was carried out at 42 °C for 60 min.

Expression studies by reverse transcription PCR

In order to analyze the tissue expression of the PoPLC-β1 mRNA, reverse transcription (RT)-PCR was performed. The 18SrRNA and β-actin genes of *P. olivaceus* were used as internal controls. All the PCR cycles were performed as follows: 94 °C for 5 min, 25–30 cycles (25 cycles for 18SrRNA and β-actin; 30 cycles for PoPLC-β1) of 94 °C for 30 s, 58 °C for 30 s, 72 °C for 20 s, and a final 7-min elongation step at 72 °C. The amplified PCR products were separated on 1.0 % agarose/TAE gels containing ethidium bromide and visualized with a Gel Doc image analysis system (Bio-Rad, USA). The resultant products were purified via agarose gel extraction (QIAquick® Gel Extraction kit) and sequenced (COSMO Co. Ltd., DNA Sequencing Service, Seoul, Korea).

Quantitative expression analysis of PoPLC-β1

Total RNA from various tissues was prepared as per the previously described method. 18SrRNA from a constitutive expression gene of β-actin was used as the internal control to verify the real-time PCR reaction. Quantitative real-time PCR for the tissue-specific expression analysis of PoPLC-β1 with the gene-specific primers was conducted using a LightCycler® 480 II SYBR Green (Roche, Switzerland). The SYBR Green RT-PCR assay was performed by the method described previously (Guan et al. 2007) using the ΔΔCt method, and the relative quantitative values were expressed in accordance with the 2–ΔΔCt method. In addition to investigating stimulation, the experiments also analyzed the expression of interleukins 1β and 6, as they are a lymphocyte mitogen and proinflammatory cytokines (Benveniste 1998).

Results

Confirmation of the PoPLC-β1 sequence

After confirming the sequence of PoPLC-β1 with those previously published sequences (Seo et al. 2011), we compared the characteristic regions of PoPLC-β1. The

deduced product was examined using a specific primer for RT-PCR and quantitative real-time PCR (Table 1).

Tissue-specific distribution of PoPLC-β1 mRNAs

The expression pattern of the PoPLC-β1 gene in the brain, eye, gills, heart, esophagus, liver, spleen, pyloric ceca, stomach, intestine, kidney, and muscle was analyzed by RT-PCR and real-time PCR. All tissues showed detectable levels of PoPLC-β1 mRNA expression, but expression was especially higher in the brain and heart tissues (Fig. 1).

Changes in PoPLC-β1 expression in response to stress
Effects of LPS, ConA, and PolyI:C stimulation on PoPLC-β1

In mammals, G protein has been shown to modulate the signal transduction of both PLC-β isoforms (Charo et al. 1994; Franci et al. 1995). To evaluate general changes in the expression levels of PLC-β1 in *P. olivaceus* related to the immune response, we used RT-PCR and real-time PCR to compare PoPLC-β1 expression to the interleukin gene expression profiles (IL-1β, IL-6) in the spleen and kidney, following stimulation of LPS, ConA, and PolyI:C, which play a major role in various disease processes in teleost fish (Figs. 2, 3, and 4).

LPS and ConA stimulation did not induce significant changes in the expression level of PoPLC-β1 (Figs. 5 and 6). However, PolyI:C injection in spleen and kidney led to a dramatic increase in the expression of PoPLC-β1 at 3 h post injection, which was increased by more than 10-fold compared to the control groups and a time dependent increase was observed until 24 h post injection in the spleen and kidney (Fig. 7). As shown in Figs. 4 and 7, PoPLC-β1 expression significantly increased along with the IL-1β and IL-6 expression in the spleen and kidney following PolyI:C stimulation. The mRNA levels of IL-1β and IL-6 in the infected spleen and kidney tissues were sharply increased and reached a peak at 3 or 6 h post injection.

Table 1 Oligonucleotide primers used for the RT-PCR and quantitative real-time PCR to analyze mRNA expression of PLC-β1

Primer name	5'-3' sequence	Description
β-actin-F	GACATGGAGAAGATCTGGCA	For RT-PCR and real-time qRT-PCR GenBank accession no AU090737
β-actin-R	ATGTCCTGCTCGAAGTCCAG	
18 s rRNA-F	GTTGGTGGAGCGATTTGTCTGG	For RT-PCR and real-time qRT-PCR GenBank accession no DQ267937
18 s rRNA-R	CATCTAAGGGCATCACAGACCTG	
IL-1β-F	AACAGCCAAGGCAAAGATTG	For RT-PCR and real-time qRT-PCR GenBank accession no AB070835
IL-1β-F	AATGTCCAGCTCCTCCTTCA	
IL-6-F	TGGGCATCAACTCGGGCTAC	For RT-PCR and real-time qRT-PCR GenBank accession no DQ267937
IL-6-R	CGAAGGCCACAGATTGGTC	
PoPLC-β1-F	CAACGTGATGGAGCAAAGAG	For RT-PCR and real-time qRT-PCR
PoPLC-β1-R	GGTGTGTGGCTCAGTCCGTT	

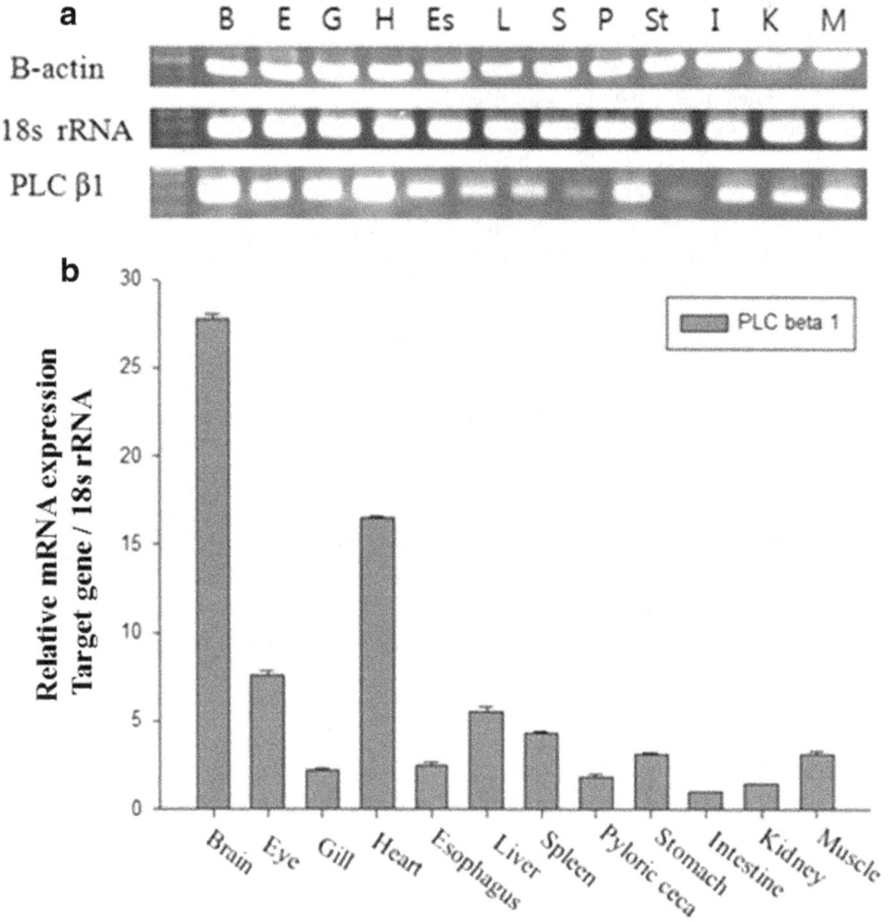

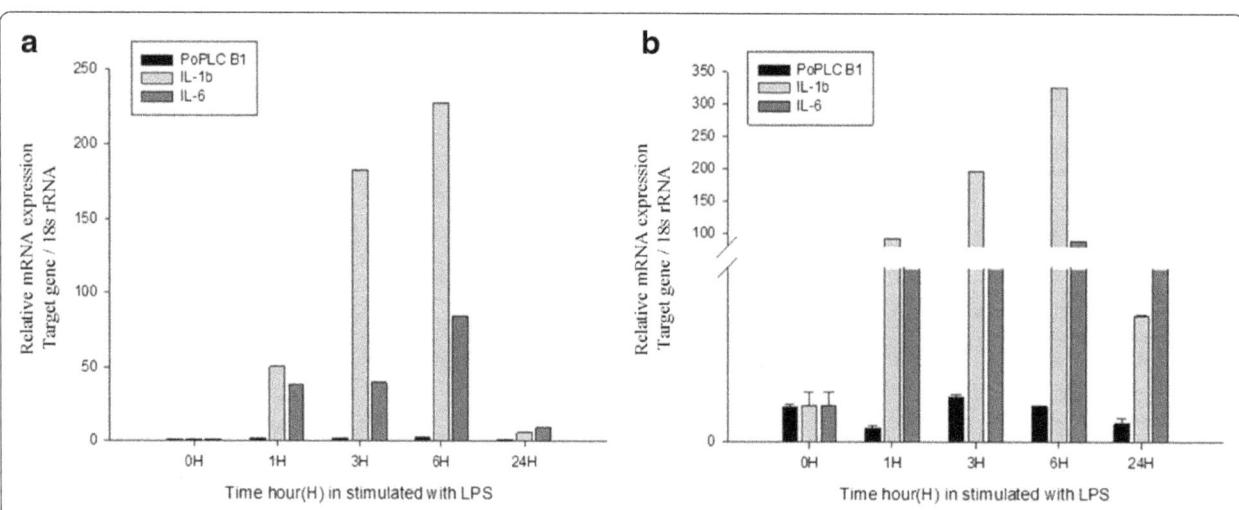

Fig. 1 Tissue-specific distribution of the PoPLC-β1 mRNA. **a** RT-PCR expression analysis of PoPLC-β1 in flounder tissues. **b** Relative expression of PoPLC-β1 in flounder tissues. The olive flounder 18SrRNA gene was used as a reference gene to normalize the expression mRNA levels between sample tissues. Each value is the average of three replicate samples and data are shown as means ± S.E

Fig. 2 Real-time PCR analysis of PoPLC-β1, IL-1β, IL-6, and 18SrRNA following stimulation with LPS from the **a** spleen and **b** kidney tissue at 0, 1, 3, 6, and 24 h

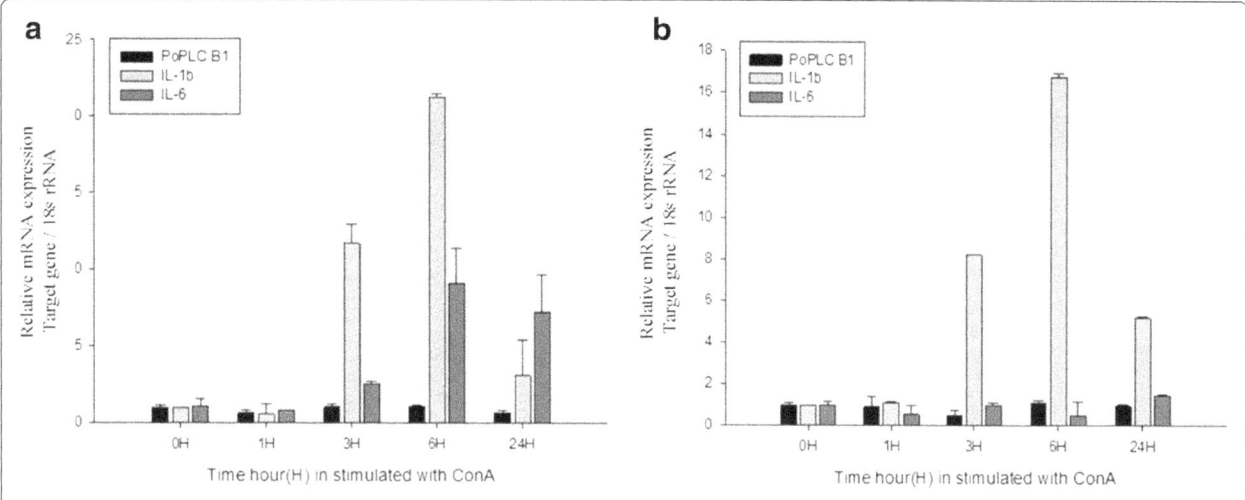

Fig. 3 Real-time PCR analysis of PoPLC-β1, IL-1β, IL-6, and 18SrRNA following stimulation with ConA from the **a** spleen and **b** kidney tissue at 0, 1, 3, 6, and 24 h

Environmental stress

Changes of water temperature and salinity both affected the expression of PoPLC-β1. To our surprise, salinity stress induced a greater effect than thermal stress. In the salinity stress condition, the expression level of PoPLC-β1 was increased by twofold in both the brain and heart and was slightly increased in the gills (Fig. 8a). Under temperature stress, PoPLC-β1 expression increased by twofold in the heart compared to controls (Fig. 8b).

Discussion

In higher vertebrates, immunity can be broadly classified into adaptive immunity and innate immunity. Innate immunity is crucial for the recruitment of non-specific immune cells, including lymphocytes, dendritic cells, granulocytes, and macrophages, to the sites of infection by microorganisms and foreign substances. After these immune cells recognize the infection and kill the foreign cells, the adaptive immune system kicks in, which involves specific antigen presentation to B and T lymphocytes to fight the pathogen (Bonizzi and Karin 2004). Phospholipid second messengers produced by PLC-β have been shown to play a significant role in T lymphocyte chemotaxis (Smit et al. 2003). Furthermore, the PLC-β subfamily appears to be essential for the generation of IP$_3$, DAG, and mobilization of Ca^{2+} (Bach et al. 2007). Moreover, PLC-β plays a critical role in activation of chemoattractant-induced integrin, cell-substrate adhesion, and in the movement and antibody production of B lymphocytes (Kawakami and Xiao 2013). Thus, PLC-βs are involved in the specialization and activation of

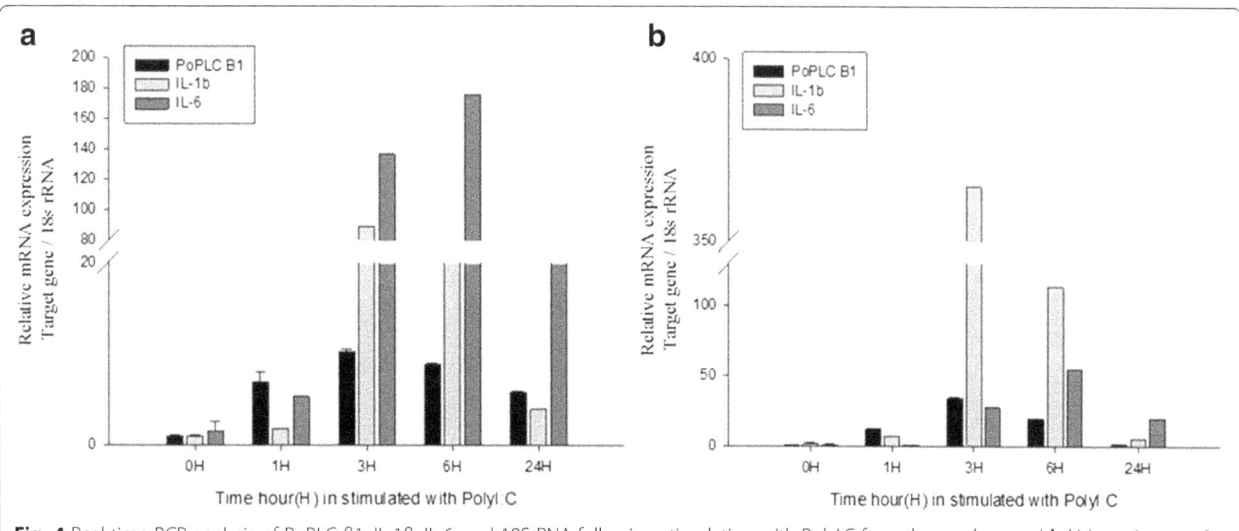

Fig. 4 Real-time PCR analysis of PoPLC-β1, IL-1β, IL-6, and 18SrRNA following stimulation with PolyI:C from the **a** spleen and **b** kidney tissue at 0, 1, 3, 6, and 24 h

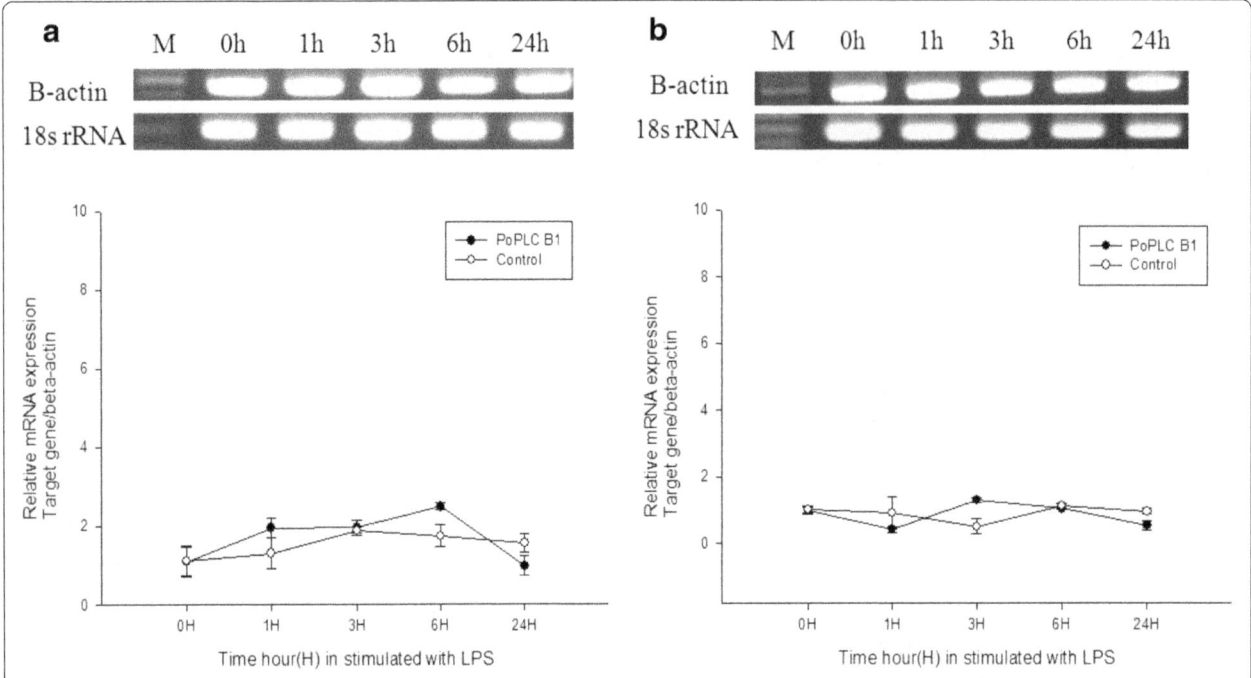

Fig. 5 RT-PCR and real-time PCR analysis of PoPLC-β1 following stimulation with LPS and PBS (control) treatment from the **a** spleen and **b** kidney tissue at 0, 1, 3, 6, and 24 h

immune cells, which regulate both the innate and adaptive immune responses (Forster et al. 1998). Previous studies have demonstrated that PLC-βs show tissue-specific expression patterns and are regulated by G protein. Indeed, in mammals, PLC-β1 is expressed in a wide range of tissues

and cell types (Kawakami and Xiao 2013), including cardiac tissues (Jalili et al. 1999) and the brain (Suh et al. 2008).

PAMPs are characteristic molecular patterns that are highly conserved in different microorganisms (Savan and Sakai 2002). PAMPs, including LPS, ConA, and PolyI:C,

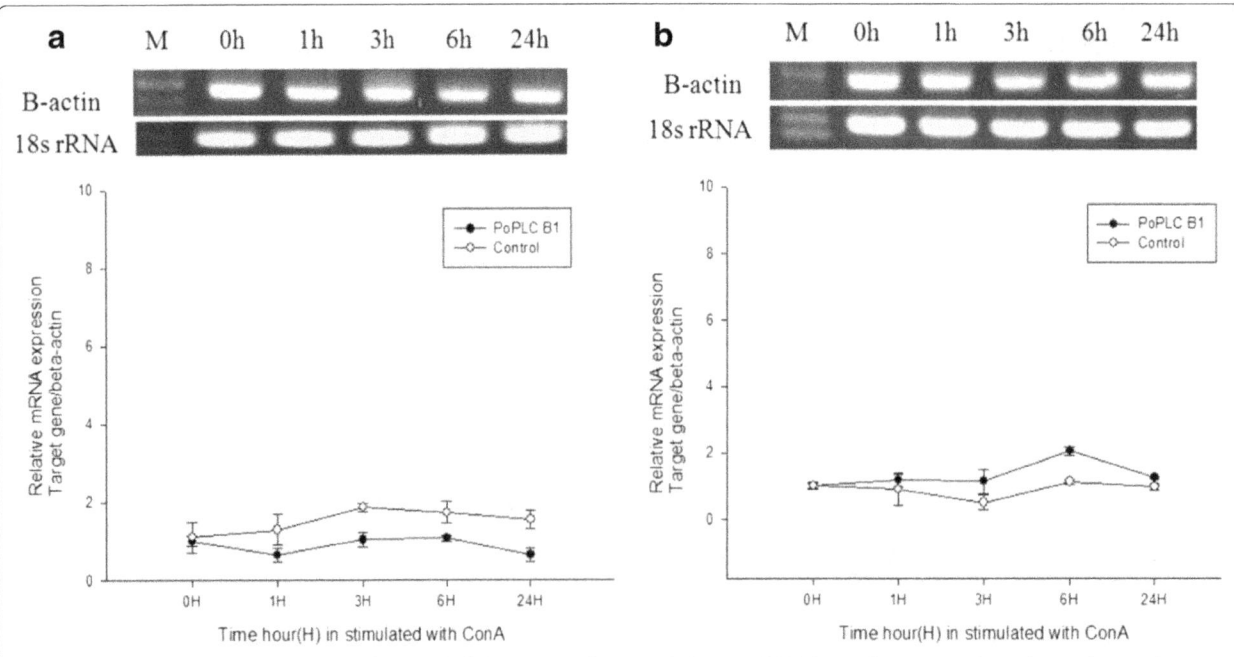

Fig. 6 RT-PCR and real-time PCR analysis of PoPLC-β1 following stimulation with ConA and PBS (control) treatment from the **a** spleen and **b** kidney tissue at 0, 1, 3, 6, and 24 h

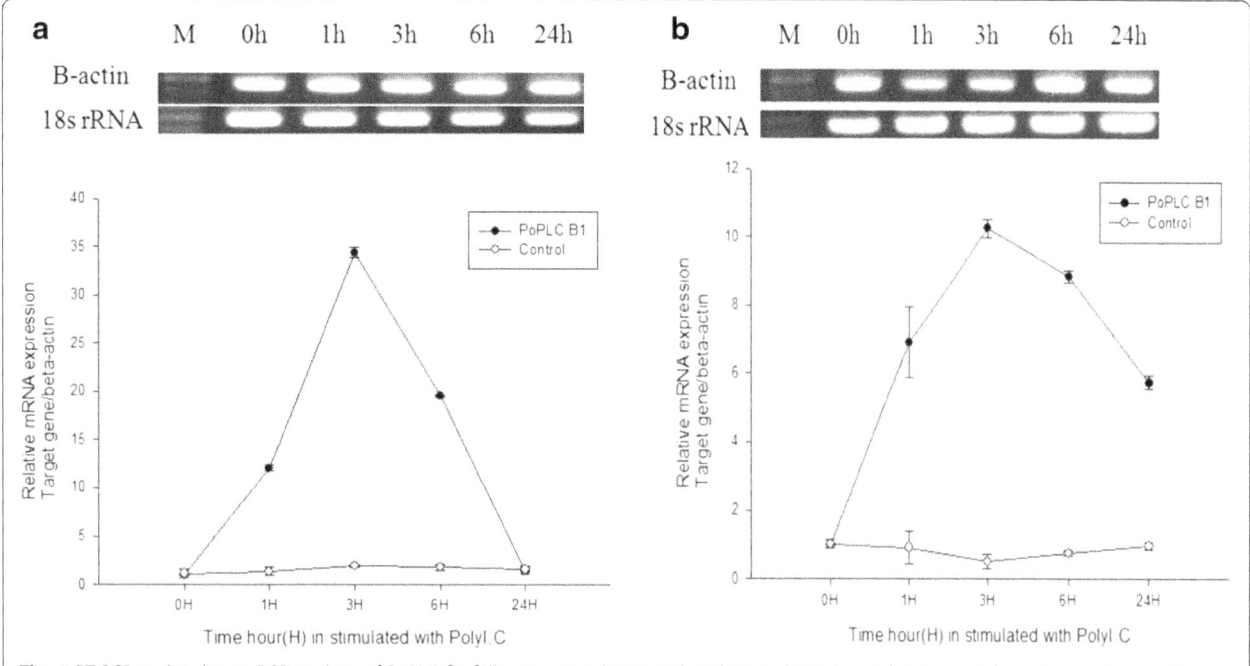

Fig. 7 RT-PCR and real-time PCR analysis of PoPLC-β1 following stimulation with PolyI:C and PBS (control) treatment from the **a** spleen and **b** kidney tissue at 0, 1, 3, 6, and 24 h

activate immune system responses in the host. In particular, LPS is a constituent of the outer membrane of gram-negative bacteria and induces the proliferation of B lymphocytes (Estepa and Coll 1992). ConA is a plant lectin isolated from jack bean (*Canavalia ensiformis)* that was originally identified as a mitogen for T lymphocytes. It has been shown to activate lymphocyte proliferation by the production of lymphokines and monokines and to activate tumor necrosis factor-alpha, macrophages, and neutrophils (Xue and Bigio 2005). ConA also triggers a mechanism similar to that induced by the specific worm allergen

(Panitch and McFarlin 1977). PolyI:C is a synthetic double-stranded RNA polymer that enhances the cytotoxicity of natural killer (NK) cells and macrophages (Biron 1997), increases interferon gamma (IFN-γ) production, and promotes lymphocyte adhesion to the endothelium (Doukas et al. 1994). Therefore, stimulation with LPS, ConA, and PolyI:C represent a challenge by bacteria, parasites, and viruses and were thus used to analyze the effects of immune challenge on PoPLC-β1.

There is significant evidence for cell-mediated responses after PolyI:C stimulation. NK cells respond to PolyI:C by

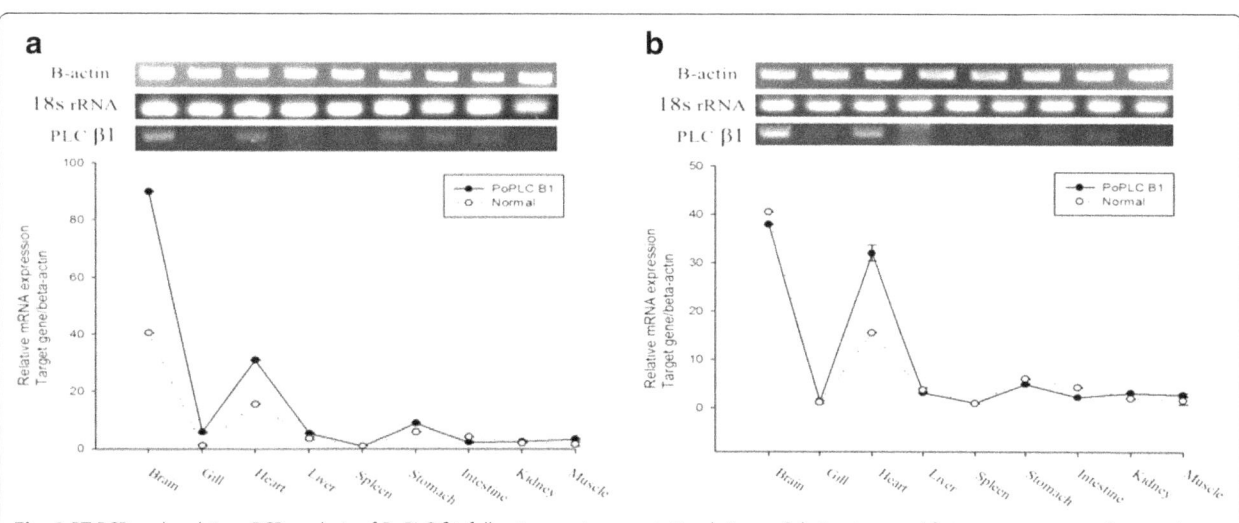

Fig. 8 RT-PCR and real-time PCR analysis of PoPLC-β1 following environment stimulation. **a** Salinity stress and **b** temperature stress from various flounder tissues

releasing proinflammatory cytokines such as IL-6 and IL-8, as well as the antiviral cytokine IFN-α (Schmidt et al. 2004; Longhi et al. 2009). In addition, PolyI:C induces inflammation and long-lasting T cell immunity (Salem et al. 2006; Trumpfheller et al. 2008; Stahl-Hennig et al. 2009). In the present study, we observed a rapid reaction in the spleen and kidney, within 3 h following PolyI:C infection. The mRNA levels of IL-1β and IL-6 from the LPS, ConA, and PolyI:C infected organs were substantially upregulated and reached a peak at 3 or 6 h post injection. Our results indicated that PLC-β1 activation by pathogens induces inflammation cytokines including IL-1β and IL-6 production in the spleen and kidney tissues. Proinflammatory cytokines such as IL-1β and IL-6 can modulate the PLC-β and PIP_2 signaling pathway (Zini et al. 2003). PLC-β activation pathway may possess the ability to modulate proinflammatory responses leading to downstream of NF-kB and interaction with G protein subunits (Townsend and Emala 2013). Also, PLC-β has been known as a central mediator of chemotaxis and plays a critical role in the recruitment of leukocyte to inflammatory tissues (Lehmann et al. 2008). Therefore, these results shed insight into the role of PLC-β1 in immune responses against pathogenic organism and provide a molecular foundation for further research, and monitoring in various teleost infection and immunity systems.

Stress factors in fish can be divided into two main categories: physical factors and chemical factors (Beckmann et al. 1990). The physical factors include salinity, temperature, density, and dissolved oxygen. In particular, salinity and temperature changes affect growth, reproduction, metabolism, osmoregulation, and immune function (Ackerman et al. 2000). Stress induced by changes in salinity has been associated with increased formation of ROS, which may considerably influence immune function and lead to oxidative stress (Paital and Chainy 2010; Shin et al. 2010). Furthermore, rapid changes in water temperature have been shown to enhance oxidative stress by increasing the amount of ROS generation in fish (Halliwell and Gutteridge 1999; Liu et al. 2007). Moreover, cells increase their antioxidant defense levels and associated enzymes in response to drastic temperature and salinity changes (Martinez-Alvarez et al. 2005). We observed increases in PoPLC-β1 in response to temperature and salinity stress in the brain and heart. There is evidence that the PLC-β1 level distinctly increases in response to ROS accumulation, which may play a role in protection against the damage induced by the physiological changes accompanying the stress response. This pattern of PLC-β1 expression may explain the process of protection from damage in these physiological responses (Yasuda et al. 2008). Since PLC-β1 can directly activate the PKCα and PKCβII isozymes, it is possible that overexpression of PLC-β1 could protect cells from oxidant-induced cell death via the activation of PKC (Lee et al. 2000).

Conclusions

In conclusion, our results showed that the PLC-β1 expression level increases in response to external stimuli mimicking immune challenge and stress in *P. olivaceus*. Research on the function of PLC-β1 and mechanisms controlling its expression in teleosts is limited, and our results provide clear evidence that the olive flounder PLC-β1 signal pathways may play a critical role in immune function at the cellular level and in inflammation reactions. In addition, PLC-β1 appears to act as an oxidative-stress suppressor to prevent cell damage in fish. Therefore, these results should serve as a foundation for further research on the piscine immune system and related signal pathways of lower vertebrates.

Abbreviations
ConA, concanavalin A; DAG, diacylglycerol; ER, endoplasmic reticulum; IP3, inositol 1,4,5-trisphosphate; LPS, lipopolysaccharide; PAMPs, pathogen-associated molecular pathogens; PH, pleckstrin homology; PIP2, phosphatidylinositol-4,5-bisphosphate PKC, protein kinase C; PLC, phosphoinositide-specific phospholipase C; PolyI:C, polyinosinic:polycytidylic acid; TIM, triose phosphate isomerase

Acknowledgements
This work was supported by a Research Grant of Pukyong National University (No 2014 0332).

Authors' contributions
SJ carried out the molecular genetic studies and manuscript writing. HY participated in the design of the study and data analysis. HH participated in the collection and assembly of the data. JK participated in its design and coordination and helped to draft the manuscript. All authors read and approved the final manuscript.

Competing interests
The authors declare that they have no competing interests.

Disclosure
The dataset(s) supporting the conclusions of this article is not included in the article.

Author details
[1]Department of Aquatic Life Medicine, Pukyong National University, Busan 608-737, South Korea. [2]Department of Biotechnology, Pukyong National University, Busan 608-737, South Korea.

References
Ackerman PA, Forsyth RB, Mazur CF, Iwama GK. Stress hormones and the cellular stress response in salmonids. Fish Physiol Biochem. 2000;23:327–36.

Akiyama Y, Stevenson GW, Schlick E, Matsushima K, Miller PJ, Stevenson HC. Differential ability of human blood monocyte subsets to release various cytokines. J Leukoc Biol. 1985;37:519–30.

An KW, Kim NN, Shin HS, Kil G-S, Choi CY. Profiles of antioxidant gene expression and physiological changes by thermal and hypoosmotic stresses in black porgy(Acanthopagrus schlegeli). Comp Biochem Physiol A Mol Integr Physiol. 2010;156:262–8.

Bach TL, Chen QM, Kerr WT, Wang Y, Lian L, Choi JK, et al. Phospholipase C β is critical for T cell chemotaxis. J Immunol. 2007;179:2223–7.

Beckmann RP, Mizzen LE, Welch WJ. Interaction of HSP70 with newly synthesized proteins: implications for protein folding and assembly. Science. 1990;248:850–4.

Benveniste EN. Cytokine actions in the central nervous system. Cytokine Growth Factor Rev. 1998;9:259–75.

Biron CA. Activation and function of natural killer cell responses during viral infections. Curr Opin Immunol. 1997;9:24–34.

Bly JE, Clem LW. Temperature and teleost immune functions. Fish Shellfish Immunol. 1992;2:159–71.

Bonizzi G, Karin M. The two NF-kB activation pathways and their role in innate and adaptive immunity. Trends Immunol. 2004;25:280–8.

Borski RJ, Yoshikawa JSM, Madsen SS, Nishioka RS, Zabetian C, Bern HA, et al. Effects of environmental salinity on pituitary growth hormone content and cell activity in the euryhaline tilapia Oreochromis mossambicus. Gen Comp Endocrinol. 1994;95:483–94.

Bowden TJ. Modulation of the immune system of fish by their environment. Fish Shellfish Immunol. 2008;25:373–83.

Britoa R, Chimal ME, Rosas C. Effect of salinity in survival, growth, and osmotic capacity of early juveniles of Farfantepenaeus brasiliensis (Decapoda: Penaeidae). J Exp Mar Biol Ecol. 2000;244:253–63.

Bunney TD, Katan M. PLC regulation: emerging pictures for molecular mechanisms. Trends Biochem Sci. 2011;36:88–96.

Charo IF, Myers SJ, Herman A, Franci C, Connolly AJ, Coughlin SR. Molecular cloning and functional expression of two monocyte chemoattractant protein 1 receptors reveals alternative splicing of the carboxyl-terminal tails. Proc Natl Acad Sci. 1994;91:2752–6.

Chen CC, Wang JK, Lin SB. Antisense oligonucleotides targeting protein kinase C-alpha, -beta I, or -delta but not –eta inhibit lipopolysaccharide-induced nitric oxide synthase expression in RAW 264.7 macrophages: involvement of a nuclear factor kappa B-dependent mechanism. J Immunol. 1998;161:6206–14.

Doukas J, Cutler AH, Mordes JP. Polyinosinic: polycytidylic acid is a potent activator of endothelial cells. Am J Pathol. 1994;145:137–47.

Estepa A, Coll JM. In vitro immunostimulants for optimal responses of kidney cells from healthy trout and from trout surviving viral haemorrhagic septicaemia virus disease. Fish Shellfish Immunol. 1992;2:53–68.

Forster R, Kremmer E, Schubel A, Breitfeld D, Kleinschmidt A, Nerl C, et al. Intracellular and surface expression of the HIV-1 coreceptor CXCR4/fusin on various leukocyte subsets: rapid internalization and recycling upon activation. J Immunol. 1998;160:1522–31.

Franci C, Wong LM, Van Damme J, Proost P, Charo IF. Monocyte chemoattractant protein-3, but not monocyte chemoattractant protein-2, is a functional ligand of the human monocyte chemoattractant protein-1 receptor. J Immunol. 1995;154:6511–7.

Guan ZB, Shui Y, Zhang SQ. Two related ligands of the TNF family, BAFF and APRIL, in rabbit: molecular cloning, 3D modeling, and tissue distribution. Cytokine. 2007;39:192–200.

Halliwell B, Gutteridge JMC. Free radicals in biology and medicine, vol. 3. Oxford: Oxford University Press; 1999. p. 86–187.

Jalili T, Takeishi Y, Walsh RA. Signal transduction during cardiac hypertrophy: the role of Gαq, PLC βI, and PKC. Cardiovasc Res. 1999;44:5–9.

Kawakami T, Xiao W. Phospholipase C-b in immune cells. Adv Biolog Regul. 2013;53:249–57.

Keller R. Concanavalin A, a model 'antigen' for the in vitro detection of cell-bound reaginic antibody in the rat. Clin Exp Immunol. 1973;13:139–47.

Kim JH, Kang JC. Oxidative stress, neurotoxicity, and non-specific immune responses in juvenile red sea bream, Pagrus major, exposed to different waterborne selenium concentrations. Chemosphere. 2015;135:46–52.

Kim JK, Lim S, Kim J, Kim S, Kim JH, Ryu SH, et al. Subtype-specific roles of phospholipase C-β via differential interactions with PDZ domain proteins. Adv Enzym Regul. 2011;51:138–51.

Lee YH, Kim SY, Kim JR, Yoh KT, Baek SH, Kim MJ, et al. Overexpression of phospholipase C beta1 protects NIH3T3 cells from oxidative stress induced cell death. Life Sci. 2000;67:827–37.

Lehmann DM, Seneviratne AMPB, Smrcka AV. Small molecule disruption of G protein βγ subunit signaling inhibits neutrophil chemotaxis and inflammation. Mol Pharmacol. 2008;73:410–8.

Liu Y, Wang WN, Wang AL, Wang JM, Sun RY. Effects of dietary vitamin E supplementation on antioxidant enzyme activities in Litopenaeus vannamei (Boone, 1931) exposed to acute salinity changes. Aquaculture. 2007;265:351–8.

Longhi MP, Trumpfheller C, Idoyaga J, Caskey M, Matos I, Kluger C, et al. Dendritic cells require a systemic type I interferon response to mature and induce CD4+ Th1 immunity with Poly IC as adjuvant. J Exp Med. 2009;206:1589–602.

Manetti R, Annunziato F, Tomasevic L, Gianno V, Parronchi P, Romagnani S, et al. Polyinosinic acid:polycytidylic acid promotes T helper type 1-specific immune responses by stimulating macrophage production of interferon-a and interleukin-12. Eur J Immunol. 1995;25:2656–60.

Martinez-Alvarez R, Morales AE, Sanz A. Antioxidant defenses in fish: biotic and abiotic factors. Rev Fish Biol Fish. 2005;15:75–88.

Otaegui D, Querejeta R, Arrieta A, Lazkano A, Bidaurrazaga A, Arriandiaga JR, et al. Phospholipase Cb4 isozyme is expressed in human, rat, and murine heart left ventricles and in HL-1 cardiomyocytes. Mol Cell Biochem. 2010;337:167–73.

Paital B, Chainy GBN. Antioxidant defenses and oxidative stress parameters in tissues of mud crab (Scylla serrata) with reference to changing salinity. Comp Biochem Physiol Part C: Toxicol Pharmacol. 2010;151:142-151.

Panitch HS, McFarlin DE. Experimental allergic encephalomyelitis: enhancement of cell-mediated transfer by concanavalin A. J Immunol. 1977;119:1134–7.

Rebecchi MJ, Pentyala SN. Structure, function, and control of phosphoinositide-specific phospholipase C. Physiol Rev. 2000;80:1291–335.

Rhee SG. Regulation of phosphoinositide-specific phospholipase C. Ann Rev Biochem. 2001;70:281–312.

Salem ML, El-Naggar SA, Kadima A, Gillanders WE, Cole DJ. The adjuvant effects of the toll-like receptor 3 ligand polyinosinic-cytidylic acid poly (I:C) on antigen-specific CD8+ T cell response are partially dependent on NK cells with the induction of a beneficial cytokine milieu. Vaccine. 2006;24:5119–32.

Sampaio LA, Bianchini A. Salinity effects on osmoregulation and growth of the euryhaline flounder Paralichthys orbignyanus. J Exp Mar Biol Ecol. 2002;269:187–96.

Sasaki T, Irie-Sasaki J, Jones TG, Oliveira-dos-Santos AJ, Stanford WL, Bolon B, et al. Function of PI3Kγ in thymocyte development, T cell activation, and neutrophil migration. Science. 2000;287:1040–6.

Savan R, Sakai M. Analysis of expressed sequence tags (EST) obtained from common carp, Cyprinus carpio L., head kidney cells after stimulation by two mitogens, lipopolysaccharide and concanavalin-A. Comp Biochem Physiol Part B: Biochem Mol Biol. 2002;131:71–82.

Schmidt KN, Leung B, Kwong M, Zarember KA, Satyal S, Navas TA, et al. APC-independent activation of NK cells by the Toll-like receptor 3 agonist double-stranded RNA. J Immunol. 2004;172:138–43.

Schreck CB, Solazzi MF, Johnson SL, Nickelson TE. Transportation stress affects performance of Coho Salmon, Oncorhynchus kisutch. Aquaculture. 1989;82:15–20.

Schütze S, Potthoff K, Machleidt T, Berkovic D, Wiegmann K, Krönke M. TNF activates NF-kB by phosphatidylcholine-specific phospholipase C-induced "acidic" sphingomyelin breakdown. Cell. 1992;71:765–76.

Seo JS, Jeon EJ, Park EM, Kim WJ, Kim NY, Lee EH, et al. Molecular cloning and characterization of PLCB1 (phospholipase C, beta 1) gene from the olive flounder, Paralichthys olivaceus. Genes Genomics. 2011;33:701–9.

Shin HS, Yoo JH, Min TS, Lee K-Y, Choi CY. The effects of quercetin on physiological characteristics and oxidative stress resistance in olive flounder, Paralichthys olivaceus. Asian-Australasian J Animal Sci. 2010;23:588–97.

Singer AU, Waldo GL, Harden TK, Sondek J. A unique fold of phospholipase C-beta mediates dimerization and interaction with G alpha q. Nat Struct Mol Biol. 2002;9:32–6.

Smit MJ, Verdijk P, van der Raaij-Helmer EM, Navis M, Hensbergen PJ, Leurs R. CXCR3-mediated chemotaxis of human T cells is regulated by a Gi- and phospholipase C-dependent pathway and not via activation of MEK/p44/p42 MAPK nor Akt/PI-3 kinase. Blood. 2003;102:1959–65.

Song C, Hu CD, Masago M, Kariyai K, Yamawaki-Kataoka Y, Shibatohge M, et al. Regulation of a novel human phospholipase C, PLCε, through membrane targeting by Ras. J Biol Chem. 2001;276:2752–7.

Stahl-Hennig C, Eisenblatter M, Jasny E, Rzehak T, Tenner-Racz K, Trumpfheller C, et al. Synthetic double-stranded RNAs are adjuvants for the induction of T helper 1 and humoral immune responses to human papillomavirus in Rhesus Macaques. PLoS Pathog. 2009;5:e1000373.

Suh PG, Park JI, Manzoli L, Peak JC, Katan M, Fukami K, et al. Multiple roles of phosphoinositide-specific phospholipase C isozymes. BMB Rep. 2008;41:415–34.

Tall E, Dormán G, Garcia P, Runnels L, Shah S, Chen J, et al. Phosphoinositide binding specificity among phospholipase C isozymes as determined by photo-cross-linking to novel substrate and product analogs. Biochem. 1997;36:7239–48.

Townsend EA, Emala CW. Quercetin acutely relaxes airway smooth muscle and potentiates β-agonist-induced relaxation via dual phosphodiesterase inhibition of PLCβ and PDE4. Am J Physiol-Lung Cell Mol Physiol. 2013;305:L396–403.

Trumpfheller C, Caskey M, Nchinda G, Longhi MP, Mizenina O, Huang Y, et al. The microbial mimic polyIC induces durable and protective CD4+ T cell immunity together with a dendritic cell targeted vaccine. Proc Natl Acad Sci U S A. 2008;105:2574–9.

Wang P, Toyoshima S, Osawa T. Concanavalin A-induced translocation of part of the GTP-binding activity from the membrane to the cytosol in murine thymocytes. J Biochem. 1998;104:169–72.

Ward SG, Cantrell DA. Phosphoinositide 3-kinases in T lymphocyte activation.
Curr Opin Immunol. 2001;13:332–8.

Xue M, Bigio MRD. Immune pre-activation exacerbates hemorrhagic brain injury
in immature mouse brain. J Neuroimmunol. 2005;165:75–82.

Yasuda E, Nagasawa K, Nishida K, Fujimoto S. Decreased expression of
phospholipase C-b1 protein in endoplasmic Reticulum stress-loaded neurons.
Biol Pharm Bull. 2008;31:719–21.

Zini N, Lisignoli G, Solimando L, Bavelloni A, Grassi F, Guidotti L, et al. IL1-β and
TNF-α induce changes in the nuclear polyphosphoinositide signalling system
in osteoblasts similar to that occurring in patients with rheumatoid arthritis:
an immunochemical and immunocytochemical study. Histochem Cell Biol.
2003;120:243–50.

Re-evaluation of the optimum dietary protein level for maximum growth of juvenile barred knifejaw *Oplegnathus fasciatus* reared in cages

Kang-Woong Kim[1], Mohammad Moniruzzaman[2], Kyoung-Duck Kim[1], Hyon Sob Han[1], Hyeonho Yun[2], Seunghan Lee[2] and Sungchul C. Bai[2*]

Abstract

We determined the optimum dietary protein level in juvenile barred knifejaw *Oplegnathus fasciatus* in cages. Five semi-purified isocaloric diets were formulated with white fish meal and casein-based diets to contain 35, 40, 45, 50, and 60 % crude protein (CP). Fish with an initial body weight of 7.1 ± 0.06 g (mean ± SD) were randomly distributed into 15 net cages (each size: 60 cm × 40 cm × 90 cm, $W \times L \times H$) as groups of 20 fish in triplicates. The fish were fed at apparent satiation level twice a day. After 8 weeks of feeding, the weight gain (WG) of fish fed 45, 50, and 60 % CP diets were significantly higher than those of fish fed 35 and 40 % CP diets. However, there were no significant differences in WG among fish fed 45, 50, and 60 % CP diets. Generally, feed efficiency (FE) and specific growth rate (SGR) showed a similar trend as WG. However, the protein efficiency ratio (PER) was inversely related to dietary protein levels. Energy retention efficiency increased with the increase of dietary protein levels by protein sparing from non-protein energy sources. Blood hematocrit content was not affected by dietary protein levels. However, a significantly lower amount of hemoglobin was found in fish fed 35 % CP than in fish fed 40, 45, 50, and 60 % CP diets. Fish fed 60 % CP showed the lowest survival rate than the fish fed 35, 40, 45, and 50 % CP diets. Broken-line analysis of WG showed the optimum dietary protein level was 45.2 % with 18.8 kJ/g diet for juvenile barred knifejaw. This study has potential implication for the successful cage culture of barred knifejaw.

Keywords: *Oplegnathus fasciatus*, Barred knifejaw, Optimum protein level, Fish growth, Fish feeds

Background

Protein is the basic component of all animal tissues, and it constitutes about 65–75 % in fish tissues on a dry matter basis (Wilson 2002). Dietary protein has great impact on the growth and body composition of fish (Lovell 1989) because it provides the essential amino acids for body protein synthesis as well as energy for growth and maintenance. However, protein is one of the most expensive major nutrients in fish feeds, and its inclusion in fish diets has significant

effects on operational costs in aquaculture (NRC 1993). It is well documented that increase of dietary protein can lead to improved fish production especially in the case of carnivorous fish species. However, the excess level of dietary protein will be used for energy and will lead to an increase in ammonia excretion which ultimately deteriorates the fish culture water that may be harmful for fish growth (Catacutan and Coloso 1995; Tibbetts et al. 2000). Furthermore, an inadequate level of dietary protein in aqua feeds can result in the stunted growth of fish. But the objective of proper feed formulation is to produce cost-effective feeds with maximum fish production at a minimum protein level (Halver and Hardy 2002). Therefore, it is imperative to determine the optimum

* Correspondence: scbai@pknu.ac.kr
[2]Department of Marine Bio-materials and Aquaculture/Feeds and Foods Nutrition Research Center, Pukyong National University, Busan 608-737, Republic of Korea
Full list of author information is available at the end of the article

dietary protein level in aquaculture diets for achieving cost-effective maximum growth of fish together with improved culture environment.

Barred knifejaw, *Oplegnathus fasciatus*, belonging to the family Oplegnathidae, is a popular food fish and an economically important cage aquaculture species in Korea as well as in East Asia (Meng et al. 1995). In 2014, its aquaculture production in Korea was approximately 884 metric tons (National Statistical Office 2014) which mostly came from cage aquaculture. It has high market value and consumer demand. The dietary protein requirements of several important aquaculture fish species have been determined including channel catfish, *Ictalurus punctatus* (Garling and Wilson 1976); Asian sea bass, *Lates calcarifer* (Catacutan and Coloso 1995); Indian major carp, *Labeo rohita* (Das et al. 1991); Nile tilapia, *Oreochromis niloticus* (El-Sayed and Teshima 1992); olive flounder, *Paralichthys olivaceus* (Kim et al. 2005); Korean rockfish, *Sebastes schlegelii* (Kim et al. 2004), and Japanese eel, *Anguilla japonica* (Okorie et al. 2007); black sea bass, *Centropristis striata* (Alam et al. 2008); black sea bream, *Sparus macrocephalus* (Zhang et al. 2010); and silver pomfret, *Pampus argenteus* (Hossain et al. 2010). For most of the cultured species, dietary protein requirement has been found to be between 30 and 55 % of the diet depending on the species, fish size, dietary protein sources, and environmental condition (Hepher 1988; NRC 1993). Therefore, it is important to estimate dietary protein requirements in fish under different conditions (Luo et al. 2004). In a previous study, Kang et al. (1998) reported that the optimum dietary protein and lipid levels for parrot fish or barred knifejaw are 46 and 16 %, respectively, in a protein and lipid ratio experiment in tanks. However, for the first time, in this study, we aimed to re-evaluate the optimum dietary protein level in diets for the juvenile barred knifejaw with a fixed dietary energy level in cage culture condition, which is widely being practiced in Korea.

Methods

Diet formulation

Composition of the experimental diets is shown in Table 1. Five experimental diets containing white fish meal and casein as the main protein sources were formulated with protein levels of 35, 40, 45, 50, or 60 % at the expense of α-potato starch and squid liver oil. The diets were formulated to be isocaloric, containing 18.8 kJ/g energy based on calculation (Garling and Wilson 1976; NRC 1993). In the diets, wheat flour and α-potato starch were used as carbohydrate sources to adjust the energy content of the experimental diets. Carboxymethylcellulose (CMC) of high viscosity was used

Table 1 Composition of the experimental diets (% of dry matter basis)

Ingredients	Protein level in the diets (%)				
	35	40	45	50	60
White fish meal[a]	25.00	20.00	15.00	10.00	5.00
Casein[b]	15.80	24.50	33.30	42.00	55.80
α-potato starch[a]	28.94	25.45	21.80	18.25	9.70
Wheat flour[c]	7.00	7.00	7.00	7.00	7.00
Squid liver oil[d]	16.21	16.00	15.85	15.70	15.45
Vitamin premix.[e]	3.00	3.00	3.00	3.00	3.00
Mineral premix.[f]	3.00	3.00	3.00	3.00	3.00
Vitamin C[g]	0.05	0.05	0.05	0.05	0.05
Carboxymethylcellulose[b]	1.00	1.00	1.00	1.00	1.00

[a]Suhyup feed Co. Ltd., Uiryeong, Korea
[b]United States Biochemical, Cleveland, OH, USA
[c]Young Nam Flour Mills Co., Busan, Korea
[d]Ewha Oil Co., Ltd., Busan, Korea
[e]Contains (as mg/kg in diets) ascorbic acid, 300; dl-calcium pantothenate, 150; choline bitartrate, 3000; inositol, 150; menadione, 6; niacin, 150; pyridoxine.HCl, 15; riboflavin, 30; thiamine mononitrate, 15; dl-α-tocopherol acetate, 201; retinyl acetate, 6; biotin, 1.5; folic acid, 5.4; B_{12}, 0.06
[f]Contains (as mg/kg in diet) Al, 1.2; Ca, 5000; Cl, 100; Cu, 5.1; Co, 9.9; Na, 1280; Mg, 520; P, 5000; K, 4300; Zn, 27; Fe, 40.2; I, 4.6; Se, 0.2; Mn, 9.1
[g]Vitamin C: L-ascorbyl-2-monophosphate, 35 % ascorbic acid activity (Hoffmann La Roche, Switzerland)

as binder. The experimental diets were also fortified with vitamin and mineral premixes (Table 1). The actual nutrients and amino acid contents in experimental diets are shown in Table 2. All ingredients were mixed and pelleted by a pelleting machine without heating using a 2-mm-diameter module (Baokyong Commercial Co., Busan, Korea). After air drying for 48 h, all the pellets were broken up, sieved into the proper pellet size, sealed, and kept at −20 °C until used.

Fish and feeding trial

Juvenile barred knifejaw, *O. fasciatus*, were transported from Geoje Marine Hatchery (Geoje, Korea) of the National Fisheries Research and Development Institute, Republic of Korea, to Youngchang Fisheries Farm (Tongyeong, Republic of Korea). Before commencement of the experiment, barred knifejaw were acclimated in a circular concrete tank containing 5000 L water where they were fed commercial diet for 2 weeks at the fish farm. A feeding trial was conducted in a rectangular concrete tank (5 m × 5 m × 3 m, $W \times L \times H$) having a flow-through system, containing 15 floating net cages (each size: 60 cm × 40 cm × 90 cm, $W \times L \times H$) in triplicates of each experimental diet. Flow rate was adjusted to ensure adequate circulation of seawater. Supplemental aeration was provided to maintain dissolved oxygen levels near saturation. Water temperature was maintained 19 °C at the beginning of the feeding trial and was 22 °C at the end of the feeding trial according to the normal changes of natural water temperature. Twenty fish weighing 7.1 ± 0.06 g (mean ±

Table 2 Proximate composition and amino acid contents of experimental diets (% of dry matter basis)

Parameters	Protein levels (%)				
	35	40	45	50	60
Moisture	25.9	26.9	25.8	27.1	26.9
Crude protein	35.0	40.6	44.8	50.2	60.1
Crude lipid	18.8	18.1	17.6	17.0	16.3
Crude ash	8.8	9.3	9.8	10.2	8.0
Estimated energy (kJ g^{-1})	18.7	18.8	18.8	18.7	18.8
P/E ratio	18.7	21.6	23.8	26.8	31.9
Amino acids					
Arg	2.11	2.36	2.43	3.52	3.61
His	1.08	1.28	1.38	1.52	1.58
Lys	3.51	4.11	4.36	5.16	5.63
Leu	3.12	3.72	3.88	4.23	4.38
Ile	2.52	2.81	3.25	3.56	3.67
Met	1.21	1.38	1.51	1.58	1.76
Cys	0.33	0.42	0.48	0.58	0.72
Phe	1.72	2.02	2.21	2.62	2.78
Tyr	1.63	1.79	2.23	2.38	2.54
Thr	1.32	1.43	2.36	2.53	2.69
Trp	0.36	0.56	0.68	0.72	0.83
Val	2.63	2.76	3.53	3.66	3.88

Values are means of duplicate samples of each diet

SD) were randomly distributed to each net cage. Fish were fed one of the experimental diets twice (0900 and 1800 h) a day to apparent satiation at the rate of 4 % of wet body weight in the first 4 weeks and 3 % in the second 4 weeks. Total body weight of the fish in each cage was determined every 2 weeks, and the feed amounts were adjusted accordingly. The experiment was conducted under the guidelines of the Animal Ethics Committee Regulations, No. 554, issued by Pukyong National University, Busan, Republic of Korea.

Sample analyses and measurements

At the end of the feeding trial, weight gain (WG), feed efficiency (FE), specific growth rate (SGR), protein efficiency ratio (PER), hepatosomatic index (HSI), condition factor (CF), protein retention efficiency (PRE), energy retention efficiency (ERE), hematocrit (percentage of packed cell volume—PCV%), hemoglobin (Hb), and survival rate of juvenile barred knifejaw were determined (Table 3). After the final weighing, five fish were randomly collected from each aquarium and blood samples were obtained using heparinized syringes from the caudal vein and pooled for blood hemoglobin and hematocrit determination (Brown 1980). For the determination of HSI, liver weight was taken by dissecting the fish. Crude protein, moisture, and ash of whole-body samples were analyzed by AOAC methods (1995). In brief, samples of diets and fish were dried to a constant weight at 135 °C for 2 h to determine moisture content.

Table 3 Growth performance and hematological characteristics of juvenile barred knifejaw fed diets with five different protein levels for 8 weeks

Parameters	Protein level in the diets (%)				
	35	40	45	50	60
WG (%)[1]	161 ± 5.7[c]	171 ± 2.5[b]	181 ± 3.1[a]	182 ± 5.7[a]	181 ± 6.1[a]
FE (%)[2]	70.9 ± 2.3[b]	75.9 ± 3.4[ab]	82.0 ± 2.4[a]	81.7 ± 2.7[a]	79.8 ± 2.9[a]
SGR (%/day)[3]	1.85 ± 0.05[c]	1.94 ± 0.02[b]	2.00 ± 0.04[a]	2.01 ± 0.04[a]	2.00 ± 0.05[a]
PER[4]	2.03 ± 0.07[a]	1.86 ± 0.06[ab]	1.48 ± 0.07[b]	1.44 ± 0.07[b]	1.33 ± 0.05[c]
HSI[5]	3.23 ± 0.05[ab]	3.32 ± 0.05[a]	3.19 ± 0.06[ab]	3.08 ± 0.04[b]	3.03 ± 0.03[c]
CF[6]	2.55 ± 0.06[b]	2.71 ± 0.07[ab]	2.83 ± 0.09[a]	2.81 ± 0.07[a]	2.78 ± 0.05[a]
PRE (%)[7]	35.1 ± 2.9[a]	28.2 ± 3.5[ab]	26.3 ± 1.8[b]	26.1 ± 2.0[b]	23.5 ± 2.3[c]
ERE (%)[8]	37.9 ± 3.1[b]	43.2 ± 3.5[a]	44.9 ± 2.9[a]	41.6 ± 28[ab]	42.9 ± 2.7[a]
Hematocrit (%)	32.4 ± 1.9	29.3 ± 2.9	28.9 ± 3.2	30.4 ± 2.1	31.1 ± 2.8
Hemoglobin (g/dl)	5.9 ± 0.4[b]	6.8 ± 0.6[a]	7.3 ± 1.0[a]	7.1 ± 0.8[a]	6.8 ± 0.6[a]
Survival rate (%)	98.3 ± 2.9[a]	98.3 ± 2.9[a]	100[a]	100[a]	95.0 ± 5.0[b]

Values are means from triplicate groups of fish where the means in each row with a different superscript are significantly different ($P < 0.05$)
[1]Percent weight gain: (final wt. − initial wt.) × 100/initial wt
[2]Feed efficiency: (wet weight gain/dry feed intake) × 100
[3]Specific growth rate: 100 × (ln final wt. − ln initial wt.)/days
[4]Protein efficiency ratio: (wet weight gain/protein intake) × 100
[5]Hepatosomatic index: (liver weight/body weight) × 100
[6]Condition factor: [fish wt. (g)/fish length (cm)^3] × 100
[7]Protein retention efficiency: [(final total body protein − initial total body protein)/total dietary protein fed] × 100
[8]Energy retention efficiency: [(final total body energy − initial total body energy)/total dietary energy fed] × 100

Ash was determined by incineration using a muffle furnace at 550 °C for 3 h. Crude lipid was determined by Soxhlet extraction unit using the Soxtec System 1046 (Foss, Hoganas, Sweden), and crude protein content was analyzed by Kjeldahl method ($N \times 6.25$) after acid digestion. Amino acids were measured with an automatic amino acid analyzer (S433; Sykam, Gilching, Germany).

Statistical analysis

All the data were subjected to one-way ANOVA using SAS version 9.1 software (SAS Institute, Cary, NC, USA) to test the effects of dietary protein levels (Zar 1984). When a significant treatment effect was observed, a least significant difference (LSD) test was used to compare the differences between treatment means ($P < 0.05$). Broken-line analysis (Robbins et al. 1979) was used to estimate the optimum level of protein requirement for juvenile barred knifejaw.

Results

The growth performance of barred knifejaw fed experimental diets at different protein levels is shown in Table 3. After 8 weeks of feeding trial, WG of the fish fed the 45, 50, and 60 % crude protein (CP) were significantly higher than those of the fish fed the 35 and 40 % CP diets ($P < 0.05$). However, there were no significant differences in WG among the fish fed the 45, 50, or 60 % CP diets. FE was significantly lower in the fish fed the 35 % CP than in the fish fed the 45, 50, and 60 % CP diets ($P < 0.05$). In contrast to WG and FE, PER and PRE decreased with increasing dietary protein levels. The highest and the lowest ERE values were observed at the 60 and 35 % CP levels, respectively. HSI was highest at the 35 % CP level, whereas the lowest HSI was observed at the 60 % CP level. CF followed the same trend like FE of fish at different protein levels. A significantly higher mortality was observed in the fish fed the 60 % CP diet compared to the fish fed the 35, 40, 45, and 50 % CP diets. In considering hematological characteristics of fish, the juvenile barred knifejaw fed the 35 % CP diet showed a significantly lower hemoglobin level than the fish fed the 40, 45, 50, and 60 % CP diets ($P < 0.05$). There were no significant differences in hematocrit levels among the fish fed the experimental diets ($P > 0.05$).

The whole-body proximate composition of juvenile barred knifejaw is shown in Table 4. The table shows that CP and crude lipid (CL) content in the whole body increased with the increase in dietary protein levels. Significantly higher whole-body CP content was found in the fish fed the 45, 50, and 60 % CP diets than in the fish fed the 35 and 40 % CP diets. Whole-body CL content was found highest in the fish fed the 50 % CP diet and lowest in the fish fed the 40 % CP diet. No significant

Table 4 Proximate composition (%) of the whole body of juvenile barred knifejaw *Oplegnathus fasciatus* fed the experimental diets for 8 weeks

Protein levels (%)	Moisture	Crude protein	Crude lipid	Crude ash
35	66.5 ± 1.1	15.9 ± 0.1[b]	9.5 ± 0.3[ab]	4.3 ± 0.3
40	65.3 ± 1.0	15.8 ± 0.3[b]	9.2 ± 0.2[b]	4.2 ± 0.1
45	66.1 ± 0.6	16.5 ± 0.3[a]	9.8 ± 0.1[a]	4.1 ± 0.2
50	65.0 ± 0.8	16.5 ± 0.3[a]	10.1 ± 0.2[a]	4.0 ± 0.3
60	65.5 ± 0.9	16.8 ± 0.2[a]	10.0 ± 0.2[a]	4.3 ± 0.2

Values are means from triplicate groups of fish where the means in each column with a different superscript are significantly different ($P < 0.05$)

differences were found in the fish fed the experimental diets in terms of whole-body ash and moisture contents.

Broken-line analysis of WG of juvenile barred knifejaw indicated that the optimum dietary protein level was 45.2 % (Fig. 1).

Discussion

After 8 weeks of the feeding trial, ANOVA showed that WG of the fish fed the 45 % CP diet was significantly higher than those of the fish fed the 35 and 40 % CP diets but not significantly different from those of the fish fed the 50 and 60 % CP diets (Table 3). However, based on broken-line analysis of WG of barred knifejaw, the optimum dietary protein was 45.2 %. Similarly, Kim et al. (2004) reported that the optimum dietary protein level for Korean rockfish was 45.1 % CP based on broken-line analysis. In contrast to the present study, Kang et al. (1998) postulated that the optimum dietary protein level should be 46 % for the same species when reared in tank condition. Lovell (1972) reported that the protein requirement of fish varies on the culture environment. Hastings and Dupree (1969) found that channel catfish fed a practical-type diet containing 40 % protein showed linear growth in terms of WG in an aquarium system, whereas the same species of fish fed the same diet with

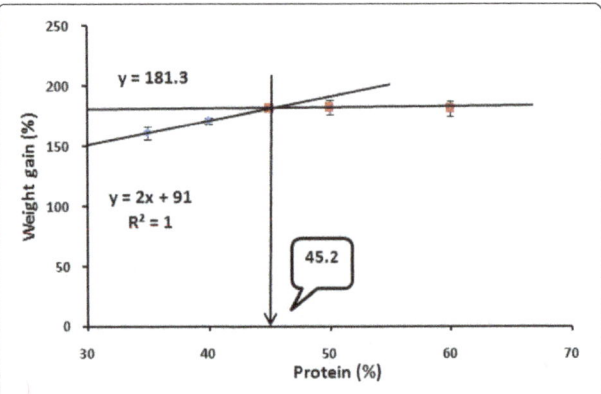

Fig. 1 Broken-line model of percent weight gain in barred knifejaw fed five different levels of dietary protein for 8 weeks

38 % protein showed linear WG in a pond. Moreover, Lovell (1972) proposed that in the cage culture system the required protein level could be 40 % for the same species of fish. Luo et al. (2004) reported that protein requirements for grouper fish species vary over a wide range between 40 and 60 % depending on the species and culture condition.

In line with our result, the protein requirement of black sea bass, *C. striata*, was found to be 45.3 % (Alam et al. 2008); 45 % for European sea bass, *Dicentrarchus labrax* (Peres and Oliva-Teles 1999); and 45 % for Florida pompano, *Trachinotus carolinus* (Lazo et al. 1998). However, in relation to the present study, the protein requirement levels were higher in olive flounder, *P. olivaceus* (46.4 %; Kim et al. 2002); grouper, *Epinephelus malabaricus* (47.8 %; Chen and Tsai 1994); and silver pomfret, *P. argenteus* (49 %; Hossain et al. 2010), and lower in hybrid striped bass, *Morone chrysops* × *Morone saxatilis* (40 %; Gatlin et al. 1994) and white bass, *M. chrysops* (41 %; Rudacille and Kohler 1998). Generally, when dietary protein levels increase, the growth of fish also increases (NRC 1993). In this experiment, WG, FE, SGR, and CF of fish improved with increasing dietary protein levels up to 45 % CP, and then, no further improvements were observed in these parameters at higher protein levels (Table 2). The trends were in agreement with Kim et al. (2004).

In the present study, there were clear reducing trends of PER and PRE with an increasing protein level in the treatment groups (Table 3). The result shows that possibly the dietary protein was efficiently utilized by fish for protein synthesis which is in agreement with Berger and Halver (1987). Similar results have been reported in other fish species (Bai et al. 1999; Kim et al. 2004; Kim et al. 2005; Hossain et al. 2010; Zhang et al. 2010). However, Kikuchi et al. (1992) and Lee et al. (2000) reported that PER values of olive flounder increased with increasing dietary protein levels. In another study, Dabrowski (1979) reported that the relationship between dietary protein and PER differs from species to species. In the present study, ERE increased with the increase of dietary protein levels which means dietary protein could be spared by nonprotein energy sources. Dietary protein sparing helps to reduce feed cost and nitrogen waste outputs (Wang et al. 2006). Ng et al. (2008) reported that lipid plays an important role for protein sparing when the dietary protein level is low in relation to the requirement.

Hematological parameters like hemoglobin (Hb) and hematocrit (PCV) concentration levels were more or less similar in all the treatment groups (Table 3). A significantly lower amount of Hb was found in the blood of the fish fed the 35 % CP diet compared with the other experimental fish fed the higher levels of CP-contained diets. However, PCV concentration level was more or less similar in all the dietary treatment groups which may have resulted in no abnormalities in health status of the experimental fish. Kim et al. (2004) also found dietary protein levels have no significant effect on the hematological and serological characteristics of juvenile Korean rockfish. Hepatosomatic index (HSI) indicates the body condition of fish. In this study, HSI of barred knifejaw decreases with the protein level increases in all the dietary treatments which may indicate higher utilization of protein levels from the diets which is in agreement with Kim and Lall (2001). Survival rates in the fish fed the 60 % CP diet showed significantly higher mortality than those of the fish fed the 35, 40, 45, and 50 % CP diets possibly due to the production of a high level of nitrogenous wastes by the fish through higher fecal output which pollutes the culture environment (Catacutan and Coloso 1995; Tibbetts et al. 2000; Alam et al. 2008).

Proximate compositions in terms of moisture and ash contents of the fish fed the experimental diets were not significantly affected by dietary protein levels (Table 4) which are in accordance with Okorie et al. (2007) for juvenile Japanese eel and Kim et al. (2004) for Korean rockfish. In this experiment, the whole-body CP content increased with the increasing dietary protein levels which agree with the result found by Kim et al. (2002). Similarly, the body lipid content generally increased as the dietary protein level increased which is in agreement with Shiau and Lan (1996) for grouper and Bai et al. (1999) for yellow puffer. On the contrary, Kim et al. (2002) reported that as the CP content of the whole body increases, the whole-body CL content decreases.

Conclusions
In conclusion, based on broken-line analysis of WG, it can be corroborated that the optimum dietary protein level could be 45.2 % in juvenile barred knifejaw to achieve maximum growth when dietary energy content was fixed at 18.8 kJ/g diet. From this present finding, we assume that a cost-effective practical feed could be developed for the sustainable production of barred knifejaw in floating net cages.

Abbreviations
CF, condition factor; CL, crude lipid; CP, crude protein; ERE, energy retention efficiency; FE, feed efficiency; HSI, hepatosomatic index; PCV, packed cell volume; PER, protein efficiency ratio; PRE, protein ratio efficiency; SGR, specific growth rate; WG, weight gain

Acknowledgements
This research was supported by a grant (R2016016) from the National Institute of Fisheries Science (NIFS) and Feeds and Foods Nutrition Research Center (FFNRC) at Pukyong National University, Republic of Korea.

Authors' contributions

KWK conducted the research, analyzed the samples, and prepared the draft manuscript; MM helped to write the draft manuscript; KDK and HSH helped in the research design and reviewed the manuscript; HY and SL helped for the statistical analysis; and SCB designed and monitored the experiment and finalized the draft manuscript. The manuscript has been read and approved by the authors, and none of its parts has been submitted and published elsewhere. The authors also declare that nobody who qualifies for authorship has been excluded from the list of authors.

Competing interests

The authors declare that they have no competing interests.

Author details

[1]Aquafeed Research Center, National Institute of Fisheries Science (NIFS), Pohang 791-923, Republic of Korea. [2]Department of Marine Bio-materials and Aquaculture/Feeds and Foods Nutrition Research Center, Pukyong National University, Busan 608-737, Republic of Korea.

References

Alam MS, Watanabe WO, Carroll PM. Dietary protein requirements of juvenile black sea bass, *Centropristis striata*. J World Aqua Soc. 2008;39:656–63.

Association of Official Analytical Chemists (AOAC). Official methods of analysis. Arlington: Association of Official Analytical Chemists, Inc; 1995.

Bai SC, Wang XJ, Cho ES. Optimum dietary protein level for maximum growth of juvenile yellow puffer. Fish Sci. 1999;65:380–3.

Berger A, Halver JE. Effect of dietary protein, lipid and carbohydrate content on the growth feed efficiency and carcass composition of striped bass, *Morone saxatills* (Walbaum), fingerlings. Aqua Res. 1987;18:345–56.

Brown BA. Hematology: principles and procedures. Philadelphia: Lea and Febiger; 1980.

Catacutan MR, Coloso RM. Effect of dietary protein to energy ratios on growth, survival, and body composition of juvenile Asian seabass, *Lates calcarifer*. Aquaculture. 1995;131:125–33.

Chen HY, Tsai JC. Optimum dietary protein level for the growth of juvenile grouper, *Epinephelus malabaricus*, fed semi-purified diets. Aquaculture. 1994;119:265–71.

Dabrowski K. Feeding requirement of fish with particular attention to common carp. A review. Polish Arch Hydrobiol. 1979;26:135–58.

Das KM, Mohanty S, Sarkar S. Optimum dietary protein to energy ratio for *Labeo rohita* fingerlings. In: De Silva S, editor. Fish nutrition research in Asia, Proceedings, Fourth Asian Fish Nutrition Workshop, Special Publication 5. Philippines: Asian Fisheries Society; 1991. p. 69–74.

El-Sayed AM, Teshima S. Protein and energy requirements of Nile tilapia, *Oreochromis niloticus*, fry. Aquaculture. 1992;103:55–63.

Garling DI, Wilson RP. Optimum dietary protein to energy ratio for channel catfish fingerlings, *Ictalurus punctatus*. J Nutr. 1976;106:1368–75.

Gatlin III DM, Brown ML, Keembiyehetty CN, Jaramillo Jr F, Nematipour GR. Nutritional requirements of hybrid striped bass (*Morone chrysops × M. saxatilis*). J World Aqua Soc. 1994;33:97–109.

Halver JE, Hardy RW. Fish nutrition. 3rd ed. San Diego: Academic; 2002.

Hastings WH, Dupree HK. Practical diets for channel catfish. Progress in sport fisheries research 1968, vol. 77. USA: U.S. Dept. of Interior, Bureau of Sport Fisheries and Wildlife Res. Public; 1969. p. 224–6.

Hepher B. Nutrition of pond fishes. Sydney: Cambridge University Press; 1988.

Hossain MA, Almatar SM, James CM. Optimum dietary protein level for juvenile silver pomfret, *Pampus argenteus* (Euphrasen). J World Aqua Soc. 2010;41: 710–20.

Kang YJ, Lee SM, Hwang HK, Bai SC. Optimum dietary protein and lipid levels on growth in parrot fish (*Oplegnathus fasciatus*). J Aquaculture (Korean). 1998;11:1–10.

Kikuchi K, Sugita H, Watanabe T. Effect of dietary protein level on growth and body composition of Japanese flounder, *Paralichthys olivaceus*. Suisanzoshoku. 1992;40:335–40.

Kim JD, Lall SP. Effects of dietary protein level on growth and utilization of protein and energy by juvenile haddock (*Melanogrammus aeglefinus*). Aquaculture. 2001;195:311–9.

Kim KW, Wang XJ, Bai SC. Optimum dietary protein level for maximum growth of juvenile olive flounder, *Paralichthys olivaceus* (Temminck et Schlegel). Aqua Res. 2002;33:673–9.

Kim KW, Wang XJ, Han K, Kang JC, Bai SC. Optimum dietary protein level and protein-to-energy ratio for growth of juvenile Korean rockfish *Sebastes schlegeli*. J World Aqua Soc. 2004;35:305–14.

Kim KW, Kang YJ, Choi SM, Wang XJ, Choi YH, Bai SC, Lee JY, Jo JY. Optimum dietary protein levels and protein to energy ratios in olive flounder *Paralichthys olivaceus*. J World Aqua Soc. 2005;36:165–78.

Lazo JP, Davies DA, Arnold CR. The effects of dietary protein level on growth, feed efficiency and survival of juvenile Florida pompano (*Trachinotus carolinus*). Aquaculture. 1998;169:225–32.

Lee SM, Cho SH, Kim KD. Effects of dietary protein and energy levels on growth and body composition of juvenile flounder, *Paralichthys olivaceus*. J World Aqua Soc. 2000;36:165–78.

Lovell RT. Protein requirements of cage-cultured channel catfish, Proceedings of the 26th annual conference of the Southeastern Association of Game and Fish Commissioners. 1972. p. 357–61.

Lovell RT. Nutrition and feeding of fish. New York: Van Nostrand-Reinhold; 1989.

Luo Z, Liu YJ, Mai KS, Tian LX, Liu DH, Tan XY. Optimal dietary protein requirement of grouper *Epinephelus coioides* juveniles fed isoenergetic diets in floating net cages. Aqua Nutr. 2004;10:247–52.

Meng QW, Su JX, Miao XZ. Fish taxonomy. Beijing: China Agriculture Press; 1995. p. 734–56.

National Research Council (NRC). Nutrient requirements of fish. Washington, D.C.: National Academy Press; 1993.

National Statistical Office. Survey on the status of fish culture. Daejeon: National Statistical Office; 2014.

Ng WK, Abdullah N, De Silva SS. The dietary protein requirement of the Malaysian mahseer, *Tor tambroides* (Bleeker), and the lack of protein-sparing action by dietary lipid. Aquaculture. 2008;284:201–6.

Okorie EO, Kim YC, Lee S, Bae JY, Yoo JH, Han K, Park GJ, Choi SM, Bai SC. Reevaluation of the dietary protein requirements and optimum dietary protein to energy ratios in Japanese eel, *Anguilla japonica*. J World Aqua Soc. 2007;38:418–26.

Peres H, Oliva-Teles A. Effect of dietary lipid level on growth performance and feed utilization by European sea bass juveniles (*Dicentrarchus labrax*). Aquaculture. 1999;179:325–34.

Robbins KR, Norton HW, Baker DH. Estimation of nutrient requirements from growth data. J Nutr. 1979;109:1710–4.

Rudacille JB, Kohler CC. Dietary protein requirement of juvenile white bass, *Morone chrysops* (book of abstract). Aquaculture. 1998;98:457–8.

Shiau SY, Lan CW. Optimum dietary protein level and protein to energy ratio for growth of grouper (*Epinephelus malabaricus*). Aquaculture. 1996;145:259–66.

Tibbetts SM, Lall SP, Anderson DM. Dietary protein requirement of juvenile American eel (*Anguilla rostrata*) fed practical diets. Aquaculture. 2000;186: 145–55.

Wang Y, Guo J, Li K, Bureau DP. Effects of dietary protein and energy levels on growth, feed utilization and body composition of cuneate drum, *Nibea miichthioides*. Aquaculture. 2006;252:421–8.

Wilson RP. Amino acids and proteins. In: Halver JE, Hardy RW, editors. Fish nutrition. San Diego: Academic; 2002. Pages 144–179.

Zar JH. Biostatistical analysis. 2nd ed. Englewood Cliffs: Prentice-Hall International, Inc.; 1984.

Zhang J, Zhou F, Wang LL, Shao Q, Xu Z, Xu J. Dietary protein requirement of juvenile black sea bream, *Sparus macrocephalus*. J World Aqua Soc. 2010;41:151–64.

Genetic identification of anisakid nematodes isolated from largehead hairtail (*Trichiurus japonicus*) in Korea

Jeong-Ho Kim[1*], Woo-Hwa Nam[1] and Chan-Hyeok Jeon[2]

Abstract

Background: The nematode species belonging to genus *Anisakis* occur at their third larval stage in numerous marine teleost fish species worldwide and known to cause accidental human infection through the ingestion of raw or undercooked fish or squids. They may also draw the attention of consumers because of the visual impact of both alive and dead worms. Therefore, the information on their geographical distribution and clear species identification is important for epidemiological survey and further prevention of human infection.

Results: For identification of anisakid nematodes species isolated from largehead hairtail (*Trichiurus japonicus*), polymerase chain reaction-restriction fragment length polymorphism (PCR-RFLP) analysis of internal transcribed spacers of ribosomal DNA were conducted. Mitochondrial cytochrome c oxidase subunit 2 gene was also sequenced, and phylogenetic analysis was conducted. From the largehead hairtail ($n = 9$), 1259 nematodes were isolated in total. Most of the nematodes were found encapsulated throughout the viscera (56.2 %, 708/1259) or moving freely in the body cavity (41.5 %, 523/1259), and only 0.3 % (4/1259) was found in the muscles. By PCR-RFLP, three different nematode species were identified. *Anisakis pegreffii* was the most dominantly found (98.7 %, 1243/1259) from the largehead hairtail, occupying 98.7 % (699/708) of the nematodes in the mesenteries and 98.1 % (513/523) in the body cavity. Hybrid genotype (*Anisakis simplex* × *A. pegreffii*) occupied 0.5 %, and *Hysterothylacium* sp. occupied 0.2 % of the nematodes isolated in this study.

Conclusions: The largehead hairtail may not significantly contribute accidental human infection of anisakid nematode third stage larvae because most of the nematodes were found from the viscera or body cavity, which are not consumed raw. But, a high prevalence of anisakid nematode larvae in the largehead hairtail is still in concern because they may raise food safety problems to consumers. Immediate evisceration or freezing of fish after catch will be necessary before consumption.

Keywords: *Anisakis pegreffii*, *Hysterothylacium* sp., Largehead hairtail, *Trichiurus japonicus*, Cutlass fish

Background

Anisakid nematodes (Nematoda: Anisakidae) infect animals belonging to almost all phyla and commonly found in aquatic vertebrates. Their life cycles involve various hosts at different levels in marine food webs. Crustaceans work as the first intermediate hosts; fish and cephalopods work as intermediate or paratenic hosts and marine mammals and fish-eating birds as definitive hosts (Mattiucci and Nascetti 2008). Since Van Thiel et al. (1960) reported that

the third stage larvae (L3) of *Anisakis* from the Atlantic herring *(Clupea harengus)* can infect human, the number of clinical cases has been increasing worldwide, with the increase of knowledge about this parasite and improvement of diagnostics, as well as more globalized customs of eating raw marine fish (Cipriani et al. 2015).

The consumption of raw or undercooked fish harboring anisakid nematode L3 can lead to human infection (Audicana and Kennedy 2008). According to Fagerholm (1991), 10 genera exist in the family Anisakidae. However, the human infection is known to be caused frequently by the L3 of the genera *Anisakis* Dujardin, 1845 and less frequently by *Pseudoterranova* Krabbe, 1878

* Correspondence: jhkim70@gwnu.ac.kr
[1]Department of Marine Bioscience, Gangneung-Wonju National University, Gangneung, Gangwon 210-702, South Korea
Full list of author information is available at the end of the article

(Sakanari and McKerrow 1989). Of the genus *Anisakis*, *A. simplex* sensu stricto (s.s.) and *A. pegreffii* are recognized as zoonotic species causing anisakiasis in Asia and Europe, respectively (Mattiucci et al. 2013; Umehara et al. 2007). Thus, not all the nematode larvae found in fish are infective to human, and a clear diagnosis is important for proper epidemiological survey and further prevention of human infection.

Identification of anisakid nematodes has been traditionally conducted by morphological observation, but it is not always sufficient to make a clear identification at the species level because morphological characters of taxonomic significance in this group are very few, particularly for the larval stages lacking reliable diagnostic features at the species level (Mattiucci and Nascetti 2008). Moreover, morphological identification of nematodes is not always efficient when a huge number of the specimens have to be identified or the specimens are physically damaged and the morphological keys are missed. DNA-based diagnostic techniques such as polymerase chain reaction-restriction fragment length polymorphism (PCR-RFLP) of internal transcribed spacer (ITS) region, direct sequencing of ribosomal DNA (rDNA), and mitochondrial DNA cytochrome c oxidase subunit 2 (mtDNA cox2) gene analysis have solved these constraints and have been widely used (Mattiucci and Nascetti 2008 and the references therein). Current taxonomy of the genus *Anisakis* based on these techniques revealed that there are two clades including nine species in this genus, and *A. simplex*, the most frequently encountered *Anisakis* species in human infection belongs to clade I (Mattiucci and Nascetti 2008).

The largehead hairtail or cutlassfish (Trichiuridae: *Trichiurus* species) distributes in circumtropical and temperate waters of the world and has been considered as one of the economically important fish species in the western North Pacific including Korea (Nakamura and Parin 1993). The annual catch of the largehead hairtail in Korea has been fluctuating between 50,000 and 90,000 t, after a sharp decrease in mid-1990 (Kim et al. 2011). Thus, importation of frozen fish from several countries has been increasing to meet the domestic demand (Cha and Kim 2009). Juveniles feed mostly on euphausiids, small pelagic planktonic crustaceans and small fishes, and adults feed mainly on fishes and occasionally on squids and crustaceans (Nakamura and Parin 1993). All of these food items are known to be the intermediate hosts of anisakid nematodes (Mattiucci and Nascetti 2008).

There have been many papers regarding anisakid nematodes isolated from *Trichiurus* species (Borges et al. 2012; Kong et al. 2015; Lee et al. 2009; Shih 2004; Umehara et al. 2010; Zhang et al. 2013). But, the identification of host species was not clearly conducted, and several scientific names (e.g., *Trichiurus lepturus*, *Trichiurus haumela*,

Trichiurus sp.) have been indiscriminately used for the largehead hairtail in their studies. However, clear identification of host species, together with the information on their geographical distribution and feeding behavior are important to understand the occurrence of anisakid nematodes and predict the likelihood of infection in a given area. In particular, the largehead hairtails are often sold as fresh or frozen fillets, which are difficult to identify since most of the morphological features are no longer available. In this case, other techniques are necessary to identify these morphologically similar and closely related species in commercial products.

Although the largehead hairtail is commercially and economically important fish species in Korea, its anisakid nematode fauna has not been clearly investigated and only fragmentary information exists. Lee et al. (2009) reported *A. pegreffii* and *Anisakis typica* from several marine fish species including *T. lepturus*, but they did neither mention the origin of the fish nor the prevalence of infection in each host fish species. Moreover, there was no information in the microhabitat of anisakid nematodes in host fish. In this study, we investigated anisakid nematode fauna from the largehead hairtail caught from Korean waters. We collected nematodes in the largehead hairtail freshly caught from coastal waters around Jeju, Korea. The host species were identified by mitochondrial DNA (mtDNA) cytochrome oxidase I (COI) gene sequencing described by Hsu et al. (2009). The collected nematodes were identified by PCR-RFLP with subsequent sequencing and mtDNA cox2 gene sequencing, to provide information on the epidemiology of *Anisakis* infection in Korea.

Methods
Nematodes collection

Fresh largehead hairtail (*n* = 9) caught by lines were purchased from local fisheries market located in Hallim, Jeju Island, off the southern coast of the Korean Peninsula, 2015. The freshly caught fish were placed on flake ice and immediately transported to the laboratory. All of the individual fish were measured and weighed. Then, they were dissected to collect nematodes. The body cavity was longitudinally opened, and the viscera were carefully examined for collecting nematodes. The muscles were sliced, placed, and pressed between transparent glass plates, then inspected with the naked eyes under the light. All of the collected live nematodes were washed repeatedly with 0.9 % NaCl solution and preserved individually in absolute alcohol for molecular analysis. Prevalence (the number of host infected with parasites divided by the number of hosts examined) and mean intensity (the average of parasite infection among the infected hosts) were used as quantitative descriptors of the parasite population as previously described by Bush et al. (1997).

DNA extraction and PCR

Before genomic DNA extraction, the nematodes were washed three times with PBS, then placed in 1.5 ml Eppendorf tube containing 400 μL TNES lysis buffer and crushed with a sterile pestle. Genomic DNA was extracted using the phenol-chloroform method with slight modification (Wasko et al. 2003). The extracted DNA was resuspended with 50 μL TE buffer (10 mM Tris-HCl, pH 7.5, 1 mM EDTA). DNA concentration and purity were measured using Nanodrop 1000 (Thermo Scientific, USA). The ITS region (ITS1-5.8S-ITS2) of rDNA was amplified using primers A (forward: 5′-GTCGAATTCGTAGGT GAACCTGCGGAAGGATCA-3′) and B (reverse: 5′-GCCGGATCCGAATCCTGGTTAGTTTCTTTTCCT-3′) and B (D'Amelio et al. 2000). Amplification was conducted using MyCyclerTM (BioRad, USA), with the following conditions. Denaturation at 94 °C for 5 min, then 35 cycles at 94 °C for 40 s, 54 °C for 40 s, 72 °C for 90 s, and post-amplification at 72 °C for 7 min.

Digestion of PCR products with restriction enzymes

Restriction enzymes *Hinf* I, *Hha* I and *Rsa* I were used in the RFLP analysis for the amplified PCR products (D'Amelio et al. 2000). Restriction endonuclease digestion was performed using 3 μL of PCR products, 0.5 μL of restriction enzymes, 1 μL of CutSmart® buffer (NEB, USA) and distilled water up to a final volume of 10 μL. The digestion with these restriction enzymes was performed at 37 °C for 90 min. The digested products were analyzed by electrophoresis in 1.5 % agarose gel containing ethidium bromide and photographed using Gel Logic 100 Imaging System (Biostep, Germany).

Mitochondrial DNA cox2 gene amplification and sequencing

For the amplification of mtDNA cox2 gene, the primers 210 (reverse: 5′-CACCAACTCTTAAAATTATC-3′) and 211 (forward: 5′-TTTCTAGTTATATA GATTGRTTYAT-3′) were used (Nadler and Hudspeth 2000). After initial denaturation at 94 °C for 5 min, 35 cycles were run at 94 °C for 40 s, 48 °C for 40 s, and 72 °C for 60 s followed by post-amplification step at 72 °C for 7 min. PCR products were purified by AccuPrep® Gel Purification Kit (Bioneer, Korea) according to the manufacturer's instructions. 10 ng/μL of purified PCR products were directly sequenced by ABI Prism 3730 XL DNA Analyzer (PE Applied Biosystems, USA).

Phylogenetic analysis

The sequence data of mtDNA cox2 gene was aligned with the published sequences in GenBank database (NCBI) using Clustal W method (Thompson et al. 1994). The phylogenetic tree was constructed by MEGA version 6 for comparing the genetic relationship among other published sequences by the neighbor-joining criteria (Tamura et al. 2013). The nucleotide sequences were registered in GenBank.

Host identification

Genomic DNA was extracted from the muscle tissue of largehead hairtail individuals according to the methods described by Wasko et al. (2003). PCR amplification of mtDNA COI gene was conducted with the primers and conditions described by Hsu et al. (2009). PCR products were purified and sequenced by the same protocols mentioned above. The obtained sequences were aligned with other published sequences including *T. lepturus*, *T. japonicus*, and *Trichiurus* sp. registered in NCBI using Clustal W method (Thompson et al. 1994). The phylogenetic tree was constructed by MEGA version 6 for comparing the genetic relationship among other published sequences of *Trichiurus* species by the neighbor-joining criteria (Tamura et al. 2013). The nucleotide sequences were registered in GenBank.

Results

We examined nine individual largehead hairtail (mean whole body length = 90.7 ± 5.2 cm, mean body weight = 462.0 ± 47.3 g). All the examined individual largehead hairtail fish harbored nematodes (100 %, 9/9). In total, 1259 nematodes were found and mean intensity was 139.9 larvae/fish (1259/9). Most of the nematodes were found embedded in the surfaces of viscera including mesenteries (56.2 %, 708/1259) or existed freely in the body cavity (41.5 %, 523/1259). Two percent of nematodes were found in the intestines (0.6 %, 7/1259) or stomach (1.4 %, 17/1259). Only 0.3 % (4/1259) of nematodes was found in the muscles. All the information is summarized in Table 1.

The amplification of the rDNA region produced approximately 1 kb fragment (data not shown). PCR-RFLP analysis using *Hinf*I, *Hha*I, and *Rsa*I restriction enzymes revealed

Table 1 Infection levels of nematodes in the largehead hairtail in this study

P[a] (%)	MI[b]	% of nematodes (number of nematodes in each microhabitat/total number of nematodes)					
		Viscera			Body cavity	Muscles	Total
		Intestines	Stomach	Mesenteries			
100.0 (9/9)	139.9 (1259/9)	0.6 (7/1259)	1.4 (17/1259)	56.2 (708/1259)	41.5 (523/1259)	0.3 (4/1259)	100 (1259/1259)

[a]Prevalence of infection
[b]Mean intensity

different banding patterns depending on the nematodes species. Digestion of the PCR product with *Hinf*I produced three different patterns, i.e., 350–300–250 bp, 700–350 bp, and 620–350–300–250 bp, respectively (Fig. 2a). Digestion using *Hha*I produced two different patterns, i.e., 550–430 bp and 350–250–200–150 bp (Fig. 2b), and digestion using *Rsa*I produced two different patterns, i.e., 550–300 bp and 650–220 bp, respectively (Fig. 2c). These banding patterns corresponded to the known patterns of *A. pegreffii* (lane 1, 2 in Fig. 1a–c), *Hysterothylacium* sp. (lane 3, 4 in Fig. 1a–c), and hybrid genotype (*A. simplex* × *A. pegreffii*) (lane 5, 6 in Fig. 1a–c), respectively.

Of 1259 nematodes, most of the nematodes (98.7 %, 1243/1259) were identified as *A. pegreffii*. The rest of them were identified as hybrid genotype, occupying 0.5 % (6/1259) and *Hysterothylacium* sp., occupying 0.2 % (2/1259) of the nematodes. PCR amplification was failed for eight individuals (0.6 %, 8/1259), probably due to the failure of DNA extraction. The results are summarized in Table 2. For *A. pegreffii*, 20 individuals were randomly selected and sequenced for mtDNA cox2 gene because a huge number of samples were collected. For other species, all the individuals were sequenced. All of the samples generated 629 bp size products (data not shown). When the phylogenetic tree was constructed with the mtDNA cox2 sequences of these samples and those previously registered in NCBI, taxonomic position of the *Anisakis* species identified in this study corresponded with the results obtained by PCR-RFLP; The mtDNA cox2 sequences of *A. pegreffii* (Genbank accession number: KU921231-921250) were clustered with those isolated from chub mackerel in

Korea and Japan (Fig. 2), with 97.9–100.0 % homology. The mtDNA cox2 sequences of *Hysterothylacium* sp. showed the highest homology (92.0 %) with *Hysterothylacium* sp. isolated from chub mackerel (Genbank accession number: KC633443). The mtDNA cox2 sequences of all the hybrid genotype specimens were found to be *A. pegreffii* (data not shown).

PCR amplification and sequencing of mtDNA COI sequences from fish samples revealed that all of the nine samples produced one single band of 630 bp (data not shown). The obtained sequences (Genbank accession number: KU963588, 963589) were clustered with those of *T. japonicus* previously registered in GenBank, with 98.4–99.8 % homology (data not shown).

Discussion

Human infection with *A. simplex* L3 accidentally occur when raw or undercooked fish contaminated with the L3 is ingested. Although live L3 can rarely develop to L4 or adults in the human gastrointestinal tracts, they can cause mild to severe epigastric pain, nausea, vomiting, and diarrhea (Audicana and Kennedy 2008). In addition, allergic reactions occur in some patients, which may occur even by the ingestion of dead *A. simplex* larvae (Audicana and Kennedy 2008). In fish, the larvae are mainly located freely in the visceral cavity or encapsulated as flat, tight spirals in and on the visceral organs; however, a minor proportion may migrate into the muscles, which may draw attention of consumers (Levsen and Berland 2012). Thus, it is necessary to have knowledge about the host range of each *Anisakis*

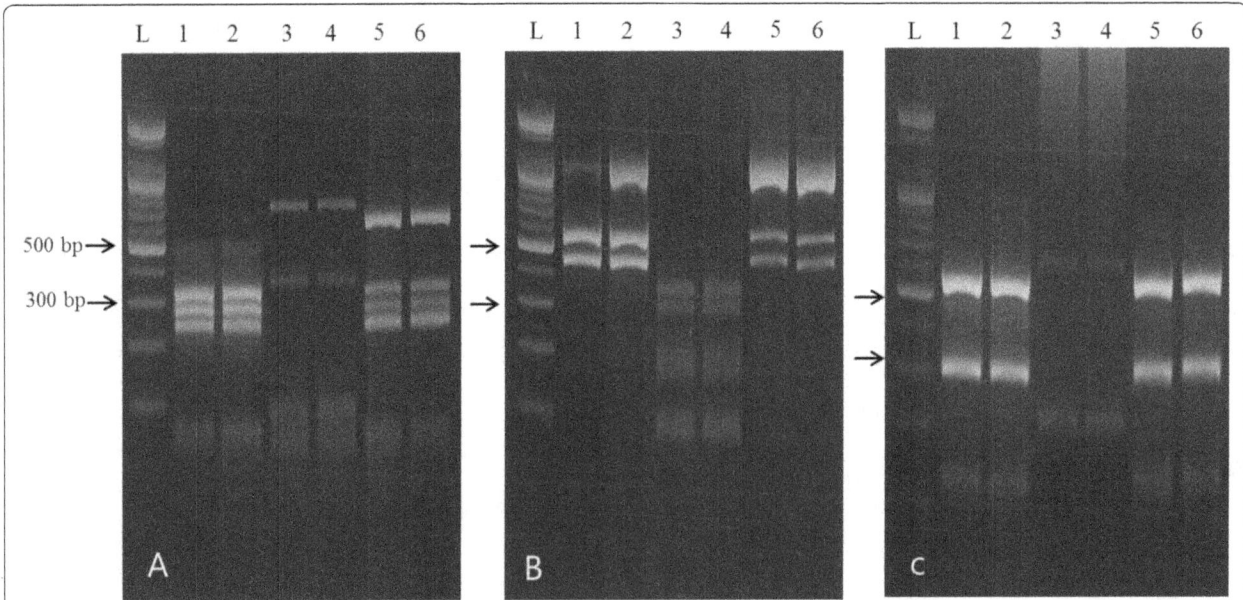

Fig. 1 PCR-RFLP profiles of the nematode species obtained by digestion of PCR products with *Hinf*I (**a**), *Hha*I (**b**), and *Rsa*I (**c**) restriction enzymes, respectively (*L* ladder, *1–2*: *A. pegreffii*, *3–4*: *Hysterothylacium* sp., *5–6*: hybrid genotype)

Table 2 Anisakid nematodes assemblage of the largehead hairtail examined in this study

Microhabitat/species name		A. pegreffii	Hybrid	Hysterothylacium sp.	Fail	Total
Viscera	Intestines	7	0	0	0	7
	Stomach	17	0	0	0	17
	Mesenteries	701	3	0	4	708
Body cavity		514	3	2	4	523
Muscle		4	0	0	0	4
	Total	1243	6	2	8	1259

species, together with the clear identification and their geographical distribution to avoid accidental human infection and warrant food safety issue.

In this study, most of the nematodes were found throughout the viscera (56.2 %) or from the body cavity of the largehead hairtail (41.5 %) (Table 1). Those from the viscera were embedded in or stick to the surfaces of the mesenteries and intestines, whereas those from the body cavity were freely moving in the body cavity. On the other hand, only 0.4 % (6/1,420) of the nematodes was found in the muscles.

Of 1259 nematodes, most of the nematodes (98.7 %, 1243/1259) were identified as *A. pegreffii* by PCR-RFLP and direct sequencing. Other species, *Hysterothylacium* sp. and hybrid genotype (*A. simplex* × *A. pegreffii*) were rarely found.

A. pegreffii Campana-Rouget and Biocca, 1955 is the dominant species of *Anisakis* in the Mediterranean Sea and also widely distributed in the austral region between 30° N and 55° S. Many pelagic and demersal fish species are known to be the intermediate/paratenic hosts caught in this area, and toothed whales (Delphinidae, Ziiphidae, Physeteridae) and baleen whales (Neobalaenidae) are known to be the definitive hosts (Mattiucci and Nascetti 2008). *A. pegreffii* has been also reported frequently at the larval stage in many fish species in East Asian waters (Bak et al. 2014 and the references therein). Recent reports in Italy suggest that *A. pegreffii* can be the etiological agent of human infection (Mattiucci et al. 2013).

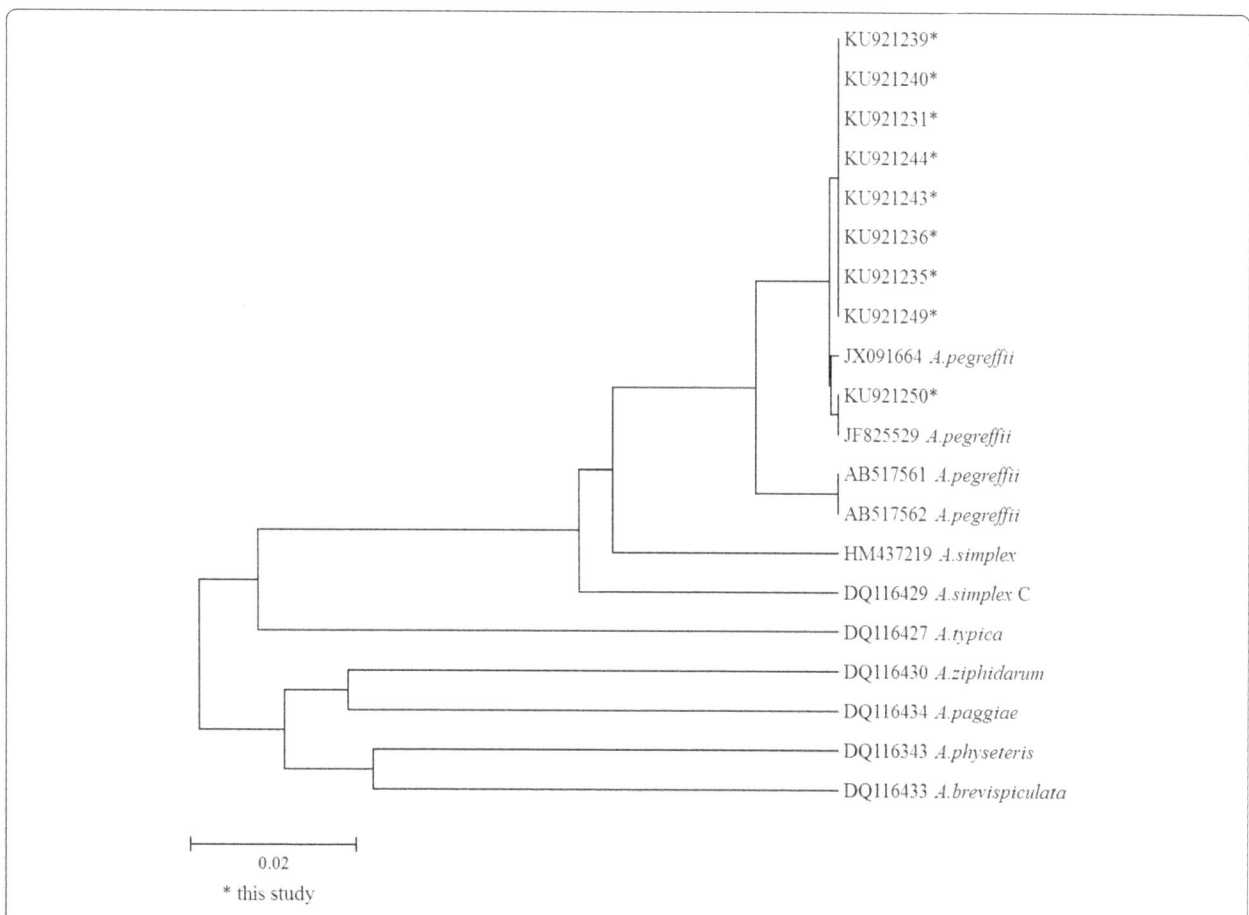

Fig. 2 Molecular phylogenetic trees showing the genetic relationships among *Anisakis* spp. based on mtDNA cox2 gene sequences. Analysis was performed using the MEGA 6 program. The *scale bar* indicates evolutionary distance

Genus *Hysterothylacium* Ward and Magath, 1917 is one of the common genera in the family Raphidascarididae (Fageholm 1991), in which there are approximately 70 species (Shamsi et al. 2013). The adult stages are usually found in the alimentary canal of predatory teleost, while the L3 are found encapsulated in the mesentery and viscera of various species of marine teleosts and also in various marine invertebrates (Keskin et al. 2015). Thus, it is not host specific in the larval stages, exhibiting a global distribution (Shamsi et al. 2013). Many species of the genus *Hysterothylacium* have been poorly described; thus, their genetic characteristics have been frequently deposited in Genbank without clear identification at the species level. Although the RFLP patterns in this study was in accordance with those of *Hysterothylacium* sp. described in Bak et al. (2014), we could not identify these specimens at the species level because the highest homology of the mtDNA cox2 sequences showed 92 % with that of *Hysterothylacium* sp. registered in GenBank. It would be necessary to conduct morphological observation and further molecular analysis, for clear identification of these specimens.

The prevalence of infection and mean intensity for anisakid nematodes L3 in *T. japonicus* calculated in this study were extremely higher than those described by Lee et al. (2009). There can be several explanations for this difference. In our study, we collected freshly caught *T. japonicus* from Korean waters, whereas they collected *T. lepturus* in a fisheries market, without information on their geographical origin.

In marine food webs, the abundance of phytoplankton is strongly related to temperature and changes in phytoplankton biomass can affect all subsequent stages, first impacting on marine crustaceans, the predominant group in zooplanktons abundance and biomass, and thereafter migratory fish, cephalopods, and predatory marine mammals further up the food chain (Utaaker and Robertson 2015). The life cycles of *Anisakis* species involve various hosts at different levels in marine food webs. The high prevalence of *Anisakis* species in fish hosts in this study is therefore thought to reflect the richness of micro/macroinvertebrates and possibly subsequently have relations with the biomass of other predators such as pinnipeds and cetaceans in waters investigated. Several papers also reported relatively a high prevalence of anisakid nematodes larvae in the largehead hairtail (Borges et al. 2012; Kong et al. 2015; Umehara et al. 2010), as in this study. The low number of the nematodes larvae in the previous study (Lee et al. 2009) may just reflect the low abundance of micro/macroinvertebrates in the areas where the largehead hairtail were caught.

Several papers described anisakid nematodes L3 from the largehead hairtail (Borges et al. 2012; Kong et al. 2015; Lee et al. 2009; Shih 2004; Umehara et al. 2010;

Zhang et al., 2013), but there was no report in which *T. japonicus* was mentioned as host species, except for our study. It is not clear if the host species mentioned was in fact *T. japonicus* or other *Trichiurus* species in their studies because the identification of host species was not clearly conducted. But except for the paper by Borges et al. (2012), they probably sampled *T. japonicus* because the known geographical distribution of *T. japonicus*, the northwestern Pacific (Hsu et al. 2009; Tzeng et al. 2007) includes their sampling areas.

After *Anisakis* L3 in the invertebrate or primary fish hosts are ingested by predatory fish, they penetrate in the intestinal wall and then may remain within the body cavity or migrate into the musculature or internal organs of the fish. The different relative abundance of these L3 among these microhabitats in fish can be affected by many factors such as the parasites species, the fish species, the fish age, and the environmental conditions (Lymbery and Cheah 2007). Although the mechanisms of larval penetration within fish hosts are still unknown, some studies revealed that L3 migrate from the visceral organs to the muscle after the death of the fish hosts and this postmortem migration is thought to be also affected by interactions among parasites, host, and external environments (Cipriani et al. 2016; Lymbery and Cheah 2007). Thus, the risk of human infection can be enhanced if the fish are ingested whole or if the fish have been kept without evisceration for a period of time.

Several methods such as postharvest handling, freezing, and adequate cooking are suggested for prevention or control measures of the human infection by ingestion of raw or undercooked fish harboring *Anisakis* L3 (Audicana and Kennedy 2008). Deep freezing and/or adequate cooking remain the most effective measures to prevent human infection (Audicana and Kennedy 2008; Lymbery and Cheah 2007). In addition, immediate evisceration or freezing of fish after catch will be necessary for the prevention of human infection (Lymbery and Cheah 2007). In case of our study, most of the *Anisakis* L3 on the largehead hairtail was found from the body cavity and viscera, which are generally not consumed raw and removed before cooking. In addition, *A. simplex* (s.s.), the most commonly involved species in human infection, was not found in this study. Thus, the largehead hairtail may not be the common source of human infection in Korea. But care should be still taken because *A. pegreffii* is also considered as a possible human pathogen (Mattiucci et al. 2013), and a high prevalence of anisakid nematode larvae in the largehead hairtail may raise food safety problems to the consumers. Immediate evisceration or freezing of fish after catch will be necessary before consumption.

Conclusions

Information on the anisakid nematodes fauna of commercially important fish species is important because the L3 of anisakid nematodes, mostly *A. simplex*, can cause accidental human infection in countries where the customs of eating raw or undercooked fish or cephalopods exist. Given the growing popularity of these customs worldwide, the risk of human infection is widespread, together with the increased number of new host records and expansion of the geographical ranges of *Anisakis* spp. The largehead hairtail, one of the commercially important fish species in Korea, was investigated for anisakid nematodes fauna. Most of the nematodes were found throughout the viscera (58.1 %) and the body cavity (41.5 %), which is not consumed raw. Only 0.3 % was found in the muscles. *A. pegreffii* was the most frequently encountered nematode species (98.7 %) from the largehead hairtail in this study. Detailed information on the distribution of different *Anisakis* species in the largehead hairtail in this study has significant implications for the biology, ecology, and epidemiology of anisakid nematodes.

Abbreviations
COI, cytochrome c oxidase subunit 1; ITS, internal transcribed spacer; L3, third stage larvae; mtDNA cox2, mitochondrial DNA cytochrome c oxidase subunit 2; RFLP, restriction fragment length polymorphism

Acknowledgements
We thank Jeong-Ho Jang, Min-Woo Kim, Chan-Goo Kim, and Hae-Young Yun for their help during the experiments.

Funding
This research was supported by the Basic Science Research Program through the National Research Foundation of Korea (NRF) funded by the Ministry of Science, ICT, and future planning (2015R1A2A2A01004406).

Authors' contributions
JHK conceived, designed the experiments, and wrote the manuscript. WHN and CHJ carried out sample collections and analyses. JHK, WHN, and CHJ interpreted results. All authors read and approved the final manuscript.

Competing interests
The authors declare that they have no competing interests.

Author details
[1]Department of Marine Bioscience, Gangneung-Wonju National University, Gangneung, Gangwon 210-702, South Korea. [2]East Coast Life Science Institute, Gangneung-Wonju National University, Gangneung 210-702, South Korea.

References
Audicana MT, Kennedy MW. *Anisakis simplex*: from obscure infectious worms to inducer of immune hypersensitivity. Clin Microbiol Rev. 2008;21:360–79. doi: 10.1128/CMR.00012-07.

Bak TJ, Jeon CH, Kim JH. Occurrence of anisakid nematode larvae in chub mackerel (*Scomber japonicus*) caught off Korea. Int J Food Microbiol. 2014; 191:149–56. doi:10.1016/j.ijfoodmicro.2014.09.002.

Borges JN, Cunha LFG, Santos HLC, Monteiro-Nato C, Santos CP. Morphological and molecular diagnosis of anisakid nematode larvae from cutlassfish (*Trichiurus lepturus*) off the coast of Rio de Janeiro, Brazil. PLOS One. 2012;7: e40447. doi:10.1371/journal.pone.0040447.

Bush AO, Lafferty KD, Lotz JM, Shostak A. Parasitology meets ecology on its own terms: Margolis et al., revisited. J Parasitol. 1997;83:575–83. doi:10.2307/3284227.

Cha YG, Kim KS. A causality analysis of the prices between imported fisheries and domestic fisheries in distribution channel. J Fish Business Admin. 2009;40: 105–26 (in Korean with English summary).

Cipriani P, Smaldone G, Acerra V, D'Angelo L, Anastasio A, Bellisario B, et al. Genetic identification and distribution of the parasitic larvae of *Anisakis pegreffii* and *Anisakis simplex* (s. s.) in European hake *Merluccius merluccius* from the Tyrrhenian Sea and Spanish Atlantic coast: implications for food safety. Int J Food Microbiol. 2015;198:1–8. doi:10.1016/j.ijfoodmicro.2014.11.019.

Cipriani P, Acerra V, Bellisario B, Sbaraglia GL, Cheleschi R, Nascetti G, et al. Larval migration of the zoonotic parasite *Anisakis pegreffii* (Nematoda: Anisakidae) in European anchovy, *Engraulis encrasicolus*: implications to seafood safety. Food Control. 2016;59:148–57. doi:10.1016/j.foodcont.2015.04.043.

D'Amelio S, Mathiopoulos KD, Santos CP, Pugachev ON, Webb SC, Picanço M, et al. Genetic markers in ribosomal DNA for the identification of members of the genus *Anisakis* (Nematoda: ascaridoidea) defined by polymerase-chain-reaction-based restriction fragment length polymorphism. Int J Parasitol. 2000;30:223–6. doi:10.1016/S0020-7519(99)00178-2.

Fageholm HP. Systematic implications of male caudal morphology in ascaridoid nematode parasites. Syst Parasitol. 1991;19:215–28. doi:10.1007/BF00011888.

Hsu KC, Shih NT, Ni IH, Shao KT. Speciation and population structure of three *Trichiurus* species based on mitochondrial DNA. Zool Studies. 2009;48:835–49.

Keskin E, Koyuncu CE, Genc E. Molecular identification of *Hysterothylacium aduncum* specimens isolated from commercially important fish species of Eastern Mediterranean Sea using mtDNA cox1 and ITS rDNA gene sequences. Parasitol Int. 2015;64:222–8. doi:10.1016/j.parint.2014.12.008.

Kim YH, Yoo JT, Lee EH, Oh TY, Lee DW. Age and growth of largehead hairtail *Trichiurus lepturus* in the East China Sea. Kor J Fish Aquat Sci. 2011;44:695–700. doi:10.5657/KFAS.2011.0695 (in Korean with English Summary).

Kong QM, Fan LF, Zhang JH, Akao N, Dong KW, Lou D, et al. Molecular identification of *Anisakis* and *Hysterothylacium* larvae in marine fishes from the East China Sea and the Pacific coast of central Japan. Int J Food Microbiol. 2015;199:1–7. doi:10.1016/j.ijfoodmicro.2015.01.007.

Lee MW, Cheon DS, Choi CS. Molecular genotyping of *Anisakis* species from Korean sea fish by polymerase chain reaction restriction fragment length polymorphism (PCR-RFLP). Food Control. 2009;20:623–6. doi:10.1016/j. foodcont.2008.09.007.

Levsen A, Berland B. *Anisakis* species. In: Woo PTK, Buchmann K, editors. Fish parasites: pathobiology and protection. CAB International: USA; 2012. p. 298–309.

Lymbery AJ, Cheah FY. Anisakid nematodes and anisakiasis. In: Murrell KD, Fried B, editors. Food-borne parasitic zoonoses: fish and plant-borne parasites. New York: Springer Science; 2007. p. 185–207.

Mattiucci M, Nascetti G. Advances and trends in the molecular systematics of anisakid nematodes, with implications for their evolutionary ecology and host-parasite co-evolutionary process. Adv Parasitol. 2008;66:47–148. doi:10. 1016/S0065-308X(08)00202-9.

Mattiucci M, Fazii P, De Rosa A, Paoletti M, Megna AS, Glielmo A, et al. Anisakiasis and gastroallergic reactions associated with *Anisakis pegreffii* infection, Italy. Emer Inf Dis. 2013;19:196–9. doi:10.3201/eid1903.121017.

Nadler SA, Hudspeth DSS. Phylogeny of the Ascaridoidea (Nematoda: Ascaridida) based on three genes and morphology: hypotheses of structural and sequence evolution. J Parasitol. 2000;86:380–93. doi:10.1645/0022-3395(2000)086[0380:POTANA]2.0.CO;2.

Nakamura I, Parin NV. FAO Species Catalogue. Vol. 15. Snake mackerels and cutlassfishes of the world (Families Gempylidae and Trichiuridae). An annotated and illustrated catalogue of the snake mackerels, snoeks, escolars, gemfishes, sackfishes, domine, oilfish, cutlassfishes, scabbardfishes, hairtails, and frostfishes known to date. FAO Fish Synop. 1993;125:99–106.

Sakanari JA, McKerrow JH. Anisakiasis. Clin Microbiol Rev. 1989;2:278–84. doi:10. 1128/CMR.2.3.278.

Shamsi S, Gasser R, Beveridge I. Description and genetic characterisation of *Hysterothylacium* (Nematoda: Raphidascarididae) larvae parasitic in Australian marine fishes. Parasitol Int. 2013;62:320–8. doi:10.1016/j.parint.2012.10.001.

Shih HH. Parasitic helminth fauna of the cutlass fish, *Trichiurus lepturus* L., and the differentiation of four anisakid nematode third-stage larvae by nuclear ribosomal DNA sequences. Parasitol Res. 2004;93:188–95. doi:10.1007/s00436-004-1095-7.

Tamura K, Stecher G, Peterson D, Filipski A, Kumar S. MEGA6: Molecular Evolutionary Genetics Analysis Version 6.0. Mol Biol Evol. 2013;30:2725–9. doi:10.1093/molbev/mst197.

Thompson JD, Higgins DG, Gibson TJ. CLUSTAL W: improving the sensitivity of progressive multiple sequence alignment through sequence weighting, position-specific gap penalties and weight matrix choice. Nucleic Acids Res. 1994;22:4673–80. doi:10.1093/nar/22.22.4673.

Tzeng CH, Chen CS, Chiu TS. Analysis of morphometry and mitochondrial DNA sequences from two *Trichiurus* species in waters of the western North Pacific: taxonomic assessment and population structure. J Fish Biol. 2007;70(B):165–76. doi:10.1111/j.1095-8649.2007.01368.x.

Umehara A, Kawakami Y, Araki J, Uchida A. Molecular identification of the etiological agent of the human anisakiasis in Japan. Parasitol Int. 2007;56: 211–5. doi:10.1016/j.parint.2007.02.005.

Umehara A, Kawakami Y, Ooi HK, Uchida A, Ohmae H, Sugiyama H. Molecular identification of *Anisakis* type I larvae isolated from hairtail fish off the coasts of Taiwan and Japan. Int J Food Microbiol. 2010;143:161–5. doi:10.1016/j.ijfoodmicro.2010.08.011.

Utaaker KS, Robertson LJ. Climate changes and food borne transmission of parasites: a consideration of possible interactions and impacts for selected parasites. Food Res Int. 2015;68:16–23. doi:10.1016/j.foodres.2014.06.051.

Van Thiel PH, Kuipers FC, Roskam RT. A nematode parasitic to herring, causing atue abdominal syndromes in man. Trop Geogr Med. 1960;2:97–113.

Wasko AP, Martins C, Oliveira C, Foresti F. Non-destructive genetic sampling in fish. An improved method for DNA extraction from fish fins and scales. Hereditas. 2003;138:161–5. doi:10.1034/j.1601-5223.2003.01503.x.

Zhang LP, Du XJ, An RY, Li L, Gasser RB. Identification and genetic characterization of *Anisakis* larvae from marine fishes in the South China Sea using electrophoretic-guided approach. Electrophoresis. 2013;34:888–94. doi:10.1002/elps.201200493.

Apparent digestibility coefficients of the extruded pellet diets containing various fish meals for olive flounder, *Paralichthys olivaceus*

Md Mostafizur Rahman[1], Hyon-Sob Han[2], Kang-Woong Kim[2], Kyoung-Duck Kim[2], Bong-Joo Lee[2] and Sang-Min Lee[1*]

Abstract

Apparent digestibility coefficients (ADCs) of dry matter, crude protein, crude lipid, energy, essential amino acids, and fatty acids in extruded pellets containing various fish meals were determined for olive flounder (*Paralichthys olivaceus*). Eight extruded pellet diets were prepared to contain different fish meals (herring fish meal, anchovy fish meal, mackerel fish meal, sardine fish meal-A, sardine fish meal-B, tuna fish meal, pollock fish meal-A, and pollock fish meal-B) designated as HM, AM, MM, SM-A, SM-B, TM, PM-A, and PM-B, respectively. Chromic oxide (Cr_2O_3) was used as an inert indicator at a concentration of 0.5 % in the diet. Feces were collected from triplicate groups of fish (151 ± 4.0 g) using a fecal collection column attached to the fish rearing tank for 4 weeks. Dry matter ADCs of the MM, SM-A, SM-B, and PM-A diets were higher than those of all the other dietary groups, and the lowest digestibility of dry matter was observed in the PM-B diet. Fish fed the MM, SM-A, and PM-A diets showed significantly higher ADC of protein than those fed the AM, SM-B, TM, and PM-B diets. Lipid ADC of PM-B was significantly lower than that of the other diets. Energy ADCs of fish fed the MM, SM-A, and PM-A diets were significantly higher than those of the other diets. The availability of essential amino acids in the MM, SM-A, and PM-A diets were generally higher than that of the other fish meal diets, while TM showed the lowest values among all the experimental diets. ADCs of fatty acids in the AM, MM, SM-A, and PM-A diets were generally higher than those of fatty acids in the other diets, and the lowest values were recorded for the PM-B diet. These results provide information on the bioavailability of nutrients and energy in various fish meals which can be used to properly formulate practical extruded feeds for olive flounder.

Keywords: *Paralichthys olivaceus*, Apparent digestibility coefficient, Fish meals

Background

Determination of the digestibility of nutrients in diets provides the first indication of their nutritional value and is considered as the first step of their quality evaluation (Allan et al. 2000; Glencross et al. 2007; Luo et al. 2009; Liu et al. 2009). Fish meal is certainly the best dietary protein source because it is quite palatable and provides an excellent balance of essential amino acids and fatty acids and some other substances (Hardy 2010). Fish meal is the preferred animal protein supplement in the diets of aquatic animals. It carries huge quantities of energy and is rich in protein, lipids, minerals, and vitamins. It also serves as the benchmark ingredient in aquaculture diets because of its high nutrient content and digestibility (Udo et al. 2012). Fish meal in animal diets increases feed consumption, feed efficiency, and growth through better feed palatability and also improves nutrient uptake, digestion, and absorption among other ingredients (Yisa et al. 2013). Some studies have investigated apparent digestibility coefficients of various fish meals in several fish species such as grower

* Correspondence: smlee@gwnu.ac.kr
[1]Department of Marine Biotechnology, Gangneung-Wonju National University, Gangneung 25457, South Korea
Full list of author information is available at the end of the article

rockfish, *Sebastes schlegeli* (Lee 2002); juvenile snakehead, *Ophiocephalus argus* (Yu et al. 2013); juvenile cobia, *Rachycentron canadum* (Zhou et al. 2004); Nile tilapia, *Oreochromis niloticus* (Köprücü and Özdemir 2005); Atlantic cod, *Gadus morhua* (Tibbetts et al. 2006); and juvenile haddock, *Melanogrammus aeglefinus* L. (Tibbetts et al. 2004). The raw materials of fish meal are processed by heating, pressing, separation, evaporation, and drying. Heating condenses the protein, breaks the fat depots, and also releases oil and water. Pressing improves the meal quality and decreases the moisture content of the press cake as much as possible. Drying process removes sufficient water from the wet and unstable mixture of press cake to form a stable fish meal.

Extrusion process can cause physical and chemical changes, such as ingredient particle size reduction and inactivation of enzymes. In addition, the heat associated with the extrusion process may also cause deactivation of anti-nutritional factors (Allan and Booth 2004) and improve the utilization of nitrogen-free extracts or other elements (Burel et al. 2000). Extrusion may also confer important benefits to the physical attributes of pellets including nutrient digestibility, palatability, pellet durability, water stability, and pellet storage life (Barrows and Hardy 2000). Extruded pellets are highly recommended for fish culture because of easy observation of feeding activity, easy management, and minimal water pollution. Cho et al. (2006) reported that extruded pellets can improve the digestibility of ingredients and they are generally well accepted by olive flounder, *Paralichthys olivaceus*.

Olive flounder is a commercially important carnivorous fish widely cultured in Eastern Asia including Korea, Japan, and China (Kim et al. 2014). Previous studies were conducted to investigate apparent digestibility coefficients of various fish meals for flounder (Deng et al. 2010; Kim et al. 2010). However, only limited information is available on the digestibility of different fish meals in flounder-extruded pellets. Therefore, the present study was conducted to determine the apparent digestibility coefficients of dry matter, crude protein, crude lipid, energy, essential amino acids, and selected fatty acids from different fish meals used in extruded diets for olive flounder.

Methods
Diet preparation
The proximate, essential amino acid and fatty acid (% of total fatty acids) compositions of the test ingredients (fish meals) are shown in Tables 1 and 2, respectively. Eight experimental diets were formulated using steam-dried herring fish meal, anchovy fish meal, mackerel fish meal, sardine fish meal-A, sardine fish meal-B, tuna fish meal, pollock fish meal-A, and pollock fish meal-B (designated as HM, AM, MM, SM-A, SM-B, TM, PM-A, and PM-B,

respectively) (Table 3). Chromic oxide (Cr_2O_3) served as the inert indicator at a concentration of 0.5 % in the diet. All dry ingredients were thoroughly mixed, and the experimental diets were manufactured using a twin-screw extruder (Model ATX-2, Fesco Precision Co., Daegu, Korea). Extrusion conditions were as follows: feeder speed, 16 to 18 rpm; conditioner temperature, 75 °C; main screw speed, 640 rpm; and barrel temperature, 100 to 115 °C. Extruder pellets were oven-dried at 60 °C for 6 h to maintain the moderate moisture content of 5 to 8 % and stored at –25 °C until use.

Fish and experimental condition
Juvenile olive flounder were obtained from a hatchery (Namhae, Korea) and acclimated to the laboratory conditions for 10 months. The experimental fish (151 ± 4.0 g) were then randomly distributed into 400-l cylindrical fiberglass tanks filled with 200 l of water at a density of 25 fish per tank. Filtered seawater was supplied at a flow rate of 3 l/min to each rearing tank. Fish rearing tanks had a sloping bottom leading to a centrally located drainage slot, and the effluent water was first directed over a fecal collection column before going to waste (Lee 2002). The water temperature was

Table 1 Proximate and amino acid compositions of the fish meals used to test diets

	Fish meals							
	HM	AM	MM	SM-A	SM-B	TM	PM-A	PM-B
Proximate analysis (% in dry matter)								
Dry matter	93.3	92.2	92.4	91.3	94.0	92.0	93.7	93.1
Crude protein	73.4	67.3	76.6	71.5	71.0	62.7	74.7	63.3
Crude lipid	10.4	8.6	6.8	10.0	10.2	10.6	5.9	5.4
Ash	16.6	19.7	16.7	16.0	14.6	20.1	15.7	26.4
Gross energy (kcal/g)	4.9	4.5	4.6	4.7	4.8	4.3	4.7	3.9
Essential amino acids (% in protein)								
Arg	6.4	6.0	6.5	7.1	6.4	6.4	7.1	7.0
His	2.8	2.0	4.5	2.5	3.0	3.3	2.5	2.5
Ile	4.4	4.0	4.5	4.1	4.7	4.2	3.9	4.2
Leu	8.0	6.6	7.9	7.8	8.3	7.6	8.0	8.0
Lys	8.4	7.2	8.6	5.8	8.9	9.3	5.7	5.3
Met + Cys	4.2	3.9	4.3	2.7	4.3	4.0	2.8	2.9
Phe + Tyr	7.6	6.3	7.5	8.0	8.0	7.2	8.3	8.3
Thr	4.8	4.8	4.7	4.2	4.9	4.8	4.9	4.3
Val	5.9	4.9	5.0	4.4	5.3	5.6	4.3	4.7

HM herring fish meal, *AM* anchovy fish meal, *MM* mackerel fish meal, *SM-A* sardine fish meal-A, *SM-B* sardine fish meal-B, *TM* tuna meal, *PM-A* pollock fish meal-A, *PM-B* pollock fish meal-B

Table 2 Fatty acid compositions (% of fatty acids) of the fish meals

	Fish meals							
	HM	AM	MM	SM-A	SM-B	TM	PM-A	PM-B
C14:0	4.6	4.2	3.6	4.7	5.4	3.8	2.2	3.3
C14:1	0.4	0.3	0.6	0.6	0.5	1.0		0.4
C16:0	21.0	21.8	19.7	21.0	22.8	26.1	17.9	23.4
C16:1	3.5	5.6	3.6	5.1	6.0	4.7	3.8	5.7
C18:0	4.5	5.9	7.9	6.9	6.0	8.1	4.8	6.6
C18:1n-9	10.8	15.1	12.9	13.9	9.6	17.1	16.9	27.2
C18:2n-6	2.1	2.2	2.4	1.5	3.3	2.1	1.7	3.1
C20:0	0.3	0.4	1.8	0.8	2.6	0.5	1.5	0.3
C20:1n-9	3.3	1.7	0.9	1.0	0.8	1.2	3.5	3.3
C18:3n-3	0.8	0.4	2.6	0.5	3.1	0.8	1.8	0.3
C20:2n-6	2.6	2.0	1.4	1.5	1.4	1.2	1.3	0.6
C22:1n-9		0.5		0.9		0.9	2.5	0.9
C20:3n-3			1.6	0.7	2.0		1.1	
C20:4n-6	1.2	0.8	1.6	1.8	1.6	2.5	1.7	0.8
C22:2n-6	0.6		0.6	0.6	1.5		0.8	
C20:5n-3	12.4	16.4	9.4	11.1	13.0	6.2	14.1	7.9
C22:3n-3	0.4	0.7	0.6	0.5		0.3		
C22:5n-3	1.3	3.2	2.5	2.5	1.4	1.2	1.6	0.9
C22:6n-3	25.2	17.5	22.5	20.9	15.3	20.4	20.1	9.3
n-3HUFA	39.3	37.7	36.5	35.8	31.7	28.1	36.9	18.1

HM herring fish meal, *AM* anchovy fish meal, *MM* mackerel fish meal, *SM-A* sardine fish meal-A, *SM-B* sardine fish meal-B, *TM* tuna meal, *PM-A* pollock fish meal-A, *PM-B* pollock fish meal-B

Table 3 Formulation and chemical composition of the experimental diets

	Diets							
	HM	AM	MM	SM-A	SM-B	TM	PM-A	PM-B
Ingredients (%)								
Herring fish meal	72							
Anchovy fish meal		72						
Mackerel fish meal			72					
Sardine fish meal-A				72				
Sardine fish meal-B					72			
Tuna meal						72		
Pollock fish meal-A							72	
Pollock fish meal-B								72
Wheat flour	14	14	14	14	14	14	14	14
α-potato-starch	5	5	5	5	5	5	5	5
Wheat gluten	2	2	2	2	2	2	2	2
Fish oil	3.7	3.7	3.7	3.7	3.7	3.7	3.7	3.7
Vitamin premix[a]	1	1	1	1	1	1	1	1
Mineral premix[b]	1	1	1	1	1	1	1	1
Stay-C (50 %)	0.3	0.3	0.3	0.3	0.3	0.3	0.3	0.3
Vitamin E (25 %)	0.2	0.2	0.2	0.2	0.2	0.2	0.2	0.2
Choline salt (50 %)	0.3	0.3	0.3	0.3	0.3	0.3	0.3	0.3
Cr_2O_3	0.5	0.5	0.5	0.5	0.5	0.5	0.5	0.5
Nutrient content (dry matter basis)								
Crude protein (%)	53.4	51.7	54.9	52.9	51.8	47.6	54.0	47.5
Crude lipid (%)	9.3	8.5	7.8	9.6	8.6	9.7	8.0	7.5
Ash (%)	12.6	16.0	13.1	13.4	12.8	16.4	12.3	20.2
NFE (%)[c]	24.7	23.8	24.2	24.1	26.8	26.3	25.7	26.6

[a]Vitamin premix contained the following ingredients (g/kg premix), which were diluted in cellulose: thiamin hydrochloride, 2.7; riboflavin, 9.1; pyridoxine hydrochloride, 1.8; niacin, 36.4; Ca-D-pantothenate, 12.7; myo-inositol, 181.8; D-biotin, 0.27; folic acid, 0.68; p-aminobenzoic acid, 18.2; menadione, 1.8; retinyl acetate, 0.73; cholecalciferol, 0.003; and cyanocobalamin, 0.003
[b]Mineral premix contained the following ingredients (g/kg premix): $MgSO_4 \cdot 7H_2O$, 80.0; $NaH_2PO_4 \cdot 2H_2O$, 370.0; KCl, 130.0; ferric citrate, 40.0; $ZnSO_4 \cdot 7H_2O$, 20.0; Ca-lactate, 356.5; CuCl, 0.2; $AlCl_3 \cdot 6H_2O$, 0.15; KI, 0.15; $Na_2Se_2O_3$, 0.01; $MnSO_4 \cdot H_2O$, 2.0; and $CoCl_2 \cdot 6H_2O$, 1.0
[c]Calculated = 100 − (crude protein + crude lipid + ash)

21.4 ± 2.10 °C, and the photoperiod followed the natural conditions during the experimental period.

Feces collection

Triplicate groups of fish were hand-fed with one of the experimental diets to apparent satiation once a day at 15.00 h. Two hours after feeding, the rearing tanks and collection column were brushed out in order to remove uneaten feed and fecal residues. The next day, feces were collected from the fecal collection columns at 9:00 h. Feces collected from the settling columns were immediately filtered with filter paper (Whatman # 1) for 60 min at 4 °C and stored at −75 °C for chemical analyses. Fecal samples from each tank were pooled at the end of the experiment.

Analytical methods

Freeze-dried feed and feces samples were finely grounded using a grinder. Fish scales were removed from the feces samples using a 300-μm sieve before analysis. Crude protein content was determined by the Kjeldahl method using an Auto Kjeldahl System (Buchi, Flawil, Switzerland). Crude lipid was determined by the ether-extraction method. Crude fiber was determined using an automatic analyzer (Fibertec, Tecator, Sweden), while ash content was determined by treatment in a muffle furnace at 600 °C for 4 h. Gross energy content was analyzed using an adiabatic bomb calorimeter (Parr, USA). For amino acid composition, samples were freeze-dried and then hydrolyzed with 6 N HCl at 110 °C for 24 h. Amino acid concentrations in the experimental diets and fecal samples were determined using an automatic analyzer (Hitachi Model 835-50, Japan) equipped with an ion exchange column (Hitachi Resin #

2619, 2.6 × 150 mm, Japan). Lipid for fatty acid analysis was extracted by a combination of chloroform and methanol (2:1, *v/v*) using the method of Folch et al. (1957). Fatty acid methyl esters were measured by transesterification with 14 % BF_3 methanol (Sigma, St Louis, MO, USA). The particular fatty acid composition was identified using a gas chromatography (PerkinElmer, Clarus 600, GC, USA) that has a flame ionization detector, equipped with SPTM-2560 capillary column (100 m × 0.25 mm i.d., film thickness 0.20 mm; Supelco, Bellefonte, PA, USA). Injector and detector temperatures were 260 °C. The column temperature was programmed from 140 to 240 °C at a rate of 5 °C/min. Helium was utilized by the carrier gas. Fatty acid composition from the samples was identified by comparison with retention times of the known standard fatty acid methyl esters (PUFA 37 component FAME Mix Supelco). Chromic oxide was determined by a wet-acid digestion method (Furukawa and Tsukahara 1966).

Apparent dry matter digestibility coefficients were calculated as $100 - (100 × (\% \ Cr_2O_3 \ in \ diet/\% \ Cr_2O_3$ in feces)).

Apparent digestibility coefficients of nutrients, energy, essential amino acids, and selected fatty acids were calculated as 100 − (100 × (% feed marker/% feces marker) × (% nutrient, energy, amino acid, or fatty acid in feces/ % nutrient, energy, amino acid, or fatty acid in feed)).

Statistical analysis
All data were subjected to one-way analysis of variance, followed by Duncan's multiple range test (Duncan 1955) at a significance level of $P < 0.05$. Linear correlations were determined between nutrient digestibility and contents of the test ingredients (fish meals). All data are presented as mean ± SE (standard error) of three replicate groups. All statistical analyses were carried out using SPSS version 20.0 (SPSS Inc., Chicago, IL, USA).

Results
The apparent digestibility coefficients (ADCs) of dry matter, crude protein, crude lipid, and energy of the extruded floating pellet diets containing various fish meals for olive flounder are shown in Table 4. The ADCs of dry matter ranged from 69 to 87 %. Dry matter ADCs of the MM, SM-A, SM-B, and PM-A diets were higher than those of the TM and PM-B diets. The dry matter ADC of PM-B was the lowest among the experimental groups.

Protein ADCs of diets ranged from 87 to 95 %. Protein ADCs of the MM, SM-A, and PM-A diets were significantly higher than those of the AM, SM-B, TM, and PM-B diets while the lowest values were observed in fish fed the TM and PM-B diets. Lipid ADCs ranged from 83

Table 4 Apparent digestibility coefficients (%) of dry matter, crude protein, crude lipid, and energy in olive flounder fed the diets containing various fish meals

Diets	Dry matter	Crude protein	Crude lipid	Energy
HM	81.5 ± 1.47^{bc}	93.2 ± 0.31^{cd}	90.5 ± 1.24^{b}	90.7 ± 0.65^{c}
AM	80.7 ± 1.71^{bc}	91.6 ± 1.47^{bc}	94.6 ± 0.71^{cd}	90.3 ± 0.24^{c}
MM	83.6 ± 0.74^{cd}	95.3 ± 0.16^{d}	94.7 ± 0.91^{cd}	93.5 ± 0.49^{d}
SM-A	84.4 ± 0.51^{cd}	95.1 ± 0.18^{d}	95.9 ± 0.06^{d}	93.0 ± 0.07^{d}
SM-B	83.5 ± 0.06^{cd}	90.8 ± 0.08^{b}	93.1 ± 0.46^{bcd}	89.3 ± 0.11^{c}
TM	77.5 ± 1.04^{b}	87.2 ± 0.70^{a}	92.4 ± 1.59^{bc}	86.2 ± 0.40^{b}
PM-A	87.0 ± 0.45^{d}	95.4 ± 0.20^{d}	93.6 ± 1.24^{bcd}	93.9 ± 0.39^{d}
PM-B	69.2 ± 2.97^{a}	87.2 ± 1.29^{a}	83.0 ± 1.82^{a}	83.5 ± 0.98^{a}

Values (mean ± SE of triplicate groups) in the same column with different superscripts are significantly different ($P < 0.05$)

to 96 %. The lipid ADCs of the PM-B diet was significantly lower than those of the other diets, and the SM-A group showed the highest value. Energy ADCs ranged from 84 to 94 %. The energy ADCs of the MM, SM-A, and PM-A diets were significantly higher than those of the other groups while the PM-B diet showed the lowest value.

Essential amino acid ADCs of diets containing various fish meals for olive flounder are shown in Table 5. In general, essential amino acid availability reflected crude protein digestibility, with fish fed the MM, SM-A, and PM-A diets showing the highest values compared to the other experimental groups. Amino acid digestibility values, for most essential amino acids, in TM were the lowest for juvenile olive flounder among the fish meals tested. Fatty acid ADCs of diets containing various fish meals for olive flounder are shown in Table 6. Among all fish meals, the ADC of selected fatty acids in PM-B was significantly lower than that of fatty acids in other fish meals.

Discussion
Dry matter ADC of various protein feedstuffs offers an estimate of overall digestibility, and a low value generally indicates that a high level of indigestible material is present in the feedstuff (Li et al. 2013). Thus, dry matter ADCs have been considered to provide a better estimate of the amount of indigestible material present in feedstuffs in comparison with digestibility coefficients for individual nutrients (Luo et al. 2008). In this study, the MM, SM-A, SM-B, and PM-A diets were equally well digested and had higher dry matter ADCs than the TM and PM-B diets. These differences can be explained by the differences in origin, quality, and chemical composition of ingredients used in the diet. We found that the dry matter digestibility was positively correlated ($r = 0.95$) with ash content of fish meals tested in the current study.

Table 5 Apparent amino acid digestibility coefficients (%) of diets containing various fish meals for olive flounder

Essential amino acids	Diets							
	HM	AM	MM	SM-A	SM-B	TM	PM-A	PM-B
Arg	94.3 ± 0.39[cd]	93.1 ± 1.46[bc]	98.1 ± 0.03[f]	96.6 ± 0.19[ef]	92.9 ± 0.27[bc]	89.8 ± 0.65[a]	96.1 ± 0.14[de]	92.0 ± 0.39[b]
His	93.2 ± 0.41[bc]	90.8 ± 1.77[a]	98.1 ± 0.13[e]	96.1 ± 0.06[de]	92.1 ± 0.15[ab]	90.3 ± 0.62[a]	95.0 ± 0.28[cd]	89.8 ± 0.75[a]
Ile	92.7 ± 0.60[cd]	90.5 ± 2.23[bc]	97.0 ± 0.13[e]	94.9 ± 0.10[de]	90.6 ± 0.28[bc]	87.6 ± 0.86[a]	94.8 ± 0.32[de]	89.0 ± 0.48[ab]
Leu	93.1 ± 0.48[cd]	91.3 ± 2.04[bc]	97.3 ± 0.11[e]	95.3 ± 0.14[de]	91.0 ± 0.21[bc]	88.0 ± 0.79[a]	95.0 ± 0.19[de]	89.3 ± 0.43[ab]
Lys	84.5 ± 0.62[de]	92.3 ± 2.08[cd]	97.8 ± 0.11[f]	96.4 ± 0.04[ef]	91.3 ± 0.12[bc]	88.6 ± 0.93[a]	95.3 ± 0.21[ef]	89.5 ± 0.50[ab]
Met + Cys	96.1 ± 0.26[d]	94.5 ± 1.23[c]	98.5 ± 0.02[e]	97.3 ± 0.01[de]	94.0 ± 0.11[c]	88.5 ± 0.36[a]	97.1 ± 0.09[de]	91.1 ± 0.37[b]
Phe + Tyr	92.2 ± 0.67[bc]	90.2 ± 1.89[ab]	96.9 ± 0.10[d]	94.8 ± 0.21[d]	90.2 ± 0.22[ab]	88.5 ± 0.71[a]	94.6 ± 0.35[cd]	88.7 ± 0.42[a]
Thr	92.0 ± 0.42[b]	90.1 ± 1.83[b]	96.7 ± 0.12[c]	94.4 ± 0.24[c]	90.1 ± 0.12[b]	86.6 ± 0.66[a]	94.7 ± 0.34[c]	87.7 ± 0.48[a]
Val	89.9 ± 0.36[c]	86.5 ± 1.94[b]	95.7 ± 0.16[e]	92.7 ± 0.24[d]	88.6 ± 0.20[bc]	83.9 ± 0.81[a]	93.4 ± 0.76[de]	86.3 ± 0.51[ab]

Values (mean ± SE of triplicate groups) in the same row with different superscripts are significantly different ($P < 0.05$)

It has been suggested that a high level of ash generally affects digestibility of dry matter and results in high waste outputs and can also cause mineral imbalances. Therefore, the low dry matter digestibility of the TM and PM-B diets may be attributed to their high ash content (20.1 and 26.4 %, respectively). Kitagima and Fracalossi (2011) reported low dry matter digestibility for fish and shrimp offal meal with high ash contents. Similar results have also been observed in rainbow trout (*Oncorhynchus mykiss*) (Bureau et al. 1999) and hybrid tilapia (*O. niloticus* × *Oreochromis aureus*) (Zhou and Yue 2012).

The protein quality of the dietary ingredients is usually the leading factor affecting fish performance and protein digestibility and is the first measure of its availability for fish (Yu et al. 2013). The ADC of protein in this study revealed that the protein of HM, MM, SM-A, and PM-A must be highly digestible by olive flounder. This indicates that each of these fish meals can be utilized efficiently as protein sources for olive flounder. The ADC of protein for the SM-A diet (95 %) is higher than that previously reported for rainbow trout (Gaylord et al. 2008). The ADC of protein for the MM diet (95 %) is higher than that reported for juvenile Pacific white shrimp, *Litopenaeus vannamei* (Lemos et al. 2009). The ADC of protein for the HM diet (93 %) is similar to that reported for herring fish meal in the Atlantic cod, *G. morhua* (Tibbetts et al. 2006), and salmonids such as the Atlantic salmon, *Salmo salar* (Anderson et al. 1997); coho salmon, *O. kisutch* (Sugiura et al. 1998); and rainbow trout (Burel et al. 2000). The ADC of protein for the AM diet (91 %) is similar to that reported for anchovy fish meal in salmonid species (Anderson et al. 1995; Sugiura et al. 1998, 2000; Thiessen et al. 2004; Glencross et al. 2005). In the present study, the protein ADCs of the TM and PM-B diets were lower than those of the other ingredients tested. The ADC of protein appeared to have a positive relationship with dry matter of the test ingredients ($r = 0.84$). The differences in ADC of protein among fish meals can be attributed to their different nutrient compositions, raw materials, species, locations, seasons of catch, and processing conditions used to produce the meal (Luo et al. 2008; Lemos et al. 2009; Terrazas-Fierro et al. 2010).

Table 6 Apparent fatty acid digestibility coefficients (%) of diets containing various fish meals for olive flounder

Fatty acids	Diets							
	HM	AM	MM	SM-A	SM-B	TM	PM-A	PM-B
C14:0	91.7 ± 0.12[b]	94.9 ± 0.12[bc]	96.1 ± 0.16[c]	97.0 ± 1.51[c]	93.0 ± 0.47[bc]	92.9 ± 0.59[bc]	93.2 ± 0.47[bc]	80.6 ± 3.12[a]
C16:0	90.0 ± 0.13[bc]	92.2 ± 0.34[bc]	94.6 ± 0.17[c]	93.1 ± 0.40[c]	89.4 ± 0.71[bc]	87.4 ± 0.77[b]	93.7 ± 0.26[c]	69.7 ± 4.70[a]
C18:0	86.8 ± 0.27[bcd]	90.7 ± 0.66[de]	91.9 ± 0.24[de]	94.1 ± 0.37[e]	84.3 ± 1.02[bc]	83.7 ± 0.82[b]	89.4 ± 0.38[cde]	68.9 ± 4.69[a]
C18:1n-9	92.5 ± 0.62[b]	95.5 ± 0.17[b]	95.8 ± 0.31[b]	95.2 ± 0.25[b]	93.8 ± 0.32[b]	95.9 ± 2.16[b]	95.2 ± 0.21[b]	80.5 ± 2.53[a]
C18:2n-6	88.9 ± 0.74[b]	96.6 ± 0.27[c]	97.1 ± 0.29[c]	96.9 ± 0.22[c]	93.3 ± 0.75[bc]	95.5 ± 0.22[bc]	93.9 ± 0.24[bc]	72.1 ± 6.24[a]
C18:3n-3	93.4 ± 0.11[cd]	96.5 ± 0.27[cd]	97.8 ± 0.80[d]	96.4 ± 0.27[cd]	90.1 ± 0.40[bc]	85.8 ± 0.60[b]	91.7 ± 0.44[bcd]	61.3 ± 6.39[a]
C20:4n-6	95.7 ± 0.19[c]	97.4 ± 0.18[c]	97.1 ± 0.12[c]	94.3 ± 0.43[bc]	92.7 ± 1.87[bc]	94.4 ± 0.63[bc]	89.9 ± 3.15[b]	82.6 ± 2.59[a]
C20:5n-3	97.2 ± 0.07[b]	98.5 ± 0.08[b]	98.7 ± 0.08[b]	97.3 ± 0.48[b]	97.5 ± 0.21[b]	98.2 ± 0.34[b]	97.9 ± 0.20[b]	92.5 ± 1.52[a]
C22:6n-3	96.7 ± 0.05[b]	98.0 ± 0.14[b]	98.0 ± 0.36[b]	97.3 ± 0.20[b]	96.8 ± 0.14[b]	96.9 ± 0.57[b]	97.4 ± 0.10[b]	88.0 ± 1.89[a]

Values (mean ± SE of triplicate groups) in the same row with different superscripts are significantly different ($P < 0.05$)

The quality of dietary protein depends on its amino acid composition and their digestibility and availability (Rollin et al. 2003). Lack of an essential amino acid leads to poor dietary protein utilization and therefore reduces growth and deceases feed efficiency. Although the data presented in this study suggest a reasonable agreement between protein and amino acid digestibilities, individual amino acid availabilities within a feed ingredient are variable. The amino acid availability coefficients of the MM, SM-A, and PM-A diets were significantly higher than those of the other experimental diets, suggesting that olive flounder can efficiently utilize these fish meals. In most of the cases, ADCs of essential amino acid in the TM diet were the lowest of all the fish meals that were tested, possibly due to lower quality of the starting raw material. Many researchers have reported that some amino acids of fish meal are inefficiently utilized or made unavailable due to differences in the processing conditions or the low quality of the raw material processed (Wilson et al. 1981; Anderson et al. 1992, Anderson et al. 1995; Yamamoto et al. 1998; Mu et al. 2000, Chu et al. 2015).

The ADC of dietary lipid usually ranges from 85 to 95 % in fish (NRC 1993). In the present study, lipid digestibilities were considered to be high (>90 %), except for PM-B (83 %). Previous studies reported ADC values of lipid in different fish meals including Peruvian fish meal (94 %) for juvenile snakehead, *O. argus* (Yu et al. 2013); white fish meal (78 %); and brown fish meal (76 %) for loach, *Misgurnus anguillicaudatus* (Chu et al. 2015). The digestibility of lipids is known to be influenced by a number of factors, including degree of unsaturation, dietary lipid level, and various other constituents (Yuan et al. 2010).

Digestibility of fatty acids is identified to be influenced by a number of factors including their chain length, degree of unsaturation, level of incorporation in dietary fat, and other constituent fatty acids and their melting points (Olsen et al. 2000; Martins et al. 2009; Oujifard et al. 2012). High specificity towards unsaturated fatty acids has commonly been found for fish digestive lipases (Caballero et al. 2002). In the present study, all diets showed high fatty acid digestibility except for PM-B. The low digestibility coefficient of fatty acids for the PM-B diet may be attributed to the poor quality of raw material processed. However, digestibility of individual fatty acids has been affected by other factors including emulsification, enzymatic hydrolysis, and micellar incorporation (Francis et al. 2007).

The ADCs of energy for the HM and SM-A diets, in the current study, are in the same range as reported in Atlantic cod (93 %) (Tibbetts et al. 2006) and rainbow trout (95 %) (Gaylord et al. 2008). It has been reported that carnivorous fish are capable of efficiently utilizing energy from animal products (Sullivan and Reigh 1995; Gaylord and Gatlin 1996; McGoogan and Reigh 1996; Lee 2002; Zhou et al. 2004). It was found that a high ash content of fish meal might reduce energy digestibility (Gomes et al. 1995).

Conclusions

The MM, SM-A, and PM-A diets showed higher dry matter, crude protein, crude lipid, and energy ADCs than the other diets. Due to variation within individual amino acid and fatty acid ADCs among diets, the use of specific amino acid and fatty acid ADCs may allow more accurate and economical formulation of the feed for olive flounder.

Abbreviations

ADCs, apparent digestibility coefficients; AM, anchovy fish meal; Arg, arginine; His, histidine; HM, herring fish meal; Ile, isoleucine; Leu, leucine; Lys, lysine; Met + Cys, methionine + cysteine; MM, mackerel fish meal; Phe + Tyr, phenylalanine + tyrosine; PM-A, pollock fish meal-A; PM-B, pollock fish meal-B; SE, standard error; SM-A, sardine fish meal-A; SM-B, sardine fish meal-B; Thr, threonine; TM, tuna fish meal; Val, valine

Acknowledgements

This work was supported by a grant from the National Institute of Fisheries Science (R2016016) in Korea.

Funding

This study was funded by a grant from the National Institute of Fisheries Science (R2016016) in Korea. The funding organization played an active role in the manufacture of the experimental feed and analyses.

Authors' contributions

MMR conducted the feeding trial and drafted the manuscript. HSH, KWK, KDK, and BJL manufactured the experimental feed and performed the analyses. SML conceived and designed the study and experimental facility and also revised the manuscript. All authors read and approved the final manuscript.

Competing interests

The authors declare that they have no competing interests.

Author details

[1]Department of Marine Biotechnology, Gangneung-Wonju National University, Gangneung 25457, South Korea. [2]Aquafeed Research Center, National Institute of Fisheries Science, Pohang 37517, South Korea.

References

Allan GL, Booth MA. Effects of extrusion processing on digestibility of peas, lupins, canola meal and soybean meal in silver perch *Bidyanus bidyanus* (Mitchell) diets. Aquac Res. 2004;35:981–91.

Allan GL, Parkinson S, Booth MA, Stone DAJ, Rowland SJ, Frances J, Warner-Smith R. Replacement of fish meal in diets for Australian silver perch, *Bidyanus bidyanus*: I. Digestibility of alternative ingredients. Aquaculture. 2000;186:293–310.

Anderson JS, Lall SP, Anderson DM, Chandrasoma J. Apparent and true availability of amino acids from common feed ingredients for Atlantic salmon (*Salmo salar*) reared in sea water. Aquaculture. 1992;108:111–24.

Anderson JS, Lall SP, Anderson DM, McNiven MA. Availability of amino acids from various fish meals fed to Atlantic salmon (*Salmo salar*). Aquaculture. 1995;138:291–301.

Anderson JS, Higgs DA, Beames RM, Rowshandeli M. Fish meal quality assessment for Atlantic salmon (*Salmo salar* L.) reared in sea water. Aquac Nutr. 1997;3:25–38.

Barrows FT, Hardy RW. Feed manufacturing technology. In: Stickney RR, editor. Encyclopedia of aquaculture. New York: Wiley; 2000. p. 354–9.

Bureau DP, Harris AM, Cho CY. Apparent digestibility of rendered animal protein ingredients for rainbow trout (*Oncorhynchus mykiss*). Aquaculture. 1999;180:345–58.

Burel C, Boujard T, Tulli F, Kaushik SJ. Digestibility of extruded peas, extruded lupin and rapeseed meal in rainbow trout (*Oncorhynchus mykiss*) and turbot (*Psetta maxima*). Aquaculture. 2000;188:285–98.

Caballero MJ, Obach A, Rosenlund G, Montero D, Gisvold M, Izquierdo MS. Impact of different dietary lipid sources on growth, lipid digestibility, tissue fatty acid composition and histology of rainbow trout, *Oncorhynchus mykiss*. Aquaculture. 2002;214:253–71.

Cho SH, Lee SM, Park BH, Lee SM. Effect of feeding ratio on growth and body composition of juvenile olive flounder *Paralichthys olivaceus* fed extruded pellets during the summer season. Aquaculture. 2006;251:78–84.

Chu ZJ, Yu DH, Yuan YC, Qiao Y, Cai WJ, Shu H, Lin YC. Apparent digestibility of selected protein feed ingredients for loach *Misgurnus anguillicaudatus*. Aquac Nutr. 2015;21:425–32.

Deng J, Mai K, Ai Q, Zhang W, Tan B, Xu W, Liufu Z. Alternative protein sources in diets for Japanese flounder *Paralichthys olivaceus* (Temminck and Schlegel): II. Effects on nutrient digestibility and digestive enzyme activity. Aquac Res. 2010;41:861–70.

Duncan DB. Multiple-range and multiple *F*-tests. Biometrics. 1955;11:1–42.

Folch J, Lees M, Sloane-Stanley GH. A simple method for the isolation and purification of total lipids from animal tissues. J Biol Chem. 1957;226:497–509.

Francis DS, Turchini GM, Jones PL, De Silva SS. Effects of fish oil substitution with a mix blend vegetable oil on nutrient digestibility in Murray cod, *Maccullochella peelii peelii*. Aquaculture. 2007;269:447–55.

Furukawa A, Tsukahara H. On the acid digestion for the determination of chromic oxide as an index substance in the study of digestibility of fish feed. Bull Jpn Soc Sci Fish. 1966;32:502–6.

Gaylord TG, Gatlin III DM. Determination of digestibility coefficients of various feedstuffs for red drum (*Sciaenops ocellatus*). Aquaculture. 1996;139:303–14.

Gaylord TG, Barrows FT, Rawles SD. Apparent digestibility of gross nutrients from feedstuffs in extruded feeds for rainbow trout, *Oncorhynchus mykiss*. J Word Aquac Soc. 2008;39:827–34.

Glencross B, Evans D, Dods K, McCafferty P, Hawkins W, Maas R, Sipsas S. Evaluation of the digestible value of lupin and soybean protein concentrates and isolates when fed to rainbow trout, *Oncorhynchus mykiss*, using either stripping or settlement faecal collection methods. Aquaculture. 2005;245:211–20.

Glencross BD, Booth M, Allan GL. A feed is only as good as its ingredients—a review of ingredient evaluation strategies for aquaculture feeds. Aquac Nutr. 2007;13:17–34.

Gomes EF, Rema P, Kaushik SJ. Replacement of fish meal by plant proteins in the diet of rainbow trout (*Oncorhynchus mykiss*): digestibility and growth performance. Aquaculture. 1995;130:177–86.

Hardy RW. Utilization of plant proteins in fish diets: effects of global demand and supplies of fishmeal. Aquac Res. 2010;41:770–6.

Kim KD, Kim DG, Kim SK, Kim KW, Son MH, Lee SM. Apparent digestibility coefficients of various feed ingredients for olive flounder, *Paralichthys olivaceus*. Kor J Fish Aquat Sci. 2010;43:325–30.

Kim KW, Kim SS, Khosravi S, Rahimnejad S, Lee KJ. Evaluation of *Sargassum fusiforme* and *Ecklonia cava* as dietary additives for olive flounder (*Paralichthys olivaceus*). Turk J Fish Aquat Sci. 2014;14:321–30.

Kitagima RE, Fracalossi DM. Digestibility of alternative protein-rich feedstuffs for channel catfish, *Ictalurus punctatus*. J World Aquac Soc. 2011;42:306–12.

Köprücü K, Özdemir Y. Apparent digestibility of selected feed ingredients for Nile tilapia (*Oreochromis niloticus*). Aquaculture. 2005;250:308–16.

Lee SM. Apparent digestibility coefficients of various feed ingredients for juvenile and grower rockfish (*Sebastes schlegeli*). Aquaculture. 2002;207:79–95.

Lemos D, Lawrence AL, Siccardi AJ. Prediction of apparent protein digestibility of ingredients and diets by in vitro pH stat degree of protein hydrolysis with

species-specific enzymes for juvenile Pacific white shrimp *Litopenaeus vannamei*. Aquaculture. 2009;295:89–98.

Li MH, Oberle DF, Lucas PM. Apparent digestibility of alternative plant-protein feedstuffs for channel catfish, *Ictalurus punctatus* (Rafinesque). Aquac Res. 2013;44:282–8.

Liu H, Wu X, Zhao W, Xue M, Guo L, Zheng Y, Yu Y. Nutrients apparent digestibility coefficients of selected protein sources for juvenile Siberian sturgeon (*Acipenser baerii* Brandt), compared by two chromic oxide analyses methods. Aquac Nutr. 2009;15:650–6.

Luo Z, Tan XY, Chen YD, Wang WM, Zhou G. Apparent digestibility coefficients of selected feed ingredients for Chinese mitten crab *Eriocheir sinensis*. Aquaculture. 2008;285:141–5.

Luo Z, Li XD, Gong SY, Xi WQ. Apparent digestibility coefficients of four feed ingredients for *Synechogobius hasta*. Aquac Res. 2009;40:558–65.

Martins DA, Valente LMP, Lall SP. Apparent digestibility of lipid and fatty acids in fish oil, poultry fat and vegetable oil diets by Atlantic halibut, *Hippoglossus hippoglossus* L. Aquaculture. 2009;294:132–7.

McGoogan BB, Reigh RC. Apparent digestibility of selected ingredients in red drum (*Sciaenops ocellatus*) diets. Aquaculture. 1996;141:233–44.

Mu YY, Lam TJ, Guo JY, Shim KF. Protein digestibility and amino acid availability of several protein sources for juvenile Chinese hairy crab *Eriocheir sinensis* H. Milne-Edwards (Decapoda, Grapsidae). Aquac Res. 2000;31:757–65.

NRC (National Research Council). Nutrient requirements of fish. Washington, DC: National Academy Press; 1993. 114 pp.

Olsen RE, Myklebust R, Ringø E, Mayhew TM. The influences of dietary linseed oil and saturated fatty acids on caecal enterocytes in Arctic char (*Salvelinus alpinus* L.): a quantitative ultrastructural study. Fish Physiol Biochem. 2000;22:207–16.

Oujifard A, Seyfabadi J, Kenari AA, Rezaei M. Growth and apparent digestibility of nutrients, fatty acids and amino acids in Pacific white shrimp, *Litopenaeus vannamei*, fed diets with rice protein concentrate as total and partial replacement of fish meal. Aquaculture. 2012;342–343:56–61.

Rollin X, Mambrini M, Abboudi T, Larondelle Y, Kaushik SJ. The optimum dietary indispensable amino acid pattern for growing Atlantic salmon (*Salmo salar* L.) fry. Brit J Nutr. 2003;90:865–76.

Sugiura SH, Dong FM, Rathbone CK, Hardy RW. Apparent protein digestibility and mineral availabilities in various feed ingredients for salmonid feeds. Aquaculture. 1998;159:177–202.

Sugiura SH, Babbitt JK, Dong FM, Hardy RW. Utilization of fish and animal by-product meals in low-pollution feeds for rainbow trout *Oncorhynchus mykiss* (Walbaum). Aquac Res. 2000;31:585–93.

Sullivan JA, Reigh RC. Apparent digestibility of selected feedstuffs in diets for hybrid striped bass (*Morone saxatilis* ♀ × *Morone chrysops* ♂). Aquaculture. 1995;138:313–22.

Terrazas-Fierro M, Civera-Cerecedo R, Ibarra-Martínez L, Goytortúa-Bores E, Herrera-Andrade M, Reyes-Becerra A. Apparent digestibility of dry matter, protein, and essential amino acid in marine feedstuffs for juvenile whiteleg shrimp *Litopenaeus vannamei*. Aquaculture. 2010;308:166–73.

Thiessen DL, Maenz DD, Newkirk RW, Classen HL, Drew MD. Replacement of fishmeal by canola protein concentrate in diets fed to rainbow trout (*Oncorhynchus mykiss*). Aquac Nutr. 2004;10:379–88.

Tibbetts SM, Lall SP, Milley JE. Apparent digestibility of common feed ingredients by juvenile haddock, *Melanogrammus aeglefinus* L. Aquac Res. 2004;35:643–51.

Tibbetts SM, Milley JE, Lall SP. Apparent protein and energy digestibility of common and alternative feed ingredients by Atlantic cod, *Gadus morhua* (Linnaeus, 1758). Aquaculture. 2006;261:1314–27.

Udo IU, Ekanem SB, Ndome CB. Determination of optimum inclusion level of some plant and animal protein-rich feed ingredients in least-cost ration for African catfish (*Clarias gariepinus*) fingerlings using linear programming technique. Int J Oceanogra Marine Ecol Sys. 2012;1:24–35.

Wilson RP, Robinson EH, Poe WE. Apparent and true availability of amino acids from common feed ingredients for channel catfish. J Nutr. 1981;111:923–9.

Yamamoto T, Akimoto A, Kishi S, Unuma T, Akiyama T. Apparent and true availabilities of amino acids from several protein sources for fingerling rainbow trout, common carp, and red sea bream. Fish Sci. 1998;64:448–58.

Yisa AG, Edache JA, Udokainyang AD, Iloama CN. Growth performance and carcass yield of broiler finishers fed diets having partially or wholly withdrawn fish meal. Int J Poultry Sci. 2013;12:117–20.

Yu HR, Zhang Q, Cao H, Wang XZ, Huang GQ, Zhang BR, Fan JJ, Liu SW, Li WZ, Cui Y. Apparent digestibility coefficients of selected feed ingredients for juvenile snakehead, *Ophiocephalus argus*. Aquac Nutr. 2013;19:139–47.

Yuan YC, Gong SY, Yang HJ, Lin YC, Yu DH, Luo Z. Apparent digestibility of selected feed ingredients for Chinese sucker, *Myxocyprinus asiaticus*. Aquaculture. 2010;306:238–43.

Zhou QC, Yue YR. Apparent digestibility coefficients of selected feed ingredients for juvenile hybrid tilapia, *Oreochromis niloticus* × *Oreochromis aureus*. Aquac Res. 2012;43:806–14.

Zhou QC, Tan BP, Mai KS, Liu YJ. Apparent digestibility of selected feed ingredients for juvenile cobia (*Rachycentron canadum*). Aquaculture. 2004;241:441–51.

Antioxidant and angiotensin I-converting enzyme inhibitory activities of northern shrimp (*Pandalus borealis*) by-products hydrolysate by enzymatic hydrolysis

Sang-Bo Kim, Na Young Yoon, Kil-Bo Shim and Chi-Won Lim[*]

Abstract

In the present study, we investigated to the antioxidant and angiotensin I-converting enzyme (ACE) inhibitory activities of the northern shrimp (*Pandalus borealis*) by-products (PBB) hydrolysates prepared by enzymatic hydrolysis. The antioxidant and ACE inhibitory activities of five enzymatic hydrolysates (alcalase, protamex, flavourzyme, papain, and trypsin) of PBB were evaluated by the 2, 2'-azino-bis [3-ethylbenzothiazoline-6-sulfonic acid] (ABTS$^+$) radical scavenging and superoxide dismutase (SOD)-like activities, reducing power and Li's method for ACE inhibitory activity. Of these PBB hydrolysates, the protamex hydrolysate exhibited the most potent ACE inhibitory activity with IC$_{50}$ value of 0.08 ± 0.00 mg/mL. The PBB protamex hydrolysate was fractionated by two ultrafiltration membranes with 3 and 10 kDa (below 3 kDa, between 3 and 10 kDa, and above 10 kDa). These three fractions were evaluated for the total amino acids composition, antioxidant, and ACE inhibitory activities. Among these fractions, the < 3 kDa and 3–10 kDa fractions showed more potent ABTS$^+$ radical scavenging activity than that of > 10 kDa fraction, while the > 10 kDa fraction exhibited the significant reducing power than others. In addition, 3–10 kDa and > 10 kDa fractions showed the significant ACE inhibitory activity. These results suggested that the high molecular weight enzymatic hydrolysate derived from PBB could be used for control oxidative stress and prevent hypertension.

Keywords: Northern shrimp, Enzymatic hydrolysate, Antioxidant, Angiotensin I-converting enzyme

Abbreviations: ACE, Angiotensin I converting enzyme; PBB, Pandalus borealis by-products; ABTS$^+$, 2,2'-Azino-bis[3-ethylbenzothiazoline-6-sulfonic acid]; SOD, Superoxide dismutase; RAS, Renin angiotensin system; OPA, o-phthaldialdehyde; DMSO, Dimethyl sulfoxide; EDTA, Ethylenediaminetetraacetic acid; K$_3$Fe(CN)$_6$, Potassium ferricyanide; TCA, Trichloroacetic acid; HHL, Hippuryl-L-histidyl-L-leucine; BSC, Benzene sulfonyl chloride; FeCl$_3$, Iron (III) chloride; MWCO, Molecular weight cut-offs; ROS, Reactive oxygen species

Background

Hypertension is one of the primary causes of cardiovascular disease which leads to stroke, coronary artery disease, and sudden cardiac death (Bhuyan and Mugesh 2011). The renin-angiotensin system (RAS) plays a key role in regulating blood volume and hypertension (Hall 1991; Hall et al. 1989). Angiotensin I-converting enzyme (ACE) is important in the functioning of the RAS. Renin converts angiotensinogen to angiotensin I, and ACE catalyzes cleavage of angiotensin I into angiotensin II which is the main active component of hypertension (Paul et al. 2006; Takahashi et al. 2011). The increased angiotensin II causes vasoconstriction along with increased blood volume and water retention (Parish and Miller 1992). Several ACE inhibitors have been tested and developed in order to inhibit angiotensin II-mediated hypertension. Most synthetic ACE inhibitors such as captopril, enalapril, lisinopril, ramipril, and benzapril on the market have been reported with their side effects including skin rash, loss of taste, and dry cough (Dr and Lisa 2012). Moreover, a recent research has found that oxidative stress is a principal factor for hypertension (Bagatini et al. 2011). Excess reactive oxygen species affects cellular functions and reduces the

* Correspondence: whasoo@korea.kr
Food Safety and Processing Research Division, National Institute of Fisheries Science, Busan 46083, South Korea

bioavailability of endothelial nitric oxide and enhances low-density lipoprotein oxidation in the vascular system (Ray et al. 2012; Toeroek 2008; Mattson 2009). Therefore, there is a necessity for the development of new ACE inhibitors with potent oxidative stress inhibition and fewer side effects from natural resources.

Northern shrimp (*Pandalus borealis*), one of the most popular shrimp species in Korea, belongs to family Pandalidae and distributes widely in the deep sea at depths of 20–1330 m with a temperature of 2–14 °C found in the water around the eastern coast of Korea (Bauer 2004). This is rich in nutrients such as proteins, minerals, and vitamins. However, the inedible parts of shrimp by-products including head, shell, and tail portions account for approximately 50 % of the catch and constitute valuable and useful bioactive materials, such as carotenoprotein, pigments, chitin, and chitosan (Chakrabarti 2002; Babu et al. 2008; Younes et al. 2015). Recently, much research has been carried out on the utilization of protein-rich fisheries by-products as nutraceuticals and nutritional supplements with high nutrient (Chae et al. 1998; Guerard et al. 2001; Arvanitoyannis and Kassaveti 2008).

The aim of the present study was to determine antioxidant and ACE inhibitory activities of enzymatic hydrolysates and its molecular weight cut-off fractions of the *P. borealis* by-products for prevention of hypertension.

Methods
Materials
P. borealis was purchased from the market of Yang yang-gun, Gangwon-do, Korea, in May, 2012. Alcalase® 2.4 L, papain, trypsin, serine, *o*-phthaldialdehyde (OPA), pyrogallol, 2, 2′-azino-bis [3-ethylbenzothiazoline-6-sulfonic acid] (ABTS⁺), dimethyl sulfoxide (DMSO), ethylenediaminetetraacetic acid (EDTA), potassium ferricyanide ($K_3Fe(CN)_6$), trichloroacetic acid (TCA), hippuryl-L-histidyl-L-leucine (HHL), angiotensin I-converting enzyme (ACE), sodium borate, sodium chloride, hydrochloric acid, pyridine, benzene sulfonyl chloride (BSC), captopril, and L-ascorbic acid were purchased from Sigma-Aldrich Chemical Co. (St. Louis, MO, USA) and iron(III) chloride ($FeCl_3$) was obtained from Junsei Chemical Co. (Tokyo, Japan). Protamex and flavourzyme were purchased from Novo Co. (Novozyme Laboratories, Copenhagen, Denmark).

Enzymatic hydrolysis and fractionation of shrimp by-product
The by-products of *P. borealis* (PBB) were lyophilized and stored at −20 °C until use. The crude protein content of PBB was 44.50 ± 0.35 % by AOAC method (AOAC, 2000) and enzymatic hydrolysis of PBB was performed using five enzymes, alcalase, papain, trypsin, protamex, and flavourzyme, under their optimal conditions (Table 1). A 100-g sample (on the basis of protein weight) and 1 % enzyme were mixed and then the mixture was incubated for 8 h at

Table 1 The proximate compositions including moisture, crude fat, ash, and crude protein of PBB

	Moisture (%)	Crude fat (%)	Ash (%)	Crude protein (%)
PBB	21.07 ± 0.22	12.19 ± 0.12	17.18 ± 0.64	44.50 ± 0.35
Protamex	7.05 ± 0.21	1.12 ± 0.15	18.35 ± 0.21	73.92 ± 0.95
< 3 K	14.04 ± 0.13	4.66 ± 0.21	15.85 ± 0.18	62.07 ± 3.23
3–10 K	7.26 ± 0.17	2.27 ± 0.25	11.75 ± 0.26	69.31 ± 2.11
> 10 K	5.48 ± 0.22	2.30 ± 0.17	12.69 ± 0.31	74.02 ± 2.63

The proximate composition of PBB was measured by AOAC methods

each optimal temperature with stirring. After incubation, the mixture was heated at 100 °C to inactivate the enzyme. The inactivated mixture was centrifuged at 2000×*g* for 20 min. After centrifugation, the supernatant was lyophilized and stored at −20 °C until use.

The protamex hydrolysate (38 g) was dissolved in 50 mL deionized water and filtered by two ultrafiltration membranes (Amicon Ultra-filter devices; Millipore, Billerica, MA, USA) with 3 and 10 kDa molecular weight cut-offs (< 3 kDa, 3–10 kDa, and > 10 kDa). The soluble fractions were prepared by centrifuging at 3000×*g* for 20 min and passed through the membrane sequentially, beginning with the largest molecular weight cut-off membrane cartridge (10 kDa). The retentate and permeate were collected separately, and the retentate was recirculated into the feed until the maximum permeate yield was reached. Permeate from the 10 kDa membrane was then filtered through the 3 kDa membrane with recirculation until the maximum permeate yield was reached.

Degree of hydrolysis (DH)
The DH of enzymatic hydrolysates of PBB was calculated by determining free amino groups with OPA (Nielsen 2001).

$$DH = h / h_{tot} \times 100$$

where h_{tot} is the total number of peptide bonds per protein equivalent, and h is the number of hydrolyzed bonds. The factor h_{tot} is dependent on the amino acids composition of the raw material (Adler-Nissen 1986).

Total amino acids contents
The total amino acids composition was determined using an amino acid analyzer (S43000; Sykam, Eresing, Germany). Samples were hydrolyzed in hydrochloric acid (6 N) in vacuum-sealed tubes at 110 °C for 24 h.

Tryptophan is measured after alkaline hydrolysis (Sato et al. 1984). Samples (9–10 mg, 0.1 mg precision) were dissolved in 10 mL of 4.2 N sodium hydroxide and hydrolysed in an oven at 110 °C for 20 h. The hydrolysates were filtered using filter paper (ADVANTEC No. 5B) and mass up 50 mL by 0.2 N sodium citrate buffer (pH

4.2). The 1–5 mL of solution was dried in a waterbath at 70 °C. The pH of the hydrolysates was adjusted to 4.2 with 6 N HCl and 0.2 N sodium citrate buffer and mass up 25 mL. The solutions were filtered through 0.2 μm membrane filter and analyzed by amino acid analyzer (Sykam 4300, Sykam, Germany).

ABTS⁺ radical scavenging activity

The $ABTS^+$ radical scavenging activity was determined using the method of Roberta et al. (1999). The ABTS solution was diluted with water to achieve an absorbance of 0.75 ± 0.03 at 734 nm. Then, 180 μL of ABTS solution was added to 20 μL of different concentrations of samples. The mixture was incubated in the dark for 10 min and measured the absorbance by spectrophotometer (BIO-TEK US/MQX 200, USA) at 734 nm. L-ascorbic acid was used as a positive control and the $ABTS^+$ radical scavenging activity of each sample was expressed as IC_{50} value.

Superoxide dismutase (SOD)-like activity

The SOD-like activity of the sample was evaluated according to the procedure of Marklund and Marklund (1974) with a slight modification. One hundred microliters of sample solutions were mixed with 100 μL of pyrogallol (7.2 mM) and 100 μL of 50 mM Tris-HCl buffer at pH 8.5 containing 0.2 mM EDTA. After 10 min, 50 μL of 1 N HCl was added to the mixture to stop the reaction and measured the absorbance at 420 nm. L-ascorbic acid was used as a positive control, and the SOD-like activity of the PBB was expressed as IC_{50} value.

Reducing power

The reducing power of shrimp shell extracts were measured by the method of Oyaizu (1986). Different concentrations of samples in 10 % DMSO were mixed with 50 μL of 0.2 M sodium phosphate buffer (pH 6.6) and 50 μL of potassium ferricyanide (10 mg/mL). The mixtures were incubated at 50 °C for 20 min. Then, 50 μL of TCA (100 mg/mL) was added and centrifuged at $2000 \times g$ for 10 min. After centrifugation, 100 μL of the supernatant was mixed with 20 μL of iron(III) chloride (1 mg/mL) and the mixture was measured at 700 nm. Reducing power was expressed as the 0.5 of absorbance (EC_{50}) and L-ascorbic acid was used as a positive control.

ACE inhibitory activity

The inhibitory activity of ACE was monitored according to the method of Li et al. (2005). A 20 μL of sample, 50 μL of 5 mM HHL, and 100 mM of sodium borate buffer (pH 8.3) containing 300 mM NaCl were pre-incubated at 37 °C for 5 min. The reaction was initiated by the addition of 10 μL of ACE solution (100 mU/mL), and the mixture was incubated at 37 °C for 30 min. The reaction was stopped by adding 100 μL of 1 M HCl, and then sodium borate buffer (320 μL), pyridine (600 μL), and BSC (200 μL) were added to the reaction mixture. After incubation at room temperature for 30 min, the absorbance of reaction mixture was measured at 492 nm and the captopril was used as a positive control.

Statistical analysis

The data were analyzed using analysis of variance through the general linear model procedure (SAS Institute, Cary, NC, USA). Duncan's multiple range test was applied to determine the significance of the differences between means ($P < 0.05$).

Results and discussion

Enzymatic hydrolysis and fractionation

Recently, the enzymatic hydrolysates have been studied and utilized as nutraceutical resources. In particular, the interest of many researchers on the enzymatic hydrolysates derived from fish and shrimp processing by-products and the various biological activities such as antioxidant, antibacterial, antiobesity, and antihypertensive activities is growing (Sila et al. 2015; Benoit et al. 2008; Cancre et al. 1999).

In the present study, PBB was hydrolyzed by five specific enzymes including alcalase, papain, trypsin, protamex, and flavourzyme for 8 h, respectively, and DH values of their hydrolysates were given in Table 2. The cleavage of peptide bonds by protease leads to decomposition of protein tertiary structure and reduction of the molecular weight of proteins (Adler-Nissen 1986). This reaction also increases in the concentration of free amino and carboxyl groups and its functional properties of proteins (Kristinsson and Rasco 2000). The DH values of PBB hydrolysates were as follows: protamex (59.85 ± 0.09 %) > papain (58.86 ± 0.08 %) > trypsin (58.31 ± 0.08 %) > alcalase (55.96 ± 0.04 %) > flavourzyme (55.68 ± 0.08 %). Among the enzymatic hydrolysates, protamex hydrolysate of PBB showed the highest DH value.

Table 2 The conditions of enzymatic hydrolysis and degree of hydrolysis of the enzymatic hydrolysates of PBB

Enzyme	Crude protein (%)	pH	Temperature (°C)	DH (%)
Alcalase	70.57 ± 2.05	8.0	50	$55.96 \pm 0.04^{1,d}$
Protamex	73.92 ± 0.95	8.0	45	59.85 ± 0.09^{a}
Flavourzyme	67.40 ± 0.16	7.0	50	55.68 ± 0.08^{e}
Papain	69.45 ± 1.33	6.0	37	58.86 ± 0.08^{b}
Trypsin	66.62 ± 0.66	8.0	37	58.31 ± 0.08^{c}

Enzymatic hydrolysates were obtained from 8 h under the optical conditions.
[1]Means within the same row with different superscripts are significantly different by Duncan's multiple range test ($P < 0.05$)

Table 3 Total amino acids composition for the molecular weight cut-off fractions of protamex hydrolysate (g/100 g)

Amino acids	< 3 kDa	3–10 kDa	> 10 kDa
Asp	3.66	4.77	3.65
Thr	1.96	2.21	1.94
Ser	1.91	2.18	1.10
Glu	6.14	7.16	6.11
Pro	2.95	0.34	3.22
Gly	5.06	4.72	5.15
Ala	4.44	4.10	4.43
Cys	0.08	0.18	0.17
Val	2.53	2.62	2.11
Met	1.44	1.24	1.74
Ile	2.21	2.54	1.88
Leu	3.80	3.68	3.73
Tyr	1.51	1.43	1.50
Phe	2.24	2.27	2.25
His	1.29	1.73	1.24
Lys	3.06	3.49	3.89
Arg	2.94	3.36	3.07
Trp	0.22	0.15	0.30
Total	47.44	48.17	47.48

Total amino acids contents

Total amino acid composition of the three MWCO fractions of protamex hydrolysate was presented in Table 3. As shown in Table 3, the total amino acid contents of three MWCO fractions were 47.44 g/100 g (< 3 kDa), 48.17 g/100 g (3–10 kDa), and 47.48 g/100 g (>10 kDa), respectively. The total amino acid compositions of three MWCO fractions were similar to each other. All MWCO fractions were rich in Glu, Gly, Asp, Ala, Leu, and Lys, while these fractions contained low levels of Cys and Trp.

Antioxidant activity

Reactive oxygen species (ROS), containing superoxide (O_2^-), hydrogen peroxide (H_2O_2), hydroxyl radical (OH^-), and singlet oxygen (O_2^-), can cause the oxidative damage to the important components such as protein, lipid, nucleic acids and have been associated with the occurrence of various diseases including hypertension (Ngo et al. 2011a, b).

The antioxidant activities of the five PBB enzymatic hydrolysates were measured by the scavenging activity on $ABTS^+$ radicals, SOD-like activity, and reducing power (Table 4). The alcalase and protamex hydrolysates of PBB exhibited noticeable $ABTS^+$ radical scavenging activity with IC_{50} value of 0.16 ± 0.02 and 0.17 ± 0.00 mg/mL, respectively. However, all hydrolysates showed lower $ABTS^+$ radical scavenging activity than that of L-ascorbic acid.

SOD, an important antioxidant defense enzyme, catalyzes the dismutation of the superoxide anion (O_2^-) into oxygen (O_2) and hydrogen peroxide (H_2O_2). The SOD-like activity is widely used for assay to measure the inhibition of pyrogallol autoxidation. The results of SOD-like activity of the five enzymatic hydrolysates were shown in Table 4. The protamex hydrolysate exerted the greatest SOD-like activity (2.04 ± 0.15 mg/mL) among hydrolysates, while the alcalase hydrolysate exhibited moderate activity.

The reducing power is the ability to donate an electron or hydrogen (Dorman et al. 2003). The electron donation capacity of the five PBB enzymatic hydrolysates were evaluated and showed in Table 4. All the PBB hydrolysates exhibited moderate reducing power.

The antioxidant activities of the three MWCO fractions of protamex hydrolysate were evaluated and showed in Table 5. The < 3 kDa and 3–10 kDa fractions showed potent $ABTS^+$ radical scavenging activity ($IC_{50} = 0.22 \pm 0.01$ mg/mL and 0.22 ± 0.00 mg/mL),

Table 4 Antioxidant and ACE inhibitory activities for the enzymatic hydrolysates of PBB

Sample	ABTS+scavenging activity IC$_{50}$ (mg/mL)[1]	SOD-like activity IC$_{50}$ (mg/mL)	Reducing power EC$_{50}$ (mg/mL)[2]	ACE inhibitory activity IC$_{50}$ (mg/mL)
Alcalase	0.16 ± 0.02[3, c]	2.82 ± 0.72[a]	9.42 ± 0.82[a]	0.11 ± 0.01[b]
Protamex	0.17 ± 0.00[b, c]	2.04 ± 0.15[a]	6.75 ± 0.94[b]	0.08 ± 0.00[c]
Flavourzyme	0.20 ± 0.00[a, b]	3.13 ± 0.51[a]	4.63 ± 0.15[c]	0.11 ± 0.00[b]
Papain	0.21 ± 0.03[a]	2.59 ± 0.48[a]	4.46 ± 0.31[c]	0.11 ± 0.01[b]
Trypsin	0.22 ± 0.02[a]	2.44 ± 0.13[a]	6.62 ± 0.06[b]	0.13 ± 0.00[a]
L-Ascorbic acid[A]	0.004 ± 0.000[d]	0.02 ± 0.00[b]	0.04 ± 0.00[d]	
Captopril[B]				0.00002 ± 0.00000[d]

[1]IC$_{50}$ (50 % inhibitory concentration) values of ABTS+ scavenging, SOD-like, and ACE inhibitory activities were expressed as a mean ± SD
[2]The reducing power was expressed as an EC$_{50}$ (concentration of the 0.5 absorbance) value
[3]Means within the same row with different superscripts are significantly different by Duncan's multiple range test ($P < 0.05$)
[A]L-ascorbic acid was used as a positive control of ABTS+ radical scavenging and SOD-like activities and reducing power
[B]Captopril was used as a positive control of ACE inhibitory activity

Table 5 Antioxidant and ACE inhibitory activities for the molecular weight cut-off fractions of the protamex hydrolysate

Sample	ABTS$^+$ scavenging activity IC$_{50}$ (mg/mL)[1]	SOD-like activity IC$_{50}$ (mg/mL)	Reducing power EC$_{50}$ (mg/mL)[2]	ACE inhibitory activity IC$_{50}$ (mg/mL)
< 3 K	0.22 ± 0.01[3, b]	>10	20.14 ± 0.39[a]	0.06 ± 0.00[a]
3–10 K	0.22 ± 0.00[b]	>10	13.84 ± 0.16[b]	0.03 ± 0.00[b]
> 10 K	0.24 ± 0.01[a]	>10	7.04 ± 0.83[c]	0.03 ± 0.00[b]
L-Ascorbic acid[A]	0.005 ± 0.000[c]	0.07 ± 0.00[d]	0.04 ± 0.00[d]	
Captopril[B]				0.00001 ± 0.0000[c]

Antioxidant and ACE inhibitory activities were measured using the molecular weight cut-off fractions of the protamex hydrolysate
[1]IC$_{50}$ values of ABTS$^+$ scavenging, SOD-like, and ACE inhibitory activities were expressed as a mean ± SD
[2]The reducing power was expressed as an EC$_{50}$ value
[3]Means within the same row with different superscripts are significantly different by Duncan's multiple range test ($P < 0.05$)
[A]L-ascorbic acid was used as a positive control of ABTS$^+$ radical scavenging and SOD-like activities, and reducing power
[B]Captopril was used as a positive control of ACE inhibitory activity

while the > 10 kDa fraction was exhibiting stronger reducing power with EC$_{50}$ value of 7.04 ± 0.83 mg/mL than those of the < 3 kDa and 3–10 kDa fractions. However, the three MWCO fractions showed not high antioxidant activities as much as protamex enzymatic hydrolysate.

ACE inhibitory activity

The inhibition of ACE, a key enzyme regulating the blood pressure, has been recognized as the most effective therapy for the treatment of hypertension. However, many synthetic ACE inhibitors including captopril, enalapril, alacepril, fosinopril, and lisinopril cause side effects such as cough, taste alterations, skin rashes, and angioneurotic edema (Alderman 1996; Cicoira et al. 2001; Vyssoulis et al. 2001). Therefore, it is necessary to develop safe and effective ACE inhibitors from natural products.

The ACE inhibitory activity of the enzymatic hydrolysates of PBB was shown in Table 4. Among the five PBB hydrolysates, the protamex hydrolysate exhibited the most potent ACE inhibitory activity with IC$_{50}$ value of 0.08 ± 0.00 mg/mL, followed by flavourzyme (IC$_{50}$ = 0.11 ± 0.00 mg/mL) > alcalase (IC$_{50}$ = 0.11 ± 0.01 mg/mL) > papain (IC$_{50}$ = 0.11 ± 0.00 mg/mL) > trypsin (IC$_{50}$ = 0.13 ± 0.00 mg/mL).

The ACE inhibitory activity of the three MWCO fractions of protamex hydrolysate was measured and shown in Table 5. Among the MWCO fractions, the > 3 kDa fraction including 3–10 kDa and > 10 kDa fractions showed significant ACE inhibitory activity with IC$_{50}$ value of 0.03 ± 0.00 mg/mL. This result also indicated that high molecular weight fraction, 3–10 kDa and > 10 kDa, exhibited potent ACE inhibitory activity compared with protamex enzymatic hydrolysate.

Recently, many researchers have reported that various bioactive peptides derived from fisheries processing by-products including crab shell (Yoon et al. 2013), Pacific cod skin (Ngo et al. 2011a, b), squid skin and muscle (Mendis et al. 2005; Rajapakse et al. 2005), and tuna (Qian et al. 2007; Je et al. 2005; Lee et al. 2010). In the present study, PBB protamex hydrolysate showed notable ABTS$^+$ radical scavenging and ACE inhibitory activities. In addition, its MWCO fractions exerted potent ACE inhibitory activities. Therefore, more detailed investigations are necessary to isolate and identify the peptides from active enzymatic hydrolysate and to clarify the mechanism of active peptides.

Conclusions

In this study, the five enzymatic hydrolysates of PBB derived from fisheries processing by-products were investigated on the antioxidant and ACE inhibitory activities. The PBB Protamex hydrolysate which showed the most potent ACE inhibitory activity than other fractions was fractionated as the below 3 kDa, between 3 and 10 kDa, and above 10 kDa fractions respectively to isolate the active materials. But these fractions showed lower antioxidant activities than those of enzymatic hydrolysate, while these fractions showed the significant ACE inhibitory activity. In addition, 3-10 kDa and >10 kDa fractions showed the better ACE inhibitory activity than <3 kDa fraction. These results suggested that the high molecular weight enzymatic hydrolysate derived from PBB could be used for prevent hypertension.

Acknowledgements
This work was supported by a grant from the National Institute of Fisheries Science (NIFS), (R2016064/RP-2016-EC-002).

Authors' contributions
SB carried out the anti-oxidant and ACE inhibitory activities assay. NY performed the enzymatic hydrolysis and fractionation. KB analyzed the total amino acids composition. CW participated in the design of the study and performed the statistical analysis. All authors read and approved the final manuscript.

Competing interests
The authors have no pecuniary or other personal interest, direct or indirect, in any matter that raises or may raise a conflict with our duties.

References

Adler-Nissen, J. 1986. Enzymic hydrolysis of food protein. Baking, UK Elsevier 9–56, 110–169

Alderman CP. Adverse effects of the angiotensin-converting enzyme inhibitors. Ann Pharmacother. 1996;30:55–61.

AOAC. 2000. Association of Official Analytical Chemists. 17th Edn., Official Method of Analysis, Washington D.C., USA.

Arvanitoyannis I, Kassaveti A. Fish industry waste: treatments, environmental impacts, current and potential uses. Int J Food Sci Tech. 2008;43:726–45.

Babu CM, Chakrabarti R, Sambasivarao KRS. Enzymatic isolation of carotenoid-protein complex from shrimp head waste and its use as a source of carotenoids. LWT-Food Sci Tech. 2008;41:227–35.

Bagatini MD, Martins CC, Battisti V, Gasparetto D, da Rosa CS, Spanevello RM, Ahmed M, Schmatz R, Rosa M, Schetinger C, Morsch VM. Oxidative stress versus antioxidant defenses in patients with acute myocardial infarction. Heart Vessels. 2011;26:55–63.

Bauer R. 2004. Remarkable shrimps, adaptations and natural history of the carideans. Animal Natural History Series. University of Oklahoma Press 7, 296.

Benoit C, Rozenn RP, Elisa C, Martine FP. Peptides from fish and crustacean by-products hydrolysates stimulate cholecystokinin release in STC-1 cells. Food Chem. 2008;111:970–5.

Bhuyana BJ, Mugesh G. Effect of peptide-based captopril analogues on angiotensin converting enzyme activity and peroxynitrite-mediated tyrosine nitration. Org Biomol Chem. 2011;9:5185–92.

Cancre I, Ravallec R, Wormhoudt AV, Stenberg E, Gildberg A, Gal YL. Secretagogues and growth factors in fish and crustacean protein hydrolysates. Mar Biotechnol. 1999;1:489–94.

Chae H, In M, Kim M. Process development for the enzymatic hydrolysis of food protein: effects of pre-treatment and post-treatments on degree of hydrolysis and other product characteristics. Biotechnol Bioproc E. 1998;3:35–9.

Chakrabarti R. Carotenoprotein from tropical brown shrimp shell waste by enzymatic process. Food Biotechnol. 2002;16:81–90.

Cicoira M, Zanolla L, Rossi A, Golia G, Franceschini L, Cabrini G, Bonizzato A, Graziani M, Anker SD, Coats AJS, Zardini P. Failure of aldosterone suppression despite angiotensin-converting enzyme (ACE) inhibitor administration in chronic heart failure is associated with ACE DD genotype. J Am Coll Cardiol. 2001;37:1808–12.

Dorman HJD, Peltoketo A, Hiltunen R, Tikkanen MJ. Characterization of the antioxidant properties of de-odorized aqueous extracts from selected Lamiaceae herbs. Food Chem. 2003;83:255–62.

Dr RCP, Lisa JM. Adverse effects of angiotensin converting enzyme (ACE) inhibitors. Drug Saf. 2012;7:14–31.

Guerard F, Dufosse L, Broise D, Binet A. Enzymatic hydrolysis of proteins from yellowfin tuna (Thunnus albacares) wastes using alcalase. J Mol Catal B: Enzym. 2001;11:1051–9.

Hall JE, Coleman TG, Guyton AC. The renin-angiotensin system, normal physiology and changes in older hypertensives. J Am Geriatr Soc. 1989;37:801–13.

Hall JE. Control of blood pressure by the renin-angiotensin-aldosterone system. Clin Cardiol. 1991;14:6–21.

Je YJ, Park PJ, Kim SK. Antioxidant activity of a peptide isolated from Alaska Pollack (Theragra chalcogramma) frame protein hydrolysate. Food Res Int. 2005;38:45–50.

Kristinsson HG, Rasco BA. Biochemical and functional properties of Atlantic salmon (Salmo salar) muscle hydrolyzed with various alkaline protease. J Agric Food Chem. 2000;48:657–66.

Lee SH, Qian ZJ, Kim SK. A novel angiotensin I converting enzyme inhibitory peptide from tuna frame protein hydrolysate and its antihypertensive effect in spontaneously hypertensive rats. Food Chem. 2010;118:96–102.

Li GH, Liu H, Shi YH, Le GW. Direct spectrophotemetric measurement of angiotensin I-converting enzyme inhibitory activity for screening bioactive peptides. J Pharm Biomed Anal. 2005;37:219–24.

Marklund S, Marklund G. Involvement of the superoxide anion radical in the auto oxidation of pyrogallol and a convenient assay for superoxide dismutase. Eur J Biochem. 1974;47:469–74.

Mattson MP. Roles of the lipid peroxidation product 4-hydroxynonenal in obesity, the metabolic syndrome, and associated vascular and neurodegenerative disorders. Exp Geront. 2009;44:625–33.

Mendis E, Rajapakse M, Byun HG, Kim SK. Investigation of jumbo squid (Dosidicus gigas) skin gelatin peptides for their in vitro antioxidant effects. Life Sci. 2005; 77:2166–78.

Nielsen PM, Petersen D, Dambmann C. Improved method for determining food protein degree of hydrolysis. J Food Sci. 2001;66:642–6.

Ngo DH, Ryu BM, Vo TS, Himaya SWA, Wijesekara I, Kim SW. Free radical scavenging and angiotensin-I converting enzyme inhibitory peptides from Pacific cod (Gadus macrocephalus) skin gelatin. Int J Biol Macromol. 2011a;49:1110–6.

Ngo DH, Wijesekara I, Vo TS, Ta QV, Kim SW. Marine food-derived functional ingredients as potential antioxidants in the food industry: an overview. Food Res Int. 2011b;44:523–9.

Oyaizu M. Studies on product of browning reaction prepared from glucose amine. Jpn J Nutr. 1986;44:307–15.

Qian ZJ, Je JY, Kim SK. Antihypertensive effect of angiotensin I converting enzyme-inhibitory peptide from hydrolysates of bigeye tuna dark muscle, Thunnus obesus. J Agric Food Chem. 2007;55:8398–403.

Parish RC, Miller LJ. Adverse effects of angiotensin converting enzyme (ACE) inhibitors. Drug Saf. 1992;7:14–31.

Paul M, Poyan Mehr A, Kreutz R. Physiology of local rennin-angiotensin systems. Physiol Rev. 2006;86:747–803.

Rajapakse N, Mendis E, Byun HG, Kim SK. Purification and in vitro antioxidative effects of giant squid muscle peptides on free radical-mediated oxidative systems. J Biochem. 2005;16:562–9.

Ray PD, Huang BW, Tsuji Y. Reactive oxygen species (ROS) homeostasis and redox regulation in cellular signaling. Cell Signal. 2012;24:981–90.

Roberta R, Nicoletta P, Anna P, Ananth P, Min Y, Catherine RE. Antioxidant activity applying an improved ABTS radical cation decolorization assay. Free Radic Biol Med. 1999;26:1231–7.

Sato D, Seino T, Kobayashi T, Murai A, Yugari Y. Determination of the tryptophan content of feed and foodstuffs by ion exchnages liquid chromatography. Agric Biol Chem. 1984;48:2961–9.

Sila A, Kamoun Z, Chlissi Z, Makni M, Moncef N, Sahnoun Z, Nedjar-Arroume N, Bougatef A. Ability of natural astaxanthin from shrimp by-products to attenuate liver oxidative stress in diabetic rats. Pharmacol Rep. 2015;67:310–6.

Takahashi H, Yoshika M, Komiyama Y, Nishimura M. The central mechanism underlying hypertension: a review of the roles of sodium ions, epithelial sodium channels, the rennin-angiotensin-aldosterone system, oxidative stress and endogenous digitalis in the brain. Hypertens Res. 2011;34:1147–60.

Toeroek J. Participation of nitric oxide in different models of experimental hypertension. Physiol Res. 2008;57:813–25.

Vyssoulis GP, Karpanou EA, Papavassiliou MV, Belegrinos DA, Giannakopoulou AE, Toutouzas PK. Side effects of antihypertensive treatment with ACE inhibitors. Am J Hypertens. 2001;14:114–5.

Yoon NY, Shim KB, Lim CW, Kim SB. Antioxidant and angiotensin I converting enzyme inhibitory activities of red snow crab Chionoecetes japonicas shell hydrolysate by enzymatic hydrolysis. Fish Aquat Sci. 2013;16:237–42.

Younes I, Nasri R, Bkhairia I, Jellouli K, Nasri M. New proteases extracted from red scorpion fish (Scorpaena scrofa) viscera: characterization and application as a detergent additive and for shrimp waste deproteinization. Food Bioprod Process. 2015;94:453–62.

6

Photoinactivation of major bacterial pathogens in aquaculture

Heyong Jin Roh, Ahran Kim, Gyoung Sik Kang and Do-Hyung Kim[*]

Abstract

Background: Significant increases in the bacterial resistance to various antibiotics have been found in fish farms. Non-antibiotic therapies for infectious diseases in aquaculture are needed. In recent years, light-emitting diode technology has been applied to the inactivation of pathogens, especially those affecting humans. The purpose of this study was to assess the effect of blue light (wavelengths 405 and 465 nm) on seven major bacterial pathogens that affect fish and shellfish important in aquaculture.

Results: We successfully demonstrate inactivation activity of a 405/465-nm LED on selected bacterial pathogens. Although some bacteria were not fully inactivated by the 465-nm light, the 405-nm light had a bactericidal effect against all seven pathogens, indicating that blue light can be effective without the addition of a photosensitizer. *Photobacterium damselae*, *Vibrio anguillarum*, and *Edwardsiella tarda* were the most susceptible to the 405-nm light (36.1, 41.2, and 68.4 J cm^{-2}, respectively, produced one log reduction in the bacterial populations), whereas *Streptococcus parauberis* was the least susceptible (153.8 J cm^{-2} per one log reduction). In general, optical density (OD) values indicated that higher bacterial densities were associated with lower inactivating efficacy, with the exception of *P. damselae* and *Vibrio harveyi*. In conclusion, growth of the bacterial fish and shellfish pathogens evaluated in this study was inactivated by exposure to either the 405- or 465-nm light. In addition, inactivation was dependent on exposure time.

Conclusions: This study presents that blue LED has potentially alternative therapy for treating fish and shellfish bacterial pathogens. It has great advantages in aspect of eco-friendly treating methods differed from antimicrobial methods.

Keywords: Photoinactivation, Blue light, Bacterial fish pathogen, Fish disease, Light-emitting diode

Abbreviations: BE, Bactericidal efficiency; BHIA, Brain and heart infusion agar; BHIB, Brain and heart infusion broth; LED, Light-emitting diode; OD, Optical density; PBS, Phosphate buffered saline; PPFD, Photosynthesis photon flux density; TSA, Tryptic soy agar

Background

Aquaculture has been the fastest-growing food-producing sector since 1970, with an average growth rate of ~9 % per year, compared with a 2.8 % growth rate of terrestrial farmed meat production over the same period (Bostock et al. 2010; Subasinghe et al. 2001). Worldwide, disease is considered to be a significant constraint on aquaculture; the economic losses caused by disease are estimated to be several billion US dollars per year (Subasinghe et al. 2001). Bacterial diseases are a major threat to aquaculture

because bacteria can survive well and reach high densities in an aquatic environment independent of their hosts, which is generally not the case in terrestrial environments (Defoirdt et al. 2011; Pridgeon and Klesius 2013). In particular, the larval stages of several farmed aquatic animals are highly susceptible to bacterial diseases (Defoirdt et al. 2011). Major bacterial pathogens include *Vibrio*, *Aeromonas*, *Edwardsiella*, and *Streptococcus* species, which affect fish such as salmon, carp, and flat fish (Baeck et al. 2006; Han et al. 2006; Milton et al. 1996; Romalde 2002; Weinstein et al. 1997; Wiklund and Dalsgaard 1998; Won and Park 2008). Inactivation of microorganisms can be accomplished with light

* Correspondence: dhkim@pknu.ac.kr
Department of Aquatic life Medicine, College of Fisheries Science, Pukyong National University, Busan, South Korea

technologies, including ultraviolet C irradiation therapy, photodynamic therapy (PDT), and blue light therapy (Arrojado et al. 2011; Yin et al. 2013). Ultraviolet (UV) irradiation has an adverse effect on fish; it causes intensive skin lesions (Ghanizadeh and Khodabandeh 2010) and reduction of goblet cells in fish skin, resulting in less mucus production and, consequently, downregulation of innate immunity (Kaweewat and Hofer 1997). The use of blue light (400–500 nm) as a mono-therapy is gaining increasing attention because of its potential antimicrobial effect and because it does not require an exogenous photosensitizer (Yin et al. 2013). Blue light is much less harmful to mammalian cells than UV irradiation (Kleinpenning et al. 2010). Light treatment has been applied in aquaculture for many years. For example, European sea bass and sole larvae showed the fastest development and the lowest degree of deformity under blue light (half-peak bandwidth = 435–500 nm) than under other wavelengths of light (Villamizar et al. 2011). Also, another study found that retina from fish exposed to blue light revealed no signs of damage as assessed by extensive histological examination (Migaud et al. 2007). In spite of this potential, there is little information on light therapy as it applies to bacterial pathogens that threaten aquaculture. The aim of this study was to determine the extent of inactivation of bacterial fish pathogens, in particular, seven species including both Gram-negative and Gram-positive bacteria carried out in in vitro experiment. The effects of light-emitting diode (LED) on different bacterial densities and the effects of different light intensities were also evaluated.

Methods

Bacterial strains and identification

Seven bacterial species were evaluated in this study. The bacterial strains were grown on tryptic soy agar (TSA) or brain and heart infusion agar (BHIA), supplemented with 1 % NaCl. A strain of *Vibrio anguillarum* isolated from diseased cod was purchased from the Korean collection for type cultures (KCTC), and *Edwardsiella tarda* KE1 and *Aeromonas salmonicida* RFAS1 originated from diseased olive flounder and black rockfish were previously used (Han et al. 2006, 2011). *Vibrio harveyi* Vh21FL, *Photobacterium damselae* Dae1-1L, *Streptococcus iniae* BS9, and *Streptococcus parauberis* SpOF3K obtained from diseased olive flounder were confirmed by polymerase chain reaction that was previously described (Table 1) (Mata et al. 2004; Osorio et al. 2000; Pang et al. 2006).

LED source

The 405- and 465-nm LEDs, each composed of 120 individual LEDs, were kindly provided by the LED-Marine Convergence Technology R&D Center (Pukyong National University). The spectra of the 405- and 465-nm LEDs as measured by a temperature-controllable integrating system (Withlight Co. Ltd., Korea) are shown in Fig. 1. The maximum irradiation of the 405- and 465-nm LED array were 250 and 516 μ mol m^{-2} s^{-1}, respectively, as calculated using a laboratory radiometer (Biospherical Instruments Inc., USA). Photosynthesis photon flux density (PPFD; μ mol m^{-2} s^{-1}) was converted to radiant flux density (mW cm^{-2}) by using the following formula:

$$\text{Radiant flux}(W) = h \times C \times NA \times PPDF(\mu \text{ mol})/\lambda \times 10^{-3}$$
$$h(\text{Plank's constant}) = 6.626 \times 10^{-34}$$
$$C(\text{Light velocity}) = 3 \times 10^{8} \text{ms}^{-1} \quad \lambda = \text{Wavelength}(nm)$$
$$NA(\text{Avogadro's constant}) = 6.02 \times 10^{23}$$

Antibacterial activity of LEDs

Approximately 10^5 CFU ml^{-1} of each culture was suspended in phosphate buffered saline (PBS; pH 7.2–7.4). Each bacterial suspension (10 ml, with a depth of 5 mm) was plated on a 30-mm petri dish on TSA (*V. anguillarum*, *V. harveyi*, *P. damselae*, *E. tarda*, and *A. salmonicida*) or BHIA (*S. iniae* and *S. parauberis*) supplemented with 1 % NaCl, exposed to 250 μ mol m^{-2} s^{-1} of the 405- or 516 μ mol m^{-2} s^{-1} of the 465-nm LED light, and placed in a 25 °C incubator for 0, 1, 3, 6, 12, 24, or 48 h. Each lamp was placed 3.5 cm above open plates containing the bacterial cultures and positioned perpendicularly. Temperature was routinely monitored during irradiation. The cultures were stirred with a sterile magnetic bar for a few seconds just before being plated, and bacterial counts were performed. A method slightly modified from a previous study (Maclean et al. 2009) was used to express the inactivation data: $\log_{10} (N/N_0)$ was plotted as a function of exposure time, where N_0 is the initial bacterial population in CFU ml^{-1} prior to inactivation and N is 10 CFU ml^{-1}. Thus, the mean bactericidal efficiency (BE) was defined as the $\log_{10}$ reduction in a bacterial population [$\log_{10}(10/N_0)$] by inactivation per unit dose in J cm^{-2}. Exposure time was deduced from the time at which bacterial populations reached 10 CFU ml^{-1}.

In order to determine the effects of initial bacterial density on the antibacterial activity of LEDs, 200 μl of six 10-fold serial dilutions (10^3, 10^4, 10^5, 10^6, 10^7, and 10^8 CFU ml^{-1}, in BHIB supplemented with 1 % NaCl) were inoculated in a 96-well microplate. The plates were exposed to a 405- or 465-nm LED at 25 °C. Optical density (OD) was measured at 630 nm after 24 h irradiation using a Sunrise™ spectrophotometer (TECAN Austria), and data was analyzed using OD of 24 h exposure group/ OD of 24 h non-exposure group × 100 (%) formula.

The data points shown in Fig. 2 and in Table 3 are expressed as mean values with standard deviations. Two-tailed Student's *t* tests and ANOVA Tukey's test were used

Table 1 Bacterial strains and primers used in this study

Bacterial strains	Isolation source	Primers	References
Gram-negative bacteria			
Vibrio harveyi Vh21FL	Diseased olive flounder	Tox F: 5'-GAAGCAGCACTCACCGAT-3' Tox R: 5'-GGTGAAGACTCATCAGCA-3'	Pang et al. (2006)
Vibrio anguillarum KCTC 2711[a]	Diseased cod	–	–
Photobacterium damselae Dae1-1L	Diseased olive flounder	Car 1: 5'-GCTTGAAGAGATTCGAGT-3' Car 2: 5'-CACCTCGCGGTCTTGCTG-3' Ure -3': 5'-CTTGAATATCCATCTCATCTGC-3' Ure -5': 5'-TCCGGAATAGGTAAAGCGGG-3'	Osorio et al. (2000)
Edwardsiella tarda KE1	Diseased olive flounder	–	Han et al. (2006)
Aeromonas salmonicida RFAS1	Disease black Rockfish	–	Han et al. (2011)
Gram-positive bacteria			
Streptococcus iniae BS9	Diseased olive flounder	Lox-1: 5'-AAGGGGAAATCGCAAGTGCC-3' Lox-2: 5'-ATATCTGATTGGGCCGTCTAA-3'	Mata et al. (2004)
Streptococcus parauberis SpOF3K	Diseased olive flounder	Spa F: 5'-TTTCGTCTGAGGCAATGTTG-3' Spa R: 5'-GCTTCATATATCGCTATACT-3'	

[a]Type strain

to determine statistically significant differences ($P < 0.05$ or $P < 0.01$) between groups exposed to blue light and controls.

Results

This study successfully demonstrates the bactericidal effects of 405- and 465-nm LEDs on selected bacterial fish and shellfish pathogens. As shown in Fig. 2, growth of the pathogens evaluated was clearly inactivated by exposure to either a 405- or 465-nm LED, although the degree of inactivation varied depending on bacterial species and sampling time point. The one exception was that a 465-nm LED was unable to inactivate *V. harveyi*, but that strain was inactivated by a 405-nm LED. Complete inactivation of *A. salmonicida* and *S. parauberis* was seen 24 h after irradiation with a 405-nm LED, whereas only 6 h were required for complete inactivation of *V. anguillarum* and *P. damselae* under the same conditions. Although *S. iniae* was more rapidly inhibited by a 465-nm LED, overall, there were no differences between 405 and 465 nm LEDs in the inactivation rate of *S. parauberis*.

BE was measured in this study using a method modified from one that was previously described (Maclean et al. 2009). Details of the inactivation parameters for all bacterial species are listed in decreasing order of BE in Table 2. We calculated BE using exposure time, which was deduced from the time at which bacterial populations reached 10 CFU ml^{-1}. *P. damselae*, *V. anguillarum*, and *E. tarda* were the most susceptible bacteria, while *S. parauberis* was the least susceptible, to exposure to a 405-nm LED. Our results show that Gram-negative bacteria, such as *P. damselae* (36.1 J cm^{-2}), *V. anguillarum* (41.2 J cm^{-2}), and *E. tarda* (68.4 J cm^{-2}), seem to be more sensitive to a 405-nm LED light than are Gram-positive bacteria like *S. parauberis* (153.8 J cm^{-2}) and *S. iniae* (90.4 J cm^{-2}) (Table 2). However, some Gram-negative bacteria such as *A. salmonicida* (98.7 J cm^{-2}) and *V. harveyi* (126.4 J cm^{-2}) have lower susceptibility than *S. iniae*.

The degree of inactivation of bacterial suspensions with varying initial population densities in BHIB + 1 % NaCl following exposure to a 405- or 465-nm LED for

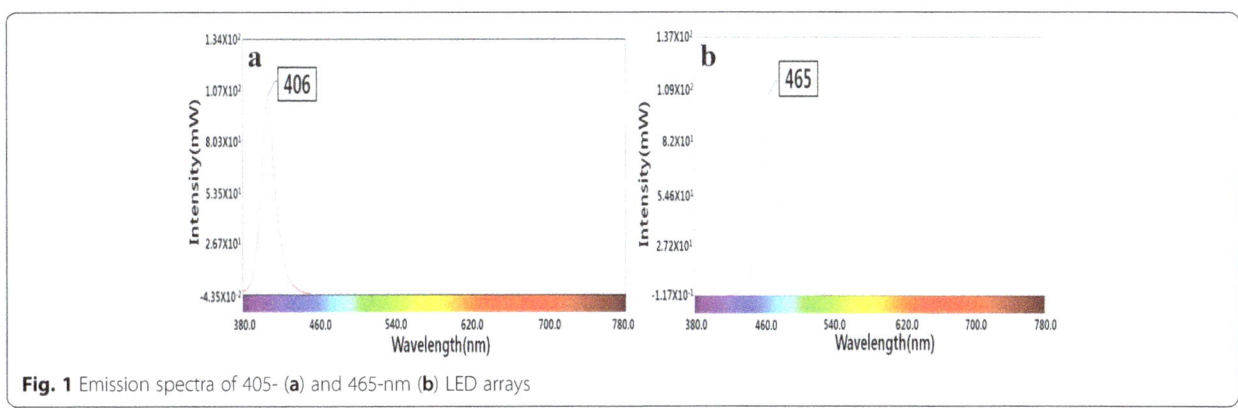

Fig. 1 Emission spectra of 405- (**a**) and 465-nm (**b**) LED arrays

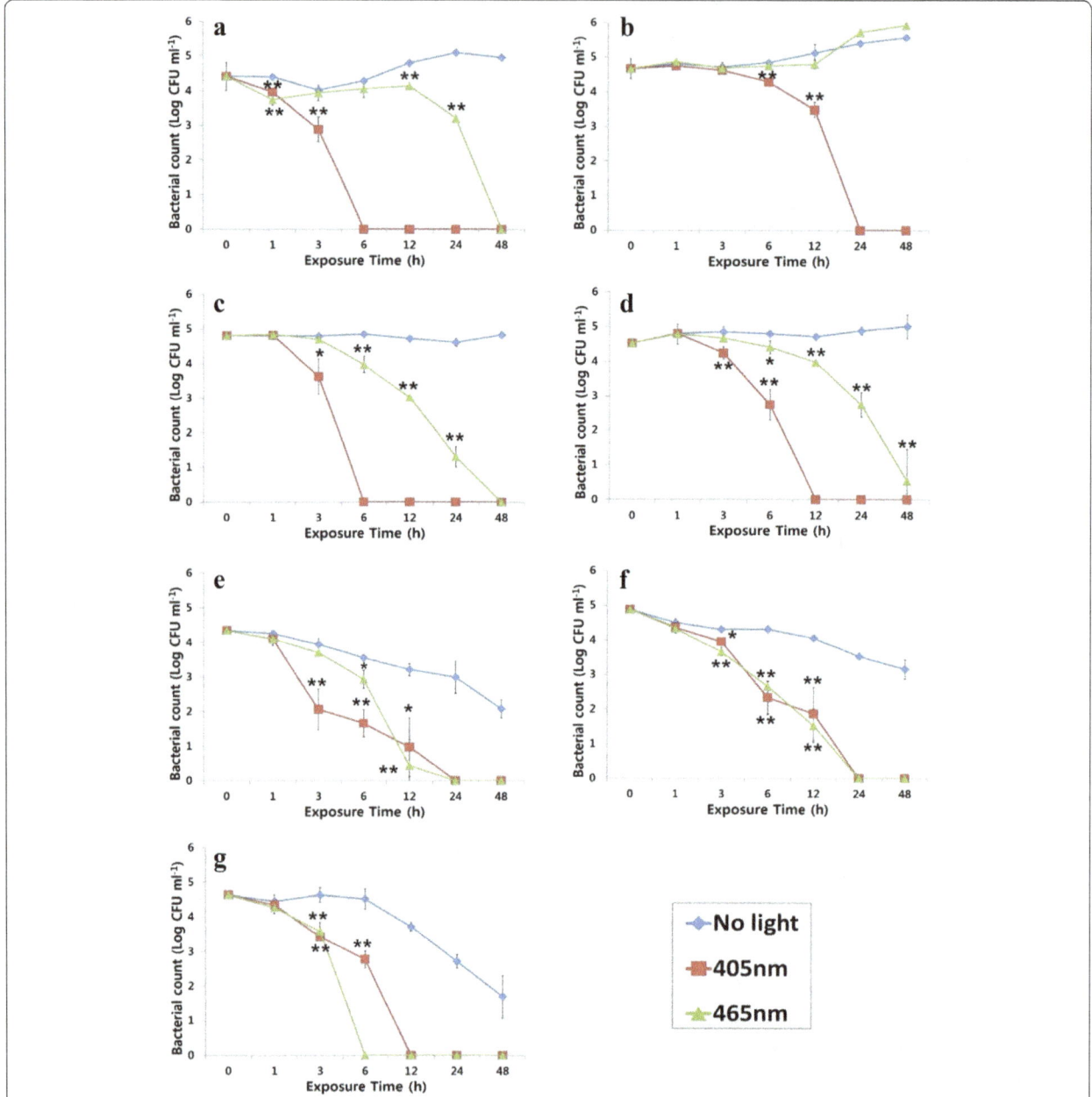

Fig. 2 Viable bacterial counts of *V. anguillarum* (**a**), *V. harveyi* (**b**), *P. damselae* (**c**), *E. tarda* (**d**), *A. salmonicida* (**e**), *S. parauberis* (**f**), and *S. iniae* (**g**) in phosphate buffered saline at 25 °C and several sampling time points (1, 3, 6, 12, 24, and 48 h) after LED exposure in a 405- or 465-nm LED (respectively, 250 μ mol m^{-2} s^{-1} or 516 μ mol m^{-2} s^{-1}) * significant difference, $P<0.05$; **significant difference, $P<0.01$

24 h is displayed in Table 3. In general, the OD values indicate that the higher starting bacterial densities were associated with lower inactivating efficacies. However, there were exceptions: unlike the other bacterial species, *P. damselae* exposed to a 405- or 465-nm LED and *V. harveyi* exposed to a 465-nm LED were not affected by their initial concentrations. *P. damselae* was able to survive a 405- or 465-nm light exposure in BHIB + 1 % NaCl, but it was much more susceptible when suspended in PBS.

Discussion

Antimicrobials are commonly used in aquaculture to prevent and treat bacterial infections in fish. Significant increases in the bacterial resistance to various antibiotics, such as oxytetracycline, quinolones, and amoxicillin, have been repeatedly found in proximity to fish farms (Defoirdt et al. 2011; Guardabassi et al. 2000; Schmidt et al. 2000). Excessive use of antimicrobials may significantly reduce their effectiveness and their usefulness in aquaculture. More importantly, studies have

Table 2 Energy levels and bactericidal efficiencies for the inactivation of bacterial species using 405- and 465-nm LEDs

Species	Wavelength (nm)	Dose (J cm^{-2})	Log$_{10}$ reduction (±SD)	Dose/log$_{10}$ reduction (±SD)	BE[a] (±SD)
Photobacterium damselae	405	137 (±4)	3.8	36.1 (±1.0)	0.028 (±0.001)
	465	1387 (±126)	3.7	374.7 (±34.1)	0.003 (±0)
Vibrio anguillarum	405	132 (±4)	3.2	41.2 (±1.3)	0.024 (±0.001)
	465	1934 (±3)	4	483.6 (±0.7)	0.002 (±0)
Edwardsiella tarda	405	260 (±10)	3.8	68.4 (±2.7)	0.015 (±0.001)
	465	2178 (±517)	4.0	544.4 (±129.2)	0.002 (±0)
Streptococcus iniae	405	262 (±5)	2.9	90.4 (±1.6)	0.011 (±0)
	465	247 (±3)	3.5	70.5 (±1)	0.014 (±0)
Aeromonas salmonicida	405	345 (±131)	3.5	98.7 (±37)	0.011 (±0.001)
	465	555 (±10)	3.3	168.1 (±40.6)	0.006 (±0)
Vibrio harveyi	405	543.7 (±4.4)	4.3	126.4 (±1.0)	0.008 (±0)
	465	NA	NA	NA	NA
Streptococcus parauberis	405	446 (±66)	2.9	153.8 (±22.8)	0.007 (±0.001)
	465	743 (±153)	2.9	256.2 (±52.7)	0.004 (±0.001)

[a]Bactericidal efficiency, calculated as log$_{10}$ (10/N_0) per radiant flux (J cm^{-2})
N_0 initial bacterial population in CFU ml^{-1}, NA not applicable, BE bactericidal efficiency

demonstrated that resistance plasmid for some antibiotics can be shared between bacterial fish pathogens, aquatic bacteria, and human pathogens, and some of them appear to have originated in the aquatic environment (Cabello et al. 2013). Thus, non-antibiotic therapies for infectious diseases are receiving considerable attention (Jori et al. 2006; Maisch 2009). It was previously demonstrated that blue light has a broad-spectrum bactericidal effect on both Gram-negative and Gram-positive bacteria (Dai et al. 2012; Maclean et al. 2009). In this study, growth of the bacterial fish and shellfish pathogens evaluated was clearly inactivated by exposure to either a 405- or 465-nm LED light. Inactivation was dependent on light intensity and exposure time. Overall, our results show that Gram-negative bacteria, such as *P damselae* (36.1 J cm^{-2}), *V. anguillarum* (41.2 J cm^{-2}), and *E. tarda* (68.4 J cm^{-2}),

Table 3 Relative growth of pathogenic bacteria with different initial population densities when exposed to 405- and 465-nm LED arrays for 24 h

Bacteria	Wavelength (nm)	Relative growth (%)[a] (± standard deviation) Initial bacterial cell counts (log CFU ml^{-1})					
		3	4	5	6	7	8
Vibrio harveyi	405	0 (±0)[b]	2 (±1)[b]	74 (±4)[c]	77 (±3)[c]	86 (±12)[c]	83 (±2)[c]
	465	70 (±61)[b]	68 (±2)[b]	72 (±2)[b]	69 (±3)[b]	78 (±18)[b]	76 (±2)[b]
Vibrio anguillarum	405	0 (±0)[b]	1 (±2)[b]	0 (±0)[b]	0 (±0)[b]	0 (±0)[b]	31 (±10)[c]
	465	0 (±0)[b]	0 (±0)[b]	0 (±0)[b]	107 (±40)[c]	107 (±30)[c]	105 (±25)[c]
Edwardsiella tarda	405	0 (±1)[b]	0 (±0)[b]	0 (±0)[b]	0 (±0)[b]	1 (±1)[b]	71 (±10)[c]
	465	1 (±1)[b]	0 (±0)[b]	76 (±18)[c]	121 (±24)[d]	86 (±6)[c]	83 (±2)[c]
Aeromonas salmonicida	405	0 (±0)[b]	0 (±0)[b]	0 (±0)[b]	0 (±0)[b]	0 (±0)[b]	0 (±0)[b]
	465	0 (±0)[b]	0 (±0)[b]	0 (±0)[b]	0 (±0)[b]	0 (±0)[b]	44 (±4)[c]
Photobacterium damselae	405	77 (±3)[b]	105 (±2)[d]	109 (±12)[d]	103 (±4)[d]	86 (±8)[b, c]	99 (±1)[c, d]
	465	103 (±1)[b]	108 (±17)[b]	95 (±8)[b]	86 (±7)[c]	66 (±6)[b, c]	88 (±3)[b, c]
Streptococcus iniae	405	0 (±0)[b]	0 (±0)[b]	0 (±0)[b]	16 (±6)[b]	84 (±2)[b]	95 (±3)[c]
	465	1 (±1)[b]	0 (±0)[b]	0 (±0)[b]	1 (±0)[c]	2 (±1)[c]	51 (±8)[d]
Streptococcus parauberis	405	0 (±0)[b]	0 (±0)[b]	0 (±0)[b]	7 (±6)[b, c]	10 (±2)[c]	44 (±5)[d]
	465	0 (±0)[b]	0 (±1)[b]	0 (±0)[b]	11 (±2)[c]	10 (±4)[c]	44 (±6)[d]

Different letters (b, c, and d) in the superscripts indicate significant differences (P<0.05)
[a]Ratio of optical density (OD) of control to OD of LED treatment group (treatment OD$_{630\ nm}$/control OD$_{630\ nm}$)

seem to be more sensitive to a 405-nm light than are Gram-positive bacteria like *S. parauberis* (153.8 J cm^{-2}) and *S. iniae* (90.4 J cm^{-2}). This result does not agree with a previous study which showed that Gram-positive bacteria such as *Staphylococcus*, *Clostridium*, and *Streptococcus* species were more susceptible to LED light than Gram-negative bacteria. Exceptions have been reported; *Enterococcus faecalis* suspensions exposed to 10 mW cm^{-2} light for up to 120 min experienced negligible inactivation (Maclean et al. 2009). Another study also found that the Gram-positive *Listeria monocytogenes* was more resistant to a 405-nm light than was the Gram-negative *Salmonella enterica* on acrylic and PVC surfaces (Murdoch et al. 2012). Taken together, it appears that Gram-positive bacteria are not always more rapidly inactivated than Gram-negative bacteria. The BE observed in this study are much lower than those seen in a previous study. This is because it took bacterial counts nine times over 200 min, which was much more frequent than in our study, where sampling was done only seven times over 48 h (Maclean et al. 2009). In addition, we used 250 μ mol m^{-2} s^{-1} (approximately 7.4 mW cm^{-2}) and 516 μ mol m^{-2} s^{-1} (approximately 13.3 mW cm^{-2}) intensities of 405- and 465-nm light, respectively, which are approximately 1.5–10 times lower than those used in previous studies (e.g., 19.5 mW cm^{-2} of 415 nm, 100 mW cm^{-2} of 415 or 455 nm, or 10 mW cm^{-2} of 405 nm) (Dai et al. 2013; Lipovsky et al. 2010; Maclean et al. 2009). This is one likely explanation as to why inactivation of pathogens in this study took longer than in previous studies. The precise mode of action of the antimicrobial effect of blue light is not yet fully understood. The commonly accepted hypothesis is that blue light excites endogenous intracellular porphyrins, which then behave as photosensitizers; photon absorption leads to energy transfer and, ultimately, the production of highly toxic reactive oxygen species (ROS) (Ashkenazi et al. 2003; Hamblin et al. 2005; Maclean et al. 2008). The differences in inactivation kinetics found in this study may be caused by organism-specific differences in porphyrin levels or porphyrin types, as suggested previously. The peak absorption wavelengths of different bacterial porphyrins may differ, and varying wavelengths may be required for their maximum photostimulation (Maclean et al. 2010). The degree of inactivation of bacterial suspensions with different initial densities was determined in order to assess LED activity on pathogens in the presence of nutrients mimicking a natural aquatic environment. *P. damselae* was able to survive a 405- or 465-nm light exposure when cultured on nutrient-enriched environment but was much more susceptible when suspended in PBS, as shown in Fig. 2. Several studies have reported that bacterial pathogens, including *Escherichia coli*, *A. salmonicida*, *Streptococcus pneumoniae*, and *V. harveyi*, produce different superoxide dismutase (SOD) and catalase isozymes inducible under certain growth conditions (Barnes et al.

1996; Flint et al. 1993; Vattanaviboon and Mongkolsuk 2001; Yesilkaya et al. 2000). However, *P. damselae* is not able to produce different SOD or catalase isozymes when exposed to oxidative stress induced by hydrogen peroxide, or under iron-depleted conditions (Díaz-Rosales et al. 2006). Also, *P. damselae*, possessing a high-affinity iron uptake system, grown under iron-limited conditions have a reduced amount of capsular material covering the cells (Do Vale et al. 2001; Naka et al. 2005). These indicate that *P. damselae* grown under nutrient-enriched conditions would be more resistant to oxidative stress (ROS) induced by LED irradiation than when grown under iron-limiting conditions (e.g., PBS). As it has been already demonstrated that blue light has caused no or very little damage to teleost (Migaud et al. 2007; Villamizar et al. 2011), it might be an alternative method to treat and prevent bacterial diseases in fish farm.

Conclusions

To the best of our knowledge, this study is the first to demonstrate that blue light is capable of inactivating major aquatic pathogens without requiring any external photosensitizer. As it is generally accepted that blue light is much less harmful to animal cells than is UV irradiation, and caused little damage to teleost that have already been demonstrated in previous studies (Migaud et al. 2007; Villamizar et al. 2011), application of blue light might be alternative to the use of antibiotics in aquaculture and would also have safety benefits. We hope our results will inspire further experiments to explore practical applications of blue light to fish and shellfish.

Acknowledgements
This research was a part of the project titled "LED-Marine Technology Convergence R&D Center" funded by the Ministry of Oceans and Fisheries, Korea.

Funding
This study was funded by the Ministry of Oceans and Fisheries, Korea.

Authors' contributions
HJ mainly conducted the acquisition, analysis, and interpretation of the data and drafted and revised the manuscript. AR and GS participated in the design and cooperation of this work including acquiring the bacterial density and OD data. DH contributed to the conception and design of this work, the interpretation of the data, and the drafting and revising of the manuscript. All authors approved the final manuscript.

Competing interests
Authors HJ, AR, GS and DH are listed as inventors on a related patent application to the Korean Intellectual Property Office entitled "Inactivation methods aquatic bacterial pathogens using blue LED light" (Patent application number: 10-2015-0087358). The datasets supporting the conclusions of this article are available for non-commercial purpose upon request to the authors.

References

Arrojado C, Pereira C, Tomé JPC, Faustino MAF, Neves MGPMS, Tomé AC, Cavaleiro JAS, Cunha Â, Calado R, Gomes NCM, Almeida A. Applicability of photodynamic antimicrobial chemotherapy as an alternative to inactivate fish pathogenic bacteria in aquaculture systems. Photochem Photobiol Sci. 2011;10:1691–700.

Ashkenazi H, Malik Z, Harth Y, Nitzan Y. Eradication of *Propionibacterium acnes* by its endogenic porphyrins after illumination with high intensity blue light. FEMS Immunol Med Microbiol. 2003;35:17–24.

Baeck GW, Kim JH, Gomez DK, Park SC. Isolation and characterization of *Streptococcus* sp. diseased flounder (*Paralichthys olivaceus*) in Jeju Island. J Vet Sci. 2006;7:53–8.

Barnes AC, Horne MT, Ellis AE. Effect of iron on expression of superoxide dismutase by *Aeromonas salmonicida* and associated resistance to superoxide anion. FEMS Microbiol Lett. 1996;142:19–26.

Bostock J, McAndrew B, Richards R, Jauncey K, Telfer T, Lorenzen K, Little D, Ross L, Handisyde N, Gatward I, Corner R. Aquaculture: global status and trends. Philos T Roy Soc B. 2010;365:2897–912.

Cabello FC, Godfrey HP, Tomova A, Ivanova L, Dölz H, Millanao A, Buschmann AH. Antimicrobial use in aquaculture re-examined: its relevance to antimicrobial resistance and to animal and human health. Envrion Microbiol. 2013;15:1917–42.

Dai T, Gupta A, Murray CK, Vrahas MS, Tegos GP, Hamblin MR. Blue light for infectious disease : *Propionibacterium acnes*, *Helicobacter pylori*, and beyond? Drug Resist Updat. 2012;15:223–36.

Dai T, Gupta A, Huang YY, Tin R, Murray CK, Vrahas MS, Sherwood ME, Tegos GT, Hamblin MR. Blue light resques mice from potentially fatal *Pseudomonas aeruginosa* burn infection: efficacy, safety, and mechanism of action. Antimicrob Agents Chemother. 2013;57:1238–45.

Defoirdt T, Sorgeloos P, Bossier P. Alternatives to antibiotics for the control of bacterial disease in aquaculture. Curr Opin Microbiol. 2011;14:251–8.

Díaz-Rosales P, Chabrillón M, Arijo S, Martinez-Manzanares E, Moriñigo MA, Balebona MC. Superoxide dismutase and catalase activities in *Photobacterium damselae* ssp. *Piscicida*. J Fish Dis. 2006;29:355–64.

Do Vale A, Ellis AE, Silva MT. Electron microscopic evidence that expression of capsular polysaccarhide by *Photobacterium damselae* subsp. *piscicida* is dependent on iron availability and growth phase. Dis Aquat Organ. 2001;44:237–40.

Flint DH, Tuminello JF, Emptage MH. The inactivation of Fe-S cluster containing hydrolases by superoxide. J Biol Chem. 1993;268:22369–76.

Ghanizadeh KE, Khodabandeh S. Effects of ultraviolet radiation on skin structure and ultrastructure in Caspian Sea salmon, *Salmo trutta caspius*, during alevin stage. Toxicol Environ Chem. 2010;92:903–14.

Guardabassi L, Dalsgaard A, Raffatellu M, Olsen JE. Increase in the prevalence of oxolinic acid resistant *Acinetobacter* spp. Observed in a stream receiving the effluent from a freshwater trout farm following the treatment with oxolinic acid-medicated feed. Aquaculture. 2000;188:205–18.

Hamblin MR, Viveiros J, Yang C, Ahmadi A, Ganz RA, Tolkoff MJ. *Helicobacter pylori* accumulates photoactive porphyrins and is killed by visible light. Antimicrob Agents Chemother. 2005;4:2822–7.

Han HJ, Kim DH, Lee DC, Kim SM, Park SI. Pathogenicity of *Edwardsiella tarda* to olive flounder, *Paralichthys olivaceus*. J Fish Dis. 2006;29:601–9.

Han HJ, Kim DY, Kim WS, Kim CS, Jung SJ, Oh MJ, Kim DH. Atypical *Aeromonas salmonicida* infection in the black rockfish, *sebastes schlegeli* Hilgendorf, in Korea. J Fish Dis. 2011;34:47–55.

Jori G, Fabris C, Soncin M, Ferro S, Coppellotti O, Dei D, Fantetti L, Chiti G, Roncucci G. Photodynamic therapy in the treatment of microbial infections: basic principles and perspective applications. Laser Surg Med. 2006;38:468–81.

Kaweewat K, Hofer R. Effect of UV-B radiation on goblet cells in the skin of different fish species. J Photochem Photobiol B. 1997;41:222–6.

Kleinpenning MM, Smits T, Frunt MHA, Van Erp PEJ, Van de Kerkhof PCM, Gerritsen RMJP. Clinical and histological effects of blue light on normal skin. Photodermatol Photo. 2010;26:16–21.

Lipovsky A, Nitzan Y, Gedanken A, Lubart R. Visible light-induced killing of bacteria as a function of wavelength: implication for wound healing. Laser Surg Med. 2010;42:467–72.

Maclean M, Macgregor SJ, Anderson JG, Woolsey G. The role of oxygen in the visible-light inactivation of *Staphylococcus aureus*. J Photochem Photobiol B. 2008;92:180–4.

Maclean M, Macgregor SJ, Anderson JG, Woolsey G. Inactivation of bacterial pathogens following exposure to light from a 405-nanometer light-emitting diode array. Appl Environ Microbiol. 2009;75:1932–7.

Maclean M, MacGregor SJ, Anderson JG, Woolsey GA, Coia JE, Hamilton K, Taggart I, Watson SB, Thakker B, Gettinby G. Environmental decontamination of a hospital isolation room using high-intensity narrow-spectrum light. J Hosp Infect. 2010;76: 247–51.

Maisch T. A new strategy to destroy antibiotic resistant microorganisms: antimicrobial photodynamic treatment. Mini-Rev Med Chem. 2009;9:974–83.

Mata AI, Gibello A, Casamayor A, Blanco MM, Domínguez L, Fernández-Garayzábal JF. Multiplex PCR assay for detection of bacterial pathogens associated with warm-water streptococcosis in fish. Appl Environ Microbiol. 2004;70:3183–7.

Migaud H, Cowan M, Taylor J, Ferguson HW. The effect of spectral composition and light intensity on melatonin, stress and retinal damage in post-smolt Atlantic salmon, *Salmo salar*. Aquaculture. 2007;270:390–404.

Milton DL, Toole RO', Horstedt P, Wolf-Watz H. Flagellin A is essential for the virulence of *Vibrio anguillarium*. J Bacteriol. 1996;178:1310–9.

Murdoch LE, Maclean M, Endarko E, MacGregor SJ, Anderson JG. Bactericidal effects of 405 nm light exposure demonstrated by inactivation of *Escherichia*, *Salmonella*, *Shigella*, *Listeria*, and *Mycobacterium* species in liquid suspensions and on exposed surfaces. Sci World J. 2012;2012:1–8.

Naka H, Hirono I, Aoki T. Molecular cloning and functional analysis of *Photobacterium damselae* subsp. *piscicida* heme receptor gene. J Fish Dis. 2005;28:81–8.

Osorio CR, Toranzo AE, Romalde JL, Barja JL. Multiplex PCR assay for ureC and 16S rRNA genes clearly discriminates between both subspecies of *Photobacterium damselae*. Dis Aquat Organ. 2000;40:177–83.

Pang L, Zhang XH, Zhong Y, Chen J, Li Y, Austin B. Identification of *Vibrio haveyi* using PCR amplification of the ToxR gene. Lett Appl Microbiol. 2006;43:249–55.

Pridgeon JW, Klesius PH. Major bacterial diseases in aquaculture and their vaccine development. In: Hemming D, editor. Anim Sci Rev 2012. UK: CAB International; 2013. p. 141–56.

Romalde JL. *Photobacterium damselae* subsp. *Piscicida*: an integrated view of a bacterial fish pathogen. Int Microbiol. 2002;5:3–9.

Schmidt AS, Bruun MS, Dalsgaard I, Pedersen K, Larsen JL. Occurrence of antimicrobial resistance in fish-pathogenic and environmental bacteria associated with four Danish rainbow trout farms. Appl Environ Microbiol. 2000;66:4908–15.

Subasinghe RP, Bondad-Reantaso MG, McGladdery SE. Aquaculture development, health and wealth. In: Subasinghe RP, Bueno P, Phillips MJ, Hough C, McGladdery SE, Arthur JR, editors. In proceedings of the Conference on Aquaculture in the Third Millennium: 20-25 February 2000. Bangkok: NACA and FAO; 2001. p. 167–91.

Vattanaviboon P, Mongkolsuk S. Unusual adaptive, cross protection responses and growth phase resistance against peroxide killing in a bacterial shrimp pathogen. *Vibrio harveyi*. FEMS Microbiol Lett. 2001;200:111–6.

Villamizar N, Blanco-Vives N, Migaud H, Davie A, Carboni S, Sánchez-Vázquez FJ. Effects of light during early larval development of some aquacultured teleosts: a review. Aquaculture. 2011;315:86–94.

Weinstein MR, Litt M, Kertesz DA, Wyper P, Rose D, Coulter M, McGeer A, Facklam R, Ostach C. Invasive infections due to a fish pathogen, *Streptococcus iniae*. New Engl J Med. 1997;28:589–94.

Wiklund T, Dalsgaard I. Occurrence and significance of atypical *Aeromonas salmonicida* in non-salmonid and salmoid fish species. Dis Aquat Organ. 1998;32:49–69.

Won KM, Park SI. Pathogenicity of *Vibrio harveyi* to culture marine fishes in Korea. Aquaculture. 2008;285:8–13.

Yesilkaya H, Kadioglu A, Gingles N, Alexander JE, Mitchell TJ, Andrew PW. Role of manganese-containing superoxide dismutase in oxidative stress and virulence of *Streptococcus pneumoniae*. Infect Immun. 2000;68:2819–26.

Yin Y, Gupta A, Hamblin MR. Light based anti-infectives: ultraviolet C irradiation, photodynamic therapy, blue light, and beyond. Curr Opin Pharmacol. 2013; 13:731–62.

7

Production optimization of flying fish roe analogs using calcium alginate hydrogel beads

Bom-Bi Ha[1], Eun-Hee Jo[1], Suengmok Cho[2] and Seon-Bong Kim[1*]

Abstract

Due to decreased supplies of marine resources and byproducts, new processing technologies for the development of analogs for natural fishery products are becoming increasingly important in the fishing industry. In the present study, we investigated the optimal processing conditions for flying fish roe analogs based on alginate hydrogels. Optimized processing of these analogs was performed by response surface methodology. The optimal processing conditions for the flying fish roe analogs (based on sphericity) were at a sodium alginate concentration of 2.41 %, calcium chloride solution curing time of 40.65 min, calcium chloride concentration of 1.51 %, and a reactor stir speed of 254×g. When the experiment was performed under these optimized conditions, the size (mm), sphericity (%), and rupture strength (kPa) of the analogs were 2.2 ± 0.12, 98.2 ± 0.2, and 762 ± 24.68, respectively, indicating physical properties similar to their natural counterparts.

Keywords: Fish roe analogs, Analog food, Alginate hydrogel, Response surface methodology, Calcium alginate

Abbreviations: ANOVA, Analysis of variance; RSM, Response surface methodology; RSREG, Response surface regression; SAS, Statistical Analysis System

Background

Encapsulation technologies featuring alginate hydrogels have been widely applied to the bioprocessing of food, cosmetics, pharmaceuticals, and other biomaterials. Alginate, a gelling polymer, is composed of linear polymers of 1–4 linked β-D-mannuronic acid (M) and α-L-guluronic acid (G) residues (Moe et al. 1995; Onsøyen 1997). In the presence of divalent cations such as Ca^{2+}, alginate forms elastic hydrogels; the "egg-box" model has been adopted as a general description of alginate gel formation (Rousseau et al. 2004; Clark and Ross Murphy 1987). Alginate hydrogels have been used for the encapsulation of highly viscous high-fat food (Blandino et al. 1999), nutrients (Chen and Subirade 2006), bioactive components (Wichchukit et al. 2013), and probiotics (Subirade et al. 2010). In addition, they can protect acid-sensitive drugs from gastric fluids allowing for controlled drug release in the small intestine and are relevant to peptic ulcers (Hwang et al. 1993). Although there has been an increased number of applications of alginate hydrogels in biotechnology (Onsøyen 1996; Skjåk-bræk and Espevik 1996), studies on their use in food analogs is limited except when used as caviar (Ji et al. 2007a,b) and *Cypselurus agoo* roe analogs (Jo et al. 2014). Regarding the rheological characteristics of alginate hydrogels, their use in encapsulation may be a useful means of producing fish roe analogs.

Flying fish roe is one of the most consumed types of fish roes in Asia. Natural flying fish roe is widely used in sushi and for other culinary purposes in Korea and Japan. However, a consistent supply of these materials is difficult to secure due to abnormal climate changes and protective fishing policies of individual nations. Owing to this imbalance between supply and demand of these raw materials, there are limitations on their sustainable production. Therefore, the demand of suitable analogs is gradually increasing. It is also necessary to develop such analogs in order to eventually replace the natural

* Correspondence: owlkim@pknu.ac.kr
[1]Department of Food Science and Technology/Institute of Food Science, Pukyong National University, 45, Yongso-ro, Nam-gu, Busan 48513, South Korea
Full list of author information is available at the end of the article

materials, by establishing systemic processes and mass production techniques regardless of resource depletion.

In the present study, we optimize the processing conditions of these roe analogs using alginate hydrogels. The optimum conditions for production of the analogs were determined based on concentrations of sodium alginate and calcium chloride, sphericity, size, and rupture strength by using response surface methodology (RSM).

Methods
Materials
Sodium alginate and anhydrous calcium chloride (Junsei Chemical Co., Ltd., Japan) were used for gelation. All other chemicals and reagents used in this study were analytical grade.

Production of flying fish roe analogs
Flying fish roe analogs were produced using a technique first reported by Jo et al. (2014), with some modifications. A sodium alginate solution was prepared at different concentrations (0.4–3.6 %, w/v) and then dropped into 0.5–2.5 % (w/v) calcium chloride at 0.06 mL/s using a peristaltic pump (Cassette tube pump SMP-23, Eyela, Japan) with a single nozzle (26G × 1/2″) connected to a silicon tube. The stirring speed of calcium chloride solution in the reactor was set at 100–500×g. The mixture was dropped every 2 min and cured for 0–60 min in solution to obtain the analogs. Afterwards, the analogs were collected with a strainer, then washed with deionized water, and stored at ambient temperature. The distance between the nozzle and the surface of the calcium chloride solution was 8 cm.

Size measurements
Sizes were measured according to the method of Jo et al. (2014). An image analyzer (Image-Pro program) coupled to an optical microscope (Bx-50, Olympus, Japan) was used, with image analysis at ×40 magnification. Five prepared analogs were randomly selected for each measurement and the average of their maximum and minimum diameters reported.

Sphericity measurements
Sphericity was measured according to the method of Jo et al. (2014) and expressed as the percent ratio of

minimum diameter to maximum diameter obtained from size measurements of the analogs.

Rupture strength measurement
Rupture strengths were measured according to the method of Jo et al. (2014), using a rheometer (Model CR-100D, Sun Scientific Co., Ltd., Japan) with the following conditions: round-disk stainless steel plunger 10 mm in diameter, 40 mm/min penetration speed, adapter area of 0.79 cm^2, sample-adapter distance of 5 mm, and 1 kN load-cell. Five samples were measured in each experiment.

Experimental design
Central composite design (CCD) was adopted in the optimization of flying fish roe analog processing. The CCD in this design consists of 2^2 factorial points, four axial points ($\alpha = 2$), and three replicates of the central point (Tables 1 and 2). Concentrations of sodium alginate (X_1, %), curing time in calcium chloride solution (X_2, min), calcium chloride concentration (X_3, %), and calcium chloride solution stirring speed in the reactor (X_4, ×g) were chosen as independent variables (IVs). The range and center point values of the four IVs were based on the results of preliminary experiments (Table 1). In order to prepare superior flying fish roe analogs in terms of external appearance, sphericity (Y_1, %) was used as the dependent variable (DV). Experimental runs were randomized in order to minimize the effects of unexpected variability in the observed responses.

Data analysis
The response surface methdology (RSREG procedure) of the MINITAB statistical software (Version 14, Minitab Inc., PA, USA) was used to fit the following second-order polynomial:

$$Y = \beta_0 + \sum_{i=1}^{4} \beta_i X_i + \sum_{i=1}^{4} \beta_{ii} X_i^2 + \sum_{i=1}^{3} \sum_{j=i+1}^{4} \beta_{ij} X_i X_j \quad (1)$$

Here, Y is the DV, β_0 is a constant, β_i, β_{ii}, and β_{ij} are regression coefficients, and X_i and X_j are levels of the IVs. The response surface plots were developed using Maple software (Maple 7, Waterloo Maple Inc., Canada) and represent a function of two independent variables

Table 1 Experimental ranges and values of the independent variables in the central composite design for manufacturing process of the flying fish roe analogs

Independent variables (IVs)	Symbol	Range and levels				
		−2	−1	0	+1	+2
Sodium alginate concentration (%, w/v)	X_1	0.4	1.2	2	2.8	3.6
Curing time in calcium chloride solution (min)	X_2	0	15	30	45	60
Calcium chloride concentration (%, w/v)	X_3	0.5	1	1.5	2	2.5
Rotation speed of calcium chloride solution (×g)	X_4	100	200	300	400	500

Table 2 Central composite design and responses of the dependent variables for the flying fish roe analogs processing to the independent variables

Run order	Coded level				Response
	X_1	X_2	X_3	X_4	Y_1
Factorial portion					
1	−1	−1	−1	−1	84.2
2	1	−1	−1	−1	91.7
3	−1	1	−1	−1	91.3
4	1	1	−1	−1	95.5
5	−1	−1	1	−1	84.2
6	1	−1	1	−1	92.0
7	−1	1	1	−1	90.0
8	1	1	1	−1	95.5
9	−1	−1	−1	1	81.0
10	1	−1	−1	1	91.7
11	−1	1	−1	1	82.6
12	1	1	−1	1	91.7
13	−1	−1	1	1	81.8
14	1	−1	1	1	94.4
15	−1	1	1	1	89.5
16	1	1	1	1	95.2
Axial portion					
17	−2	0	0	0	68.4
18	2	0	0	0	95.5
19	0	−2	0	0	91.7
20	0	2	0	0	94.7
21	0	0	−2	0	88.9
22	0	0	2	0	85.7
23	0	0	0	2	95.0
24	0	0	0	2	86.4
Center portion					
25	0	0	0	0	97.5
26	0	0	0	0	98.2
27	0	0	0	0	97.7

Y (sphericity, %), X_1 (sodium alginate concentration, %), X_2 (curing time in calcium chloride solution, min), X_3 (calcium chloride concentration, %), X_4 (rotation speed of calcium chloride solution, ×g)

while keeping the other two independent variables at their optimal values.

Statistical analysis

All the results were conducted to analysis of variance at a level of $P < 0.05$, and the means were set apart using Duncan's multiple range tests ($\alpha = 0.05$). The analysis of data was subjected through the RSREG procedure of SAS software.

Results and discussion
Optimization for the production of the analogs using RSM

Process optimization was used to determine the conditions needed to obtain analogs most closely resembling natural ones. Sphericity, which indicates the degree of similarity between the analogs and a sphere, was defined as the ratio of minimum diameter to maximum diameter of the prepared analogs. The most important factors affecting sphericity (Y, sphericity, %) were determined in the preliminary study, with the central point and ranges determined by CCD (Box and Wilson 1951). The independent variables and central point in the analog production were the sodium alginate concentration (2 %, w/v), curing time in calcium chloride solution (30 min), calcium chloride concentration (1.5 %, w/v), and stir speed of calcium chloride in the reactor (300×g) (Table 1). A total of 27 intervals were included in the experiment, and sphericity results of the analogs prepared in each interval are shown in Table 2. RSM was performed using the data obtained by the RSREG procedure of SAS software. Based on these results, the statistical significance each associated with the linear term (X_1, X_2, X_3, X_4), quadratic term (X_1X_1, X_2X_2, X_3X_3, X_4X_4), and interaction term was determined by t-statistics. Furthermore, the statistical significance of the second-order polynomial model was measured using the estimated coefficient of each model as well as analysis of variance (ANOVA) (Table 3). For the linear coefficient, X_1 ($P < 0.001$) and X_4 ($P = 0.0254$) exhibited statistical significance at $P < 0.05$ but X_2 ($P = 0.0175$) and X_3 ($P = 0.6162$) did not show any statistical significance.

For the quadratic coefficient, there was statistical significance found in X_1X_1 ($P < 0.001$), X_3X_3 ($P = 0.001$), and X_4X_4 ($P = 0.020$), but not X_2X_2 ($P = 0.133$). For the interaction coefficient, no significance was exhibited in X_1X_2 ($P = 0.217$), X_1X_3 ($P = 0.993$), X_1X_4 ($P = 0.249$), X_2X_3 ($P = 0.633$), X_2X_4 ($P = 0.369$), and X_3X_4 ($P = 0.194$). In order to develop fitted response surface model equations, all insignificant terms ($P > 0.05$) were eliminated and the resulting fitted models are shown in Table 4. The coefficient of determination in the polynomial R^2 was 0.924 at $P = 0.000$. Such a high coefficient of determination and significance arise from experimental conditions designed by preliminary studies. ANOVA was employed to evaluate the statistical significance of the second-order polynomial model, and ANOVA based on the dependent variables was applied to express a Y (sphericity, %) response model. In the ANOVA results, all linear terms ($P < 0.001$) and quadratic terms ($P < 0.001$), except the cross-product term ($P = 0.453$), had probabilities >99 % (Table 5). CCD was used to identify the optimum sphericity conditions in the prepared analogs. In preliminary studies, the optimum ranges of sodium alginate concentration (X_1, 2 %), curing time in calcium chloride solution

Table 3 Estimated coefficients of the fitted quadratic polynomial equation for different responses based on t-statistic

Parameter	Y (sphericity, %)	
	Coefficient	P value
Constant	97.80	0.000
X_1	4.89	0.000
X_2	1.51	0.018
X_3	0.27	0.633
X_4	−1.40	0.026
X_1X_1	−3.76	0.000
X_2X_2	−0.94	0.133
X_3X_3	−2.42	0.001
X_4X_4	−1.57	0.020
X_1X_2	−0.88	0.217
X_1X_3	0.01	0.993
X_1X_4	0.82	0.249
X_2X_3	0.33	0.633
X_2X_4	−0.63	0.369
X_3X_4	0.93	0.194

Y (sphericity, %), X_1 (sodium alginate concentration, %), X_2 (curing time in calcium chloride solution, min), X_3 (calcium chloride concentration, %), X_4 (rotation speed of calcium chloride, $\times g$)

(X_2, 30 min), calcium chloride concentration (X_3, 1.5 %), and stirring speed of the calcium chloride solution in the reactor (X_4, 300 × g) were determined (Table 1). RSREG resulted in saddle points having eigenvalues that show positive and negative values. Additionally, the optimum conditions (coded values) of the analogs based on RSM were 0.51 % for the sodium alginate concentration (X_1), 0.71 min for the calcium chloride solution curing time (X_2), 0.02 % for calcium chloride concentration (X_3), and –0.46×g for the calcium chloride solution stirring speed (X_4). When these results were substituted into

Table 4 Analysis of variance (ANOVA) for response of dependent variables

Response	Source		DF	SS	MS	F value	P value
Y	Regression	Linear	4	677.288	169.322	23.15	0.000
		Quadratic	4	343.637	85.909	11.74	0.000
		Cross-product	6	45.159	7.526	1.03	0.453
		Total model	14	1066.084	76.149	10.41	0.000
	Residual	Lack of fit	10	87.527	8.753	67.33	0.015
		Pure error	2	0.260	0.130		
		Total error	12	87.787	7.316		
	Total		26	1153.87			

DF degrees of freedom, *SS* sum of square, *MS* mean square

equation (1), X_1, X_2, X_3, and the uncoded value of X_4 were 2.41 %, 40.65 min, 1.51 %, and 254×g, respectively. When the experiment was carried out with the uncoded values as calculated above, the expected sphericity (Y, %) of the analogs was 99.9 % compared to an experimental value of 98.2 %, representing minimal difference (Table 6).

In general, alginates undergo gelling to yield hydrogels in the presence of metallic divalent cations (Smidsrød and Haug 1972; Grant et al. 1973); for example, calcium ions make stronger alginate gels than potassium and sodium ions (Montero and Perez-Mateos 2002). When alginate solutions are added into calcium chloride solutions, calcium ions bind to the carboxylic groups of G-block alginate molecules to yield alginate gels (Moe et al. 1995; Sabraa 2005). However, M-block rich alginates have weak binding with calcium and form more elastic gels (Sabraa 2005). Alginates with high M/G ratios produce beads with smaller sizes than in the case of low M/G ratio alginate beads (Kendal Jr. et al. 2004; Mørch et al. 2006), as alginates of lower M/G ratio form more porous gels (Simpson et al. 2003). Curing of alginate gels in calcium chloride solution is promoted by increased calcium concentration, which benefits the sphericity and physical strength of the gels.

Response surface plots and impact factors

Figure 1 shows a three-dimensional graph describing the effects of the independent variables (X_1, X_2, X_3, X_4) on the dependent variable (Y) using Maple software. Among the different analog production processes, the present study focused on the effects of sodium alginate concentration (X_1, %), curing time in calcium chloride solution (X_2, min), calcium chloride concentration (X_3, %), and calcium chloride solution stirring speed in the reactor (X_4, ×g) as key factors in analog production. The graph represents the correlation between X_1, X_2, X_3, and X_4. All response surface graphs indicate that the sphericity of the analogs increased as the coded values of IVs approached zero. In particular, the sphericity of the analogs (DV) decreased remarkably as X_1 approached –2. On the other hand, X_2 was not considerably influential on the sphericity of the analogs compared to the other independent variables. Therefore, based on Fig. 1, all independent variables except for the curing time in calcium chloride solution influenced the sphericity of the analogs, and the concentration of sodium alginate played an especially important role. Jo et al. (2014) described the effects of heat, salt, and hydrocolloids on analog formation, and Ji et al. (2007a,b) made caviar analogs using calcium alginate gel capsules. The shape transition of the Ca-alginate beads was

Table 5 Optimal conditions for the flying fish roe analogs manufacturing process from sodium alginate

Response			Y (sphericity, %)	
Optimal conditions	X_1	Coded value	0.51	
		Actual value (%)	2.41	
	X_2	Coded value	0.71	
		Actual value (min)	40.65	
	X_3	Coded value	0.02	
		Actual value (%)	1.51	
	X_4	Coded value	−0.46	
		Actual value (×g)	254	
Stationary point			Saddle point	
Predicted value of response Y			99.9	
Experimental value of response Y			98.2	

X_1 (sodium alginate concentration, %), X_2 (curing time in calcium chloride solution, min), X_3 (calcium chloride concentration, %), X_4 (rotation speed of calcium chloride, ×g)

distinguished into three phases based on collection distance and was affected by the combined influence of the solution properties, the collection distance and the drop size (Chan et al. 2009).

Conclusions

Demand for analogs of crab meat and caviar food is gradually increasing, due to recent abnormal climate issues and fishery policy changes. Thus, improved

Table 6 Comparison of physical properties of natural flying fish roe and the analogs manufactured using the optimal processing condition

Factors	Flying fish roe analogs	Natural flying fish roe
Size (mm)	2.2 ± 0.12^b	1.88 ± 0.02^a
Sphericity (%)	98.2 ± 0.2^a	98.19 ± 1.55^a
Rupture strength (kPa)	762 ± 24.68^a	803.6 ± 126.37^a

The same superscripts in a raw are not significantly different each other at $P < 0.05$

development of such analogs is essential to substitute for natural products. Alginate was adopted for the production of flying fish roe analogs in this study because it easily forms gels and is easy to handle, non-toxic to humans, and inexpensive. The study elucidated the production of flying fish roe analogs under optimized conditions using RSM. The optimum conditions were 2.41 % sodium alginate concentration, 40.65 min curing time in calcium chloride solution, 1.51 % calcium chloride concentration, and a $254 \times g$ stirring speed of the calcium chloride solution in the reactor. When performed at these optimum conditions, this process yielded a high sphericity of 98.2 %. The size (mm),

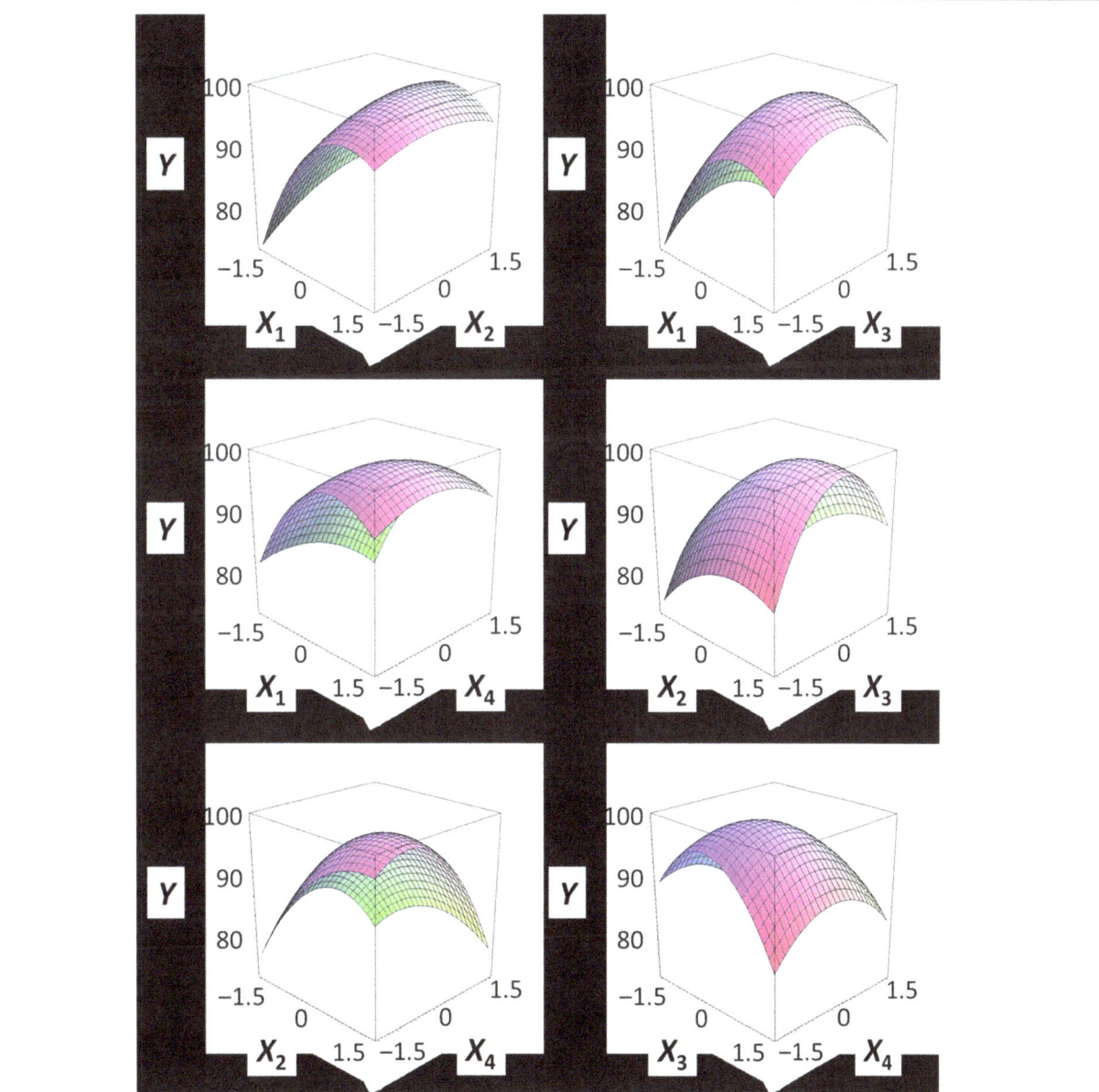

Fig. 1 Response surface plots for flying fish roe analogs manufacturing process from sodium alginate and calcium chloride. Y (sphericity, %), X_1 (sodium alginate concentration, %), X_2 (curing time in calcium chloride solution, min), X_3 (calcium chloride concentration, %), X_4 (rotation speed of calcium chloride, $\times g$)

sphericity (%), and rupture strength (kPa) of the analogs produced using these optimum conditions were 2.2 ± 0.12, 98.2 ± 0.2, and 762 ± 24.68, respectively; these physical properties are quite similar to those of natural roe. With subsequent improvements to palatability and mass production, these roe analogs may be viable substitutes for natural roe.

Funding
This work was supported by a Research Grant of Pukyong National University (2016).

Authors' contributions
B-BH and E-HJ carried out the preparation of the analogs and measurement of size, sphericity, and rupture strength. B-BH and S-MC carried out the RSM design using SAS system (Version 9.01, SAS Institute Inc., USA) and Maple software (Maple 7. Waterloo Maple Inc., Canada). B-BH and S-BK conceived of the study and participated in its design and coordination and helped to draft the manuscript. All authors read and approved the final manuscript.

Competing interests
The authors declare that they have no competing interests.

Author details
[1]Department of Food Science and Technology/Institute of Food Science, Pukyong National University, 45, Yongso-ro, Nam-gu, Busan 48513, South Korea. [2]Research Group of Innovative Functional Foods, Korea Food Research Institute, Seongnam 13539, South Korea.

References
Blandino A, Macias M, Cantero D. Formation of calcium alginate gel capsules: influence of sodium alginate and $CaCl_2$ concentration on gelation kinetics. J Biosci Bioengin. 1999;88:686–89.

Box GEP, Wilson KB. On the experimental attainment of optimum conditions. J Royal Stat Soc (Ser B). 1951;13:1–45.

Chan ES, Lee BB, Ravindra P, Poncelet D. Prediction models for shape and size of Ca-alginate macrobeads produced through extrusion-dipping method. J Colloid Interf Sci. 2009;338:63–72.

Chen LY, Subirade M. Effect of preparation conditions on the nutrient release properties of alginate-whey protein granular microspheres. Eur J Pharm Biopharm. 2006;65:354–62.

Clark AH, Ross Murphy SB. Structural and mechanical properties of biopolymer gels. Adv Polym Sci. 1987;83:57–192.

Grant GT, Morris ER, Rees DA, Smith JC, Thom D. Biological interactions between polysaccharides and divalent cations: the egg-box model. FEBS Let. 1973;32:195–98.

Hwang SJ, Rhee GJ, Jo HB, Lee KM, Kim CK. Alginate beads as controlled release polymeric drug delivery system. J Korean Pharm Sci. 1993;23:19–26.

Ji CI, Cho SM, Gu YS and Kim SB. The processing optimization of caviar analogs encapsulated by calcium-alginate gel membranes. Food Sci Biotechnol. 2007a;16:557-64.

Ji CI, Cho SM, Yoon YS, Kim SB. Optimization of physical conditions for caviar analog preparation using calcium-alginate gel capsules. J Fish Sci Technol. 2007b;10:103-12.

Jo EH, Ha BB, Kim SB. Effects of Heat, Salt and hydrocolloid treatments on flying fish Cypselurus agoo roe analogs prepared using calcium alginate hydrogels. J Fish Aquat Sci. 2014;17:203-7.

Kendall Jr WF, Darrabie MD, El-shewy HM, Opara EC. Effect of alginate composition and purity on alginate microspheres. J Microencap. 2004;21:821–28.

Moe ST, Draget KI, Skjåk-bræk G, Smidsrød O. Alginates. In: Food polysaccharides and their applications, vol. 9. New York: Marcel Dekker; 1995. p. 245–86.

Montero P, Perez-mateos M. Effects of Na^+, K^+, and Ca^{2+} on gels formed from fish mince containing a carrageenan or alginate. Food Hydrocolloids. 2002;16:375–82.

Mørch YA, Donati I, Strand BL, Skjåk-bræk G. Effect of Ca^{2+}, Ba^{2+}, and Sr^{2+} on alginate microbeads. Biomacromol. 2006;7:1471–80.

Onsøyen E. Commercial applications of alginates. Carbohydr Eur. 1996;14:26–31.

Onsøyen E. Alginates. In A. Imeson (Ed.). Thickening and gelling agents for food. London: Blackie Academic & Professional; 1997. p. 22-24.

Rousseau I, Le cerf D, Picton L, Argillier JF, Muller G. Entrapment and release of sodium polystyrene sulfonate (SPS) from calcium alginate gel beads. Eur Polym J. 2004;40:2709–15.

Sabraa W. Alginate: a polysaccharide of industrial interest and diverse biological functions. In: Polysaccharides: structural diversity and functional versatility. 2nd ed. New York: Marcel dekker; 2005. p. 515–33.

Skjåk-bræk G, Espevik T. Application of alginate gels in biotechnology and biomedicine. Carbohydr Eur. 1996;14:19–25.

Simpson NE, Grant SC, Blackband SJ, Constantinidis I. NMR properties of alginate microbeads. Biomater. 2003;24:4941–48.

Smidsrød O, Haug A. Dependence upon gel-sol state of the ion-exchange properties of alginates. Acta Chem Scand. 1972;26:2063–74.

Subirade M, Hebrand G, Hoffart V, Beyssac E, Cardoy JM, Alic M. Coated whey protein/alginate microparticles as oral controlled delivery systems for probiotic yeast. J Microencap. 2010;27:292–302.

Wichchukit S, Oztop MH, Mccarthy MJ, Mccarthy KL. Whey protein/alginate beads as carries of a bioactive component. Food Hydrocolloids. 2013;33:66–73.

8

Effects of extruded pellet and moist pellet on growth performance, body composition, and hematology of juvenile olive flounder, *Paralichthys olivaceus*

Seunghan Lee[1], Mohammad Moniruzzaman[1], Jinho Bae[1], Minji Seong[1], Yu-jin Song[1], Bakshish Dosanjh[2] and Sungchul C. Bai[1*]

Abstract

A feeding trial was conducted to evaluate the effects of two different sizes of extruded pellets (EP) (EP$_1$ - 3 mm or EP$_2$ - 5 mm) and a moist pellet (MP) in olive flounder, *Paralichthys olivaceus*, reared in semi-recirculation system. A total of 450 fish with an average initial weight of 5.0 ± 0.2 g (mean $\pm$ SD) were fed one of the three experimental diets in triplicate groups. At the end of a 6-week feeding trial, weight gain, specific growth rate, and feed efficiency of fish fed EP diets were significantly higher than those of fish fed MP ($P < 0.05$). Water quality parameters like turbidity, total ammonia nitrogen, and total phosphorous from tanks of fish fed EP$_1$ and EP$_2$ were significantly lower than those from tanks of fish fed MP. Blood plasma glutamic oxaloacetic transaminase and glucose concentration were significantly higher in fish fed MP diet compared to fish fed EP diets ($P < 0.05$). Whole body crude protein contents in fish fed EP diets were higher than those from the fish fed MP diet. Whole body amino acid content like threonine, aspartic acid, serine, tyrosine, and cystine were found to be significantly higher in fish fed EP diets than those in fish fed MP diet. In considering overall performance of olive flounder, EP$_2$ diet could be recommended for the successful aquaculture of this important fish species.

Keywords: Extruded pellet, Moist pellet, Growth, Hematology, Juvenile olive flounder

Background

Fish feeding is one of the most important factors in commercial fish farming because feeding regime may have consequences on both growth performance and feed wastage (Tsevis et al. 1992; Azzaydi et al. 2000). During the last decade, there has been a marked increase in the use of extruded pellet (EP) for feeding fish. It has been well documented that EP diets have superior water stability, better floating properties, and higher energy content than other pelleted diets (Hilton et al. 1981; Johnsen and Wandsvik 1991; Ammar 2008). However, the size of feed pellets and the rate at which they are delivered may affect the amount of feed that an individual fish can ingest over a period of time. Undesirable size of pellets or a high amount of pellets may cause feed wastage, as fish may be unable to ingest the required amount of feeds (Bailey et al. 2003).

Olive flounder, *Paralichthys olivaceus*, is one of the most commercially important marine aquaculture species in Korea. Production of the olive flounder was 42,133 metric tons, ranked it first among Korean mariculture finfish species in 2014 (KOSTAT 2015). The suitable type and size of pellets for different age groups of olive flounder are very important for maximum growth. Most of the flounder production has been sourced from use of frozen raw fish (sardine or mackerel) or raw fish-based moist pellets (MP) composed of frozen raw fish and commercially available binder meal at a certain ratio (Cho and Cho 2009).

There are several studies that have been done in the context of nutrient requirements and feeding technology

---

* Correspondence: scbai@pknu.ac.kr
[1]Department of Marine Bio-materials and Aquaculture/Feeds and Foods Nutrition Research Center (FFNRC), Pukyong National University, Busan 608-737, Republic of Korea
Full list of author information is available at the end of the article

of olive flounder. However, information on the effects of extruded pellets and their size for juvenile olive flounder is scarce. Therefore, the present experiment was conducted to evaluate the effects of two different sizes of extruded pellets and a moist pellet on the water quality, growth performance, body composition, hematological characteristics, and gut histology of juvenile olive flounder, *P. olivaceus*.

Methods
Experimental diets
In this study, the MP diet was prepared from frozen sardine and commercial wheat flour at the ratio of 3:1 (wet weight basis) and the EP_1 (3 mm) and EP_2 (5 mm) feeds were provided by EWOS Canada Ltd. EP was designed to maintain the same level of protein (56 %) and lipid (10 %). Proximate compositions of the experimental diets are shown in Table 1. Diets were stored at −20 °C (wet pellets) until use.

Experimental fish and feeding trial
Juvenile olive flounder, *P. olivaceus*, was obtained from Tong-yeong, Republic of Korea. Prior to the feeding trial, the fish were fed different experimental diets for 2 weeks to allow them to adjust to the experimental diets and conditions. Feeding trial was conducted in a semi re-circulating system with 250-L aquaria each having a water flow rate of 1.5 L/min. Supplemental aeration was provided to maintain dissolved oxygen near saturation. Water temperature was maintained at 21 ± 1 °C (mean ± SD). Salinity was maintained at 31 ± 1 g/L (mean ± SD). Fish averaging 5.0 ± 0.2 g (mean ± SD) randomly distributed to each aquarium as groups of 50 fish and fed the experimental diets in triplicate at a rate of 2.5~4.5 % wet body weight per day for 6 weeks. Total fish weight in each aquarium was determined every 2 weeks, and the amount of diets fed to fish was adjusted accordingly.

Table 1 Proximate composition of the experimental diet (percentage dry matter basis)

Items	Diets		
	EP_1^a	EP_2^a	MP^b
Moisture (%)	7.39	7.17	65.15
Crude ash (%)	10.91	11.07	12.08
Crude lipid (%)	10.18	10.04	10.40
Crude protein (%)	56.79	56.91	57.07
Crude fiber (%)	2.38	2.30	1.61
Carbohydrate (%)	14.35	14.34	13.47
Gross energy (kcal/g)	3.76	3.75	3.76

[a]*EP* extruded pellet; EWOS Canada Ltd.
[b]*MP* moist pellet; composed of frozen sardine and wheat flour (3:1)

Water quality analysis
Water samples from the fish tanks were monitored just after 2 h of feeding. Turbidity, total ammonia nitrogen (TAN), and total phosphorous (TP) were determined from the water of the each experimental tank. The concentration of turbidity, TAN, and TP were recorded according to the standard methods for marine environmental analysis (Ministry of Land Transport and Maritime Affairs 2010).

Sample collection and analysis
At the end of the feeding trial, fish were starved for 24 h and they were counted and weighed to calculate the weight gain (WG), specific growth rate (SGR), feed efficiency (FE), and survival rate. After the final weighing, three fish from each aquarium were analyzed for whole body proximate composition. Proximate composition of the experimental diets and fish bodies were performed by the standard methods of AOAC (1995). To determine the moisture content of diets and fish, they were dried to maintain constant weight at 135 °C for 2 h. Ash content was determined using a muffle furnace (550 °C for 4 h). Crude lipid content was determined by the soxhlet extraction method by using Soxtec system 1046 (Foss, Hoganas, Sweden) and crude protein content by Kjeldahl method (N × 6.25) after acid digestion, distillation, and titration of samples. Fiber content was analyzed with a fiber analyzer (FT122 Fibertec™, Foss, Hillerød, Denmark). Carbohydrate content was calculated by subtracting total percentage of nutrient contents from 100 %. Gross energy of experimental diets were calculated based on the calculation of 16.7, 16.7, and 37.7 kJ/g for protein, carbohydrate, and lipid, respectively (Halver and Hardy 2002). Blood samples were taken by using heparinized syringes from the caudal vein of five randomly chosen fish per tank. Plasma was collected after centrifugation at 3000 rpm for 10 min and stored at −70 °C in order to analyze glutamic oxaloacetic transaminase (GOT), glutamic pyruvic transaminase (GPT), glucose, total protein (T-protein), cholesterol, and triglycerides. Plasma analyses were performed at the National Fisheries Research and Development Institute (NFRDI), Gijang, Busan, Korea, by using the kits of DRI-CHEM 4000i- Fuji Dri-Chem Slide- 3150 (Minato-ku, Tokyo, Japan). Amino acid analysis of edible body parts was performed by ninhydrin method (Sykam Amino Acid Analyzer S433, Sykam, Eresing, Germany).

Statistical analysis
All data were analyzed by one-way ANOVA (Statistix 3.1; Analytical Software, St. Paul, MN, USA) to test the effects of the dietary treatments. When a significant treatment effect was observed, an LSD test was used to compare means. Treatment effects were considered at $P < 0.05$ level of significance.

Results and discussion

Water quality has been acknowledged to have profound effects on the growth performance and health of aquaculture fish species. In the present experiment, water quality parameters were affected by the experimental diets (Table 2). The observed water quality parameters especially turbidity was significantly lower among the group of fish fed extruded pellet diets. Turbidity caused by suspended solids has been reported to have great effects on fish metabolism in terms of fish growth and survival. After 2 h of feeding, turbidity, TAN, and TP were recorded to be significantly higher in the group of fish fed MP diet than those of fish fed EP diets. These results may indicate that MP diet was easily soluble in water before consumption by fish, whereas extruded pellets were more stable in water and their leaching rate in water was comparatively prolonged. Folke and Kautsky (1989) reported that water pollution by fish feeding is caused largely by increasing turbidity, as well as ammonia and phosphorus loading through uneaten feeds and feces. From an on-farm experiment with flounder, Kim and Lee (2000) reported that the excretion of nitrogen (N) ranged from 48 to 70 g and phosphorus (P) from 10 to 12 g per kilogram weight gain. However, under practical feeding conditions, flounder excreted much higher N (114 g) and P (28 g) per kilogram weight gain, suggesting a substantial waste of feed (Kim et al. 2002). Likewise, similar findings have been reported by Cha et al. (2008) in their experiment with olive flounder fed MP and chitosan-based extruded pellets.

In the present study, significantly higher growth performance was observed for the group of fish fed EP_2 than those of fish fed MP diet (Table 3). At the end of experiment, WG, SGR, and FE of fish fed EP_2 diet were significantly higher than those of fish fed EP_1 and MP diet. However, there were no significant differences in these parameters among the fish fed EP_1 and MP diets. Survival rate ranged from 94 to 97 % without any statistical differences among different treatments. Likewise, Cho and Cho (2009) reported that an extruded pellet is more recommendable than a moist pellet for the growth

Table 2 Water quality parameters after 2 h of feeding of juvenile olive flounder fed with the different experimental diets for 6 weeks

Parameters	Diets			Pooled SEM
	EP_1	EP_2	MP	
Turbidity (NTU)	0.57[b]	0.56[b]	2.63[a]	0.34
TAN (mg/L)	0.33[b]	0.31[b]	0.53[a]	0.04
TP (mg/L)	0.04[b]	0.04[b]	0.07[a]	0.01

Value are means from groups ($n = 3$) of samples where the values in each row with different superscripts are significantly different ($P < 0.05$)
TAN total ammonia nitrogen, *TP* total phosphorus, *Pooled SEM* pooled standard error of mean (SD/√n)

Table 3 Growth performances of juvenile olive flounder fed with the different experimental diets for 6 weeks

Variables	Diets			Pooled SEM
	EP_1	EP_2	MP	
WG[§]	130.8[b]	142.1[a]	128.4[b]	2.01
SGR[¶]	2.78[b]	2.95[a]	2.75[b]	0.03
FE[□]	101.1[b]	103.1[a]	101.4[b]	0.41
SR*	97.3	96.0	94.0	1.15

Values are means from groups ($n = 3$) of fish where the values in each row with different superscripts are significantly different ($P < 0.05$)
Pooled SEM pooled standard error of mean (SD/√n)
[§]Weight gain (%): (final wt. − initial wt.) × 100/initial wt.
[¶]Specific growth rate (%/day): (ln final wt. − ln initial wt.) × 100/days
[□]Feed efficiency (%): (wet weight gain/dry feed intake) × 100
*Survival rate = (total fish − dead fish) × 100/total fish

performance in flounder aquaculture. In our study, the results of lower feed efficiency in fish fed MP diet could be due to high leaching properties of the MP diet before the ingestion of feed by the fish. It is well documented that extruded pellets have superior water stability, better floating properties, and a higher energy content among the pelleted diets (Hilton et al. 1981; Johnsen and Wandsvik 1991; Ammar 2008). Aqua feed technology is moving in tandem with the aquaculture growth with the usage of extrusion procedures for the improvement of digestibility (Umar et al. 2013). Chang and Wang (1999) stated the advantages of extrusion cooking process for aquaculture feed production including improved feed conversion ratio, control of pellet density, greater feed stability in water, better production efficiency, and versatility. During extrusion cooking, various reactions take place including thermal treatment, gelatinization, protein denaturation, hydration, texture alteration, partial dehydration, and destruction of microorganisms and other toxic compounds (Kannadhason et al. 2011). According to Chang and Wang (1999), the gelatinization that occurs during extrusion process improves durability of the feed rations and digestibility of starch. In the present study, result for the fish fed EP_2 concordantly supported the various reports that extruded pellets are having a better efficiency over MP diet for juvenile olive flounder growth. However, it is difficult to attribute any reason for the observed lower weight gain for the group of fish fed EP_1 diet in the present experiment.

The present experiment clearly demonstrated the beneficial effects of pellet size of extruded pellets on the performance of olive flounder. Interestingly, we observed a lower growth rate for the group of fish fed EP_1 than did fish fed EP_2 diet. Feed pellet size will obviously have an effect on fish performance, and there are indications of this effect presented in the study. Usually, for a pellet that is larger than the mouth gape of fish, handling time becomes a limiting factor in the fish's ability to ingest enough pellets to maintain good growth which will clearly have negative effects. However, in our study, EP_2 feed was well accepted by the fish even though its size was larger than EP_1 probably because EP_2 was more suitable in relation to mouth gape size of fish than those of EP_1 diet.

It is recommended that pellet size should be approximately 20–30 % of the size of the fish species mouth gape (Craig 2009). Feeding too small a pellet results in inefficient feeding because more energy is used in finding and eating more pellets. Conversely, pellets that are too large will depress feeding and, in the extreme, cause choking. Therefore, it is better to select the largest sized feed that the fish will actively eat. Smith et al. (1995) reported that the length and diameter of pellets influence the detectability and/or attractiveness of pellets to salmonids. In another report, Irwin et al. (2002) reported that smaller sized turbot prefer to accept a bigger size of pellet (pellet size, 40 % of mouth gape) whereas the larger fish group prefers a smaller pellet size (pellet size, 20 % of mouth gape) which is higher than the preferred pellet size of salmonid species (Wankowski and Thorpe 1979; Brannas and Alanara 1992). The results may support the findings of the present study. However, feed range selectivity may be governed by hunger levels of fish (Croy and Hughes 1991). Ellis et al. (1997) reported that farmed turbot prefer pellets and due to their extended jaws they can engulf large prey items (Holmes and Gibson 1986). Some workers (Hjertnes et al. 1993; Tuene and Nortvedt 1995) have used larger pellet sizes in experiments with halibut than those recommended for Atlantic salmon, possibly because halibut have a larger mouth gape than salmonids of the same weight. In contrast, Stradmeyer et al. (1988) reported that adult salmon showed a more immediate response to larger pellets but that these were more likely to be rejected than pellets of a shorter length. However, the texture and hardness of pellets is an important issue. It has been seen that juvenile salmon can handle larger size of soft pellets than hard pellets (Mearns 1990). Tuene and Nortvedt (1995) fed 9-mm pellets to 90–662-g halibut and concluded that the high intra-individual (day-to-day) coefficient of variation of feed intake may have been caused by the large pellet size since the average consumption at each meal was less than two pellets per fish.

Whole body proximate composition data revealed significantly lower whole body crude protein contents for the group of fish fed MP diet than those of fish fed EP diet, whereas whole body crude lipid content was significantly higher among the group of fish fed MP diet (Table 4). Moisture contents for fish fed EP_2 were significantly lower than those of fish fed all other diets. However, there was no significant difference in whole body moisture content among the group of fish fed EP_1 and MP diets. Similar findings have been reported in

Table 4 Whole body proximate composition of juvenile olive flounder fed with the different experimental diets for 6 weeks (percentage of DM basis)

Items	Diets			Pooled SEM
	EP_1	EP_2	MP	
Moisture	76.87[a]	75.36[b]	77.11[a]	0.30
Crude ash	17.53	16.96	18.45	0.47
Crude lipid	3.25[b]	5.34[ab]	7.50[a]	0.58
Crude protein	74.70[a]	74.16[a]	71.91[b]	0.53

Value are means from groups ($n = 3$) of fish where the values in each row with different superscripts are significantly different ($P < 0.05$)
Pooled SEM pooled standard error of mean (SD/√n)

various previous experiments. For instance, Cho and Cho (2009) reported from their experiments that proximate composition of whole body of flounder with and without liver, except for moisture content of liver, was not significantly affected by the different diets (extruded pellets, semi-moist pellets, and moist pellets). Results for whole body amino acids (Table 5) showing only four amino acids viz. aspartic acid (Asp), threonine (Thr), serine (Ser), and tyrosine (Tyr) were significantly lowest for the fish fed MP diet than those of fish fed other

Table 5 Whole body amino acid composition of juvenile olive flounder fed with the different experimental diets for 6 weeks (percentage of DM basis)

Amino acids (AA)	Diets			Pooled SEM
	EP_1	EP_2	MP	
Essential (EAA)				
Methionine	1.64	1.50	1.37	0.05
Leucine	5.07	4.84	4.82	0.05
Isoleucine	3.38	2.90	2.98	0.11
Arginine	4.37	4.39	4.01	0.06
Histidine	1.87	1.78	1.90	0.04
Lysine	5.56	5.19	5.25	0.06
Phenylalanine	2.72[a]	2.64[ab]	2.54[b]	0.03
Threonine	2.90[a]	2.82[a]	2.18[b]	0.09
Valine	3.39	3.27	3.35	0.03
Non-essential (NEAA)				
Alanine	4.59	4.71	4.33	0.07
Glycine	5.31	6.06	5.24	0.18
Aspartate	7.25[a]	7.14[a]	6.27[b]	0.13
Proline	3.44	3.69	3.33	0.07
Serine	2.93[a]	2.87[a]	1.50[b]	0.20
Glutamate	10.32	10.00	9.96	0.09
Tyrosine	1.94[a]	1.83[a]	1.08[b]	0.12
Cystine	1.29[a]	1.15[ab]	1.00[b]	0.04

Value are means from groups ($n = 3$) of fish where the values in each row with different superscripts are significantly different ($P < 0.05$)
Pooled SEM pooled standard error of mean (SD/√n)

experimental diets. Although significant differences were recorded in whole body amino acids for other 13 amino acids, no clear trend could be drawn among different treatments. MP diet appeared to affect distinctly only these four amino acids. Due to the lack of reports on whole body amino acid contents in similar studies, it is difficult to compare the present observation with others.

Hematological characteristics can be used as an index of the health status of fish (Blaxhall 1972). Hematological changes have been detected following different types of stress conditions like exposure to pollutants, diseases, and hypoxia (Duthie and Tort 1985). Hence, it could be suggested that any unhealthy condition caused by poor nutrition could affect the hematological characteristics of fish. Plasma glucose concentration is one of the stress indicators in fish (Menezes et al. 2006) which may vary greatly depending on the physiological status of the animal (de Andrade et al. 2007). Mommsen et al. (1999) reported that plasma glucose values can increase, decrease, or keep constant under a high plasma cortisol level. Plasma GOT and GPT activities may give information on liver injury or dysfunction (Wells et al. 1986). They are also used as valuable diagnostic means of stress responses in several fish species (Almeida et al. 2002; Choi et al. 2007). The present study has been indicated that plasma GOT and glucose in the group of fish fed MP diet were significantly higher than those of fish fed EP diet because fish might be always in stress of feed competition (Table 6). However, no significant differences were found in GPT, T-protein, cholesterol, and triglyceride levels among the fish fed EP or MP diet.

In salmon aquaculture, MP diets were used due to their better acceptance with soft texture and relatively low cost compared to dry diet (Ghittino 1979). However, in yellow tail and flounder culture, MP diet has demerits in causing water pollution from leftover feed which ultimately increases production cost by decreasing water quality and quantity of fish (Kim and Shin 2006; Kim et al. 2007). In this instance, extruded pellet diet could

be a right choice to minimize water pollution and increase total production in flounder aquaculture.

Conclusions
The results from the present study demonstrated the beneficial effects of EP and their diameter over commonly used MP diets in promoting growth of olive flounder, suggesting the need to revise the feed and feeding technology for flounder aquaculture. In the present experiment, the results evidenced that fish fed EP_2 (5 mm) had the better growth and water quality parameters than did fish fed MP diet in juvenile olive flounder.

Abbreviations
CL: Crude lipid; CP: Crude protein; EAA: Essential amino acids; EP: Extruded pellet; FE: Feed efficiency; GOT: Glutamic oxaloacetic transaminase; GPT: Glutamic pyruvic transaminase; MP: Moist pellet; SGR: Specific growth rate; SR: Survival rate; WG: Weight gain

Acknowledgements
We would like to acknowledge the members of the Feeds and Foods Nutrition Research Center (FFNRC), Pukyong National University, Busan, Republic of Korea, for their assistance.

Funding
This research work has been financially supported by EWOS Canada Ltd.

Authors' contributions
SL conducted the research, analyzed the samples, and prepared the draft manuscript. MM helped to write the draft manuscript. JB, MS, and YS helped in the research conduction and statistical analysis. BKD helped in the research design and reviewed the manuscript. SCB designed and monitored the experiment and finalized the draft manuscript. All authors read and approved the final manuscript.

Competing interests
The authors declare that they have no competing interests.

Author details
[1]Department of Marine Bio-materials and Aquaculture/Feeds and Foods Nutrition Research Center (FFNRC), Pukyong National University, Busan 608-737, Republic of Korea. [2]EWOS Canada Ltd., 7721-132nd Street Surrey, Vancouver, British Columbia, Canada.

Table 6 Hematological parameters of juvenile olive flounder fed with the different experimental diets for 6 weeks

Parameters	Diets			Pooled SEM
	EP_1	EP_2	MP	
GOT (U/L)	34.0[b]	34.7[b]	44.9[a]	1.96
GPT (U/L)	5.0	5.3	5.0	0.23
Glucose (mg/dL)	62.0[b]	61.0[b]	86.3[a]	4.18
T-protein (g/dL)	3.4	3.3	3.0	0.06
Cholesterol (mg/dL)	238.0	265.5	238.6	10.5
Triglycerides (mg/dL)	259.4	308.2	325.4	28.0

Value are means from groups ($n = 3$) of samples where the values in each row with different superscripts are significantly different ($P < 0.05$)
Pooled SEM pooled standard error of mean (SD/$\sqrt{n}$), *GOT* glutamic oxaloacetic transaminase, *GPT* glutamic pyruvic transaminase

References
Almeida JA, Diniz YS, Marques SG, Faine LA, Ribas BO, Burneiko RC, Novelli EB. The use of the oxidative stress responses as biomarkers in Nile tilapia (*Oreochromis niloticus*) exposed to in vivo cadmium contamination. Env Int. 2002;27:673–9.
Ammar AA. Effect of extruded and trash fish diets on growth performance and pond productivity of sea bream, *Sparus aurata*, the sea bass, *Dicentrarchus labrax* and the flat head grey mullet, *Mugil cephalus* reared in polyculture system in earthen ponds. Egyptian J Aqua Biol Fish. 2008;12:43–58.
AOAC (Association of Official Analytical Chemists). Official methods of analysis. 16th ed. Arlington: AOAC International; 1995.

Azzaydi M, Martines FJ, Zamora S, Sanchez-Vazquez FJ, Madrid JA. The influence of nocturnal vs. diurnal feeding condition under winter condition on growth and feed conversion of European sea bass, *Dicentrarchus labrax*, *L*. Aquaculture. 2000;182:329–38.

Bailey J, Alanara A, Crampton V. Do delivery rate and pellet size affect growth rate in Atlantic salmon (*Salmo salar* L.) raised under semi-commercial farming conditions? Aquaculture. 2003;224:79–88.

Blaxhall PC. The haematological assessment of the health of freshwater fish-a review of selected literature. J Fish Biol. 1972;4:593–604.

Brannas E, Alanara A. Feeding behaviour of the Arctic charr in comparison with the rainbow trout. Aquaculture. 1992;105:53–9.

Cha SH, Lee JS, Song CB, Lee KJ, Jeon YJ. Effects of chitosan-coated diet on improving water quality and innate immunity in the olive flounder, *Paralichthys olivaceus*. Aquaculture. 2008;278:110–8.

Chang YK, Wang SS. Advances in extrusion technology: aquaculture animal feeds and foods. Lancaster: Technomic publishing company, Inc.; 1999.

Cho YJ, Cho SH. Compensatory growth of olive flounder, *Paralichthys olivaceus*, fed the extruded pellet (EP) with different feeding regimes. J World Aqua Soc. 2009;40:505–12.

Choi YS, et al. Identification of *Pseudomonas aeruginosa* genes crucial for hydrogen peroxide resistance. J Micro Biotech. 2007;17:1344–52.

Craig S. Understanding fish nutrition, feeds, and feeding. Virginia Polytechnic Institute and State University, Virginia Cooperative Extension, Publication 420-256. 2009.

Croy MI, Hughes RN. The influence of hunger on feeding behaviour and on the acquisition of learned foraging skills by the fifteen-spined stickleback, *Spinachia spinachia* L. Anim Behav. 1991;41:161–70.

de Andrade JI, Ono EA, Menezes GC, Brasil EM, Roubach R, Urbinati EC, Tavares-Dias M, Marcon JL, Affonso EG. Influence of diets supplemented with vitamins C and E on pirarucu (*Arapaima gigas*) blood parameters. Comp Biochem Physiol Part A: Mol & Int Physiol. 2007;146:576–80.

Duthie GG, Tort L. Effects of dorsal aortic cannulation on the respiration and haematology of Mediterranean living *Scyliorhinus canicula*. Comp Biochem Physiol Part A: Physiol. 1985;81:879–85.

Ellis T, Howell BR, Hughes RN. Comparative feeding efficiency of wild and hatchery-reared turbot, *Scophthalmus maximus*. Bergen: The first International Symposium on Stock Enhancement and Sea Ranching; 1997. p. 140.

Folke C, Kautsky N. The role of ecosystems for the development of aquaculture. Ambio. 1989;18:234–43.

Ghittino P. Formulation and technology of moist feed. Pages 37-40 In: J.E. Halver and K. Tiews, eds. Finfish nutrition and fish feed technology. Berlin: Heinemann; 1979.

Halver JE, Hardy RW. Fish nutrition. 3rd ed. California: Academic; 2002. p. 8.

Hilton JW, Cho SJ, Slinger CY. Effect of extrusion processing and steam pelleting diets on pellet durability pellet water absorption and physiological response of rainbow trout, (*Salmo gairdneri* R.). Aquaculture. 1981;25:185–94.

Hjertnes T, Gulbrandsen KE, Johnsen F, Opstvedt J. Effect of dietary protein, carbohydrate and fat levels in dry feed for juvenile halibut (*Hippoglossus hippoglossus* L.). In: Kaushik SJ, Luquet P, editors. Fish nutrition in practice. Versailles: INRA Editions; 1993.

Holmes RA, Gibson RN. Visual cues determining prey selection by the turbot, *Scophthalmus maximus* L. J Fish Biol. 1986;29:49–58.

Irwin S, O'Halloran J, FitzGerald RD. Mouth morphology and behavioural responses of cultured turbot towards food pellets of different sizes and moisture content. Aquaculture. 2002;205:77–88.

Johnsen FA, Wandsvik A. The impact of high energy diets on pollution control in the farming industry. Proceedings of the First International Symposium on Nutrition Strategies in Management of Aquaculture Waste. Ontario: University of Guelph; 1991. p. 51–63.

Kannadhason S, Muthukumarappan K, Rosentrater KA. Effect of starch sources and protein content on extruded aquaculture feed containing DDGS. Food Bio Tech. 2011;4:282–94.

Kim JD, Lee SB. Effects of dietary growth, feed utilization and pollution load of Japanese flounder (*Paralichthys olivaceus*). Ann Anim Res Sci. 2000;11:75–84.

Kim JD, Shin SH. Growth, feed utilization and nutrient retention of juvenile olive flounder (*Paralichthys olivaceus*) fed moist, semi-moist and extruded diets. Asian-Aust J Anim Sci. 2006;19:720–6.

Kim JD, Shin SH, Cho KJ, Lee SM. Effect of daily and alternate day feeding regimes on growth and food utilization by juvenile flounder *Paralichthys olivaceus*. J Aquac. 2002;15:15–21.

Kim KD, Kim KM, Kang YJ. Influences of feeding frequency of extruded pellet and moist pellet on growth and body composition of juvenile flounder *Paralichthys olivaceus* in suboptimal water temperature. Fish Sci. 2007;73:745–9.

KOSTAT (Korea National Statistical Office). KOSIS statistical database for fisheries production. Daejeon: KOSTAT; 2015.

Mearns KJ. The behavioural approach in identifying feeding stimulants for fish and its application in aquaculture. In: Kjorsvik E, editor. Application of behavioural studies in aquaculture. Proceedings from the Mini symposium on Ethology in Aquaculture, Trondheim, 22 October 1989. Bergen: Norwegian Society for Aquacultural Research; 1990. p. 69–74.

Menezes GC, Tavares-Dias M, Ono EA, Andrade JI, Brasil EM, Roubach R, Urbinati EC, Marcon JL, Affonso EG. The influence of dietary vitamin C and E supplementation on the physiological response of pirarucu, *Arapaima gigas*, in net culture. Comp Biochem Physiol Part A: Physiol. 2006;145:274–9.

Ministry of Land Transport and Maritime Affairs (MLTM). Standard methods for marine environmental analysis. Ministry of Land, Transport and Maritime Affairs, Seoul, Korea; 2010.

Mommsen TP, Vijayan MM, Moon TW. Cortisol in teleosts: dynamics, mechanisms of action, and metabolic regulation. Rev Fish Biol Fish. 1999;9:211–68.

Smith IP, Metcalfe NB, Huntingford FA. The effects of food pellet dimensions on feeding responses by Atlantic salmon (*Salmo salar* L.) in a marine net pen. Aquaculture. 1995;130:167–75.

Stradmeyer L, Metcalfe NB, Thorpe JE. Effect of food pellet shape and texture on the feeding response of juvenile Atlantic salmon. Aquaculture. 1988;73:217–28.

Tsevis N, Klaoudatos S, Conides A. Food conversion budget in sea bass *Dicentrarchus labrax*, fingerlings under two different feeding frequency patterns. Aquaculture. 1992;101:293–304.

Tuene S, Nortvedt R. Feed intake, growth and feed conversion efficiency of individual Atlantic halibut (*Hippoglossus hippoglossus*) fed a commercial fish feed. Aqua Nutr. 1995;1:27–35.

Umar S, Kamarudin MS, Ramezani-Fard E. Physical properties of extruded aqua feed with a combination of sago and tapioca starches at different moisture contents. Anim Feed Sci Tech. 2013;183:51–5.

Wankowski JWJ, Thorpe JE. The role of food particle size in the growth of juvenile Atlantic salmon (*Salmo salar* L.). J Fish Biol. 1979;14:351–70.

Wells RM, McIntyre RH, Morgan AK, Davies PS. Physiological stress responses in big game fish after exposure: observation on plasma chemistry and blood factors. Comp Biochem Physiol Part A: Physiol. 1986;64:565–71.

Cloning, expression, and activity of type IV antifreeze protein from cultured subtropical olive flounder (*Paralichthys olivaceus*)

Jong Kyu Lee[1] and Hak Jun Kim[2]*

Abstract

Antifreeze proteins (AFPs) lower the freezing point but not the melting point of aqueous solutions by inhibiting the growth of ice crystals via an adsorption-inhibition mechanism. However, the function of type IV AFP (AFP IV) is questionable, as its antifreeze activity is on the verge of detectable limits, its physiological concentration in adult fish blood is too low to function as a biological antifreeze, and its homologues are present even in fish from tropic oceans as well as freshwater. Therefore, we speculated that AFP IV may have gained antifreeze activity not by selective pressure but by chance. To test this hypothesis, we cloned, expressed, and assayed AFP IV from cultured subtropical olive flounder (*Paralichthys olivaceus*), which do not require antifreeze protein for survival. Among the identified expressed sequence tags of the flounder liver sample, a 5′-deleted complementary DNA (cDNA) sequence similar to the *afp4* gene of the longhorn sculpin was identified, and its full-length cDNA and genome structure were examined. The deduced amino acid sequence of flounder AFP IV shared 55, 53, 52, and 49 % identity with those of *Pleuragramma antarcticum*, *Myoxocephalus octodecemspinosus*, *Myoxocephalus scorpius*, and *Notothenia coriiceps*, respectively. Furthermore, the genomic structure of this gene was conserved with those of other known AFP IVs. Notably, the recombinant AFP IV showed a weak but distinct thermal hysteresis of 0.07 ± 0.01 °C at the concentration of 0.5 mg/mL, and ice crystals in an AFP IV solution grew star-shaped, which are very similar to those obtained from other polar AFP IVs. Taken together, our results do not support the hypothesis of evolution of AFP IV by selective pressure, suggesting that the antifreeze activity of AFP IV may have been gained by chance.

Keywords: *Paralichthys olivaceus*, Subtropical fish, Type IV antifreeze protein, Thermal hysteresis

Background

Antifreeze proteins (AFPs) lower the freezing point but not the melting point of aqueous solutions by inhibiting the growth of ice crystals via an adsorption-inhibition mechanism (DeVries and Wohlschlag 1969; Raymond and DeVries 1977; Jia and Davies 2002). Binding of AFP to the surface of an ice crystal inhibits further growth of the crystal, which creates a temperature gap between the melting and freezing points of the aqueous solution. This gap, called thermal hysteresis (TH), is used to quantitatively express the activity of AFP. AFPs are essential for polar fish to survive at the subfreezing temperature of seawater (DeVries and Wohlschlag 1969;

DeVries 1971). The freezing point of fish blood can be lowered down to −0.7 °C by solutes in the serum, and an additional 1 to 1.2 °C depression is achieved by AFPs. The total freezing point depression of approximately 1.9 °C allows the fish blood to remain unfrozen even at −1.9 °C, the temperature of polar seawater during winter (DeVries 1971; Davies and Hew 1990; Raymond and DeVries 1977; Fletcher et al. 2001). To date, four fish AFPs with distinct TH activities have been identified: antifreeze glycoproteins (AFGPs) and AFP types I, II, and III (Davies and Hew 1990). Their TH activities are typically approximately 1 °C, which is sufficient to protect the fish from freezing.

Recently, AFPs and antifreeze activity have also been reported from fish living in regions other than polar regions such as north-temperate and temperate areas (Deng et al 1997; Gauthier et al. 2008; Nishimiya et al.

* Correspondence: kimhj@pknu.ac.kr
[2]Department of Chemistry, College of Natural Sciences, Pukyong National University, Busan 48513, Republic of Korea
Full list of author information is available at the end of the article

2008; Kim 2015). Conventionally, north-temperate fish off the east coast (50° N) of Canada and/or Hokkaido (43.06° N) have been suggested to possess AFPs as a result of convergent evolution (Cheng 1998; Cheng and Chen 1999; Cheng 1998; Cheng et al. 2003; Graham et al. 2008, 2013), as sea ice forms in these coastal areas during winter similar to that in the austral winter at the Antarctic. Conversely, it is not likely that temperate and/ or subtropical fish require antifreeze activity for survival. However, contrary to our expectation, we observed antifreeze activity in some temperate and subtropical fish from the ice-free East Sea of Korea (Kim 2015). Among nine fish examined in that region, three fish, *Gymnocanthus herzensteini*, *Zoarces gillii*, and *Kareius bicoloratus*, exhibited antifreeze activity. Furthermore, during the investigation, tissues of *Paralichthys olivaceus* did not display discernible antifreeze activity although analysis of its expressed sequence tags (ESTs) revealed an EST encoding a putative type IV antifreeze protein (AFP IV) from a liver complementary DNA (cDNA) library.

AFP IV was first identified in the longhorn sculpin, *Myoxocephalus octodecemspinosus*, which inhabits the north-temperate coastal waters of North America (Deng et al. 1997; Deng and Laursen 1998). However, unlike other types of AFPs, the function of AFP IV remains questionable. This uncertainty arises as the AFP IV THs from longhorn and shorthorn sculpins are on the verge of detectable limits, the physiological concentration of this protein in adult fish blood is too low to function as a biological antifreeze (Deng and Laursen 1998; Zhao et al. 1998; Gauthier et al. 2008), and because its homologues are present in many fish from all the oceans ranging from tropic to polar as well as in freshwater fish, many of which do not require an antifreeze protein. In addition, AFP IV shares 20 % sequence identity with the apolipoprotein superfamily (Cheng 1998; Gauthier et al. 2008). Based on the initial characterization of two AFP IVs from longhorn and shorthorn sculpins, Gauthier et al. (2008) speculated that AFP IV may not have been selected and evolved as a main antifreeze owing to the presence of the more efficient type I AFP. In addition, very recently, a Chinese group reported that two AFP IV genes, *afp4a* and *afp4b*, from the tropical fish gibel carp (*Carassius auratus gibelio*) and zebrafish (*Danio rerio*) are essential for epiboly progression and convergent movement during zebrafish embryo gastrulation (Liu et al. 2009; Lee et al. 2011; Xiao et al. 2014). This evidence is consistent with previous findings that some apolipoproteins, which are homologues of AFP IV, play a significant role in embryonic morphogenesis and organogenesis in these fish (Xia et al. 2008; Choudhury et al. 2009; Zhang et al. 2011; Wang et al. 2013; Xiao et al. 2014). Together, these results imply that the main function of AFP IV is not as an antifreeze and that its

gain of ice-binding capacity might not be the consequence of selective pressure of a cold environment but may have arisen by chance. If the ice-binding affinity of AFP IV found in polar and other northern fish had evolved simply by selective pressure from the cold environment, then its homologues in temperate and tropical fish would not be expected to possess a similar affinity. To address this issue, we cloned and analyzed cDNA of the subtropical fish *P. olivaceus*, expressed its recombinant protein in *Escherichia coli*, and measured its TH activity.

Methods

Isolation and sequence analysis of *P. olivaceus afp4* cDNA and genomic DNA

An adult olive flounder cultured in Korea was purchased from a local fish market, and cDNA libraries were constructed for the liver and muscle tissues. Briefly, the tissues were ground using a mortar and pestle in liquid nitrogen. mRNA was isolated from the ground liver and muscle tissues using the Micro-FastTract™ 2.0 Kit (Invitrogen, Carlsbad, CA, USA) according to the manufacturer's instructions. Unidirectional cloning of cDNA was performed using a ZAP-cDNA synthesis kit (Stratagene, Santa Clara, CA, USA) as described in the user manual. Mass phagemid vector excision was performed, and the mixture was plated on a Luria-Bertani (LB) agar plate. The excised colonies were randomly selected and incubated separately in 2 mL LB broth with ampicillin (50 µg/mL). The cultured cells were harvested by centrifugation, and the plasmids were isolated using a standard phenol method. The plasmids were then digested with XhoI and EcoRI and electrophoresed on a 1.5 % agarose gel. The plasmids containing the 1-kb insert were selected and sequenced with the SK primer (5′-CGC TCT AGA ACT AGT GGA TC-3′) for 5′-directional analysis. The trimming of vector and linker sequences from the raw sequences was performed manually using Chromas software (http://technelysium.com.au/wp/chromas.html). NCBI databases were then searched for sequences homologous to each EST using BlastN and BlastX.

To obtain 5′ end sequences of the *afp4* gene, the CapFishing™ full-length cDNA Premix Kit (Seegene, Korea) was used. For 5′-rapid amplification of cDNA ends (RACE), the target primer R1 (5′-CAA GCC GCT CCA TGA ACC T-3′) was designed based on the 3′-untranslated region (UTR) of the *afp4* gene. The polymerase chain reaction (PCR) product was isolated, cloned using the TOPO TA Cloning Kit (Invitrogen), and sequenced.

For genome structure analysis, specific forward (5′-ATG AAA TTC TCC CTC ATT GC-3′) and reverse (5′-GGC TTG TTT AGT TGG AGA TG-3′) primers were designed based on the *afp4* cDNA sequence. The amplified PCR product of 1.2 kb was sequenced and used as a probe to screen for bacterial artificial chromosome (BAC) clones containing the target genomic region. A BAC library of olive flounder (Dr. Aoki, Tokyo University of Fisheries) was arrayed on nylon membranes in a 3×3 grid pattern using a Biomek 2000 Workstation (Beckman Coulter, Brea, CA, USA) and hybridized with the ^{32}P-labeled PCR product. The plasmids of six positive clones (A1–A6) were isolated using the phenol method. The plasmid of A3 was digested with EcoRI, KpnI, BamHI, PstI, or HindIII, and each digested DNA was electrophoresed on a 0.7 % agarose gel and transferred to a positively charged nylon membrane. The membrane was then hybridized with the 1.2-kb digoxigenin-labeled probe previously amplified by PCR (Roche, Roswell, GA, USA). The plasmids of A2–A6, except for A5, were full-digested with EcoRI and electrophoresed on a 0.7 % agarose gel, transferred to a nylon membrane, and also hybridized with the above probe. The 2.9-kb fragment of the A3 plasmid digested with EcoRI was eluted from the agarose gel and ligated into an EcoRI-digested pUC19 vector. The plasmids were then transformed into bacteria, isolated, and sequenced.

Expression and purification of *P. olivaceus* AFP IV from *E. coli*
To construct the expression vector harboring the flounder *afp4* gene, the mature coding region was amplified from the cDNA library using the forward primer, 5′-CAT ATG CAA GAT GCT GCT GAT CTG-3′, including an NdeI site and a reverse primer, 5′-TCT AGA TTA GTT GGA GAT GCT GCG-3′, including an XbaI site using *Pfu* polymerase. After A-tailing using Taq polymerase, the PCR product was cloned into the TOPO TA cloning vector (Invitrogen). The plasmids were isolated and confirmed by sequencing. The insert DNAs digested with NdeI and XbaI were subsequently ligated into pCold I expression vectors (TaKaRa, Otsu, Japan), digested with the same restriction enzymes. The resulting plasmid was transformed into *E. coli* BL21.

For protein expression, the *E. coli* transformants were inoculated in LB medium with 50 μg/mL ampicillin and grown at 37 °C overnight. The seed culture was diluted 100 times in the LB medium with 50 μg/mL ampicillin and grown at 37 °C until OD_{600} reached 0.5, at which time the culture was transferred to a 15 °C shaker and further incubated for 1 h. Then, isopropyl β-D-thiogalactopyranoside (IPTG) was added to the culture medium at a final concentration of 0.5 mM to induce the AFP IV protein. The cells were incubated further for 15–20 h at 15 °C in a shaker. The recombinant AFP IV was subsequently purified from the soluble supernatant using Ni-affinity chromatography. The AFP IV-overexpressed cell pellet was suspended in 20 mM Kpi, pH 7.4, 50 mM NaCl, and 5 mM imidazole and lysed using sonication. The supernatant of the cell lysate was

loaded onto a Ni-NTA column pre-equilibrated with a lysate buffer. The AFP IV was eluted with a buffer containing 20 mM Kpi, pH 7.4, 50 mM NaCl, and 500 mM imidazole. The fractions containing the AFP IV protein were pooled and dialyzed against a lysis buffer and concentrated using ultrafiltration. The molecular mass of the recombinant AFP was determined using matrix-assisted laser desorption/ionization time-of-flight (MALDI-TOF) using sinapinic acid (3,5-dimethoxy-4-hydroxy cinnamic acids) as a matrix with the Voyager-DE STR Biospectrometry Workstation (Applied Biosystems, Foster City, CA, USA).

TH activity of recombinant *P. olivaceus* AFP IV
TH activity and ice crystal morphology were examined using a nanoliter osmometer (Otago Osmometers, Dunedin, New Zealand) connected to a cold well stage mounted on a light microscope equipped with a Canon digital camera (Tokyo, Japan). A droplet containing a few nanoliters of the sample to be assayed was layered into a well filled with oil. The sample well was placed on the stage and frozen rapidly at approximately −20 °C. The temperature was raised slowly until a single ice crystal remained. Then, the temperature was lowered again slowly while the ice crystal morphology was maintained. The ice crystal image was pictured at ×40 magnification. The melting point of the sample was taken as the temperature at which a single ice crystal formed, and the freezing point was the temperature at which rapid growth of the ice crystal was observed. The difference between melting and freezing points is considered the TH value. While measuring the TH values, changes of ice shape were also observed.

Results
Cloning and sequence analysis of *P. olivaceus* type IV AFP cDNA
A total of 149 plasmids that contained relatively large inserts were isolated and sequenced from liver and muscle cDNA libraries of *P. olivaceus* (data not shown). Of these, one of the liver ESTs displayed high similarity to the *afp4* gene of longhorn sculpin (*M. octodecimspinosus*) (Deng et al. 1997; Deng and Laursen 1998; Zhao et al. 1998). The full-length cDNA was obtained using RACE-PCR. The nucleotide sequence of the full-length cDNA consisted of 51 nt 5′-UTR, 375 nt open reading frame (ORF), and 174 nt 3′-UTR. The polyadenylation signal (ATTAAA) was found between 15 and 19 nt upstream of the polyadenylation site. The ORF encodes a protein of 124 amino acid residues. The SignalP 3.0 program (Bendtsen et al. 2004) predicted the first 20 residues in the N-terminus as a signal peptide for secretion. As shown in Fig. 1, mature AFP IV amino acid sequence alignment demonstrated high identities with functional AFP IVs from polar fish, 55 % identity with *Pleuragramma*

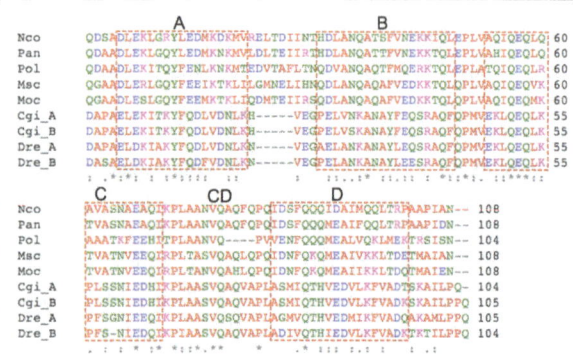

Fig. 1 Multiple sequence alignment of AFP IVs. Sequence alignment was carried out using ClustalW2 (Larkin et al. 2007). Presented are only mature protein sequences from *Paralichthys olivaceus* (Pol, GenBank Accession. No. Q8JI37); *Myoxocephalus octodecemspinosus* (Moc, ABA41379); *Myoxocephalus octodecemspinosus* (Moc, ABA41379); *Notothenia coriiceps* (Nco, ADU02181); *Pleuragramma antarcticum* (Pan, ADU02180); *Danio rerio* (Dre_A, NP_001038953 and Dre_B, XP_697091); and *Carassius gibelio* (Cgi_A, AHZ08737 and Cgi_B, AAR12991). Amino acid consensus is shown below the alignment via three kinds of consensus symbols. An *asterisk* (*) indicates a position that has a single conserved residue; a *colon* (:) indicates a position that exhibits high similarity (scoring >0.5); and a *period* (.) indicates a position that shows low similarity (scoring ≤0.5). Asterisks (*) *above* protein sequences note the predicted α-helical regions. *Dotted boxes* indicate the four helices (*A*, *B*, *C*, and *D*) of the antiparallel bundle model of AFP IV. *CD* is a linker between the *C* and *D* helices

antarcticum, 53 % with *M. octodecemspinosus*, 52 % with *Myoxocephalus scorpius*, and 49 % with *Notothenia coriiceps*. However, it showed relatively lower identities (less than 30 %) to AFP IVs from tropical fish such as zebrafish (*D. rerio*) and gibel carp (*C. gibelio*). The calculated molecular masses of the precursor and the mature *P. olivaceus* AFP IV were 13,945 and 11,941 Da, respectively.

Genomic structure of *P. olivaceus afp4*

We identified six positive BAC clones, A1–A6, by hybridization screening, all of which except for the A5 clone contained the *afp4* gene, which was further confirmed by Southern blotting (data not shown). To obtain the genomic sequences, an approximately 2.9-kb genomic fragment of A3 was subcloned and sequenced. However, this fragment did not contain about 740 bp after the first exon; we therefore amplified the gap region by PCR using specific primers based on the partial genome sequence. We determined that the genomic structure of this gene consisted of four exons and three introns (Fig. 2a) comprising a total of 2808 bp. The intron/exon boundaries were well conserved according to the GT-AG rule. Based on the cDNA and genomic sequences, the *afp4* gene contains three translated exons (exons 2–4) and one non-translated exon (exon 1). The length of intron 1 is 1389 bp. Intron 2 (117 bp) is significantly shorter than the other introns. Exon 1 and 20 nt

of exon 2 encodes the 5′-UTR. In exon 2, the ATG translation initiation site is located 21 bp from the 5′ end of exon 2. Exons 2, 3, and 4 encode 16, 41, and 67 amino acids, respectively.

Purification and TH activity of recombinant AFP IV of olive flounder

The mature AFP of olive flounder was expressed in the pCold I vector with an additional N-terminal peptide consisting of a translation-enhancing element, factor X cleavage site, and 6 His-tag. Although the recombinant AFP was expressed as a mixture of soluble and insoluble forms, soluble protein was successfully purified to relative homogeneity using His-tag affinity chromatography (Fig. 3a, b). Although as shown in Fig. 3b there were a few contaminants visible in the eluted fractions and concentrated solution, because there is no intrinsic antifreeze activity from *E. coli* and any antifreeze activity observed was solely from the recombinant AFP IV, we did not attempt to remove the contaminants completely. The calculated molecular mass of AFP with the additional N-terminal sequence was 14,097 Da. The purified protein migrated faster by sodium dodecyl sulfate-polyacrylamide gel electrophoresis (SDS-PAGE); however, its molecular mass was confirmed by MALDI-TOF. The TH activity and ice crystal morphology of AFP were detected for both the soluble fraction of the cell lysate and the purified protein. TH activity of the cell lysate was approximately 0.04 ± 0.01 °C, which indicated that the soluble fraction contained active AFP IV. The AFP IV was purified but concentrated only up to 0.5 mg/mL using ultrafiltration because it tended to aggregate at higher concentrations of greater than 0.5 mg/mL. The TH value was 0.07 ± 0.01 °C at the concentration of 0.5 mg/mL. As the purified AFP IV was more concentrated than the soluble cell lysate, the TH value was higher. Similar to other AFP IVs, the olive flounder AFP produced a star-shaped ice morphology (Fig. 3c).

Discussion

The reason underlying the presence of a gene for AFP IV in *P. olivaceus* (Pol *afp4*) is unclear, as it is a subtropical fish that does not require AFP for survival. In addition, *P. olivaceus* did not initially demonstrate an antifreeze activity (Kim 2015). However, its EST analysis revealed that it carried a transcribable *afp4* gene, which further raised the question of the function of this transcript. To test the hypothesis that the ice-binding ability of AFP IV was gained specifically by the selective pressure of cold environments, we cloned the Pol *afp4* gene and evaluated its TH activity.

Previously, Cheng (2003) proposed that the longhorn sculpin AFP IV had evolved from a domain duplication of apolipoproteins through the selection pressure of a

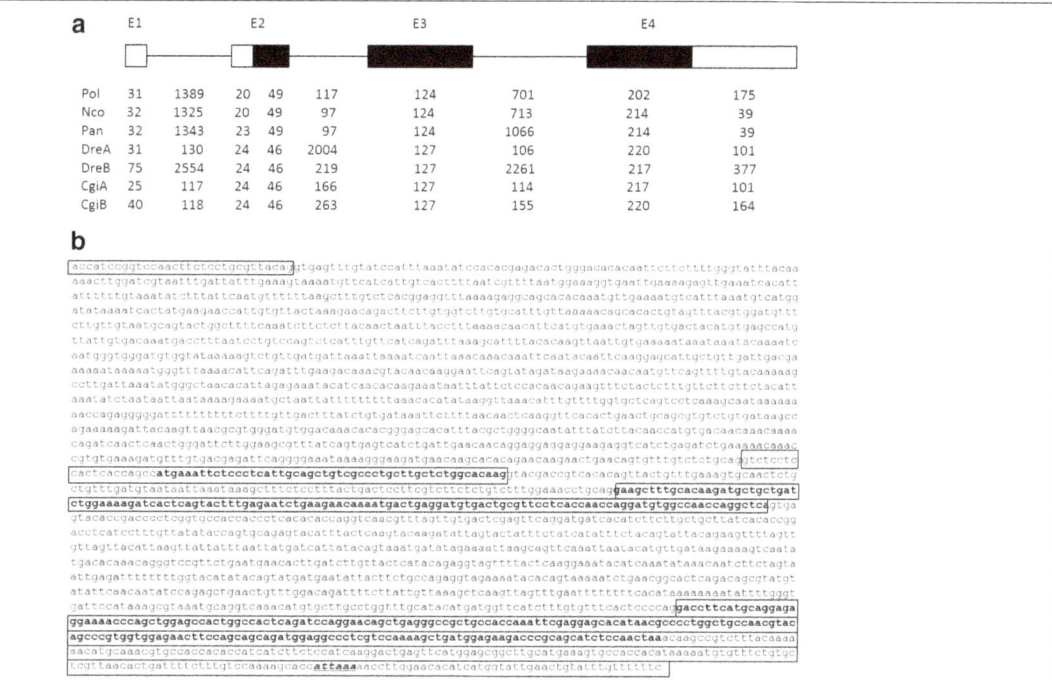

Fig. 2 Genomic structure (**a**) and genomic sequence (**b**) of Pol AFP IV. *Pol Paralichthys olivaceus, Nco Notothenia coriiceps, Pan Pleuragramma antarcticum, Dre Danio rerio,* and *Cgi Carassius gibelio.* In **b**, the exons are indicated as *boxes* starting from exon 1; the poly A signal is *underlined* and *italicized*

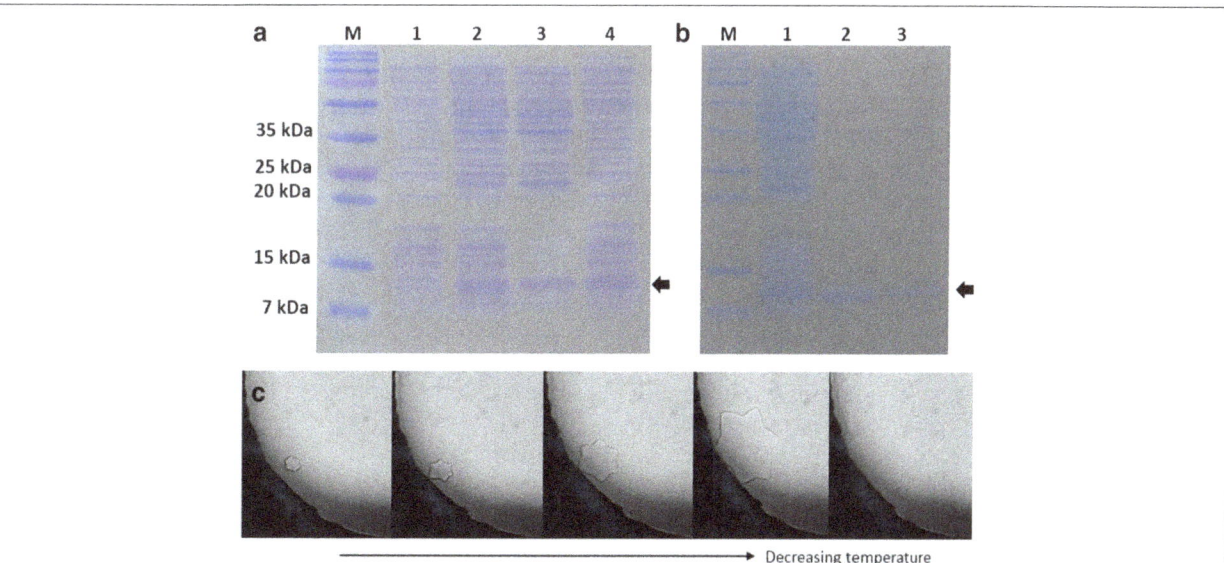

Fig. 3 SDS-PAGE of expressed (**a**) and purified (**b**) Pol AFP IV and its TH activity and ice morphology (**c**). **a** *M* protein size marker, *lane 1* lysates of uninduced bacteria, *lane 2* lysates of bacteria induced with IPTG and *lanes 3* and *4* soluble fraction and insoluble fractions of total protein from the induced bacteria, respectively. **b** *M* protein size marker, *lane 1* elution fraction 1, and *lane 2* elution fraction 2. *Arrows* indicate Pol AFP IV. **c** Antifreeze activity of recombinant Pol AFP IV. The ice morphology of the Pol AFP IV solution was examined using a nanoliter osmometer under a light microscope

subfreezing marine environment. On the other hand, its homologues have been identified in many fish isolated from polar to tropic seawater as well as from freshwater, and their gene organization is also very similar (Gauthier et al. 2008; Lee et al. 2011; Xiao et al. 2014). A comparison of the genome structures of known fish AFP IV proteins including the *P. olivaceus afp4* gene characterized in the current study showed that the exon and intron positions are highly conserved (Fig. 2a), supporting the suggestion that fish *afp4* genes resembled each other regardless of their habitats. However, the sequences and sizes of the introns differ among fish. Notably, the genomic organization of the *P. olivaceus afp4* gene exhibits greater similarity to those of AFP IV from two Antarctic fish, *P. antarcticum* and *N. coriiceps* (Lee et al. 2011), than to tropical carp and zebrafish. In particular, the two *afp4* genes found in both carp and zebrafish are arranged in a head-to-tail tandem manner. This tandem arrangement is believed to be due to gene duplication (Xiao et al. 2014). However, only a single *afp4* gene is present in the genomes of two Antarctic fish and *P. olivaceus*, with no evidence of gene duplication being evident. Together, this evidence renders the establishment of the evolution of this gene more challenging.

The sequence alignment showed that the four additional AFP IVs with a known antifreeze activity share a 48–53 % amino acid sequence identity with Pol AFP IV (Fig. 1). Deng and Laursen (1998) proposed a four-helix bundle model for AFP IV based on the high sequence identity to an apolipoprotein. Notably, as shown in Fig. 1, Pol AFP IV carries a four-residue deletion that can be mapped onto the linker between C and D helices (CD linker) linker of the four-helix bundle model (Gauthier et al. 2008; Lee et al. 2011). However, despite the deletion in the linker, Pol AFP IV exhibits almost the same TH activity as other AFP IV proteins (Deng et al. 1997; Deng and Laursen 1998; Zhao et al. 1998; Gauthier et al. 2008; Lee et al. 2011), which strongly indicates that residues important for ice-binding are conserved in Pol AFP IV and not located in the CD linker. This result also implies that the deletion is not likely to disrupt the four-helix bundle structure, because the characteristics of recombinant Pol AFP IV are similar to those of other recombinant AFP IVs. This may hold positive implications for AFP IVs from fresh- and warmwater fish as well (Xiao et al. 2014). Thus, even though the AFP IVs from *C. gibelio* and *D. rerio* each contain a deletion in the loop connecting helices A and B (Xiao et al. 2014), they might therefore have very similar three-dimensional folds as there are no drastic changes in the sequences corresponding to the four helices.

In particular, although very weak compared to AFPs I, II, and III at the same concentrations, recombinant Pol AFP IV exhibits TH activity of 0.07 ± 0.01 °C at 0.5 mg/mL. This activity is comparable to those of the Antarctic AFP IVs (Lee et al. 2011) but is slightly lower than those of recombinant shorthorn and longhorn sculpin counterparts (Deng et al. 1997; Deng and Laursen 1998; Gauthier et al. 2008; Zhao et al. 1998). Although the natural AFP IV of longhorn sculpin begins to aggregate at concentrations above 25 mg/mL (Deng et al. 1997), recombinant AFP IVs expressed in *E. coli* were less soluble and prone to aggregate at concentrations as low as 0.4 mg/mL for *M. octodecemspinosus* AFP IV, 0.5 mg/mL for *P. antarcticum* and *N. coriiceps* AFP IVs, and 10 mg/mL for *M. scorpius* AFP IV (Gauthier et al. 2008; Lee et al. 2011). Similar to these AFPs, Pol AFP IV was also only able to be concentrated up to 0.5 mg/mL. However, this low concentration is comparable to the physiological concentration (0.05–0.1 mg/mL) of AFP IV in the blood plasmas of shorthorn sculpin as detected by immunoblotting (Gauthier et al. 2008). At this low concentration, the TH value of the natural longhorn sculpin AFP IV is similar to those of other recombinant AFP IVs (Deng et al. 1997; Deng and Laursen 1998), suggesting that if the recombinant proteins were able to be concentrated, their TH values would be quite close to that of the longhorn sculpin protein. The TH values from AFP IVs are not enough to keep fish from freezing at subzero temperature, addressing the question of their real function in vivo.

The ice morphology shaped by Pol AFP IV was also similar to those produced by the Antarctic fish and shorthorn sculpin AFP IVs (Gauthier et al. 2008; Lee et al. 2011). The star-shaped ice crystals represent weak antifreeze activity, which is also consistent with the low TH values. Notably, if longhorn sculpin AFP IV had been evolved only through the selective pressure of the cold Antarctic environment approximately 10–15 million years ago, then there would likely be a very slim chance of comparable AFP IV evolution in temperate, subtropical, and/or tropical fish such as *P. olivaceus*, which might have experienced very recent temporal glaciation approximately 2.5 to 3 million years ago. Instead, the abundant expression of AFP IVs in carp and zebra fish observed in oocytes and embryos strongly support its association with early development (Goetz et al. 2006; Liu et al. 2009; Brenton et al. 2012).

Conclusions

Taken together, our results therefore do not support the hypothesis of evolution of AFP IV by selective pressure, suggesting that the antifreeze activity of AFP IV may have been gained by chance.

Funding

This work was supported by a Creative Research Grant (2014) of Pukyong National University.

Authors' contributions

JKL carried out the study and drafted the manuscript. HJK conceived of and participated in the study, and finalized the manuscript. Both authors read and approved the final manuscript.

Competing interests

The authors declare that they have no competing interests.

Author details

[1]Department of Microbiology, College of Natural Sciences, Pukyong National University, Busan 48513, Republic of Korea. [2]Department of Chemistry, College of Natural Sciences, Pukyong National University, Busan 48513, Republic of Korea.

References

Bendtsen JD, Nielsen H, Heijne GV, Brunak S. Improved prediction of signal peptides: SignalP 3.0. J Mol Biol. 2004;340:783–95.

Breton TS, Anderson JL, Goetz FW, Berlinsky DL. Identification of ovarian gene expression patterns during vitellogenesis in Atlantic cod (Gadus morhua). Gen Comp Endocrinol. 2012;179:296–304.

Cheng CH. Evolution of the diverse antifreeze proteins. Curr Opin Genet Dev. 1998;8:715–20.

Cheng CC. Freezing avoidance in polar fishes. In: Gerday C, editor. Encyclopedia of life support systems (EOLSS)—theme 6.73 extremophiles, developed under the auspices of the UNESCO. Oxford: Eolss Publishers; 2003. http://www.eolss.net.

Cheng CH, Chen L. Evolution of an antifreeze glycoprotein. Nature. 1999;401:443–4.

Cheng CH, Chen L, Near TJ, Jin Y. Functional antifreeze glycoprotein genes in temperate-water New Zealand notothenioid fish infer an Antarctic evolutionary origin. Mol Biol Evol. 2003;20:1897–908.

Choudhury M, Yamada S, Komatsu M, Kishimura H, Ando S. Homologue of mammalian apolipoprotein A-II in non-mammalian vertebrates. Acta Biochim Biophys Sin (Shanghai). 2009;41:370–8.

Davies PL, Hew CL. Biochemistry of fish antifreeze proteins. FSSEB J. 1990;4:2460–8.

Deng G, Laursen RA. Isolation and characterization of an antifreeze protein from the longhorn sculpin, Myoxocephalus octodecimspinosis. Biochim Biophys Acta. 1998;1388:305–14.

Deng G, Andrews DW, Laursen RA. Amino acid sequence of a new type of antifreeze protein, from the longhorn sculpin Myoxocephalus octodecimspinosis. FEBS Lett. 1997;402:17–20.

DeVries AL. Glycoproteins as biological antifreeze agents in Antarctic fishes. Science. 1971;172:1152–5.

DeVries AL, Wohlschlag DE. Freezing resistance in some Antarctic fishes. Science. 1969;163:1073–5.

Fletcher GL, Hew CL, Davies PL. Antifreeze proteins of teleost fishes. Annu Rev Physiol. 2001;63:359–90.

Gauthier SY, Scotter AJ, Lin FH, Baardsnes J, Fletcher GL, Davies PL. A re-evaluation of the role of type IV antifreeze protein. Cryobiology. 2008;57:292–6.

Goetz FW, McCauley L, Goetz GW, Norberg B. Using global genome approaches to address problems in cod mariculture. ICES J Marine Sci. 2006;63:393–9.

Graham LA, Lougheed SC, Ewart KV, Davies PL. Lateral transfer of a lectin-like antifreeze protein gene in fishes. PLoS One. 2008;3:e2616.

Graham LA, Hobbs RS, Fletcher GL, Davies PL. Helical antifreeze proteins have independently evolved in fishes on four occasions. PLoS One. 2013;8:e81285.

Jia Z, Davies PL. Antifreeze proteins: an unusual receptor-ligand interaction. Trends Biochem Sci. 2002;27:101–6.

Kim HJ. Antifreeze activity in temperate fish from the East Sea, Korea. Fisheries Aquatic Sci. 2015;18:137–42.

Larkin MA, Blackshields G, Brown NP, Chenna R, McGettigan PA, McWilliam H, et al. ClustalW and ClustalX version 2. Bioinformatics. 2007;23:2947–8.

Lee JK, Kim YJ, Park KS, Shin SC, Kim HJ, Song YH, et al. Molecular and comparative analyses of type IV antifreeze proteins (AFPIVs) from two Antarctic fishes, Pleuragramma antarcticum and Notothenia coriiceps. Comp Biochem Physiol B Biochem Mol Biol. 2011;159:197–205.

Liu JX, Zhai YH, Gui JF. Molecular characterization and expression pattern of AFPIV during embryogenesis in gibel carp (Carassiu auratus gibelio). Mol Biol Rep. 2009;36:2011–8.

Nishimiya Y, Yasuhiro M, Hirano Y, Kondo H, Miura A, Tsuda S. Mass preparation and technological development of an antifreeze protein. Synthesiology (Engl ed). 2008;1:7–14.

Raymond JA, DeVries AL. Adsorption inhibition as a mechanism of freezing resistance in polar fishes. Proc Natl Acad Sci U S A. 1977;74:2589–93.

Wang Y, Zhou L, Li Z, Li W, Gui J. Apolipoprotein C1 regulates epiboly during gastrulation in zebrafish. Sci China Life Sci. 2013;56:975–84.

Xia JH, Liu JX, Zhou L, Li Z, Gui JF. Apo-14 is required for digestive system organogenesis during fish embryogenesis and larval development. Int J Devel Biol. 2008;52:1089–98.

Xiao Q, Xia JH, Zhang XJ, Li Z, Wang Y, Zhou L, et al. Type-IV antifreeze proteins are essential for epiboly and convergence in gastrulation of zebrafish embryos. Int J Biol Sci. 2014;10:715–32.

Zhang T, Yao S, Wang P, Yin C, Xiao C, Qian M, et al. ApoA-II directs morphogenetic movements of zebrafish embryo by preventing chromosome fusion during nuclear division in yolk syncytial layer. J Biol Chem. 2011;286: 9514–25.

Zhao Z, Deng G, Lui Q, Laursen RA. Cloning and sequencing of cDNA encoding the LS-12 antifreeze protein in the longhorn sculpin, Myoxocephalus octodecimspinosis. Biochim Biophys Acta. 1998;1382:177–80.

Galatheoid squat lobsters (Crustacea: Decapoda: Anomura) from Korean waters

Jung Nyun Kim[1*], Mi Hyang Kim[2], Jung Hwa Choi[3] and Yang Jae Im[1]

Abstract

Ten species of Galatheoidea (squat lobsters), belonging to two families, were collected in the Korean exclusive economic zone: *Galathea balssi* Miyake and Baba, 1964, *Galathea orientalis* Stimpson, 1858, *Galathea pubescens* Stimpson, 1858, and *Galathea rubromaculata* Miyake and Baba, 1967 belonging to Galatheidae; *Bathymunida brevirostris* Yokoya, 1933, *Cervimunida princeps* Benedict, 1902, *Munida caesura* Macpherson and Baba, 1993, *Munida japonica* Stimpson, 1858, *Munida pherusa* Macpherson and Baba, 1993, and *Paramunida scabra* (Henderson, 1885) belonging to Munididae. The present study comprises the morphological description of these ten species, including drawings and color photographs, a brief review of their regional records, and a key for their identification. Although all species are common in Japanese waters, *G. balssi*, *G. rubromaculata*, *B. brevirostris*, *C. princeps*, *M. caesura*, and *M. pherusa* are new to Korean marine fauna.

Keywords: Galatheidae, Munididae, Squat lobster, Korean fauna

Background

Based on a phylogenetic study, Ahyong et al. (2010) revised the classification of the superfamily Galatheoidea to include four families: Galatheidae, Munididae, Munidopsidae, and Porcellanidae. The former three families, together with the superfamily Chirostyloidea, which includes Chirostylidae, Eumunidae, and Kiwaidae, are known as "squat lobsters," as their abdomen is tucked under the thorax, giving a "crouching or squatting" appearance (Baba et al. 2009). Squat lobsters are distributed worldwide and play an important role in the marine food webs of coastal marine zones. Some species inhabiting the eastern Pacific and eastern Atlantic are of commercial importance (Poore et al. 2011). Although galatheoid squat lobsters comprise 877 species belonging to 45 genera (World Register of Marine Species 2015), in Korean waters, only four species belonging to three genera of two different families have been reported (Kim and Kim 1997): *Galathea orientalis* Stimpson, 1858 and *Galathea pubescens* Stimpson, 1858 in Galatheidae; *Munida japonica* Stimpson, 1858 and *Paramunida*

scabra (Henderson, 1885) (as *Munida*) in Munididae (as Galatheidae).

The present study provides morphological descriptions and illustrations for ten species of squat lobsters (four Galatheidae and six Munididae), collected in the Korean exclusive economic zone during a fisheries resources investigation conducted by the National Institute of Fisheries Science (Korea), from 2003 to 2013. Of these, six species are recorded for the first time in Korean waters: *Bathymunida brevirostris* Yokoya, 1933, *Cervimunida princeps* Benedict, 1902, *Galathea balssi* Miyake and Baba, 1964, *Galathea rubromaculata* Miyake and Baba, 1967 *Munida caesura* Macpherson and Baba, 1993, and *Munida pherusa* Macpherson and Baba, 1993. Eight of the ten Korean species occur in the West Pacific and *B. brevirostris* is endemic to the Korea and Tsushima Straits. Two other species, *G. pubescens* and *Paramunida scabra* are widely distributed in the Indo-West Pacific. A key for their identification is also provided.

Methods

The specimens examined in this study were deposited in the National Institute of Fisheries Science, Korea. Individuals were collected by SCUBA diving for *G. orientalis* and bottom otter trawls (mesh size at the cod end = 0.98 × 0.98 mm), operating during daytime for 30–

---

* Correspondence: crangonk@korea.kr
[1]West Sea Fisheries Research Institute, National Institute of Fisheries Science, Incheon 22383, South Korea
Full list of author information is available at the end of the article

60 min at 3.4 knots, on average for the other species. All samples were frozen on board shortly after capture and maintained at –80 °C until their morphological identification in the laboratory. Specimens were photographed before their identification, and then preserved in 70–90 % ethyl alcohol.

Carapace length (CL) was determined measuring the distance from the posterior margin of the orbit to the posterior mid-dorsal margin of the carapace, and used to indicate specimen size. The terminology used in descriptions mainly follows that of Baba et al. (2009). Species are arranged in alphabetical order.

Results and discussion
Superfamily Galatheoidea Samouelle, 1819
(Korean name: Sae-u-but-i-sang-gwa)
Family Galatheidae Samouelle, 1819
(Korean name: Sae-u-but-i-gwa)
Genus *Galathea* Fabricius, 1793
(Korean name: Sae-u-but-i-sok)
***Galathea balssi* Miyake and Baba, 1964**
(New Korean name: Bal-seu-sae-u-but-i)
(Figs. 1 and 11a)

Galathea australiensis Balss, 1913: 13, Fig. 13 (not *G. australiensis* Stimpson, 1858).

Galathea balssi Miyake and Baba, 1964: 205, Figs. 1, 2 (type locality: East China Sea, 120–122 m); Haig, 1973: 278, Fig. 2a–f; Macpherson and Robainas-Barcia, 2015: 56 (full synonymy).

Material examined
One male (CL 6.1 mm), one ovigerous female (CL 7.5 mm), 33° 05.8′N, 126° 47.2′E, southern Jeju Island,

Korea, 101 m depth, bottom otter trawl, RV *Tamgu 1*, 25 Apr. 2004.

Description
Rostrum (Fig. 1a) about 1.5 times longer than wide, with four lateral teeth. Carapace (Fig. 1a) with several interrupted transverse striae on its anterior half; epigastric and parahepatic spines present; lateral margin with small second anterolateral spine, anterior branchial margin with two spines. Pterygostomian flap unarmed. Antennular peduncle (Fig. 1b) with basal segment bearing three distal spines. Ischium of third maxilliped (Fig. 1c) with one flexor spine on distal margin; merus with two flexor spines; carpus unarmed. First pereopod fingers (Fig. 1d) strongly gapped in male and slightly gapped in female, with large proximal tooth. Second pereopod carpus (Fig. 1e) bearing dorsal row of spines; propodus with six strong flexor corneous spines; dactylus about half length of propodus, biunguiculate, with row of teeth on flexor margin. Epipods absent on all pereopods.

Coloration
Carapace dull light brown; rostrum light reddish brown; first pereopod light brown, with light reddish brown band; second to fourth pereopods with pale brownish bands (Fig. 11a).

Distribution
Korea, Japan, East China Sea, South China Sea, Philippines, Indonesia, southwestern and eastern Australia, and Vanuatu; depth 31-222 m (Balss 1913, Miyake and Baba 1964, Haig 1973, Macpherson and Robainas-Barcia 2015).

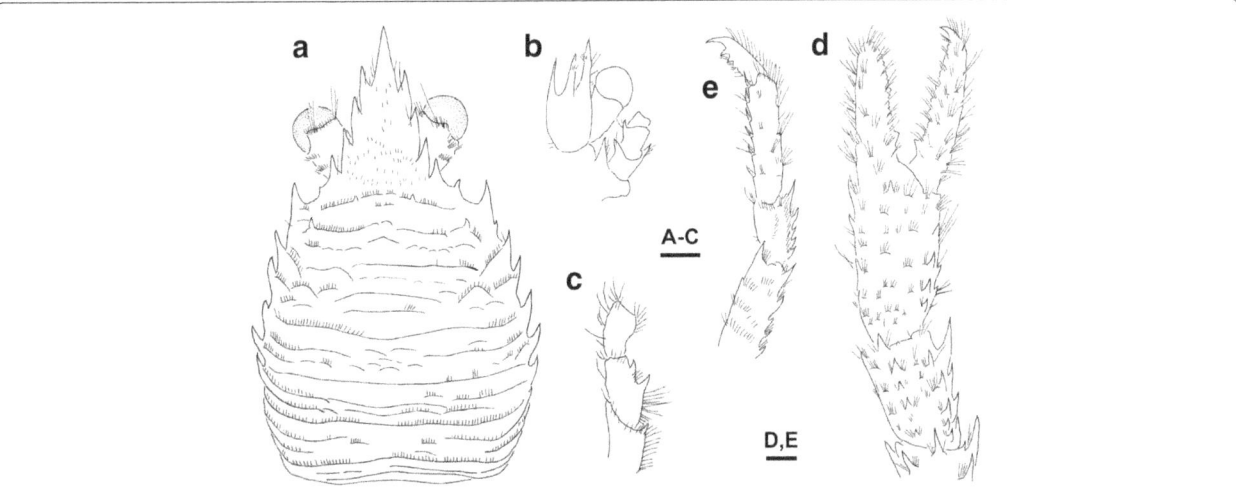

Fig. 1 *Galathea balssi* Miyake and Baba, 1964. Ovigerous female (CL 7.5 mm) from southern Jeju Island. **a** Carapace and eyes, dorsal. **b** Left eye, antennular and antennal peduncles, ventral. **c** Ischium, merus and carpus of right third maxilliped, lateral. **d** Carpus and chela of left first pereopod, dorsal. **e** Left second pereopod, lateral. Scales = 1.0 mm

Remarks

This species is easily distinguishable from its Korean congeners by the parahepatic spine on the carapace.

Galathea orientalis Stimpson, 1858
(Korean name: Sae-u-but-i)
(Fig. 2)

Galathea orientalis Stimpson, 1858: 252 (type locality: Ly-i-moon passage near Hong Kong, depth 46 m); Miers, 1879: 51; Ortmann, 1892: 252, pl. 11, Figs. 10, 10a, 10i; Miyake and Baba, 1967: 232, Fig. 5; Kim, 1973: 175, Text-Fig. 19, pl. 64, Figs, 5a, 5b; Miyake, 1982: 145, with one figure, pl. 49, Fig. 1; Kim and Kim, 1987: 232; Baba et al. 2009: 120, Figs. 100–103; Macpherson and Robainas-Barcia, 2015: 206 (full synonymy).

Material examined

Two males (CL 3.0, 4.1 mm), two ovigerous females (CL 4.4, 5.2 mm), Oryuk Islet, Busan, Korea, 0–5 m depth, SCUBA diving, 23 Jul. 2014; six males (CL 2.2–2.9 mm), four females (CL 2.3–3.1 mm), Oryuk Islet, Busan, Korea, 10–15 m depth, SCUBA diving, 3 Dec. 2014; two males (CL 3.0, 3.2 mm), one female (CL 3.9 mm), Oryuk Islet, Busan, Korea, 0–5 m depth, SCUBA diving, 10 Mar. 2015; one male (CL 2.3 mm), two ovigerous females (CL 3.9, 4.0 mm), Oryuk Islet, Busan, Korea, 10–15 m depth, SCUBA diving, 14 Aug. 2015.

Description

Rostrum (Fig. 2a) about 1.5 times longer than wide, with four lateral teeth. Carapace (Fig. 2a) with lateral orbital angle strongly produced; epigastric spine present; third stria between anteriormost spines on branchial lateral margin interrupted by cervical groove; lateral margin with moderately large second anterolateral spine; anterior branchial margin with two spines. Pterygostomial flap (Fig. 2c) with spine on anterior part. Thoracic sternum (Fig. 2b) about 0.8 times longer than broad; third sternite heart shape, anterior margin medially concave. Second pleomere (Fig. 2a) with two transverse ridges. Antennular peduncle (Fig. 2c) with basal segment bearing three distal spines. Third maxilliped merus (Fig. 2d) bearing two flexor and two extensor spines, carpus with three extensor spines. First pereopod (Fig. 2e) with fingers spooned distally. Epipods present only on first pereopod.

Coloration

Body light brown; appendages and carapace carinae and spines light reddish brown. However, color substantially varies in this species (see Baba et al. 2009).

Distribution

Korea, Japan, East China Sea, Hong Kong, Taiwan, and western Australia; shore to 549 m depth (Stimpson 1858, Miers 1879, Ortmann 1892, Miyake and Baba 1937, Kim 1973, Miyake 1982, Baba et al. 2009, Macpherson and Robainas-Barcia 2015).

Remarks

G. orientalis is easily distinguished from its Korean congeners by the small spine on the anterior part of the pterygostomian flap. This species is relatively widely distributed in Korea (Kim 1973, Kim and Kim 1987).

Galathea pubescens Stimpson, 1858
(Korean name: Tel-bo-sae-u-but-i)
(Figs. 3 and 11b)

Galathea pubescens Stimpson, 1858: 90 (type locality: Hakodate and Amami-oshima, Japan, depth 46–60 m); Balss, 1913: 11, Figs. 11, 12; Yokoya, 1933: 57; Makarov, 1938: 88, Fig. 32, 33; Miyake, 1982: 145, pl. 49, Fig. 3;

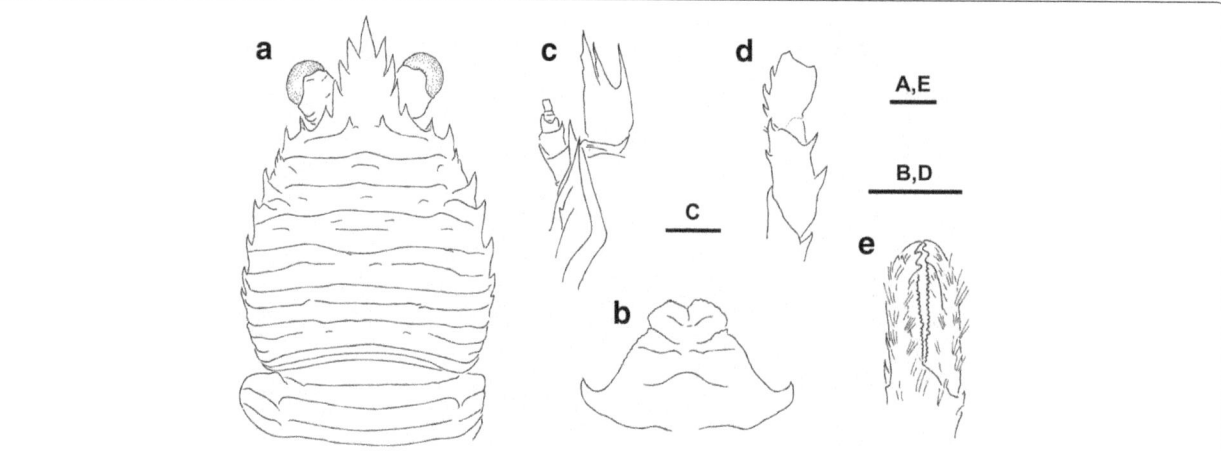

Fig. 2 *Galathea orientalis* Stimpson, 1858. Ovigerous female (CL 5.2 mm) from Oryuk Islet, Busan. **a** Carapace and eyes, dorsal. **b** Third and fourth thoracic sternites, ventral. **c** Right part of cephalothorax, showing antennular and antennal peduncles, ventral. **d** Ischium, merus, and carpus of right third maxilliped, lateral. **e** Distal part of chela of right first pereopod, ventral. Scales = 1.0 mm

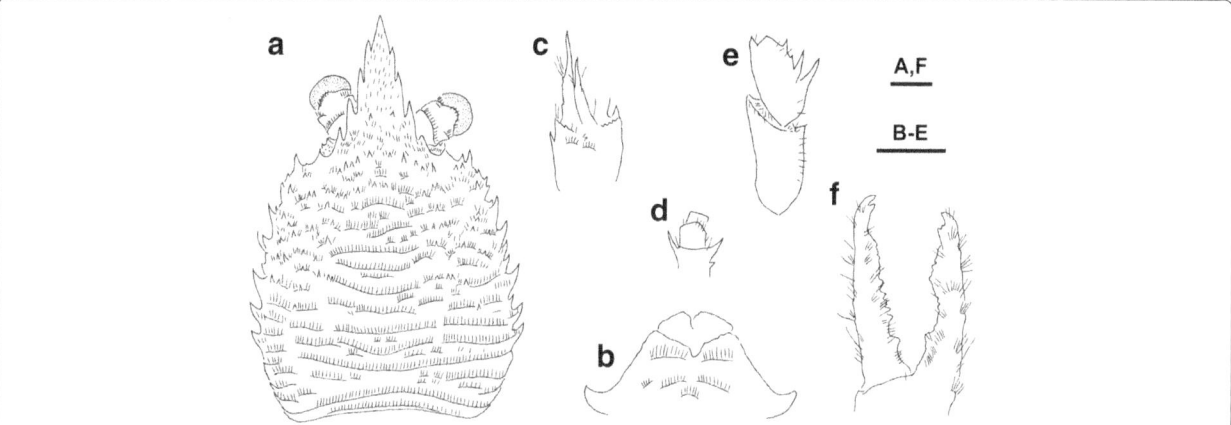

Fig. 3 *Galathea pubescens* Stimpson, 1858. Male (CL 6.0 mm) from eastern Jeju Island. **a** Carapace and eyes, dorsal. **b** Third and fourth thoracic sternites, ventral. **c** Right antennular peduncle, ventral. **d** Right antennal peduncle, ventral. **e** Ischium and merus of right third maxilliped, lateral. **f** Distal part of chela of left first pereopod, ventral. Scales = 1.0 mm

Baba, 1969: 48, Fig. 5; Kim, 1973: 176, Text-Fig. 20, pl. 65, Figs. 6a, 6b; Kim and Kim, 1987: 233; Baba et al., 2009:125, Figs. 105, 106; Macpherson and Robainas-Barcia, 2015: 266, Fig. 120A (full synonymy).

Material examined

One male (CL 6.0 mm), one female (CL 4.2 mm), 33° 09.6′ N, 127° 28.4′E, eastern Jeju Island, Korea, 112 m depth, bottom otter trawl, RV *Tamgu 1*, 25 Apr. 2004; one ovigerous female (CL 3.9 mm), 33° 24.8′N, 127° 43.5′E, eastern Jeju Island, Korea, 117 m depth, bottom otter trawl, RV *Tamgu 1*, 26 Oct. 2006; one ovigerous female (CL 7.6 mm), 32° 17.7′N, 126° 15.3′E, southern Jeju Island, Korea, 97 m depth, bottom otter trawl, RV *Tamgu 1*, 13 Apr. 2008.

Description

Rostrum (Fig. 3a) about 1.5 times longer than wide, lateral margin with four teeth. Carapace (Fig. 3a) pubescent dorsally, with many small spines on its anterior half; lateral orbital angle with small spine; epigastric region with row of small spines arranged in concentric arc; lateral margin with three to four small spines between anterolateral spine and anterior most branchial spine; branchial margin with three to four spines in its anterior half. Pterygostomian flap unarmed. Second and third pleomeres with four transverse ridges each. Thoracic sternum (Fig. 3b) about 0.7 times longer than wide; third sternite heart shaped, anterior margin medially excavated. Antennular peduncle (Fig. 3c) with basal segment bearing well-developed distodorsal and distolateral spines, distomesial spine minute. First segment of antennal peduncle (Fig. 3d) bearing one distal spine on lateral margin and two mesial spines. Third maxilliped merus (Fig. 3e) with three flexor spines, occasionally one small additional spine between distal and second spines; extensor margin with two distal spines. First pereopod (Fig. 3f) with fingers distally spooned, cutting edge with intermeshing teeth. Epipods only on first pereopod.

Coloration

Carapace and anterior half of abdomen orange, whitish along midline; first pereopod orange with whitish spine tips (Fig. 11b).

Distribution

Widely distributed in the Indo-West Pacific: Korea, Japan, Taiwan, East China Sea, South China Sea, Philippines, New Caledonia, Central Queensland, Western Australia, Tanzania, and South Africa; depth 45-494 m (Stimpson 1858, Balss 1913, Yokoya 1933, Makarov 1938, Miyake 1982, Baba 1962, Kim 1987, Baba et al. 2009 Macpherson and Robainas-Barcia 2015).

Remarks

The present species is easily distinguished from its Korean congeners by the pubescent carapace and the numerous small spines on the anterior part of the carapace.

Galathea rubromaculata Miyake and Baba, 1967
(New Korean name: Jem-bak-i-sae-u-but-i)
(Fig. 4)

Galathea rubromaculata Miyake and Baba, 1967: 236, Figs. 7, 8 (type locality: East China Sea, 32° 24.8′N, 129° 24.7′E, depth 173 m); Baba, 2005: 245; Poore et al., 2011: 333, pl. 11G-H; Macpherson and Robainas-Barcia, 2015: 271 (full synonymy).

Material examined

One male (CL 3.9 mm), 33° 15.4′N, 127° 39.8′E, eastern Jeju Island, Korea, 132 m depth, bottom otter trawl, RV *Tamgu 20*, 28 Mar 2013.

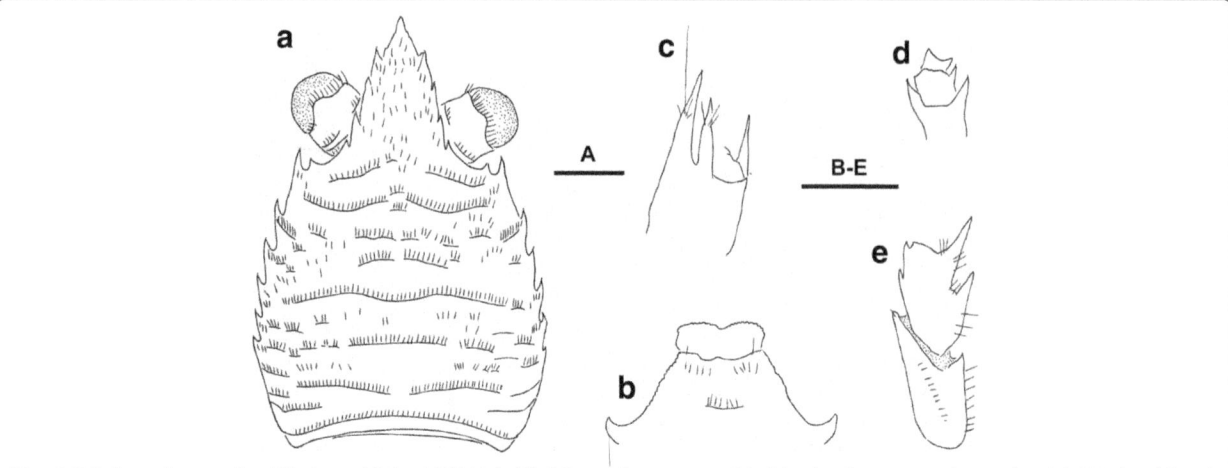

Fig. 4 *Galathea rubromaculata* Miyake and Baba, 1967. Male (CL 3.9 mm) from eastern Jeju Island. **a** Carapace and eyes, dorsal. **b** Third and fourth thoracic sternites, ventral. **c** Right antennular peduncle, ventral. **d** Right antennal peduncle, ventral. **e** Ischium and merus of right third maxilliped, lateral. Scales = 1.0 mm

Description
Rostrum (Fig. 4a) broadly triangular with four shallowly incised lateral teeth. Carapace (Fig. 4a) weakly striated, without gastric spines; lateral margin without second anterolateral spine; anterior branchial margin with two spines. Pterygostomian flap unarmed. Thoracic sternum (Fig. 4b) about 0.8 times longer than wide; third sternite rectangular, posterior margin nearly straight. Antennular peduncle (Fig. 4c) with basal segment bearing three terminal spines. First segment of antennal peduncle (Fig. 4d) bearing one distal spine on each lateral and mesial margins; second segment with one distal spine on mesial margin. Merus of third maxilliped (Fig. 4e) with two stout ventral and two small dorsal spines. Epipods absent from all pereopods.

Coloration
Red spots bilaterally arranged on white carapace and pleon.

Distribution
Korea, Japan, and Philippines; depth 132-500 m (Miyake and Baba 1967, Baba 2005, Poore et al. 2011, Macpherson and Robainas-Barcia 2015).

Remarks
This species differs from its Korean congeners by having weakly striated carapace and lacking gastric spines. Color of live individuals is also distinctive: red spots on a carapace and pleon.

Family Munididae Ahyong, Baba, Macpherson and Poore, 2010
(New Korean name: Ba-neul-i-ma-sae-u-but-i-gwa)
Genus *Bathymunida* Balss, 1914
(New Korean Name: Jjalb-eun-ba-neul-i-ma-sae-u-but-i-sok)
Bathymunida brevirostris Yokoya, 1933

(New Korean name: Jjalb-eun-ba-neul-i-ma-sae-u-but-i)
(Figs. 5 and 11c)

Munida brevirostris Yokoya, 1933: 64, Fig. 28 (type locality: north of Goto Island, Japan, depth 106 m).
Bathymunida brevirostris: Baba, 1970: 59, Figs. 1, 2; Baba and de Saint Laurent, 1996: 450, Fig. 8 (full synonymy); Baba, 2005: 239; Baba et al., 2008: 56.

Material examined
One female (CL 5.1 mm), 33° 24.8′N, 127° 43.5′E, eastern Jeju Island, Korea, 117 m depth, bottom otter trawl, RV *Tamgu 1*, 26 Oct. 2006.

Description
Rostrum (Fig. 5a) anteriorly narrowed with well-developed supraocular spines distinctly exceeding minute median rostral spine. Carapace (Fig. 5b) without continuous ridges; minute scale-like striae on dorsal surface, especially numerous on gastric region; gastric and cardiac processes strongly compressed laterally; postcervical spines pronounced; anterolateral margin with small spine, branchial marginal spines relatively strong. Eyestalks (Fig. 5a) without tubercles; eyelashes barely discernible. Merus and ischium of third maxilliped (Fig. 5b) bearing distal spine on flexor margin. Second pereopod propodus (Fig. 5c) with row of small corneous spines on ventral margin; dactylus slender, slightly shorter than propodus, slightly curved in distal half, flexor margin with one spiniform seta at its midlength, approximately.

Coloration
Carapace and abdomen pale brown with several scattered orange-red spots; chelipeds pale brown with red

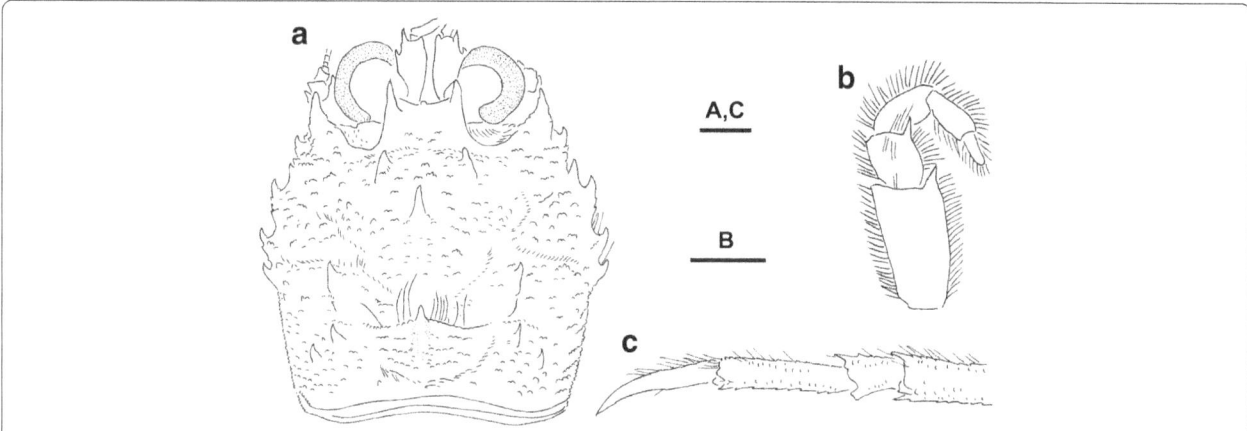

Fig. 5 *Bathymunida brevirostris* Yokoya, 1933. Ovigerous female (CL 5.1 mm) from eastern Jeju Island. **a** Carapace and cephalic appendages, dorsal. **b** Right third maxilliped, lateral. **c** Merus to dactylus of right second pereopod, lateral. Scales = 1.0 mm

marks on fingers but not on distal and proximal portions (Fig. 11c).

Distribution
Korea and Japan; depth 105-173 m (Yokoya 1933, Baba 1970, Baba and de Saint Laurent 1996, Baba 2005, Baba et al. 2008).

Remarks
This species is the sole representative of *Bathymunida* known in Korean waters and it is restricted to the Korea and Tsushima Straits.

Genus *Cervimunida* Benedict, 1902
(New Korean name: Sa-seum-sae-u-but-i-sok)
Cervimunida princeps Benedict, 1902
(New Korean name: Sa-seum-sae-u-but-i)
(Figs. 6 and 11d)

Cervimunida princeps Benedict, 1902: 249, Fig. 3 [type locality: off Honshu, Japan, "Albatross" St. 3698 (Sagami Bay, Manazuru Zaki), depth 280 m]; Balss, 1913: 18, Fig. 15, pl. 1, Fig. 1; Makarov, 1938: 100, Fig. 37; Miyake, 1982: 149, pl. 50, Fig. 4; Baba, 1986: 167, 288, Fig. 118; Baba et al., 2008: 59, Fig. 3I; Baba et al., 2009: 92, Figs. 75–77 (full synonymy).

Material examined
One male (CL 30.2 mm), southern Geoje Island, Korea, 112 m depth, bottom otter trawl, TS *Kaya* (Pukyong National University), 23 Sep. 2002; one male (CL 31.3 mm), 33° 34.5′N, 125° 19.5′E, southern Gageo Island, Korea, 89 m depth, bottom otter trawl, 9 Nov. 2006; one female (CL 22.9 mm), 35° 05.3′N, 129° 24.1′E, southern Busan, Korea, 110 m depth, bottom otter trawl, RV *Tamgu 3*, 24 Mar. 2004.

Description
Rostrum (Fig. 6a, b) laterally compressed, strongly arched; dorsal margin with two large and five additional small teeth; ventral margin with one strong tooth; supraocular spines curving dorsally, slightly reaching rostrum midlength. Carapace (Fig. 6a, b) striated; dorsal surface with six epigastric, two parahepatic, two

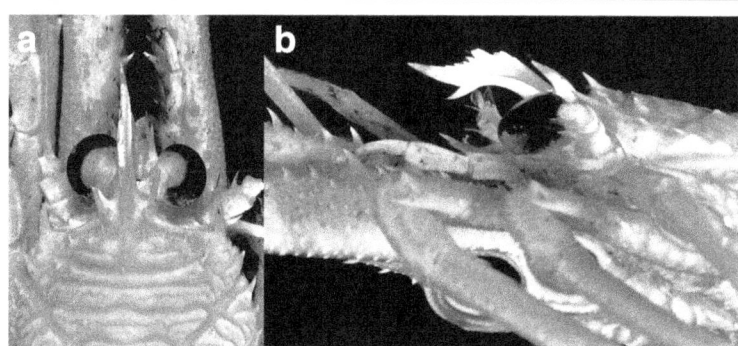

Fig. 6 *Cervimunida princeps* Benedict, 1902. Male (CL 30.2 mm) from southern Busan. **a** Carapace, cephalic and thoracic appendages, dorsal. **b** Same, lateral

postcervical, and two anterior branchial spines; cervical groove distinct; lateral margin with seven spines including two anterior branchial spines; anterolateral spine well developed. Second to fourth pleomeres with eight spines each on anterior ridge. Epipods absent from all pereopods.

Coloration

Body coloration orange to orange-red; rostrum deep red at distal and proximal portions and white in between; fingers and mesial margin of chelipeds orange-red (Fig. 11d).

Distribution

Korea, Japan, East China Sea, Taiwan, and Philippines; depth 76-452 m (Benedict 1902, Balss 1913, Makarov 1938, Miyake 1982, Baba 1986, Baba et al. 2008, Baba et al. 2009).

Remarks

The genus *Cervimunida* is represented by two species: *C. princeps* from the western Pacific *and Cervimunida johni* Porter, 1903 from the southeastern Pacific.

Genus *Munida* Leach, 1820
(Korean name: Ba-neul-i-ma-sae-u-but-i-sok)
Munida caesura Macpherson and Baba, 1993
(New Korean name: Kkeun-son-ba-neul-i-ma-sae-u-but-i)
(Figs. 7 and 11e)

Munida caesura Macpherson and Baba, 1993: 388, Fig. 3 (type locality: Tosa Bay, Japan, depth 250–300 m); Wu et al., 1998: 108, Figs. 20, 21H; Baba et al., 2009: 245, Figs. 27–29 (full synonymy).

Material examined

Four males (CL 13.1–16.7 mm), one female (CL 12.0 mm), 35° 05.3′N, 129° 24.1′E, southern Busan, Korea, 110 m depth, bottom otter trawl, RV *Tamgu 3*, 24 Mar. 2004.

Description

Carapace (Fig. 7a) with numerous transverse striae; intestinal region with scale-like stria; frontal margin oblique; anterolateral spines slightly overreaching sinus between rostral and supraocular spines. Pleomeres unarmed. Sternal plastron (Fig. 7b) with numerous striae; third sternite with anterior margin slightly bilobed, posterior margin broader than anterior margin of fourth sternite; fourth and fifth sternites with several oblique striae. Eyes (Fig. 7a) large. Antennular peduncle (Fig. 7c) with basal segment bearing subequal terminal spines. Second segment of antennal peduncle (Fig. 7d) with distomesial spine reaching distal margin of fourth segment and one additional small spine on mesial margin. Merus of third maxilliped (Fig. 7e) bearing one distal extensor spine and two flexor spines. Second pereopod (Fig. 7f) with dactylus proximally stout, proximal flexor margin convex with corneous spines on proximal three quarters and unarmed on distal quarter.

Coloration

Generally orange red; pereopods with numerous red dots; fingers of chelipeds reddish with whitish tips (Fig. 11f).

Distribution

Korea, Japan, Taiwan, South China Sea, Philippines, and Indonesia; depth 110-1211 m (Macpherson and Baba 1993, Wu et al. 1998, Baba et al. 2009).

Remarks

M. caesura can be distinguished from its Korean congeners based on the following characters: (1) second

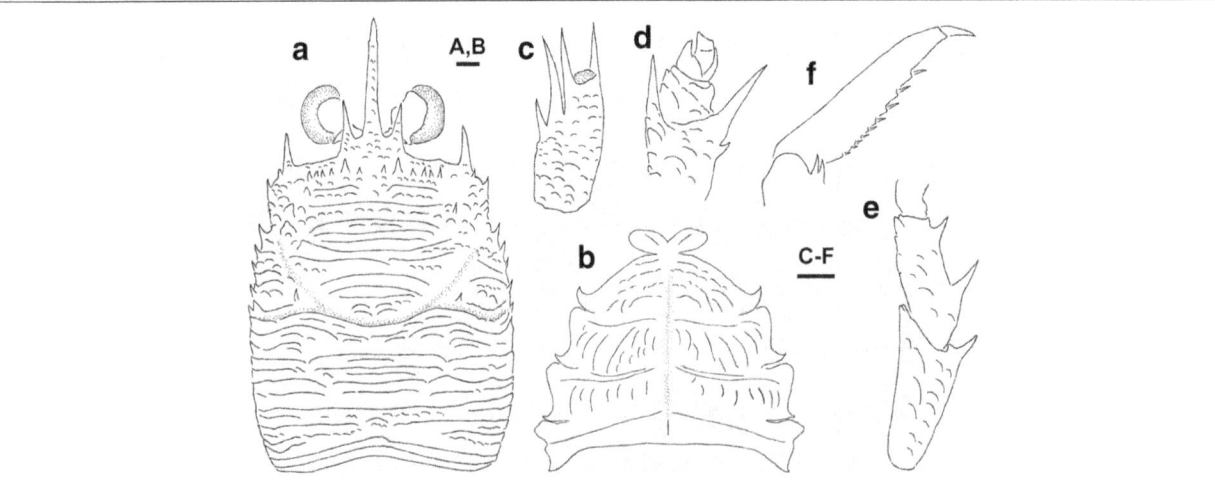

Fig. 7 *Munida caesura* Macpherson and Baba, 1993. Male (CL 16.7 mm) from southern Busan. **a** Carapace and eyes, dorsal. **b** Sternal plastron, ventral. **c** Right antennular peduncle, ventral. **d** Right antennal peduncle, ventral. **e** Ischium and merus of right third maxilliped, lateral. **f** Dactylus of right second pereopod, lateral. Scales = 1.0 mm

pleomere unarmed on anterior ridge; (2) anterior margin of the third thoracic sternite weakly bilobed; and (3) posterior margin of the third thoracic sternite broader than the anterior margin of the fourth sternite.

Munida japonica Stimpson, 1858
(Korean name: Ba-neul-i-ma-sae-u-but-i)
(Figs. 8 and 11f)

Munida japonica Stimpson, 1858: 252 (type locality: Kagoshima Bay, Japan, depth 36 m; type material extinct); Miyake and Baba, 1967: 240, Figs. 11, 12; Macpherson and Baba, 1993: 399, Fig. 9 [neotype designated (type locality: Makura-zaki, Kagoshima Pref., Japan, depth 145 m)]; Wu et al., 1998: 115, Figs. 24, 26F, G; Komai et al., 2002: 55; Baba et al., 2009: 163, Figs. 142–148 (full synonymy).

?*Munida japonica*: Kim, 1973: 178, pl. 65, Fig. 7; Miyake, 1982: 146, pl. 49, Fig. 4.

Material examined
One male (CL 9.1 mm), 33° 00.9′N, 125° 13.9′E, western Jeju Island, Korea, 90 m depth, bottom otter trawl, RV *Tamgu 1*, 26 Mar 2003; one male (CL 6.9 mm), 33° 00.0′N, 126° 08.6′E, western Jeju Island, Korea, 114 m depth, bottom otter trawl, 29 Mar. 2003; one female (CL 7.8 mm), 33° 30.0′N, 127° 50.6′E, northeastern Jeju Island, Korea, 112 m depth, bottom otter trawl, RV *Tamgu 3*, 5 Apr. 2003; one female (CL 8.9 mm), 33° 09.6′N, 127° 28.4′E, eastern Jeju Island, Korea, 112 m depth, bottom otter trawl, RV *Tamgu 1*, 25 Apr. 2004; one male (CL 7.6 mm), 33° 24.8′N, 127° 43.5′E, eastern Jeju Island, Korea, 117 m depth, bottom otter trawl, RV *Tamgu 1*, 26 Oct. 2006.

Description
Carapace (Fig. 8a) with frontal margin slightly oblique; anterior spine not reaching sinus level between rostrum and supraocular spine; posterior branchial margin with five spines; no scale-like or short stria on intestinal region. Second pleomere (Fig. 8a) with two spines on each side of anterior transverse stria. Sternal plastron (Fig. 8b) with few striae on fourth sternite; anterior margin of fourth sternite contiguous to median part of posterior margin of third sternite. Eyes (Fig. 8a) large. Basal segment of antennular peduncle (Fig. 8c) with terminal spines subequal. First segment of antennal peduncle (Fig. 8d) bearing distomesial spine almost reaching distal margin of third segment; second segment bearing small mesial spine and prominent distomesial spine overreaching distal margin of fourth segment. Merus of third maxilliped (Fig. 8e) bearing distinct distal spine on extensor margin; flexor margin with three spines. First pereopod (Fig. 8f) with fingers as long as palm, each bearing proximal spines fairly dorsal, and between subterminal and basal spines. Flexor margin of dactylus in second pereopod (Fig. 8g) bearing corneous spines in proximal two thirds and unarmed on distal portion.

Coloration
Generally orange red; chelipeds with whitish lateral margins and reddish mesial margins; ambulatory legs with slight reddish brown and white bands (Fig. 11g).

Distribution
Korea, Japan, East China Sea, Taiwan, Philippines, and Indonesia; depth 22-732 m (Stimpson 1858, Miyake and Baba 1967, Macpherson and Baba 1993, Wu et al. 1998, Komai et al. 2002, Baba et al. 2009).

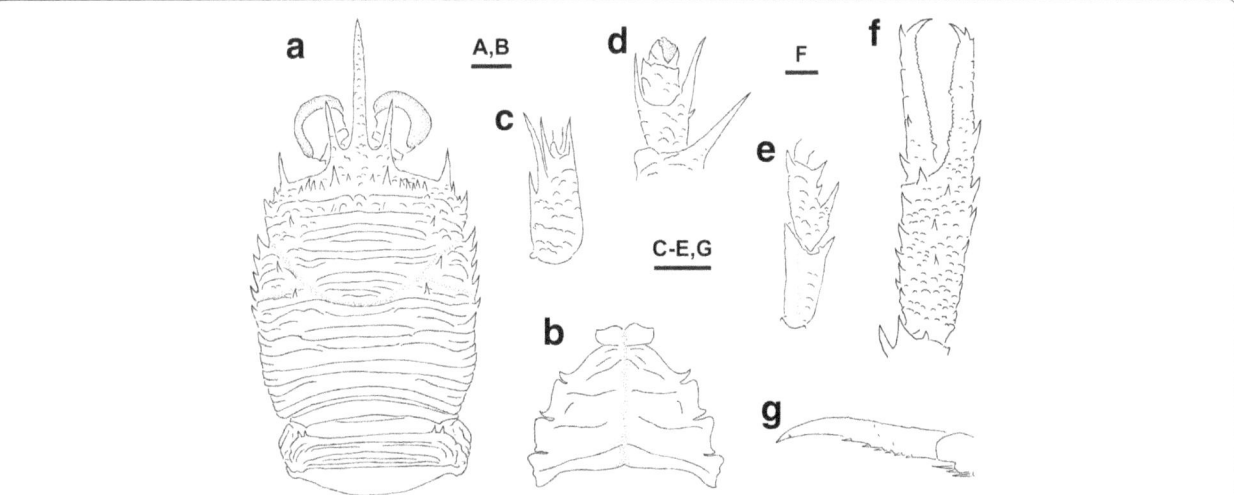

Fig. 8 *Munida japonica* Stimpson, 1858. Male (CL 6.9 mm) from western Jeju Island. **a** Carapace and eyes, dorsal. **b** Sternal plastron, ventral. **c** Right antennular peduncle, ventral. **d** Right antennal peduncle, ventral. **e** Ischium and merus of right third maxilliped, lateral. **f** Chela of right first pereopod, dorsal. **g** Dactylus of left second pereopod, lateral. Scales = 1.0 mm

Remarks

M. japonica is easily distinguished from its two Korean congeners, as it has two spines on each side of anterior ridge of the second pleomere and few striae on the fourth and fifth thoracic sternites. Macpherson and Baba (1993) clarified the taxonomy of *M. japonica* designating a neotype, as previous records of this species comprised incorrect identifications. Although Kim (1973) recorded *M. japonica* from Korea, it is difficult to fully assess the identity of the materials used by this author without reexamining them. Thus, Kim's (1973) record is questionably included in the synonymy. Nevertheless, the present study confirms the occurrence of *M. japonica* in Korea. This species is widely distributed in Korean waters from Busan to western Jeju Island.

Munida pherusa Macpherson and Baba, 1993
(New Korean name: Ju-reum-ba-neul-i-ma-sae-u-but-i)
(Fig. 9)

Munida pherusa Macpherson and Baba 1993: 408, fig. 15 (type locality: Philippines, 13° 56.5′N, 120° 20.7′E, depth 136–152 m); Wu et al., 1998: 122, Figs 28, 35B; Komai et al. 2002: 55; Baba et al., 2009: 178, Figs. 158, 159 (full synonymy).

Material examined

One female (CL 6.9 mm), 33° 30.0′N, 127° 50.6′E, northeastern Jeju Island, Korea, 112 m depth, bottom otter trawl, RV *Tamgu 3*, 5 Apr. 2003; two males (CL 3.0, 8.9 mm), 33° 24.8′N, 127° 43.5′E, eastern Jeju Island, Korea, 117 m depth, bottom otter trawl, RV *Tamgu 1*, 26 Oct. 2006.

Description

Carapace (Fig. 9a) with median gastric spine; branchial margin with five spines. Pleomeres unarmed. Sternal plastron (Fig. 9b) with numerous striae; third sternite with anterior margin producing two distinct lobes, posterior margin slightly narrower than anterior margin of fourth sternite. Eyes (Fig. 9a) large. Basal segment of antennular peduncle (Fig. 9c) bearing two subequal spines. First segment of antennal peduncle (Fig. 9d) bearing distomesial spine almost reaching distal margin of third segment; distomesial spine of second segment reaching distal margin of fourth peduncular segment. Merus of third maxilliped (Fig. 9e) with three spines on flexor margin; extensor margin with distal spine. Fingers of first pereopod (Fig. 9f) about two thirds of length of propodus; cutting edges strongly gapped in males, slightly gapped in female, each with proximal spine. Second pereopod (Fig. 9g) with dactylus bearing corneous spines on proximal two thirds of flexor margin and unarmed in distal portion.

Distribution

Korea, Japan, Taiwan, Philippines, and Indonesia; 73-167 m (Macpherson and Baba 1993, Wu et al. 1998, Komai et al. 2002, Baba et al. 2009).

Remarks

M. pherusa is distinguished from the closely related *M. caesura* by the two distinctly produced lobes in the anterior margin of the third thoracic sternite, and by showing a narrower posterior margin than the anterior margin of the fourth sternite.

Genus *Paramunida* Baba, 1988
(New Korean name: Ga-si-sae-u-but-i-sok)
Paramunida scabra Henderson, 1885
(Korean name: Ga-si-sae-u-but-i)
(Figs. 10 and 11)

Munida scabra Henderson, 1885: 409 (type locality: off the Ki (Kei) Island, 05° 49′15″S, 132° 14′15″E, depth 236 m); Henderson, 1888: 134, pl. 15, Figs. 4, 4a, 4b; Miyake and Baba, 1967: 242, Fig. 13; Kim, 1973: 178;

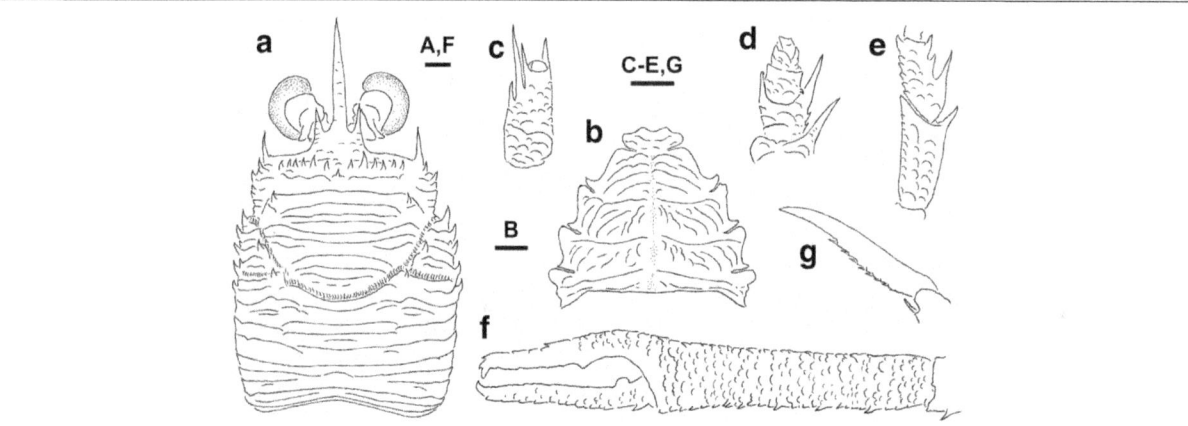

Fig. 9 *Munida pherusa* Macpherson and Baba, 1993. Male (CL 8.9 mm) from eastern Jeju Island. **a** Carapace and eyes, dorsal. **b** Sternal plastron, ventral. **c** Right antennular peduncle, ventral. **d** Right antennal peduncle, ventral. **e** Ischium and merus of right third maxilliped, lateral. **f** Chela of right first pereopod, dorsal. **g** Dactylus of left second pereopod, lateral. Scales = 1.0 mm

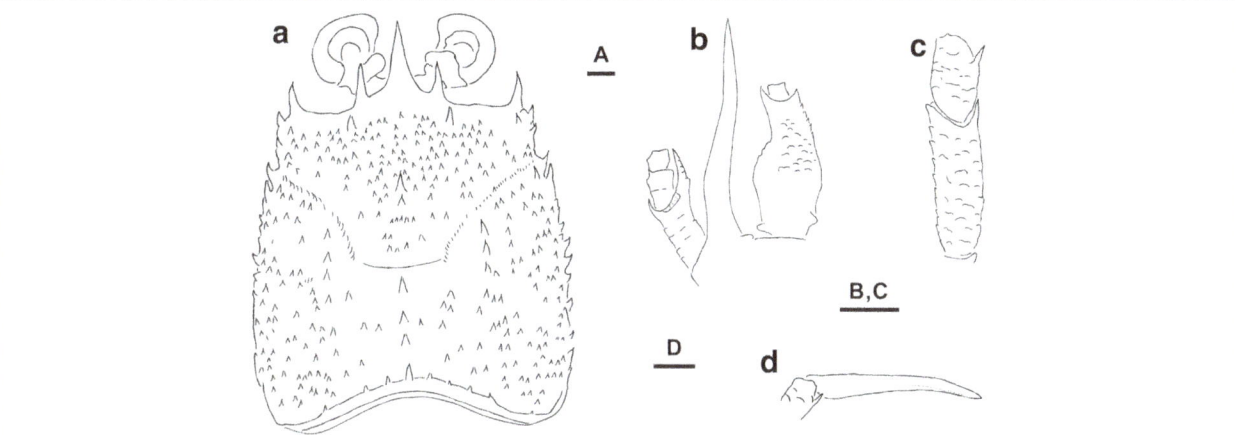

Fig. 10 *Paramunida scabra* Henderson, 1885. Female (CL 11.1 mm) from southern Busan. **a** Carapace and eyes, dorsal. **b** Right antennular and antennal peduncles, ventral. **c** Ischium and merus of right third maxilliped, lateral. **d** Dactylus of right second pereopod, lateral. Scales = 1.0 mm

Miyake, 1982: 149, pl. 50, Fig. 2; Baba, 1986: 175, 292, Fig. 125; Kim and Kim, 1987: 233.

Paramunida scabra: Macpherson 1993: 462, Fig. 8; Baba et al., 2009: 281, Figs. 257, 258 (full synonymy).

Material examined

Two males (CL 11.0, 11.1 mm), one female (CL 11.1 mm), 34° 49.9′N, 129° 04.3′E, southern Busan, Korea, 93 m, bottom otter trawl, RV *Tamgu 3*, 24 Sep. 2008.

Description

Rostrum (Fig. 10a) with thin dorsal carina; supraocular spine reaching rostrum midlength. Carapace (Fig. 10a) median gastric region with two spines; cardiac region with row of four spines on midline. Posterior ridge of fourth pleomere with distinct single median spine. Sternal plastron with striae. Antennal peduncle (Fig. 10b) with prolongation of first segment overreaching distal margin of antennular peduncle; second segment with

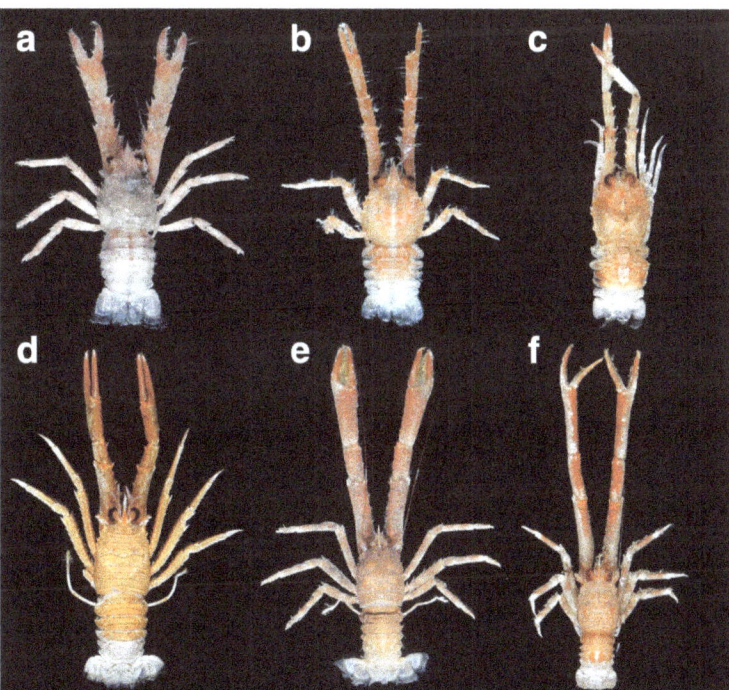

Fig. 11 The six Galatheoidea described in the present study. **a** *Galathea balssi* Miyake and Baba, 1964: ovigerous female (CL 7.5 mm) from southern Jeju Island. **b** *Galathea pubescens* Stimpson, 1858: male (CL 6.0 mm) from eastern Jeju Island. **c** *Bathymunida brevirostris* Yokoya, 1933: ovigerous female (CL 5.1 mm) from eastern Jeju Island. **d** *Cervimunida princeps* Benedict, 1902: female (CL 22.9 mm) from southern Busan. **e** *Munida caesura* Macpherson and Baba, 1993: male (CL 16.7 mm) from southern Busan. **f** *Munida japonica* Stimpson, 1858: male (CL 6.9 mm) from western Jeju Island

Table 1 Geographic ranges of the galatheoid species inhabiting Korean waters recorded in the present study

Species	Japan	Taiwan	Philippines	Indonesia	Australia	Indian Ocean	Depth (m)
[a]Galathea balssi	+		+	+	+		31–222
Galathea orientalis	+	+			+		0–549
Galathea pubescens	+	+	+		+	+	45–494
[a]Galathea rubromaculata	+		+				132–500
[a]Bathymunida brevirostris	+						105–173
[a]Cervimunida princeps	+	+	+				76–452
[a]Munida caesura	+	+	+	+			110–1211
Munida japonica	+	+	+	+			22–732
[a]Munida pherusa	+	+	+	+			73–167
Paramunida scabra	+	+	+	+		+	70–1630
Total	10	7	8	5	3	2	

[a]Newly recorded in Korea

well-developed distomesial spine nearly reaching distal margin of fourth peduncular segment. Merus of third maxilliped (Fig. 10c) bearing one distal spine on flexor margin. Second pereopod (Fig. 10d) with flexor margin of dactylus slightly curved distally, unarmed on flexor margin.

Coloration

Carapace and abdomen generally light pink or reddish brown; epigastric and cardiac regions reddish; first to fourth pereopods whitish, with red bands and a few small red spots; first pereopod fingers whitish with reddish distal and proximal parts.

Distribution

Korea, Japan, Taiwan, Hong Kong, Philippines, Indonesia, East and South China Sea, and Eastern Africa; depth 70-1630 m (Henderson 1885, 1888, Miyake and Baba 1967, Kim 1973, Miyake 1982, Baba 1986, Macpherson 1993, Baba et al. 2009).

Remarks

This species is the sole representative of the genus *Paramunida* known from Korean waters.

Conclusions

Biogeography

The geographic ranges of the 10 Galatheoidea species recorded from Korean waters are summarized in Table 1. Although all species are commonly found in Japanese waters, six species were recorded for the first time in Korean waters. Seven species are also found in Taiwan, eight in the Philippines, five in Indonesia, and three in Australia. Two species, *G. pubescens* and *P. scabra* are widely distributed in the Indo-West Pacific. *B. brevirostris* is endemic to Korea and Tsushima Straits.

Key to the Galatheoidea found in Korean waters

1. Rostrum broad, flattened, and subtriangular; supraocular spines present or absent .... Galatheidae, *Galathea* 2
   – Rostrum slender and spiniform; supraocular spines always present .... Munididae 5
2. Carapace with parahepatic spine .... *G. balssi*
   – Carapace without parahepatic spine 3
3. Epipod absent on first pereopod .... *G. rubromaculata*
   – Epipod present on first pereopod 4
4. Pterygostomian flap with spine on anterior part .... *G. orientalis*
   – Pterygostomian flap without spine on surface .... *G. pubescens*
5. Rostrum with dorsal and ventral spines .... *Cervimunida princeps*
   – Rostrum without dorsal and ventral spines 6
6. Rostrum spiniform, well developed; carapace with distinct transverse ridges or striae .... *Munida* 8
   – Rostrum short; carapace without distinct transverse ridges or striae 7
7. Carapace with prominent gastric and cardiac spines; first segment of antennal peduncle with short process .... *Bathymunida brevirostris*
   – Carapace covered with granules or spinules, lacking prominent gastric and cardiac spines; first segment of antennal peduncle with elongate, anteriorly directed process .... *Paramunida scabra*
8. Second pleomere with two spines on each side of anterior ridge .... *M. japonica*
   – Second pleomere unarmed on anterior ridge 9
9. Third thoracic sternite with anterior margin slightly bilobed, posterior margin broader than anterior margin of fourth sternite .... *M. caesura*

– Third thoracic sternite with anterior margin producing two distinct lobes, posterior margin narrower than anterior margin of fourth sternite .... *M. pherusa*

Acknowledgements
This work was supported a grant by the National Institute of Fisheries Science, Korea (R2016032).

Authors' contributions
JNK and JHC carried out the sampling and identification each species. JNK and MHK participated in drafted the manuscript. YJI conceived of the study and participated in its design and coordination. All authors read and approved the final manuscript.

Competing interests
The authors declare that they have no competing interests.

Author details
[1]West Sea Fisheries Research Institute, National Institute of Fisheries Science, Incheon 22383, South Korea. [2]Korea Inter-University Institute of Ocean Science, Pukyong National University, Busan 48513, South Korea. [3]Southeast Fisheries Research Institute, National Institute of Fisheries Science, Tongyeong 53085, South Korea.

References
Ahyong ST, Baba K, Macpherson E, Poore GCB. A new classification of the Galatheoidea (Crustacea: Decapoda: Anomura). Zootaxa. 2010;2676:57–68.

Baba K. Chirostylids and galatheids from dredgings and trawlings operated in the East China Sea by the Japanese Fisheries Research Vessel Kaiyo Maru in 1967. OHMU. 1969;2:41–57.

Baba K. Redescription of *Bathymunida brevirostris* (Yokoya, 1933) (Crustacea, Decapoda, Galatheidae). Mem Fac Educ, Kumamoto Univ, Sect 1 (Nat Sci). 1970;18:59–62.

Baba K. Reptantia Macrura, Anomura and Brachyura. In: Baba K, Hayashi K, Toriyama M, editors. Japan Fisheries Resource Conservation Association, Decapod Crustaceans from continental shelf and slope around Japan. Tokyo: Japan Fisheries Resource Conservation Association; 1986. p. 148–231. 279–316.

Baba K. Deep-sea chirostylid and galatheid crustaceans (Decapoda: Anomura) from the Indo-West Pacific, with a list of species. Galathea Rep. 2005;20:1–317.

Baba K, de Saint Laurent M. Crustacea Decapoda: Revision of the genus *Bathymunida* Balss, 1914, and description of six new related genera (Galatheidae). In: Résultats des Campagnes MUSORSTOM, vol. 15. Crosnier A. ed. Mém Mus Natl Hist Nat Paris. 1996;168:433–502.

Baba K, Macpherson E, Poore GCB, Ahyong ST, Bermudez A, Cabezas A, Lin CW, Nizinski M, Rodrigues C, Schnabel KE. Catalogue of squat lobsters of the world (Crustacea: Decapoda: Anomura-families Chirostylidae, Galatheidae and Kiwaidae). Zootaxa. 2008;1905:1–220.

Baba K, Macpherson E, Lin CW, Chan TY. Crustacean Fauna of Taiwan. Squat Lobsters (Chirostylidae and Galatheidae). Keelung, PRC: National Taiwan Ocean University; 2009.

Balss H. Ostasiatische Decapoden. I. Die Galatheidae und Paguridae, in Beiträge zur Naturgeschichte Ostasiens, herausgegeben von Pr. F. Doflein. Abhand. K. Bayerischen Akad Wiss Math Physik Klasse. 1913;Suppl 2:1–85.

Benedict JE. Description of a new genus and forty six new species of crustaceans of the family Galatheidae with a list of the known marine species. Proc Biol Soc Wash. 1902;26:243–334.

Haig J. Galatheidae (Crustacea, Decapoda, Anomura) collected by the F.I.S. Endeavour. Rec Aust Mus. 1973;28:269–89.

Henderson JR. Diagnoses of new species of Galatheidae collected during the "Challenger" expedition. Ann Mag Nat Hist. 1885;16:407–21.

Henderson JR. Report on the Anomura collected by H.M.S. Challenger during the years 1873–76. Report on the Scientific Results of the Voyage of H.M.S. Challenger during the years 1873–76. Zoology. 1888;27:1–221.

Kim HS. Anomura and Brachyura. Illustrated Encyclopedia of Fauna and Flora of Korea, Vol, 14. Seoul, KR: Ministry of Education; 1973.

Kim HS, Kim CB. The anomuran crabs (including thalassinideans) of Cheju Island and its adjacent waters, Korea (Crustacea: Decapoda). Korean J Syst Zool. 1987;3:225–36.

Kim HS, Kim W. Order Decapoda. In: List of Animals in Korea (excluding insects), The Korean Society of Systematic Zoology, ed. Seoul, KR: Academy Publishing Co.; 1997. p. 212–23.

Komai T, Ohtsuka S, Nakaguchi K, Go A. Decapod crustaceans collected from the southern part of the Sea of Japan in 2000–2001 using TRV Toyoshio-maru. Nat Hist Res. 2002;7:19–73.

Macpherson E. Crustacea Decapoda: species of the genus *Paramunida* Baba, 1988 (Galatheidae) from the Philippines, Indonesia and New Caledonia. In: Résultats des Campagnes MUSORSTOM, vol. 10. Crosnier A. ed. Mém Mus Nat'l Hist Nat, Paris. 1993;156:443–73.

Macpherson E, Baba K. Crustacea Decapoda: *Munida japonica* Stimpson, 1858, and related species (Galatheidae). In: Résultats des Campagnes MUSORSTOM, vol. 10. Crosnier A. ed. Mém Mus Natl Hist Nat Paris. 1993;156:381–420.

Macpherson E, Robainas-Barcia A. Species of the genus *Galathea* Fabricius, 1793 (Crustacea, Decapoda, Galatheidae) from the Indian and Pacific Oceans, with descriptions of 92 new species. Zootaxa. 2015;3913:1–335.

Makarov VV. [Crustacés Décapodes anomures] In: Fauna SSSR (n. ser.), Shtakelberg AA. ed. Akademii Nauk SSSR, Moscow and Leningrad, RU, 1938;16: 1–324.

Miers EJ. On a collection of Crustacea made by Capt. H.C. Dt. John RN, in the Corean and Japanese Seas. Part 1, Podophthalmia. With an appendix by Capt. St. John HC. Proc Zool Soc London. 1879;1879:18–61.

Miyake S. Japanese Crustacean Decapods and Stomatopods in Color, vol. I. Macrura. Hoikusha, Oksaka, JP: Anomura and Stomatopoda; 1982.

Miyake S, Baba K. Two new species of *Galathea* from Japan and the East China Sea. J Fac Agr Kyushu Univ. 1964;13:205–11.

Miyake S, Baba K. Galatheids of the East China Sea (Chirostylidae and Galatheidae, Decapoda, Crustacea). J Fac Agr Kyushu Univ. 1967;14:225–46.

Ortmann A. Die Decapoden-Krebse des Strassburger Museums, mit besonderer Berücksichtigung der von Herrn Dr. Döderlein bei Japan und bei den Liu-Kiu-Inseln gesammelten und z. Z. im Strassburger Museum aufbewahrten Formen. IV. Die Abtheilungen Galatheidea und Paguridea. Zool Jahr Abth Syst Geogr Biol Theire. 1892;65:241–326.

Poore G, Ahyong ST, Taylor J. In: Poore G, Ahyong ST, Taylor J, editors. The Biology of Squat Lobsters. 2011. CSIRO Publishing, Melbourne, AU and CRC Press, Boca Raton, CA, US.

Stimpson W. Prodromus descrioptionis animalium evertebratorum, quae in expeditione ad oceanum Pacificum septentrionalem, a Republica Federate missa, Cadwaldaro Ringgold et Johanne Roders ducibus, obseravit et descripsit. VII. [Preprint (December 1858)]. Proc Acad Nat Sci Philadelphia. 1858;1858:225–52.

World Register of Marine Species (2015) Galatheoidea. World Register of Marine Species Accessed 11 Oct 2015. http://marinespecies.org/aphia.php?p=taxdetails&id=106685.

Wu MF, Chan TY, Yu HP. On the Chirostylidae and Galatheidae (Crustacea: Decapoda: Galatheidea) of Taiwan. Ann Taiwan Mus. 1998;40:75–153.

Yokoya Y. Macrura of Mutsu Bay. Report on the Biological Survey of Mutsu Bay, 16. Sci Rep Tohoku Imp Univ Sendai. 1930;5:525–48.

11

A new record of the Axiid shrimp *Balssaxius habereri* (Balss, 1913) (Crustacea: Decapoda: Axiidea) in Korean waters

Jung Nyun Kim[1*], Jung Hwa Choi[2], Yang Jae Im[1] and Hyun-su Jo[3]

Abstract

Balssaxius habereri (Balss, 1903) has been newly reported in Korean waters. This species was previously known from the Pacific coast of northern Japan, Korea Strait, Yellow Sea, and Japanese coast of the East Sea. Specimens were collected from western Jeju Island using otter trawls at depths of 65–85 m. Regarding Korean axiid shrimps, a single species, *Boasaxius princeps* (Boas, 1880), previously known as *Axiopsis princeps* in Korea, has been recorded. *B. habereri* is easily distinguished from *B. princeps* as it does not have tufts of setae on its body and males do not have the first pleopod. Morphological descriptions and color photos of the specimens are provided.

Keywords: *Balssaxius habereri*, Axiidae, Korea, Distribution record

Background

The family Axiidae currently includes 193 species of 62 genera globally (Poore 2015). They can be morphologically characterized by the combination of a prominent rostrum ending in an acute tip, lacking the linea thalassinica, similar laminar rami on the second to fifth pleopods, and an oval uropodal endopod with a distal flap. They usually occur worldwide in temperate to tropical regions from intertidal habitats to water depths of more than 2000 m (Dworschak 2015). In Korea, only a single species, *Boasaxius princeps* (Boas, 1880), previously known as *Axiopsis princeps*, has been reported in the family to date (Kim and Kim 1997). During a taxonomic study on decapod crustaceans in Korean waters, three specimens of the axiid shrimp *Balssaxius habereri* (Balss, 1913) were collected from western Jeju Island, at depths of 65–85 m by a bottom otter trawl mounted on R/V *Tamgu 1* and R/V *Tamgu 20* of the National Institute of Fisheries Science, Korea. *B. habereri* was originally described by Balss (1913) as *Axius habereri* from Sagami Bay in the central Pacific coast of Japan. Subsequently, the species has been recorded from the Pacific coast of northern Japan, Korean Strait, Yellow Sea, and Japanese

coast of the East Sea. The present material is the first record of *B. habereri* in Korea. The aim of the present report was to provide morphological descriptions and color photographs of this species.

Methods

The species were collected using sampling gear comprising bottom otter trawls (mesh size at the cod end 0.98 × 0.98 mm). A net with otter boards was towed during the daytime for 30–60 min at a mean 3.4 knots. All the samples collected were frozen onboard shortly after capture and maintained at −80 °C until laboratory identification. Before identification, samples were photographed and preserved in 70–90 % ethyl alcohol.

Carapace length (CL), from the posterior margin of the orbit to the posterior middorsal margin of the carapace, was used to indicate the size of the specimens. Terminology mainly followed Sakai (2011).

The examined specimens have been deposited in the National Institute of Fisheries Science, Korea (NIFS).

Results and discussion
Family Axiidae Huxley, 1879
 Genus *Balssaxius* Sakai, 2011
 (New Korean name: Nam-bang-ga-jae-a-jae-bi-sok)
 ***Balssaxius habereri* (Balss, 1913)**
 (Figure 1)

* Correspondence: crangonk@korea.kr
[1]West Sea Fisheries Research Institute, National Institute of Fisheries Science, Incheon 22383, South Korea
Full list of author information is available at the end of the article

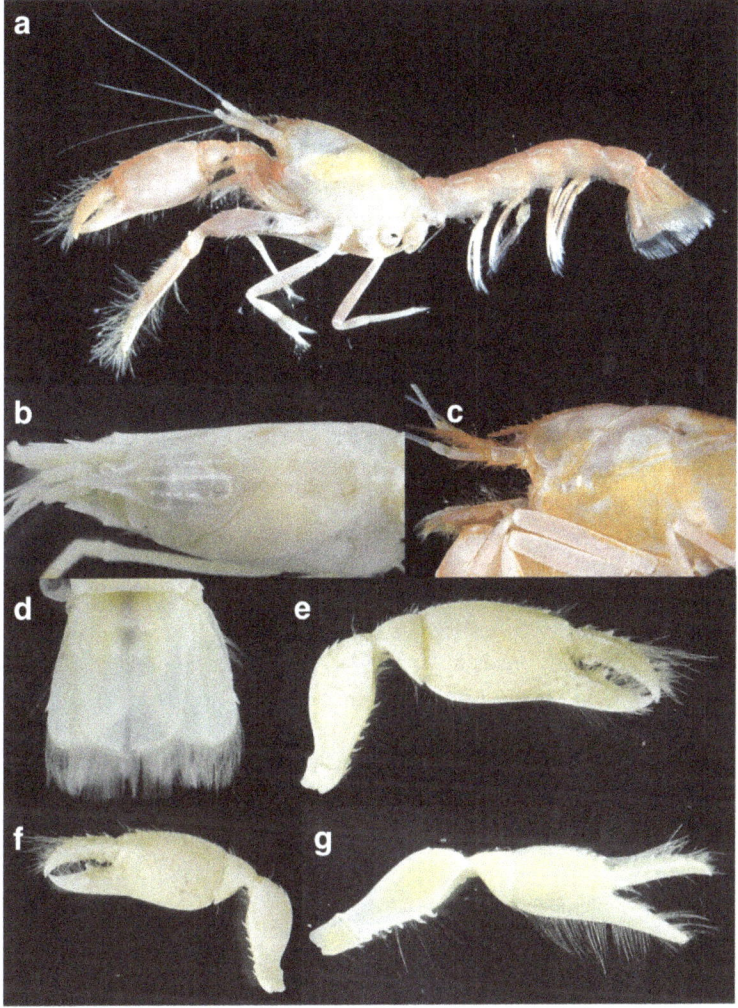

Fig. 1 *Balssaxius habereri* (Balss, 1913). **a**, **b**, **d**, **g**, male (CL 18.0 mm) from western Jeju Island; **c**, **e**, **f**, male (CL 14.1 mm) from west-northwestern Jeju Island. **a** entire animal, lateral; **b** carapace, cephalic, and thoracic appendages, dorsal; **c** same, lateral; **d** tail fan, dorsal; **e** major cheliped, lateral; **f** same, mesial; **g** minor cheliped, lateral

(New Korean name: Nam-bang-ga-jae-a-jae-bi)

Axius habereri Balss, 1913, p. 238 (type locality: Sagami Bay, Japan); 1914, p. 85, text-figs 46, 47; Yokoya 1933, p. 49.

Axiopsis (*Axiopsis*) *habereri*: De Man, 1925, p. 70; Miyake 1982, p. 89, pl. 30, fig. 3.

Calocarides habereri: Sakai and de Saint Laurent, 1989, p. 83.

Balssaxius habereri: Sakai, 2011, p. 74, fig. 13A, B.

Material examined

Jejudo: 1 ♂ (CL 18.0 mm) (western Jeju Island, 32° 59.9′ N 124° 26.0′ E, 65 m, otter trawl, *Tamgu 1*: March 14, 2002); 1 ♂ (CL 14.1 mm) (west-northwestern Jeju Island, 33° 46.1′ N 124° 49.4′ E, 85 m, otter trawl, *Tamgu 20*:

April 8, 2009); 1 ♀ (CL 19.2 mm) (west-southwestern Jeju Island, 32° 5.4′ N 125° 15.5′ E, 76 m, otter trawl, *Tamgu 20*: March 25, 2015).

Description

Rostrum (Fig. 1a–c) acutely triangular, reaching slightly beyond first segment of antennular peduncle or reaching midlength of second segment, concave dorsally, lateral margin with four small teeth. Carapace (Fig. 1a–c) without supraocular spine; gastric median carina with three teeth anteriorly; submedian carina distinct, with row of 8–10 teeth; lateral carina smooth, continuous lateral margin of rostrum; gastric region with sparse short setae; cervical groove distinct and deep; postcervical carina and pair of posterior protrusions present. Pleonal pleura all rounded marginally. Telson (Fig. 1d) longer than wide; bearing proximal lobe with spine posteriorly, followed by two denticles on lateral margin; dorsal surface with median groove, one pair of proximal submedian spines, two pairs of lateral spines; posterior margin rounded posteriorly, with small median tooth in males, without median tooth in female. Eyestalk (Fig. 1b, c) falling slightly short of midlength of rostrum. Antenna with elongate scaphocerite, reaching slightly beyond midlength of penultimate segment of antennular peduncle. Cheliped unequal. Major cheliped (Fig. 1e, f) with ischium bearing one spine on ventrodistal margin; merus with two subdistal spines on dorsal margin and four ventral spines; chela compressed laterally, with one or two subdistal spines on dorsal margin of palm; dactylus shorter than palm, laterally denticulate, with row of several spines on dorsal margin. Minor cheliped (Fig. 1g) with one ventrodistal spine on ischium; merus with one dorsodistal spine, ventral margin with five spines; palm slightly shorter than dactylus, with four anterior dorsal spines. Uropodal endopod (Fig. 1d) with longitudinal median carina bearing eight teeth, lateral margin with one or two subdistal spines; exopod with transverse suture bearing row of 10 small teeth; lateral margin with five to seven teeth including posterolateral tooth and movable spinule, mesial to base of posterolateral tooth.

Coloration
Entirely pale orange (Fig. 1a).

Distribution
Korea, Japan; 65–160 m.

Remarks
A. habereri originally described by Balss (1913) was included in the genus *Axiopsis* by De Man (1925) and later in *Calocarides* by Sakai and de Saint Laurent (1989). However, it does not belong to *Calocarides* due to the presence of sharp postcervical carina on the carapace. In contrast, the carapace of the type species of *Calocarides*, *Calocarides coronatus* (Trybom, 1904), bears no postcervical carina. Sakai (2011) established the new genus *Balssaxius* for the present species.

The present material generally agrees with the original description by Balss (1913) and the subsequent description by Sakai (2011), who examined the female lectotype deposited in Zoologische Staatssammlung in Munich. However, a few minor discrepancies, which could fall within the range of intraspecific variation, are apparent: the minor cheliped with merus has one dorsodistal and five ventral spines in the present specimens, whereas the type specimens have two dorsodistal and six ventral spines.

Conclusions
B. habereri is easily distinguished from *B. princeps* since it lacks tufts of setae and the first pleopod is lacking in males. The present species rarely occurred near Jeju Island in sandy mud bottom habitats, whereas *B. princeps* is distributed from Busan to Dokdo Island in rocky bottom habitats. Although this species is endemic in East Asian waters, i.e. Korea Strait, Yellow Sea, Japanese coast of the East Sea, and Pacific coast of Japan, it is never reported in Korean waters.

Acknowledgements
This research was funded by the National Institute of Fisheries Science, Korea (R2016032).

Authors' contributions
JNK and JHC carried out the sampling and identification the species. JNK participated in drafted the manuscript. YJI and HJ conceived of the study and participated in its design and coordination. All authors read and approved the final manuscript.

Competing interests
The authors declare that they have no competing interests.

Author details
[1]West Sea Fisheries Research Institute, National Institute of Fisheries Science, Incheon 22383, South Korea. [2]Southeast Sea Fisheries Research Institute, National Institute of Fisheries Science, Tongyeong 53085, South Korea. [3]Department of Marine Science and Production, Kunsan National University, Kunsan 54150, South Korea.

References
Balss H. Ostasiatische Decapoden. I. Die Galatheidae und Paguridae, in Beiträge zur Naturgeschichte Ostasiens, herausgegeben von Pr. F. Doflein. Abhand K Bayerischen Akad Wiss Math Physik Klasse. 1913;Suppl 2:1–85.

De Man JG. The Decapoda of the Siboga Expedition. Part VI. The Axiidae collected by the Siboga-Expedition. Siboga Exped. 1925;39a5:1–127.

Dworschak PC. Methods collecting Axiidea and Gebiidea (Decapoda): a review. Ann Naturhist Mus Wien B. 2015;117:5–21.

Kim HS, Kim W. Order Decapoda. In: The Korean Society of Systematic Zoology, editor. List of animals in Korea (excluding insects). Seoul: Academy Publ. Co; 1997. p. 212–23.

Miyake S. Japanese crustacean decapods and stomatopods in color. Vol. I. Macrura, Anomura and Stomatopoda. 1st ed. Osaka: Hoikusha, JP; 1982.

Poore G. 2015. Axiidea de Saint Laurent, 1979. World Register of Marine Species at http://www.marinespecies.org/aphia.php?p=taxdetails&id=477324 on 2015-09-30.

Sakai K. Axioidea of the world and a reconsideration of the Callianassoidea (Decapoda, Thalassinidea, Callianassida). Leiden: Brill; 2011.

Sakai K and de Saint Laurent M. A check list of Axiidae (Decapoda, Crustacea, Thalassinidea, Anomura), with remarks and in addition descriptions of one new subfamily, eleven new genera and two new species. Tokushima: Naturalists Publ Tokushima Biol Lab Shikoku Women's Univ; 1989. 3:1–104.

Yokoya Y. On the distribution of decapod crustaceans inhabiting the continental shelf around Japan, chiefly based upon the materials collected by S.S. Sôyô-Maru, during the years 1923-1930. J Coll Agr Toyko Imp Univ. 1933;12:1–222.

12

Miuraea migitae, a new record of the order Bangiales (Bangiophyceae, Rhodophyta) from Korea

Young Ho Koh, Hyung Woo Lee and Myung Sook Kim[*]

Abstract

We found specimens of foliose Bangiales from the subtidal zone of Udo, Jeju Island, Korea. In molecular analyses of *rbc*L sequences, these Korean specimens were almost identical to *Miuraea migitae* from Osaka, Japan. In the morphological comparison, Korean specimens were consistent with habitat, color, and vegetative characteristics with the description of *M. migitae*. This is the first record of *M. migitae* outside the type locality and Nagasaki in Japan. This study confirms that new or unrecorded species of the order Bangiales may be discovered from subtidal habitats.

Keywords: Bangiales, Korea, *Miuraea migitae*, New record, *rbc*L

Background

The order Bangiales (Nägeli 1874) is an order of distinctive and morphologically simple red algae that represents an ancient lineage (Butterfield 2000). The foliose Bangiales includes the most highly valued seaweeds grown in aquaculture from Korea, Japan, China, and Southeast Asia for the past several hundred years (Mumford and Miura 1988). This order has been considered monophyletic with the single family Bangiaceae. Traditionally, *Bangia* Lyngbye and *Porphyra* C. Agardh have been recognized on the basis of gametophyte morphology: unbranched uni- to multiseriate filaments in the genus *Bangia* and blade in the genus *Porphyra* (Oliveira et al. 1995). Recently, the revision of the order Bangiales, based on the phylogeny using nrSSU and *rbc*L genes, has greatly improved the taxonomic understanding of this group (Sutherland et al. 2011). According to Sutherland et al. (2011), the order Bangiales was split into eight of foliose (*Boreophyllum*, *Clymene*, *Lysithea*, *Miuraea*, *Porphyra*, *Pyropia*, and *Wildemania*) and seven of filamentous genera.

The genus *Miuraea* was established by Sutherland et al. (2011) based on molecular analyses and composed of only one species, *M. migitae* (N.Kikuchi, S.Arai, G.Yoshida & J.A.Shin) N.Kikuchi, S.Arai, G.Yoshida, J.A.Shin & M.Miyata. This species was originally described as *Porphyra migitae*, based on the specimen collected from Misaki town in Osaka Prefecture, Japan (Kikuchi et al. 2010). *M. migitae* was distinguished from other foliose Bangiales species by growing on the subtidal habitat and having fire red to pink color (Kikuchi et al. 2010). The distribution of *M. migitae* was reported only from Japan (Osaka and Nagasaki), attached to the dead bivalve shell and rope (Sutherland et al. 2011).

In Korea, the study of *Porphyra* has been carried out on morphology, physiology, flora, and culture by several authors (Hwang and Lee 2001; Kim and Kim 2011). To date, seven species of *Porphyra* and 13 species of *Pyropia* J.Agardh (previously known *Porphyra* from Korea) have been reported from Korea, mostly collected in the intertidal zone (Kim et al. 2013). Two species, *Pyropia koreana* (M.S. Hwang & I.K. Lee) M.S. Hwang, H.G. Choi, Y.S. Oh & I.K. Lee and *Pyropia yezoensis* (Ueda) M.S. Hwang & H.G. Choi, are known from subtidal habitats (Hwang and Lee 2001; Kim and Kim 2011). However, the specimens of the order Bangiales from subtidal habitats have been scarcely explored because of the difficulty of sampling in this environment, despite the possibility of discovering new or unrecorded species.

In this study, we collected foliose Bangiales specimens from a depth of 15 m in Udo, Jeju Island, Korea. We conducted morphological observations and *rbc*L sequence analysis in order to confirm the taxonomic position of these subtidal specimens. This is the first record

* Correspondence: myungskim@jejunu.ac.kr
Department of Biology, Jeju National University, Jeju 63243, South Korea

of the occurrence of the order Bangiales species from under 15 m depth of subtidal habitats in Korean coasts.

Methods

Four samples were collected by SCUBA diving from a depth of 15 m of Udo (33° 30′ N, 126° 55′ E), Jeju Island, Korea, in 9 May 2015 (E15019) and 17 May 2016 (E16064, E16065, E16066). Thalli were mounted on herbarium sheets as voucher specimens. Some parts of thallus were cutoff to stain with 1% aqueous aniline blue acidified with a drop of 1% HCL and to make permanent slide in a solution of 50% Karo syrup for microscopic observations. Photomicrographs were taken with a Canon EOS 600D digital camera mounted on the Olympus BX43

microscope. Voucher specimens were deposited in the herbarium of Jeju National University, Korea (JNUB).

Genomic DNA was extracted from the specimens using the DNeasy Plant Mini Kit (Qiagen, Hilden, Germany) following the manufacturer's instruction. The primer pairs for *rbc*L gene were *rbc*LJNF1-*rbc*LJNR1 and *rbc*LJNF2-*rbc*LJNR2 (Kang and Kim 2013). Amplification condition for *rbc*L consisted of 7 min at 97 °C for pre-denaturation, followed by 45 cycles of 1 min at 97 °C, 1 min 47 °C, and 2 min at 72 °C, with a final 7 min extension cycle at 72 °C, and a soak cycle at 4 °C. PCR products were purified using the AccuPrep® PCR Purification Kit (Bioneer, Daejeon, Korea) following the manufacturer's instructions. Nucleotide sequences of *rbc*L were determined on strands of PCR

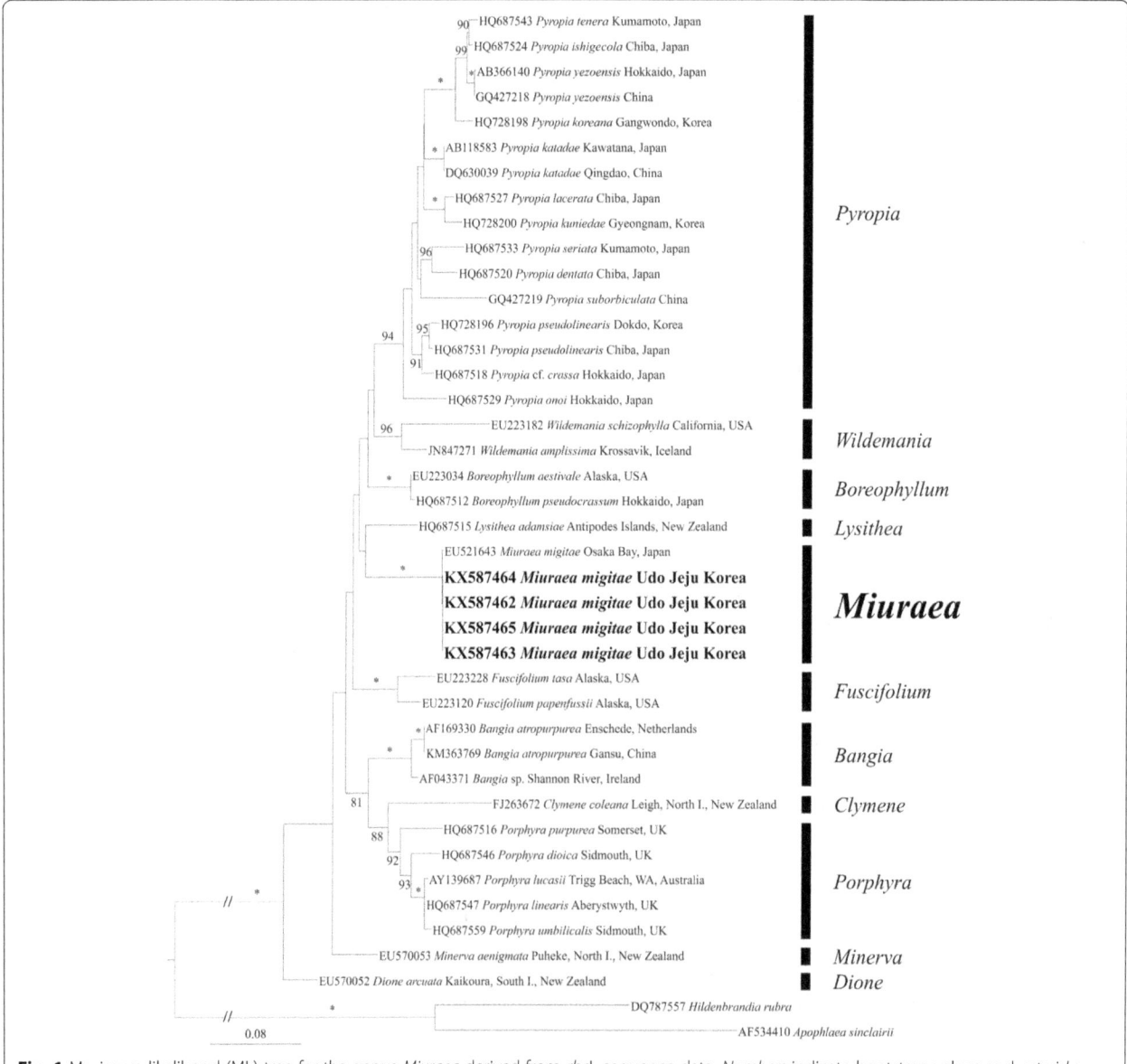

Fig. 1 Maximum likelihood (ML) tree for the genus *Miuraea* derived from *rbc*L sequence data. *Numbers* indicate bootstrap values and *asterisks* denote full support. *Scale bar* indicates number of substitutions per site

amplification products at the Macrogen sequencing facility (Macrogen Inc., Seoul, Korea). The sequences were edited using Mega ver. 7.0.14 (Tamura et al. 2013), and multiple sequence alignments were constructed with 39 taxa of the order Bangiales from GenBank with two taxa of Hildenbrandiales as out-group.

Maximum likelihood (ML) analysis was conducted using RAxML software (Stamatakis 2006) with the GTR + I + Γ evolutionary model to confirm the taxonomic position of Korean specimens. We used 1000 independent tree inferences using the "-#" option to generate bootstrap values for phylogenetic analysis with the algorithm "a" of "-f" option to search the best-scoring tree in one program run. The other options were used as a default set of the software.

Results

Molecular analyses

In total, 1174 base pairs of *rbc*L were aligned with 444 variable positions (37.8%) and 353 phylogenetically informative positions (30.0%). The Korean specimens (GenBank accession number KX587462-KX587465) were almost identical with *M. migitae* from Japan (EU521643), showed 0.2% sequence divergence (Fig. 1). In addition, *M. migitae* was clearly separated from other Korean specimens, *Pyropia koreana* (HQ728198), *Py. kuniedae* (HQ728200), *Py. pseudolinearis* (HQ718196), and *Py. yezoensis* (AB366140 and GQ427218), by 10.2–10.5%, 10.1–10.3%, 9.4–9.7% and 11–11.1% sequence divergences, respectively.

The clade of the genus *Miuraea* formed a monophyletic group with strong support, whereas there was a sister genus *Lysithea* without support. The monophyly of each genus was supported by strong bootstrap values (100% for *Bangia*, *Boreophyllum*, *Fuscifolium*, and *Miuraea*; 96% for *Wildemania*; 94% for *Pyropia*; and 92% for *Porphyra*). *Pyropia* formed a clade with *Wildemania* and *Boreophyllum*, but not supported. *Porphyra* was related to *Clymene* (a bladed species) and *Bangia* with filamentous morphology by moderate support values.

Miuraea migitae (N.Kikuchi, S.Arai, G.Yoshida & J.A.Shin) N.Kikuchi, S.Arai, G.Yoshida, J.A.Shin & M.Miyata

Basionym: *Porphyra migitae* N.Kikuchi, S.Arai, G.Yoshida & J.A. Shin
Type locality: Misaki town, Osaka Prefecture, Japan.
Holotype: SAP105477 (Kikuchi et al. 2010).
Korean name: Chambunhonggim (참분홍김).

Thalli are membranaceous, monostromatic, elliptic, obovate, and circular shape with cordate and rotund base, fire red to pink in color, and undulated entire margin (Fig. 2a, b). Thalli are up to 13 cm long, 6 cm wide, and 25–40 μm thick in the central portion (Fig. 2a, e).

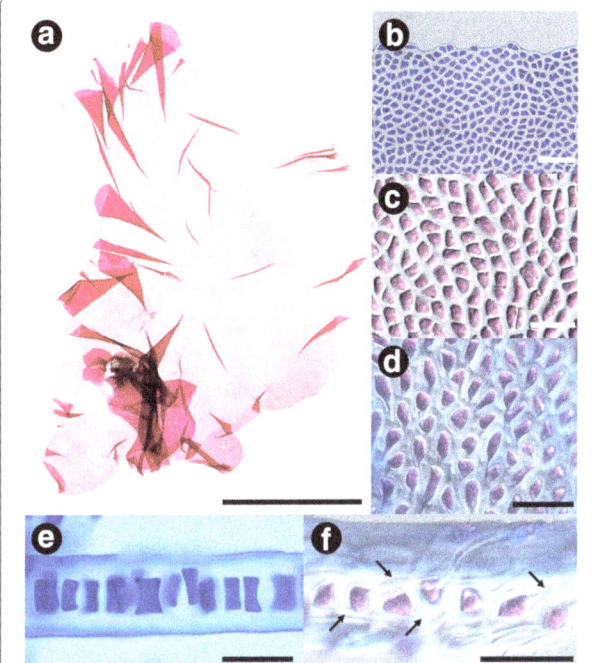

Fig. 2 *Miuraea migitae* (N.Kikuchi, S.Arai, G.Yoshida & J.A.Shin) N.Kikuchi, S.Arai, G.Yoshida, J.A.Shin & M.Miyata. **a** Vegetative thallus of *M. migitae* (E15019) collected from Udo, Jeju Island, Korea, 9 May 2015. **b** Surface view of thallus with entire margin. **c** Surface view of the central portion of thallus. **d** Surface view of the basal portion with rhizoidal cells. **e** Sectional view of vegetative cells in the central portion. **f** Sectional view of rhizoidal filaments toward both sides of thallus (*arrows*). *Scale bars* represent **a** 4 cm; **b**, **d**, **f**, 50 μm; **c**, **e**, 25 μm

Vegetative cells are oblong, triangular, polygonal with rounded angles in the surface view, and 10–17 μm long × 7–10 μm broad in size (Fig. 2b, c). In the sectional view, cells are oblong and 15–20 μm high × 7–15 μm broad in size (Fig. 2e). Basal cells are 15–35 μm long × 6–20 μm broad in the surface view (Fig. 2d) and having rhizoidal filaments in both sides of thallus in the sectional view (Fig. 2f).

To date, *M. migitae* has been only collected from the subtidal zone in Udo, Jeju Island, Korea. The habitat is an area in a strong current and is covered with sand and a rhodolith bed.

Discussion

Miuraea migitae has been known as an endemic species in the subtidal zone of Japan. It is characterized by monostromatic thallus elliptic and obovate in shape, with fire red to pink color and 25 cm length × 13 cm width in size (Kikuchi et al. 2010). According to the culture study by Kikuchi et al. (2010), *M. migitae* can be distinguished from other subtidal species of Bangiales in Japan by the presence of asexual reproductive subcycles involving archeospores and neutral spores on the foliose thallus. Although our specimens were not observed any archeospores and neutral spores on thallus, Korean specimens were consistent with habitat, color, and vegetative characteristics in the description of *M. migitae*.

In recognizing the species of *Pyropia*, the vegetative features are viewed as important diagnostic characters, such as color, length, thickness, and projection of rhizoidal filaments including habitat (Table 1). *M. migitae* was found in the subtidal habitat like *Py. koreana* and *Py. yezoensis* with monostromatic thallus, but they can be distinguished from *M. migitae* by color, basal cell size, and the projection of rhizoidal filaments (both sides in *M. migitae* and one side in *Py. koreana*) (Table 1). *Porphyra oligospermatangia*, *Pyropia katadae*, and *Py. tenera* are having elliptic to obovate thallus with cordate base (Kim and Kim 2011), but they are different from *M. migitae* in length (24–47 cm in *M. migitae* and 30–56 cm in *P. oligospermatangia*), the thickness of thallus (25–40 μm in *M. migitae*, 40–50 μm in *Py. katadae* and 35–60 μm in *Py. yezoensis*), the habitats (subtidal zone in

M. migiate and intertidal zone in *P. oligospermatangia*, *Py. katadae*, and *Py. tenera*), and the color (fire red to pink in *M. migitae*, yellowish to greenish in *P. oligospermatangia*, *Py. tenera*, and *Py. yezoensis*, and purple in *Py. katadae*) (Hwang and Lee 1994; Kim and Kim 2011).

The molecular analyses demonstrated that the *rbc*L sequences from Korean specimens and *M. migitae* from Japan (EU521643) were almost identical with 0.2% genetic distances. Lindstrom and Fredericq (2003) mentioned that the sequence divergences of *rbc*L within species of *Porphyra* from different populations differed by 0.1% in *Porphyra fallax* from British Columbia, Canada, to 0.9% in *Porphyra pseudolanceolata* and *P. pseudolinearis* from Alaska, USA. The sequence divergence between Korean and Japanese specimens fell into the intraspecific variation, showing the specimens to be conspecific. In the phylogenetic tree, *Miuraea* formed a clade with four genera, *Pyropia*, *Wildemania*, *Boreophyllum*, and *Lysithea*, without supporting value. These

Table 1 Morphological features of *Miuraea migitae* and five foliose Bangiales species having similar morphology or habitat from Korea

	Miuraea migitae	Porphyra oligospermatangia	Pyropia kuniedae	Py. koreana	Py. tenera	Py. yezoensis
Habitat	6–15-m depth of subtidal zone	Intertidal zone	Low intertidal zone	Low intertidal to 5-m depth of subtidal zone	Upper to middle intertidal zone	Middle intertidal to 5-m depth of subtidal zone
Color	Fire red, pink	Yellowish brown, pale-green toward the base	Purple, greenish red, brownish red	Bright red, brownish red	Yellowish red, brownish red, greenish red	Brownish red, greenish red
Shape	Narrow elliptic, elliptic, obovate, circular	Elliptical, linear, lanceolate	Round, ovate	Elliptical, obovate	Elliptical, round, ovate, obovate	Round, ovate, obovate, oblanceolate, reniform
Basal shape	Cordate, rotund, navel	Round, cordate	Cordate	Cuneate, round	Cuneate, round, cordate	Cuneate, round, cordate, umbilicate
Dentation	Absent	Absent	Absent	Absent	Absent	Absent
Undulation	Present	Present	Present	Present	Present	Present
Length (cm)	13–25	20–40 (−55)	7–9.5	6–10	4.5–20 (−40)	3.5–13
Width (cm)	6–13	5–15	6.5–9	4–8	1.5–9	1–9
Thickness (μm)	24–47	30–56	40–50	20–27	20–40	35–60
Vegetative cell shape	Oblong, triangular, polygonal	Circle, rectangle	Circle, rectangle, triangle	Rectangle	Rectangle	Rectangle
Vegetative cell size (μm)	$8–17 \times 6–10$	$7–11 \times 6–9$	$12–15 \times 7–10$	$13–17 \times 9–10$	$15–21 \times 9–12$	$13–19 \times 7–14$
Basal cell shape	Capitate	Elliptic	Elliptic, capitate	Elliptical	Circle, elliptic	Circle, elliptical
Basal cell size (μm)	$15–35 \times 6–20$	$18–42 \times 13–21$	$22–27 \times 13–16$	$22–25 \times 9–12$	$22–33 \times 13–21$	$22–33 \times 10–23$
Projection of rhizoidal filaments	Both sides	Both sides	Both sides	One side	Both sides	Both sides
Division formula of spermatangia	$4 \times 4 \times 8$	$4 \times 2 \times 8$	$4 \times 2–4 \times 4–8$	$4 \times 2–4 \times 4$	$4 \times 2–4 \times 4–8$	$4 \times 2–4 \times 8–16$
Division formula of zygotosporangia	$2 \times 2 \times 4$	$2 \times 2 \times 2–4$	$2 \times 2 \times 4$	$2 \times 1 \times 2$	$2–4 \times 1–2 \times 2–4$	$2 \times 2 \times 4$
Reference	This study, Kikuchi et al. 2010	Kim and Kim 2011	Hwang and Lee 2001, Kim and Kim 2011	Hwang and Lee 2001, Kim and Kim 2011	Hwang and Lee 2001, Kim and Kim 2011	Hwang and Lee 2001, Kim and Kim 2011

phylogenetic relationships were different from two previous studies: one is that the genus *Miuraea* formed a basal node of the order Bangiales except for two filamentous genera *Dione* and *Minerva* (Sutherland et al. 2011), in the other is that the genus *Miuraea* formed a sister clade composed of *Boreophyllum*, *Fuscifolium*, *Pyropia*, *Wildemenia*, and an undescribed filamentous genus (Sánchez et al. 2014). In addition, the taxonomic confusion still remained on the phylogenetic relationships of three genera, *Porphyra*, *Bangia*, and *Clymene*, even though they had moderate support value in the *rbc*L tree (Sutherland et al. 2011; Sánchez et al. 2014).

The new record of *M. migitae* from Korea extends the distribution range of this species previously restricted to only two localities, Osaka and Nagasaki in Japan (Guiry and Guiry 2016). Udo of Jeju Island is distant from the type locality of *M. migitae*, Osaka, Japan. Hwang (2008) mentioned that Udo had been connected with Osaka by shipping previously (before 1945). Although the origin of this species is unclear, this report indicated that this species might be occurred in Korea prior to that time. Lack of the records on the distribution of *M. migitae* from Udo before our finding might be due to the difficulty of collecting in its subtidal habitat where there are swift currents. Since the foliose thallus of *M. migitae* is growing up to 28 cm as the maximum length and producing asexual reproductive cells during summer season (Kikuchi et al. 2010), it can be a potential genetic resource for plant breeding to increase the production in seaweed aquaculture.

Conclusions

In conclusion, we collected the foliose Bangiales specimens from the subtidal zone of Udo, Jeju Island. Based on molecular and morphological analyses, these specimens were identified as *M. migitae*, previously considered endemic in Japan. This is a new record for Korea and indicates that collections from the subtidal zone are necessary to increase understanding of the biodiversity of seaweeds including potential genetic resources in Korea. In further studies, we are performing morphological and molecular investigations of the order Bangiales to understand the accurate phylogenetic relationships.

Acknowledgements
We thank all members of the Molecular Phylogeny of Marine Algae laboratory at Jeju National University for collecting the samples. This work was supported by a grant from the National Institute of Biological Resources (NIBR), funded by the Ministry of Environment (MOE) of Korea (NIBR201624202 for the Graduate Program of the Undiscovered Taxa from Korea).

Authors' contributions
YHK carried out the research and drafted the manuscript. HWL collected the samples from the subtidal zone by SCUBA diving. MSK designed this study

and revised the manuscript. All authors read and approved the final manuscript.

Competing interests
The authors declare that they have no competing interests.

References

Butterfield NI. *Bangiomorpha pubescens* n. gen., n. sp.: implications for the evolution of sex, multicellularity, and the Mesoproterozoic/Neoproterozoic radiation of eukaryotes. Paleobiology. 2000;26:386–404. http://dx.doi.org/10.1666/0094-8373(2000)026<0386:BPNGNS>2.0.CO;2.

Guiry MD and Guiry GM. *AlgaeBase*. World-wide electronic publication, National University of Ireland, Galway. 2016. http://www.algaebase.org. Accessed 19 July 2016.

Hwang KS. The searching and tourism resource plan of Jeju-Japan sea route before independence. J Korean Reg Dev Assoc. 2008;20:113–32.

Hwang MS, Lee IK. Two species of *Porphyra* (Bangiales, Rhodophyta), *P. koreana* sp. nov. and *P. lacerata* Miura from Korea. Algae. 1994;9:169–77.

Hwang MS, Lee IK. Taxonomy of the genus *Porphyra* (Bangiales, Rhodophyta) from Korea. Algae. 2001;16:233–73.

Kang JC and Kim MS. A novel species *Symphyocladia glabra* sp. nov. (Rhodomelaceae, Rhodophyta) from Korea based on morphological and molecular analyses. Algae. 2013;28:149–160. http://dx.doi.org/10.4490/alga.2013.28.2.149.

Kikuchi N, Arai S, Yoshida G, Shin JA, Broom JE, Nelson WA, et al. *Porphyra migitae* sp. nov. (Bangiales, Rhodophyta) from Japan. Phys Chem Chem Phys. 2010;49:345–54. http://dx.doi.org/10.2216/09-82.1.

Kim H-S and Kim S-M. *Algal Flora of Korea. Vol. 4, No. 1, Rhodophyta: Stylonematophyceae, Compopogonophyceae, Bangiophyceae. Primitive red algae*. National Institute of Biological Resources, Ministry of Environment, Incheon, Korea, 142 pp. 2011.

Kim H-S, Boo SM, Lee IK and Son CH. *National list of species of Korea. Marine Algae*. National Institute of Biological Resources, Ministry of Environment, Incheon, Korea, 336 pp. 2013.

Lindstrom SC, Fredericq S. *rbc*L gene sequences reveal relationships among north-east Pacific species of *Porphyra* (Bangiales, Rhodophyta) and a new species, *P. aestivalis*. Phycol. Res. 2003;51:211–24. http://dx.doi.org/10.1046/j.1440-1835.2003.00312.x.

Mumford TF, Miura A. *Porphyra* as food: cultivation and economics. In: Lembi CA, Waaland JR, editors. Algae and human affairs. London: Cambridge University Press; 1988. p. 87–117.

Nägeli C. *Die neuern Algensysteme und Versuch zur Begründung eines eigenen Systems der Algae und Florideen*. Schulthess, Zurich, Switzerland, 275 pp., pls I-X. 1874.

Oliveira MC, Kurniawan J, Bird CJ, Rice EL, Murphy CA, Singh RK, et al. A preliminary investigation of the order Bangiales (Bangiophyceae, Rhodophyta) based on sequences of nuclear small-subunit ribosomal RNA genes. Phycol Res. 1995;43:71–9. http://dx.doi.org/10.1111/j.1440-1835.1995.tb00007.x.

Sánchez N, Vergés A, Peteiro C, Sutherland JE and Brodie J. Diversity of bladed Bangiales (Rhodophyta) in western Mediterranean: recognition of the genus *Themis* and descriptions of *T. ballesterosii* sp. nov., *T. iberica* and *Porphyra parva* sp. nov. J. Phycol. 2014;50: 908–929. http://dx.doi.org/10.1111/jpy.12223.

Stamatakis A. RAxML-VI-HPC: maximum likelihood-based phylogenetic analyses with thousands of taxa and mixed models. Bioinformatics. 2006;22:2688–90. http://dx.doi.org/10.1093/bioinformatics/btl446.

Sutherland JE, Lindstrom S, Nelson WA, Brodie J, Lynch MDJ, Hwang MS, et al. A new look at an ancient order: generic revision of the Bangiales (Rhodophyta). J Phycol. 2011;47:1131–51. http://dx.doi.org/10.1111/j.1529-8817.2011.01052.x.

Tamura K, Stecher G, Peterson D, Filipski A, Kumar S. MEGA6: Molecular Evolutionary Genetics Analysis version 6.0. Molec Biol Evol. 2013;30:2725–9. http://dx.doi.org/10.1093/molbev/mst197.

Appropriate feeding for early juvenile stages of eunicid polychaete *Marphysa sanguinea*

Kyeong Hun Kim[1], Byoung Kwon Kim[2], Sung Kyun Kim[2], War War Phoo[2], B. A. Venmathi Maran[3] and Chang-Hoon Kim[1,2,3*]

Abstract

Survival rate (SR) and growth rate (GR) were tested with various feed sources to identify an appropriate feed to improve the productivity in the early life stage of the rock worm *Marphysa sanguinea* (Montagu, 1813) (Eunicidae: Polychaeta). In addition, feed supply rates were also examined. Three experiments were performed to identify the appropriate feed for the juvenile stages of *M. sanguinea*. Experiment 1 was done using seven different feed sources and without feed as well for the first 20 days of *M. sanguinea* culture. Decapsulated *Artemia* and extruded pellet for shrimp were showed with high SR and GR in the experiment 1. Experiment 2 was performed with five different feed sources. Two feeds were selected from experiment 1 in addition to eel feed, mixed micro-algae, and benthic diatom. Four different quantities of each feed were supplied to 3000 individuals of early juvenile stage of *M. sanguinea*. High quantity of decapsulated *Artemia* and shrimp feed resulted with a relatively good SR and GR. In experiment 3, we provided 20, 50, and 75 mg of decapsulated *Artemia* and shrimp feed to 3000 individuals of *M. sanguinea*. Our results demonstrated that after 3 months, decapsulated *Artemia* showed high survival rate and 75 mg/3000 inds provided the best quantity of feed in the earlier life stage culture of *M. sanguinea*.

Keywords: Polychaete, Feed, Feeding rate, *Marphysa sanguinea*, Decapsulated *Artemia*

Background

Polychaetes with a relatively short life cycle and a strong reproductivity not only play a role of secondary consumers in the ocean but also sometimes purify deposit by changing the organic ingredients through feeding behavior (Clark 1977; Paik 1989; Heo 2011). Polychaetes are considered as an indicator organism of marine pollution (Belan 2003; Giangrande et al. 2005; Samuelson 2001), as the major prey of benthic fish and as bait for angling becoming a target species of fishermen's sideline (Gambi et al. 1994; Olive 1994, 1999; Younsi et al. 2010). Among the polychaete species, especially, the rockworm *Marphysa sanguinea* (Montagu, 1813) (Eunicidae), is a commercially important species for aquaculture.

M. sanguinea lives in a rock block or between gravels mixed in tender deposit of upper and low intertidal region in the whole coast of South Korea and is well distributed around the world (Glasby and Hutchings 2010; Hutchings et al. 2012). The studies on the breeding of *M. sanguinea* were done by Imai (Imai 1975, 1976, 1981). Besides, the studies on feeding habit and inhabiting environment of adults (Prevedelli et al. 2007), on the early larval development (Imai 1982; Prevedelli et al. 2007), and on the salinity tolerance of juvenile (Garcês and Pereira 2011) were reported. However, the study on early nursery-stock cultivation for mass production of *M. sanguinea* was meager.

The aquaculture of *M. sanguinea* required the definite technologies for mass production and by supply of appropriated feed in each production stage. Unfortunately, no feed is placed on the market and only a pellet type of shrimp or finfish feed is used in the culture of *M. sanguinea*. It is not clear that efficiency of such feed is proper as a food source. It is uncertain how such feed affects growth and survival of *M. sanguinea*. Especially, survival of larva-

* Correspondence: chkpknu@hanmail.net
[1]Interdisciplinary Program of Biomedical Engineering, Pukyong National University, Busan 48513, Republic of Korea
[2]Department of Marine Bio-materials and Aquaculture, Pukyong National University, Busan 48513, Republic of Korea
Full list of author information is available at the end of the article

juvenile until the third month is crucial to determine a seed production of *M. sanguinea*, and it is necessary to establish a reasonable food source and feeding rate to prevent a high mortality in early life stage. Hence, the purpose of this study was to examine survival rate and growth rate by testing various feeds to know appropriate feed, feeding rate, and supply rate in an early stage of nursery-stock production, the biggest fatal stage of *M. sanguinea* aquaculture.

Methods

To investigate an appropriate feed for early life stage of *M. sanguinea*, three different experiments were performed.

Experiment 1: preliminary survey with eight different feeds (20 days)

To investigate a suitable food source in early life stage of *M. sanguinea*, 15 ml of seawater filtered by membrane filter (GF/C) was put in 6-hole well plate (SPL Life Sciences Inc.), and 100 individuals of larvae were introduced in each hole. The larvae were produced in Fisheries Science and Technology Center of Pukyong National University. One microliter of diluted solution of 0.1 g of mud with 10 ml of filtered seawater was put in each hole as substrate to induce the planktonic larvae to do metamorphosis. Water temperature was maintained as 20 °C, and water was changed in 10 ml every 2 days. Without food source (control), *Chaetoceros* sp. (1×10^4 cells/ml), benthic diatom sp. (1×10^4 cells/ml), *Chlorella* powder (1 mg), *Tetraselmis suecica* (1×10 cells/ml), sea mustard (1 ml), decapsulated *Artemia* (1 mg), and extruded pellet (EP) for shrimp (1 mg) were used as food sources. Among them, decapsulated *Artemia* and EP for shrimp were started to supply after 6 days of introduction of larvae when they formed (Kim 2015). Feeds were supplied in every 2 days after renewal of seawater. The feeding tests were triplicated, and the experimental period was 20 days. The survival and growth rate were calculated at the end of the experiment. The living worms were counted, and the length was measured by taking 10 juveniles randomly in each experimental group for calculating the growth rate. The Dixi image program (Ver 2.89, Dixi optics) was used to measure the length of juveniles.

Experiment 2: investigation on the source and the amount of feed for early juvenile culture (for 2 months)

Five feed sources with four quantities in each (total 20 conditions) were tested. Decapsulated *Artemia* and shrimp feed showing the positive results in experiment 1 were also used in addition to eel feed. For microalgae, mixed microalgae solution sold on the market and benthic diatom species generally used for larva of invertebrate as a feed were selected (Table 1). For shrimp feed, eel feed, and

Table 1 Nutritional contents of feed used in this experiment and their ingredients

Ingredient	Protein	Lipid	Fiber	Moisture	Others
Shrimp feed	Min 52%	Min 14.5%	Max 3%	Max 10%	–
Decapsulated *Artemia*	Min 46%	Min 5%	Max 5.8%	Max 13%	–
Eel feed	Min 54%	Min 10%	Max 15%	–	Ca 1.5%, P 2.7%
Mixed microalgae	–	–	–	–	*Isochrysis Pavlova, Tetraselmis*
Benthic diatom	–	–	–	–	*Navicula, Nitzschia,* etc.

decapsulated *Artemia*, four different quantities such as 5 mg/3000 inds, 10 mg/3000 inds, 50 mg/3000 inds, and 100 mg/3000 inds were supplied. For mixed microalgae and benthic diatom, 1×10^4 cells/3000 inds, 1×10^5 cells/3000 inds, 1×10^6 cells/3000 inds, and 1×10^7 cells/3000 inds were supplied. Nutritional contents of used feed are presented in Table 1. The feeding was done for every 2 days after 7 days from installation of experiment. Filtered seawater was renewed constantly, and remained food was not removed. The experimental periods were 60 days by reflecting the high mortality in early juvenile period. Number of setigers and survival rate of worms were examined at the end of experiment. A rearing tank (Fig. 1) was constituted with three reserves of 45 cm × 9 cm × 5 cm which allowed three replicates of experiments. Mud was used as a substrate with 1-cm depth. Experimental installations are shown in Fig. 2. The conditions of seawater were maintained at 20 ± 2 °C, 32–33 psu, pH 8.0, and 6–8 ppm of DO during the experimental period.

Experiment 3: more detailed investigation on the source and the amount of feed for juvenile cultivation (3 months of experiment)

Decapsulated *Artemia* and shrimp feed could be considered as a good efficient feed through the results of experiment 2; hence, a closer feeding amount ranges were investigated in the third experiment for 3 months. Three quantities such as 25 mg/3000 inds, 50 mg/3000 inds, and 75 mg/3000 inds of two feeds were reinstalled. The experimental conditions were similar to experiment 2.

Statistical analysis

Growth rate (GR) was measured with growth of body length in experiment 1.

GR (growth rate) = [(final body length – initial body length)/initial body length] × 100

For experiments 2 and 3, GR was measured with number of setigers.

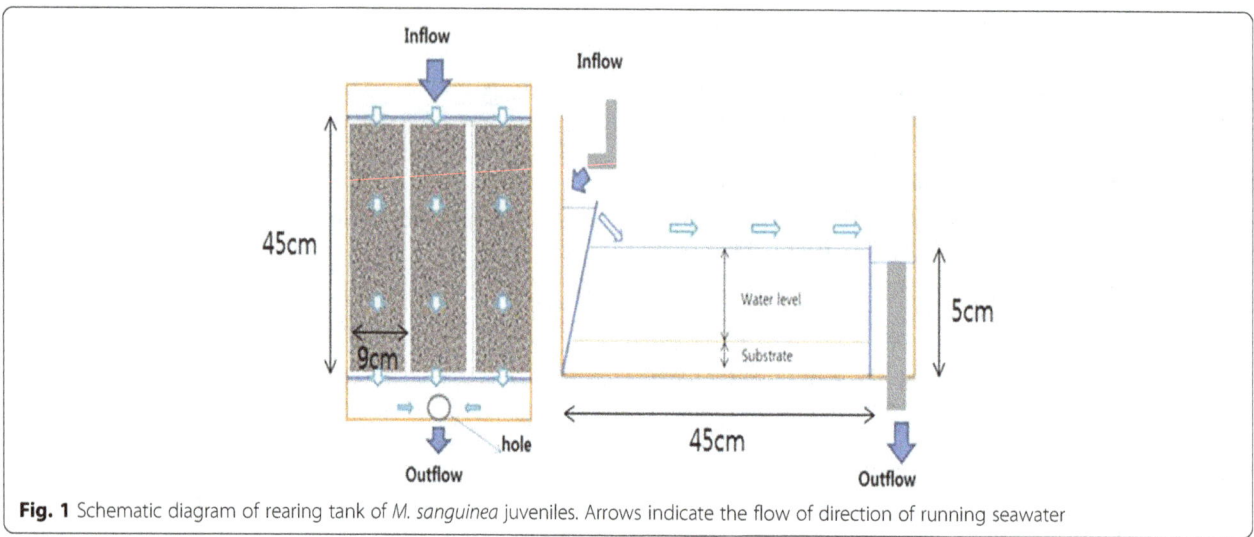

Fig. 1 Schematic diagram of rearing tank of *M. sanguinea* juveniles. Arrows indicate the flow of direction of running seawater

Survival rate was calculated as

$$\text{SR (survival rate)} = (\text{final population number} / \text{initial population number}) \times 100$$

The validation of each study section was made by two-way ANOVA test with squared transformed data, and the significant test was conducted with a p value of 0.05. The statistical analysis was performed with the Systat v.9 package.

Results

Experiment 1: a preliminary survey about appropriate feed source of an early larval/juvenile cultivation for nursery-stock cultivation

The decapsulated *Artemia* and shrimp feed showed the high growth (Fig. 3) and survival rate (Fig. 4). The growth rate (GR) of shrimp feed group was 451.5 ± 13.29% which was the highest, and decapsulated *Artemia* was 411.3 ± 6.94% at the end of 20-day experiment. The other experimental groups were showed similar or lower growth rate than the control group (without feed).

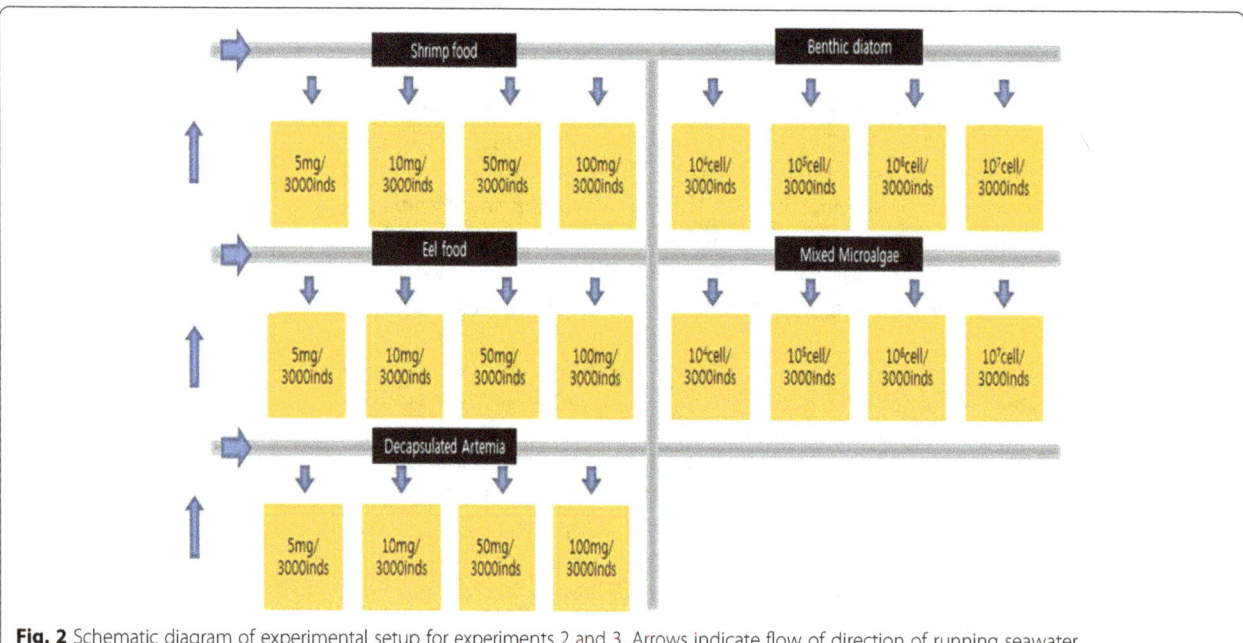

Fig. 2 Schematic diagram of experimental setup for experiments 2 and 3. Arrows indicate flow of direction of running seawater

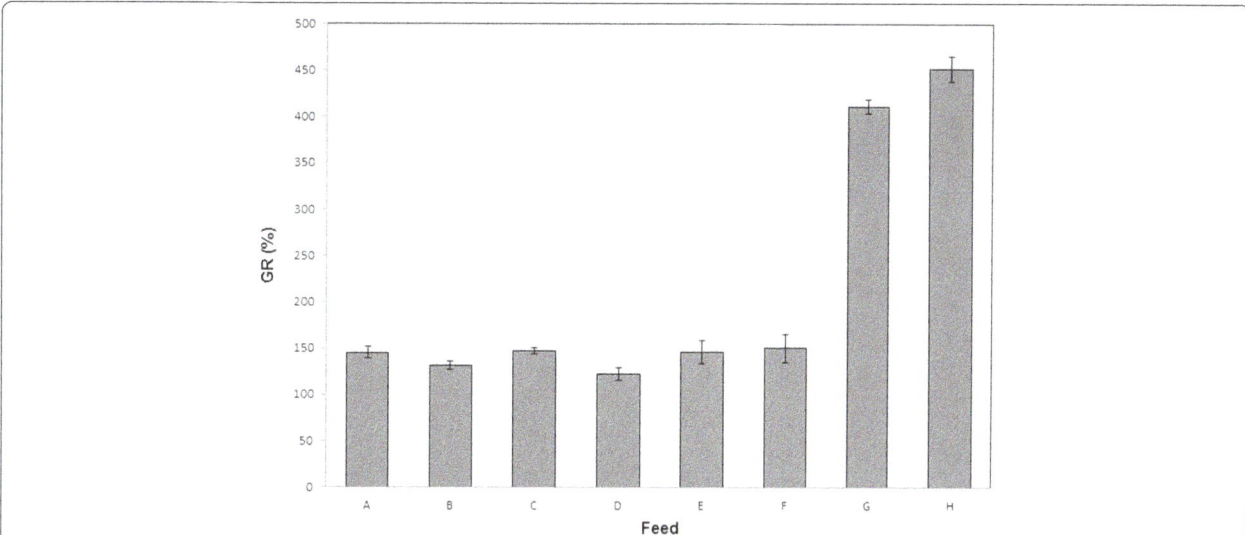

Fig. 3 Growth rate (GR) of early larval culture with eight different food sources after 20 days. A, without food source (control); B, *Chaetoceros* sp. (1×10^4 cells/ml); C, benthic diatom sp. (1×10^4 cells/ml), D, *Chlorella* powder (1 mg); E, *Tetraselmis suecica* (1×10 cells/ml); F, sea mustard (1 ml); G, decapsulated *Artemia* (1 mg); and H, extruded pellet (EP) for shrimp (1 mg)

The survival rate (SR) was the highest in decapsulated *Artemia* with 61 ± 3.2%. Shrimp feed was followed with 51 ± 3%. The control group was 35 ± 4.6%. The lowest survival rate was 28 ± 3% of sea mustard which is lower than control (Fig. 4). The results of GR and SR were shown that decapsulated *Artemia* and shrimp feed were good candidate for feed of early juvenile stage.

Experiment 2: first survey for appropriate feed source of an early juvenile cultivation for nursery-stock cultivation (2 months of experiment)

Growth and survival rates of juveniles at the end of the experiment according to feeding amount with different feed types are shown in Table 2 and Figs. 5 and 6. In the survival rate of different feed sources, decapsulated

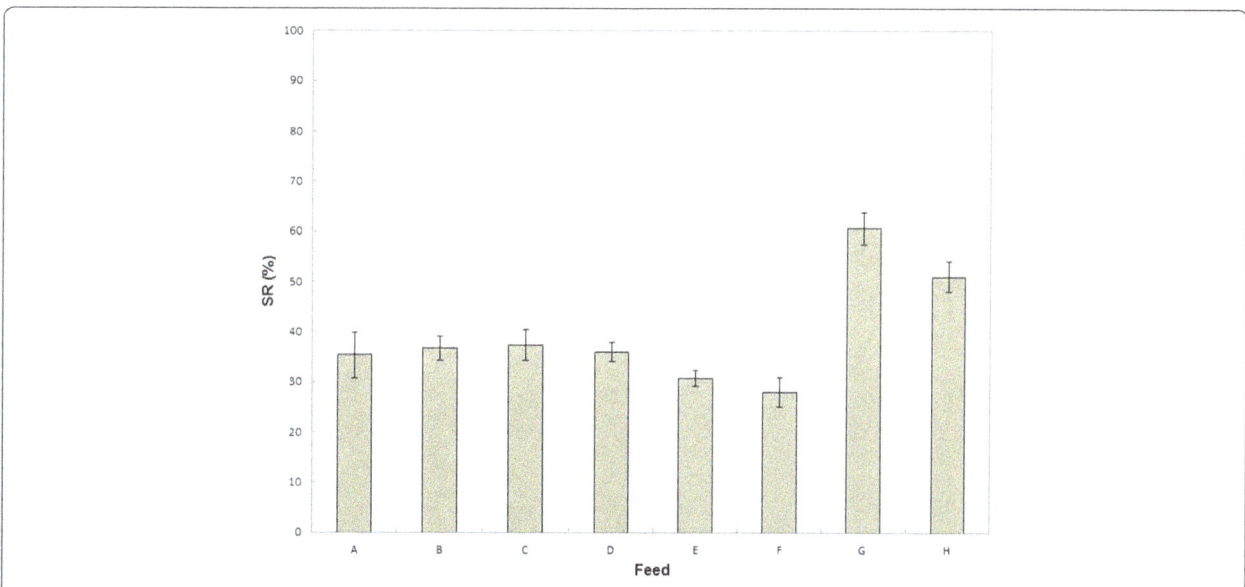

Fig. 4 Survival rates of early larval culture with eight different food sources after 20 days. A, without food source (control); B, *Chaetoceros* sp. (1×10^4 cells/1 ml); C, benthic diatom sp. (1×10^4 cells/1 ml); D, *Chlorella* powder (1 mg); E, *Tetraselmis suecica* (1×10 cells/1 ml); F, wakame (1 ml); G, decapsulated *Artemia* (1 mg); and H, extruded pellet (EP) for shrimp (1 mg)

Table 2 Survival rate (SR) and number of setigers of *M. sanguinea* juvenile by feeds (2 months)

Trials	Group	Amount of feed (per 3000 individuals)	SR (%)	Number of setigers
Shrimp feed	A	5 mg	1.00 ± 0.01	23.66 ± 1.25
	B	10 mg	0.45 ± 0.30	23.58 ± 2.58
	C	50 mg	11.81 ± 1.77	24.26 ± 3.78
	D	100 mg	3.65 ± 0.29	31.13 ± 4.95
Eel feed	A	5 mg	0.21 ± 0.01	26.10 ± 2.19
	B	10 mg	3.37 ± 1.40	20.33 ± 1.41
	C	50 mg	11.92 ± 3.28	21.80 ± 1.17
	D	100 mg	6.70 ± 1.82	22.46 ± 0.80
Decapsulated *Artemia*	A	5 mg	1.67 ± 0.54	29.66 ± 3.65
	B	10 mg	9.45 ± 1.33	23.10 ± 2.59
	C	50 mg	33.88 ± 2.54	25.03 ± 0.32
	D	100 mg	11.62 ± 2.98	33.43 ± 3.20
Benthic diatom	A	1×10^4 cell	0	
	B	1×10^5 cell	0	
	C	1×10^6 cell	0	
	D	1×10^7 cell	0.2 ± 0.1	22.00 ± 3.07
Mixed microalgae	A	1×10^4 cell	0	
	B	1×10^5 cell	0	
	C	1×10^6 cell	0	
	D	1×10^7 cell	0.03	18.67 ± 3.06

Artemia showed the highest survival rate, followed by shrimp feed and eel feed. They showed higher SR than other two feed sources such as benthic diatom and mixed microalgae. Fifty milligrams of decapsulated *Artemia* showed the highest average survival rate with 33.88 ± 2.54%. Followed by 50 mg of eel feed 11.92 ± 3.28% and 50 mg of shrimp feed 11.81 ± 1.77%, 100 mg of decapsulated *Artemia* showed 11.62 ± 2.98%.

In feeding amount, 50 mg showed the highest survival rate in general. Benthic diatom and mixed microalgae

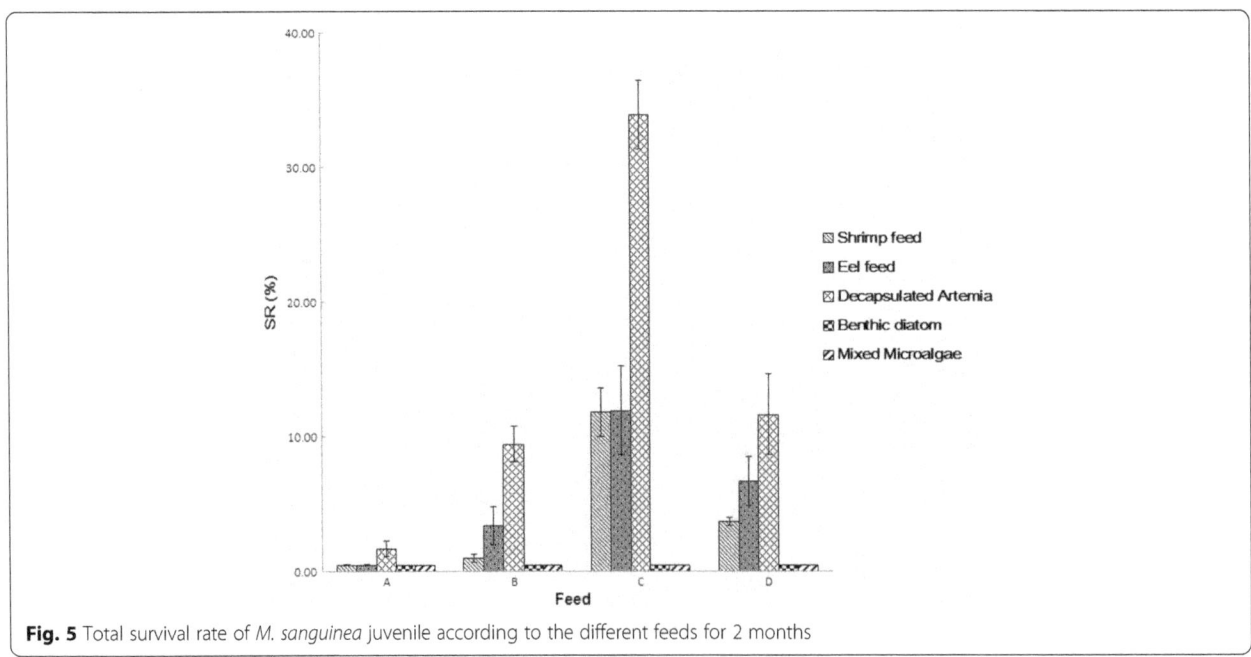

Fig. 5 Total survival rate of *M. sanguinea* juvenile according to the different feeds for 2 months

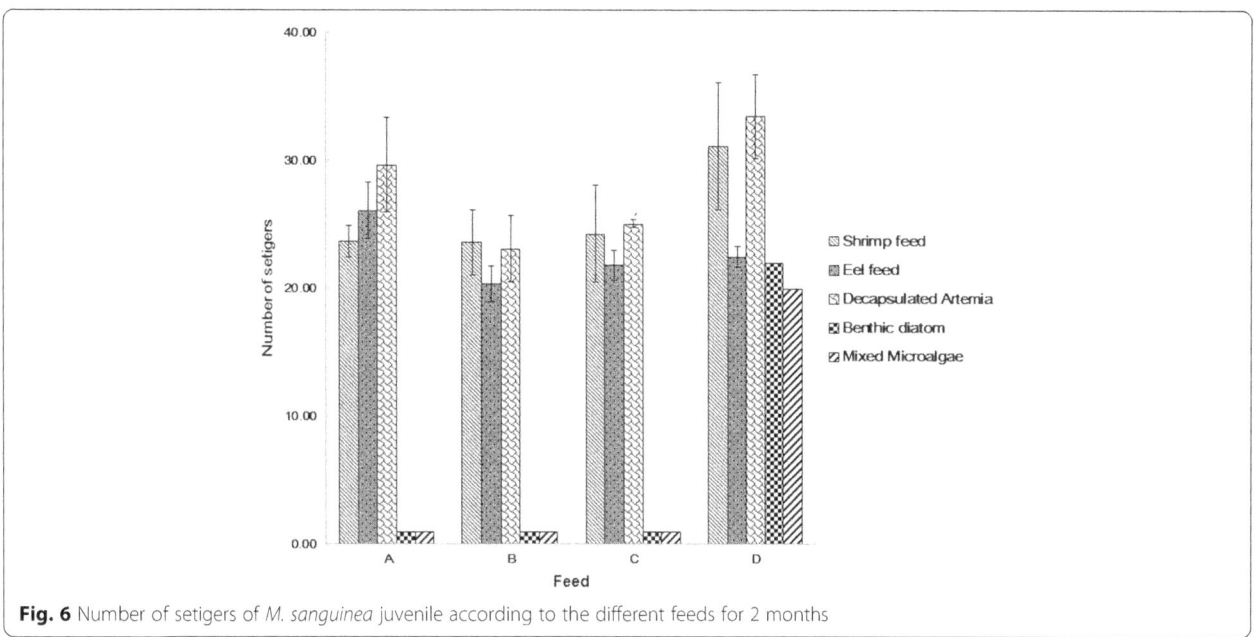

Fig. 6 Number of setigers of *M. sanguinea* juvenile according to the different feeds for 2 months

did not show survival individuals except in 1×10^7 cells (Fig. 5). Two-way ANOVA test showed the significant effect ($p < 0.05$) of feed and quantity of feed for survival rate of juvenile (Table 3). In growth presented with number of setigers, decapsulated *Artemia* and shrimp feed were higher than eel feed. Benthic diatom and mixed microalgae were showed 22.00 ± 3.07 setigers and 18.67 ± 3.06 setigers, respectively; however, they could not survive until 20 experimental days (Fig. 6). In decapsulated *Artemia* section with the highest growth rate, 100 mg/3000 inds section was 33.43 ± 3.2 setigers, and 5 mg/3000 inds section was 29.66 ± 3.65 setigers, but 50 mg/3000 inds section with the highest survival rate showed a low growth as 25.03 ± 0.32 setigers. In all sections, 100 mg/3000 inds section showed the highest growth and 50 mg/3000 inds section showed the highest survival rate (Table 2). Considering survival and growth, decapsulated *Artemia* 50 mg/3000 inds section showed good result.

Two-way ANOVA test showed the non-significant effect ($p < 0.05$) of feed and quantity of feed for increase of setiger numbers of juvenile (Table 4). So the effects of feed and feed quantity were significant for survival rate, not for juvenile's growth. Two-way ANOVA test was performed only with shrimp feed, eel feed, and decapsulated *Artemia*

which give statistically enough numbers at 20 experimental days.

Experiment 3: final survey for appropriate feed source of an early juvenile cultivation for nursery-stock cultivation (3 months of experiment)

Decapsulated *Artemia* group showed higher survival rate than shrimp feed group on average (Table 5); 75 mg/3000 inds of decapsulated *Artemia* section showed $7.91 \pm 1.28\%$ of survival rate. Next, decapsulated *Artemia* 50 mg/3000 inds showed $4.36 \pm 0.81\%$, and shrimp feed 75 mg/3000 inds section showed $3.99 \pm 1.19\%$ (Fig. 7). In feeding amount, 75 mg/3000 inds showed the highest survival rate in general. Two-way ANOVA test showed the non-significant effect ($p > 0.05$) of feed, significant effect ($p < 0.05$) of quantity of feed, and significant effect ($p < 0.05$) of interaction of two factors for survival rate of juvenile (Table 6).

In growth presented with number of setigers, decapsulated *Artemia* had more setigers than shrimp feed. Decapsulated *Artemia* showed the highest growth in 75 mg/3000 inds with 42.2 ± 8.31 setigers, followed by shrimp feed 75 mg/3000 inds section with 40.33 ± 5.94 setigers. In low amount of feed, growth was similar to shrimp

Table 3 Two-way ANOVA test results of survival rate (SR) cultured in different feeds for *M. sanguinea* juveniles

Source	Sum-of-squares	df	Mean square	F ratio	P
[a]Feed	20.191	1	20.191	10.174	0.003
[b]Qte of feed	30.851	1	30.851	18.463	0.000
Feed × Qte of feed	47.995	1	47.995	41.135	0.000

[a]Feed: various kinds of feeds
[b]Qte of feed: various quantities of feeds

Table 4 Two-way ANOVA test results of setiger numbers cultured in different feeds for *Marphysa sanguinea* juveniles

Source	Sum-of-squares	df	Mean square	F ratio	P
[a]Feed	0.145	1	0.145	0.576	0.456
[b]Qte of feed	0.257	1	0.257	1.020	0.320
Feed × Qte of feed	0.275	1	0.275	1.092	0.303

[a]Feed: various kinds of feeds
[b]Qte of feed: various quantities of feeds

Table 5 Survival rate (SR) and number of setigers of *M. sanguinea* juvenile after 3 months

Feed	Group	Amount of feed (per 3000 inds) (mg)	[1]SR (%)	Number of setigers (growth)
Shrimp feed	A	25	2.08 ± 0.29	35.36 ± 5.17
	B	50	2.37 ± 0.60	37.10 ± 4.98
	C	75	3.94 ± 1.19	40.33 ± 5.94
Decapsulated *Artemia*	A	25	2.12 ± 0.25	35.90 ± 7.98
	B	50	4.36 ± 0.81	39.56 ± 8.31
	C	75	7.91 ± 1.28	42.20 ± 5.23

Table 6 Two-way ANOVA test results of survival rate (SR) cultured in different feeds for *M. sanguinea* juveniles

Source	Sum-of-squares	df	Mean square	F ratio	P
[a]Feed	0.987	1	0.987	4.232	0.056
[b]Qte of feed	2.646	1	2.646	20.454	0.000
Feed × Qte of feed	4.026	1	4.026	93.244	0.000

[a]Feed: various kinds of feeds
[b]Qte of feed: various quantities of feeds

feed and decapsulated *Artemia*. With high amount of feed, growth was higher in decapsulated *Artemia* than in shrimp feed (Fig. 8). Two-way ANOVA test showed that the non-significant effect ($p > 0.05$) of feed and significative effect ($p < 0.05$) of quantity of feed and significant effect ($p < 0.05$) of interaction of two factors for setiger numbers of juvenile (Table 7).

Discussion and conclusions

Imai (1975) reported that mucilage secreted after 10–13 days from fertilization, later it was well agreed to the report of Kim and Jang (2008). In this experiment, we made an attempt of food supply from the seventh day of larval release, as per the report of Kim (2015), and observed that larvae ate organic matter with jaw plate from the seventh day. In the larval stage, like other lecitotrophic larvae, the worm cannot feed and only looking for appropriate settling place (Thorson 1950; Jablonski and Lutz 1983). The other

polychaetes *Nereis pelagica* and *Nereis grubei* showed polyphagia, *Perinereis cultrifera* shows phytophagy feeding seaweed, and *Perinereis nuntia* showed polyphagia placing too much emphasizing on creophagy. *Pereneris nuntia* showed more than 80% of feeding efficiency with mixed feed of eel feed as powder in the Japanese aquaculture farm (Yoshida 1976). But only few reports are available on the early juvenile stage of *M. sanguinea* (Kim 2015).

In the pilot survey for 20 days, creophagy- and phytophagy-type feeds were tested. It was very clear that decapsulated *Artemia* and shrimp feed were reasonably good for early life stage of *M. sanguinea* culture and the other feed were shown similar level to the control group (without food) (Fig. 3). Similar result was reported by Nielsen et al. (1995), filter-feeding *Nereis diversicolor* grew on a diet of suspended algal cells, but the maximum specific growth rate was lower than the feeding experiments with shrimp meat. In the case of experiment 1, growth and survival rate of shrimp feed and decapsulated *Artemia* were higher than other feeding sources. Generally, 50 mg/3000 inds sections showed good survival rate in feeding amount of shrimp, eel, and

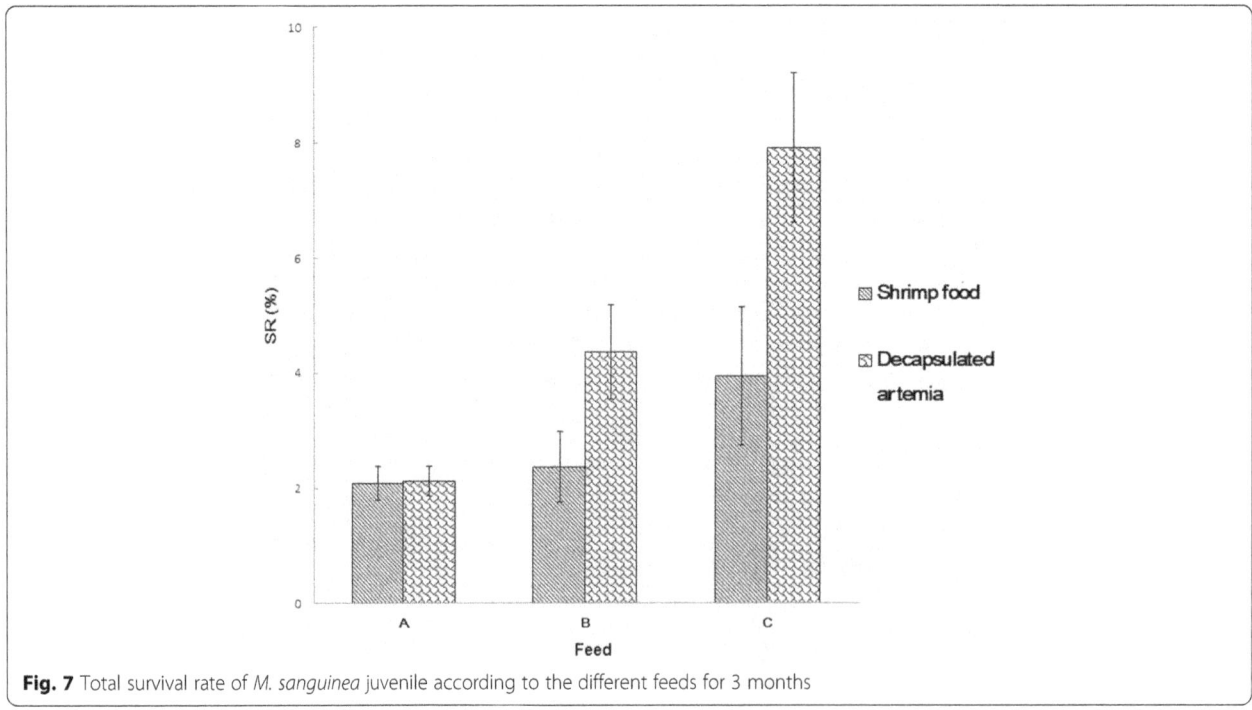

Fig. 7 Total survival rate of *M. sanguinea* juvenile according to the different feeds for 3 months

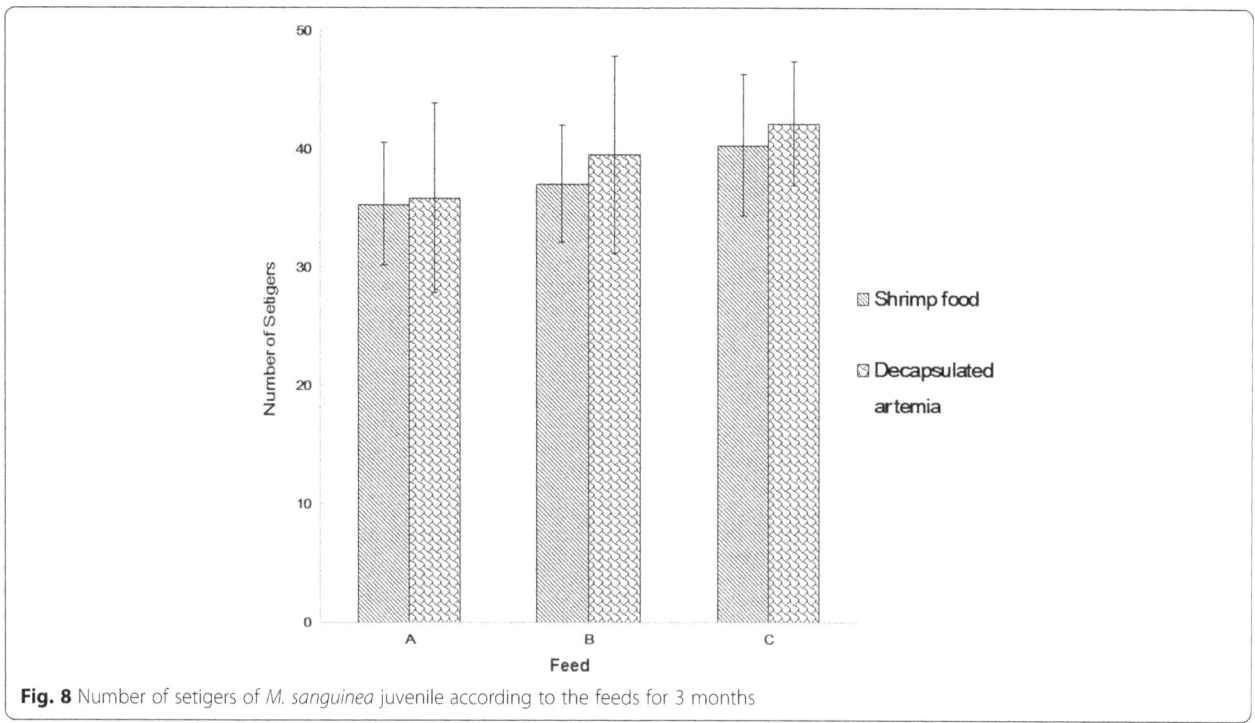

Fig. 8 Number of setigers of *M. sanguinea* juvenile according to the feeds for 3 months

decapsulated *Artemia* feeds. Decapsulated *Artemia* showed the highest growth rate. Studies performed by Vedel and Riisgård (1993) showed that *N. diversicolor* fed algal cells (*Rhodomonas* sp.) obtained specific growth rates of 3.1% comparable to 3.9% measured in worms in glass tubes placed 15 cm above the bottom in the eutrophicated Odense Fjord. The maximum specific growth rate of the facultative *N. diversicolor* fed algae is lower than 9% found for the obligate suspension-feeding blue mussel *Mytilus edulis* grown in nature in net bags (Riisgård and Poulsen 1981).

However, 50 mg/3000 inds section with the highest survival rate showed lower growth than other feeding amount sections. Both 5 mg/3000 inds and 10 mg/3000 inds sections were showed very low survival rate by insufficient food sources. It seems that there was a cannibalization in the low feeding amount sections. The other section 100 mg/3000 inds was also found with relatively low survival rate due to the degradation of water quality by the accumulation of unused feeds. In addition, there was already no survival being in adhesive

diatom and microalgae sections, and it showed a very low survival rate below 0.2% of survival rate; thus, they could not play a role of food source for *M. sanguinea*. Hence, 50 mg/3000 inds of decapsulated *Artemia* was the best food and feeding amount for the first 2 months of nursery-stock cultivation.

Two-way ANOVA test showed significant effect of feed and feed quantity for survival rate. However, there was no significant difference in growth rate. The good survival rate was showed in decapsulated *Artemia*. Eel feed showed slightly higher survival rate than shrimp feed for *M. sanguinea* larvae. But growth rate was higher in shrimp feed and easier to control in water quality than in eel feed (Kim 2015). This is similar to the result of *N. diversicolor*, where the maximum specific growth rate was higher in shrimp feed (Nielsen et al. 1995). Hence, decapsulated *Artemia* and shrimp feed were chosen for experiment 3. The quantity of feed supplied was differentiated more and less 25 mg/3000 inds in the base of 50 mg/3000 inds. In the experiment 3, our results showed that enough amount of feed can help survival and growth rates, no matter what kind of feed is provided. However, decapsulated *Artemia* showed high survival rate and 75 mg/3000 inds section provided the best quantity of feed in the earlier life stage culture of *M. sanguinea*. Since there were no much studies on the larvae of *M. sanguinea* in Korea and other countries, our results were not extensively compared. However, further comparative studies are necessary to clarify the extent of feeding in the larvae of *M. sanguinea*.

Table 7 Two-way ANOVA test results of setiger numbers cultured in different feeds for *Marphysa sanguinea* juveniles

Source	Sum-of-squares	df	Mean square	F ratio	P
[a]Feed	0.075	1	0.075	1.579	0.227
[b]Qte of feed	0.618	1	0.618	45.722	0.000
Feed × Qte of feed	0.607	1	0.607	42.775	0.000

[a]Feed: various kinds of feeds
[b]Qte of feed: various quantities of feeds

Funding
This work was supported by a Research Grant of Pukyong National University (2015 year).

Authors' contributions
KHK, BKK, and CHK manufactured the experimental feed and drafted the manuscript. KHK, BKK, SKK, and WWP conducted the feeding trial and performed the analyses. KHK and CHK conceived and designed the study and experimental facility and also revised the manuscript. All authors read and approved the final manuscript.

Competing interests
The authors declare that they have no competing interests.

Author details
[1]Interdisciplinary Program of Biomedical Engineering, Pukyong National University, Busan 48513, Republic of Korea. [2]Department of Marine Bio-materials and Aquaculture, Pukyong National University, Busan 48513, Republic of Korea. [3]Fisheries Science and Technology Centre, Pukyong National University, Goseong-gun, Gyeongsangnam-do 52957, Republic of Korea.

References
Belan TA. Marine environmental quality assessment using polychaete taxocene characteristics in Vancouver harbour. Mar Environ Res. 2003;57:89–101.

Clark RB. Reproduction, speciation and polychaete taxonomy. Essay on polychaetous annelids in memory of Dr. Olga Hartman. 1977. p. 477–502.

Gambi MC, Castelli A, Giangrande A, Lanera P, Prevedelli D, Zunarelli-Vandini R. Polychaetes of commercial and applied interest in Italy: an overview. In: Actes de la 4ème conférence internationale des Polychètes. Mem Mus natn Hist nat, Dauvin, JC, Laubier L and Reish DJ, Eds. Muséum National d'Histoire Naturelle, Paris. 1994;162:593–603.

Garcês JP, Pereira J. Effect of salinity on survival and growth of Marphysa sanguinea Montagü (1813) juveniles. Aquacult Int. 2011;19:523–30.

Giangrande A, Licciano M, Musco L. Polychaetes as environmental indicators revisited. Mar Poll Bull. 2005;50:1153–62.

Glasby CJ, Hutchings PA. A new species of Marphysa Quatrefages, 1865 (Polychaeta: Eunicida: Eunicidae) from northern Australia and a review of similar taxa from the Indowest Pacific, including the genus Nauphanta Kinberg 1865. Zootaxa. 2010;2352:29–45.

Heo CH. Larval development and effect of substrates on juvenile growth of polychaete Marphysa sanguinea, M. Fish. Thesis, Pukyong National University, Busan, Korea. 2011.

Hutchings P, Glasby CJ, Wijnhoven S. Note on additional diagnostic characters for Marphysa sanguinea (Montagu, 1813) (Annelida: Eunicida: Eunicidae), a recently introduced species in the Netherlands. Aquat Invasions. 2012;7:277–82.

Imai T. On the development and the spawning behaviour of polychaete worm, Marphysa sanguinea. Aquaculture. 1975;23(1):14–20.

Imai T. On the growth of polychaete worm, Marphysa sanguinea. Aquaculture. 1976;24(2):61–7.

Imai T. Feeding and excreting of Marphysa sanguinea (Montagu) Annelida Polychaeta. Bull Kanagawa Pref Fish Exp St. 1981;3:9–14.

Imai T. The early development and breeding of Marphysa sanguinea (Montagu). Benthos Res. 1982;23:36–41.

Jablonski D, Lutz RA. Larval ecology of marine benthic invertebrates; paleobiological implications. Biol Rev. 1983;58:21–89.

Kim BK. Feed and feeding rate in early stage on seed production of the rockworm polychaete Marphysa sanguinea. M. Eng. Thesis, Pukyong National University, Busan, Korea. 2015.

Kim CH, Jang SW. Effects of rearing conditions on the artificial seed production of a polychaete Marphysa sanguinea. J Aquacul. 2008;21:34–40.

Nielsen AM, Eriksen NT, Iversen JJL, Riisgård HU. Feeding, growth and respiration in the polychaetes Nereis diversicolor (facultative filter-feeder) and N. virens (omnivorous)—a comparative study. Mar Ecol-Prog Ser. 1995;125:149–58.

Olive PJW. Polychaete aquaculture and polychaete science: a mutual synergism. Hydrobiologia. 1999;402:175–83.

Olive PJW. Polychaeta as a world resource: a review of patterns of exploitation as sea angling baits and the potential for aquaculture based production. In: Actes de la 4ème conférence internationale des Polychètes. Mem Mus natn Hist nat, Dauvin, JC, Laubier L and Reish DJ, Eds. Muséum National d'Histoire Naturelle, Paris. 1994; 162:603–610.

Paik EI. Illustrated encyclopedia of fauna & flora of Korea vol 31. Polychaeta. Dept. Culture and Education. Seoul, Korea. 1989.

Prevedelli D, Massamba N, Siala G, Ansaloni I, Simonini R. Life cycle of Marphysa sanguinea (Polychaeta: Eunicidae) in the Venice lagoon (Italy). Mar Ecol. 2007; 28:384–93.

Riisgård HU, Poulsen E. Growth of Mytilus edulis in net bags transferred to different localities in a eutrophicated Danish fjord. Mar Pollut Bull. 1981;12:272–6.

Samuelson GM. Polychaetes as indicators of environmental disturbance on subarctic tidal flats, Iqaluit, Baffin Island, Nunavut territory. Mar Poll Bull. 2001; 42:733–41.

Thorson G. Reproductive and larval ecology of marine bottom invertebrates. Biol Rev. 1950;25:1–45.

Vedel A, Riisgård HU. Filter-feeding in the polychaete Nereis diversicolor: growth and bioenergetics. Mar Ecol-Prog Ser. 1993;100:145–52.

Yoshida M. The effects of worm density and amount of food for the culture of the Nereid worm Perinereis nuntia var. vallata: Nippon Suisan Gakkaishi. 1976; 42(11): 1193–8.

Younsi M, Daas T, Daas D, Scaps P. Polychaetes of commercial interest from the Mediterranean East Coast of Algeria. Medit Mar Sci. 2010;11:185–8.

Taurine supplementation in diet for olive flounder at low water temperature

Joo-Min Kim[1], G. H. T. Malintha[2], G. L. B. E. Gunathilaka[2], Chorong Lee[2], Min-Gi Kim[2], Bong-Joo Lee[3], Jeong-Dae Kim[4] and Kyeong-Jun Lee[2]*

Abstract: The objective of this study was to examine the effect of dietary supplementation of taurine for juvenile olive flounder (*Paralichthys olivaceus*) at low water temperature (16.4 ± 0.36 °C). Fish meal (FM)-based diet was used as the control diet. Four other experimental diets were prepared by adding taurine to FM-based diet at 0.25, 0.50, 1.00, and 1.50% (T1, T2, T3, and T4, respectively). Each experimental diet was fed to triplicate groups of fish (initial mean body weight, 19.5 g) for 10 weeks. At the end of the feeding trial, growth performance and feed utilization, hematological parameters, non-specific immune responses, whole-body proximate composition, and liver mRNA expression of insulin-like growth factor-1 (IGF-1) were investigated. Feed conversion ratio was significantly reduced while protein efficiency ratio was significantly increased in taurine-supplemented groups. Hematocrit and hemoglobin were also significantly increased while plasma cholesterol levels were decreased in taurine-supplemented groups than those in the control group. Nitro-blue-tetrazolium, myeloperoxidase and lysozyme activities, and plasma immunoglobulin level were significantly increased by taurine supplementation. These results suggest that dietary taurine supplementation is effective in improving growth performances, feed utilization, and innate immunity of olive flounder in low water temperature season.

Keywords: Taurine, Olive flounder, Innate immunity, Water temperature

Background

Taurine (2-aminoethane sulfonic acid) is an amino acid synthesized from methionine via cysteine by a series of enzymatic reactions (Oja and Kontro 1983). It is considered as a conditionally indispensable amino acid for marine or cold water fish species. The ability of fish to synthesize taurine is species or developmental stage dependent (Yokoyama et al. 2001; Kim et al. 2003, 2005). The reason for such differences in their ability to synthesize taurine might be due to various activities of L-cysteine sulfinate decarboxylase, an enzyme required for the conversion from cysteine to taurine (De la Rosa and Stipanuk 1985; Takeuchi et al. 2001; Yokoyama et al. 2001). Although taurine is nonessential, its inclusion in diets is often recommended because of its auxiliary actions such as membrane protection, antioxidation and detoxification in mammals (Wright et al. 1986). In addition, it plays a role in osmoregulation in invertebrates (Schaffer et al. 2000), acts as a carrier for lipid-soluble vitamins in mammals (Petrosian and

Haroutounian 2000) and induces bile salt production in fish (Van Waarde 1988). Due to its unique amino acid properties (i.e. low molecular weight, nitrogen content and water solubility), taurine can also act as a feed stimulant for fish (Carr 1982). It has been reported that taurine is effective as a feed attractant by stimulating the olfactory system of arctic char (*Salvelinus alpinus*), grayling (*Thymallus thymallus*) (Doving et al. 1980) and rainbow trout (*Oncorhynchus mykiss*) (Hara et al. 1984).

Effects of water temperature on nutrient utilization have been studied in many fish species (Olsen and Ringn 1998; Peres and Oliva-Teles 1999). It has been reported that the optimum temperature for growth of Japanese flounder is 20–25 °C, even though they can be fed at low temperatures ranging from 10 to 20 °C (Iwata et al. 1994). In most studies, nutrient requirements for flounder have been determined at moderate water temperatures (18–22 °C) (Lee et al. 2000, 2002).

Olive flounder is the most important marine fish species in Korea with production exceeding 60% of annual fish production (Ministry of Maritime Affairs and Fisheries 2015). Olive flounder can be exposed to suboptimal temperatures during the culture period in Korea. Thus, knowledge on nutrient utilization in different water

* Correspondence: kjlee@jejunu.ac.kr
[2]Department of Marine Life Sciences, Jeju National University, Jeju 63243, South Korea
Full list of author information is available at the end of the article

temperatures would be useful in optimizing dietary compositions or feeding conditions throughout the year for the growth of olive flounder. However, limited information is available on nutrient utilization in suboptimal water temperatures for this species. Therefore, the objective of this study was to determine the effects of taurine supplementation on growth, feed utilization, whole-body composition, innate immunity, and liver mRNA expression of IGF-1 of juvenile olive flounder at low water temperatures.

Methods

Experimental diets

Five experimental diets (Table 1) were formulated to contain taurine in fish meal-based diets at 0, 0.25, 0.5, 1.0, and 1.5% (Control, T1, T2, T3, and T4) for juvenile olive flounder. The five diets were formulated to be isonitrogenous (49.0% crude protein) and isocaloric (18.1 kcal/kg diet). All dry ingredients were thoroughly mixed with 10–15% double distilled water after addition of fish oil, pelleted through a pellet machine (SP-50, Gumgang Engineering, Daegu, Korea) at 6 mm in diameter, dried for 24 h and stored at –20 °C until used.

Table 1 Formulation and chemical analysis of the five experimental diets for olive flounder (P. olivaceus) (%, dry matter)

Ingredients	Experimental diets				
	Control	T1	T2	T3	T4
Brown fish meal	50.0	50.0	50.0	50.0	50.0
Soybean meal	5.00	5.00	5.00	5.00	5.00
Corn gluten meal	5.00	5.00	5.00	5.00	5.00
Wheat flour	27.0	27.0	27.0	27.0	27.0
Fish oil	8.00	8.00	8.00	8.00	8.00
Starch	1.00	1.00	1.00	1.00	1.00
Mineral mix[a]	1.00	1.00	1.00	1.00	1.00
Vitamin mix[b]	1.00	1.00	1.00	1.00	1.00
Choline chloride	0.50	0.50	0.50	0.50	0.50
Glycine	1.50	1.25	1.00	0.50	0.00
Taurine	0.00	0.25	0.50	1.00	1.50
Chemical analysis (%, dry matter)					
Moisture	6.9	7.2	7.1	7.0	7.1
Crude protein	48.9	48.8	49.2	49.0	49.0
Crude lipid	13.2	13.0	13.1	13.0	13.2
Crude ash	8.0	8.1	8.1	8.2	8.2
Taurine	0.12	0.33	0.42	0.92	1.34

[a]Mineral premix (g/kg mixture): $MgSO_4.7H_2O$, 80.0; $NaH_2PO_4.2H_2O$, 370.0; KCl, 130.0; Ferric citrate, 40.0; $ZnSO_4.7H_2O$, 20.0; Ca-lactate, 356.5; $CuCl_2$, 0.2; $AlCl_3$. $6H_2O$, 0.15; $Na_2Se_2O_3$, 0.01; $MnSO_4.H_2O$, 2.0; $CoCl_2.6H_2O$, 1.0
[b]Vitamin premix (g/kg mixture): L-ascorbic acid, 121.2; DL-a tocopheryl acetate, 18.8; thiamin hydrochloride, 2.7; riboflavin, 9.1; pyridoxine hydrochloride, 1.8; niacin, 36.4; Ca-D-pantothenate, 12.7; myo-inositol, 181.8; D-biotin, 0.27; folic acid, 0.68; p-aminobezoic acid, 18.2; menadione, 1.8; retinyl acetate, 0.73; cholecalficerol, 0.003; cyanocobalamin, 0.003

Fish and feeding trial

Olive flounder juveniles were purchased from a private hatchery and transported to Marine and environmental Research Institute, Jeju National University, Jeju, South Korea. Fish were adapted to experimental conditions and facilities for 2 weeks. At the end of the acclimation period, total 525 fish (initial mean body weight, 19.5 g) were randomly distributed into 15 150 L capacity polyvinyl circular tanks with three replicate groups per diet treatment. The tanks were supplied with filtered seawater at a flow-rate of 3 L/min and aeration to maintain enough dissolved oxygen. Each tank was designated as one of three replicates for a diet group. They were fed with the experimental diets to apparent satiation two times a day (08:30 and 17:00 h) for 10 weeks. Water temperature was naturally maintained at 16.4 ± 0.36 °C during the whole duration. Growth measurement was carried out at every 2 weeks.

Sample collections and analysis

At the end of the feeding trial, all the fish in each tank were bulk-weighed and counted for calculation of weight gain (WG), specific growth rate (SGR), feed conversion ratio (FCR), protein efficiency ratio (PER), and survival. Four fish per tank (4 fish/tank, 12 fish/diet) were randomly captured 24 h after the last meal, anesthetized with 2-phenoxyethanol solution (200 mg/L), and blood samples were taken from the caudal vein with heparinized syringe for determination of hematocrit, hemoglobin, and nitro-blue-tetrazolium (NBT) activity. Then, plasma samples were separated by centrifugation at $5000 \times g$ for 10 min and stored at –70 °C for determination of plasma chemicals and total immunoglobulin (Ig) level. Another set of blood samples (4 fish/tank, 12 fish/diet) were taken using non-heparinized syringe and allowed to clot at room temperature for 30 min. The serum was separated by centrifugation for 10 min at $5000 \times g$ and stored at –70 °C for the analysis of nonspecific immune responses. Three intact fish per tank (3 fish/tank, 9 fish/diet) were randomly selected and kept at –70 °C for whole-body composition analysis. Hematocrit was determined in four individual fish per tank by a microhematocrit method (Brown 1980). Hemoglobin, glucose, total protein and total cholesterol were determined in the same four fish by using the automated blood analyzer (SLIM, SEAC Inc., Florence, Italy). Oxidative radical production by phagocytes during respiratory burst was measured through the NBT activity assay described by Anderson and Siwicki (1995). Serum myeloperoxidase (MPO) activity was measured by Quade and Roth (1997). Superoxide dismutase (SOD) activity was measured by percentage reaction inhibition rate of enzyme with water soluble tetrazolium dye substrate and xanthine oxidase using a SOD assay kit

(Sigma- aldrich, 19,160, St. Louis, USA) according to the manufacturer's instructions. Plasma total Ig level was determined by Siwicki and Anderson (1993). Serum lysozyme level was measured using turbidometric assay described by Hultmark et al. (1980). The serum anti-protease activity was measured by Ellis (1990) with slight modifications (Magnadottir et al. 1999). Moisture and ash content were analyzed by the standard procedures (AOAC 2005). Crude protein were measured by using automatic Kjeltec Analyzer Unit 2300 (FOSS, Hillerød, Sweden) and crude lipid was analyzed by Folch et al. (1957). Amino acid composition of the diets was determined using a Sycom S-433D automatic amino acid analyzer (Sykam, Eresing, Germany). Hydrolysis of the samples was performed in 6 N HCl at 110 °C for 24 h under nitrogen atmosphere. Identification and quantifi-cation of each amino acid were achieved by comparing the retention times of the peaks with those of standards.

Expression levels of liver IGF-1 mRNA

Liver samples were taken from two fish per tank and total RNA was isolated using trizol reagent following the manufacturers' protocol. The quantity of the RNA was calculated using the absorbance at 260 nm. The integrity and relative quantity of RNA were checked by gel elec-trophoresis. PrimeScript RT reagent Kit with gDNA Eraser (Perfect Real Time) (TaKaRa Code DRR047) was used to remove genomic DNA and reverse transcription. Levels of IGF-1 transcript were measured by real-time PCR (SYBR Green I), using 18S rRNA as a housekeeping gene. Primers for real-time PCR were designed based on the previously cloned sequence for IGF-1 (NCBI Gen-bank accession no: AF061278) and 18S rRNA (NCBI Genbank accession no: EF126037) in *P. olivaceus*. Rela-tive expression ratio of IGF-1 was calculated based on the PCR efficiency (E) and the Ct of a sample versus the control (FM treatment) and expressed in comparison with the reference gene (18S rRNA); according to Pfaffl's mathematical model (Pfaffl 2001).

$$\text{Ratio} = \left[\left(E_{\text{IGF-1}}\right)^{\Delta GF(\text{control-sample})}\right] / \left[\left(E_{\text{actin}}\right)^{Ct(\text{control-sample})}\right]$$

Statistical analysis

The diets were assigned by a completely randomized design. Data were analyzed by one-way analysis of vari-ance (ANOVA) in SPSS version 11.0 (SPSS Inc., Chicago, IL, USA). When ANOVA identifies differences among groups, the difference in means was made with Tukey's HSD multiple range test. Statistical significance was deter-mined at $P < 0.05$ for means of treatments. Percentage data were arcsine transformed before statistical analysis.

Results

Growth performance and feed utilization of juvenile olive flounders are shown in Table 2. All the taurine-supplemented groups showed numerically increased growth compared to the control even though it was not significant. The highest growth rate was found in fish fed T3 diet. However, feed utilization was significantly influenced by the taurine supplementation reducing FCR and increasing PER. FCR was significantly lower in fish fed all the taurine-supplemented diets than that of fish fed the control diet. Higher supplementation (T3 and T4) of taurine exhibited even significantly lower FCR than that of fish fed T1 and T2 diets. PER was signifi-cantly higher in fish fed T2, T3, and T4 diets compared to that of fish fed the control and T1 diets.

Hematological responses of fish fed the diets are shown in Table 3. Hematocrit and hemoglobin levels were significantly elevated by taurine supplementation. T3 group had the highest hematocrit and hemoglobin levels among all the groups. Total cholesterol level was significantly decreased in T4 group compared to the control group.

Results of non-specific immune response of fish fed the diets are shown in Table 4. All the immune parameters showed promising results in fish fed taurine-supplemented diets. All the taurine supplementation groups showed significantly higher NBT, MPO, Ig, and antiprotease activ-ities compared to the control group. In SOD and lysozyme activities, significantly higher values were observed in taurine supplementation over 0.5% in diets except for T4 group. The highest NBT and lysozyme activities were shown in T3 group while the highest MPO, SOD, and anti-protease activities were in T2 and T3 groups. The highest Ig level was observed in T2 group.

Results of proximate compositions in whole-body sam-ples are shown in Table 5. No significant effects were observed by the supplementation of taurine. Relative ex-pression levels of IGF-1 mRNA is shown in Fig. 1. There was not much of an influence on expression, however, T4 group showed a significantly higher peak compared to other groups including the control group.

Discussion

Increased feed intake is a primary response of fish to in-creased water temperature (Bureau et al. 2002). Improved feed intake at high water temperature generally results in higher growth and feed efficiency (Kim et al. 2006). According to this phenomenon, reduced growth of fish in cold water is due to reduced feed intake. It was reported that fish fed an extruded feed at suboptimal water temper-atures had relatively lower growth performance than that of fish fed raw fish or Oregon moist pellet (Satoh et al. 2003). This phenomenon is called winter syndrome. To solve the problem of winter syndrome, it is important to

Table 2 Growth performance and feed utilization of olive flounder (*P. olivaceus*, initial BW: 19.5 g) fed the five experimental diets for 10 weeks. The diets were added with graded levels of taurine by 0, 0.25, 0.5, 1.0, and 1.5% (control, T1, T2, T3, and T4, respectively)

| | Experimental diets | | | | |
	Control	T1	T2	T3	T4
FBW[1]	38.1 ± 0.66	41.7 ± 3.70	39.8 ± 7.89	46.4 ± 3.45	44.6 ± 5.10
WG[2]	96.5 ± 3.39	113 ± 18.9	103 ± 40.3	137 ± 17.6	128 ± 26.0
SGR[3]	1.01 ± 0.03	1.13 ± 0.13	1.04 ± 0.30	1.29 ± 0.11	1.22 ± 0.17
FI[4]	30.6 ± 1.26	35.1 ± 5.85	30.8 ± 12.13	33.3 ± 4.04	30.6 ± 6.12
FCR[5]	$1.63 ± 0.02^a$	$1.58 ± 0.00^b$	$1.52 ± 0.01^c$	$1.24 ± 0.01^d$	$1.22 ± 0.03^d$
PER[6]	$1.32 ± 0.01^c$	$1.37 ± 0.00^c$	$1.42 ± 0.01^b$	$1.74 ± 0.01^a$	$1.77 ± 0.04^a$
Survival (%)	60.0 ± 5.71	57.2 ± 14.3	63.8 ± 13.2	66.7 ± 5.95	57.5 ± 15.9

Values are mean of triplicate groups and presented as mean ± S.D. Values with different superscripts in the same row are significantly different ($P < 0.05$). The *lack of superscript letter* indicates *no significant* differences among treatments
[1]Final body weight (g)
[2]Weight gain (%) = 100 × (final mean body weight – initial mean body weight)/initial mean body weight
[3]Specific growth rate (%/day) = [(ln final body weight – ln initial body weight)/days] × 100
[4]Feed intake (g fish^{-1}) = dry feed consumed/fish number
[5]Feed conversion ratio = dry feed fed/wet weight gain
[6]Protein efficiency ratio = wet weight gain/total protein given

improve feed intake or feed utilization to obtain better growth performance at low water temperature. Therefore, the present study was conducted to determine whether taurine supplement could improve feed intake of olive flounder at low water temperature.

It has been reported that taurine supplementation can improve growth performance of many fish species including olive flounder, parrot fish (*Oplegnathus fasciatus*), red seabream (*Pagrus major*) and yellow tail (*Seriola quinqueradiata*) (Kim et al. 2007; Matsunari et al. 2008; Takagi et al. 2008; Lim et al. 2013; Takagi et al. 2013; Hanini et al. 2013; Han et al. 2014; Khaoian et al. 2014). Kim et al. (2007) reported that the taurine requirement of olive flounder is 1.0% for juvenile and fingerling stages in optimal water temperature. Han et al. (2014) reported that taurine and glutamine supplementations can significantly improve growth, feed intake and FCR of olive flounder at optimal water temperature. However, studies on the effect of taurine supplementation at suboptimal water temperature

are very limited. Water temperature varies depending on seasons for many regions. Fish shows normal behavior and best growth performance only in optimal water temperature because fish is a poikilotherm. In suboptimal water temperature, feed intake of fish usually decreases resulting in decreased growth performance. To solve the problem of winter syndrome, various methods using several feed stuffs such as enzyme treated fish meal, krill extract, or krill meal have been evaluated (Satoh 2003; Satoh et al. 2003). Satoh et al. (2003) have reported that daily growth rate and apparent protein digestibility of fish-fed diet with protease are worse than those fed with MP at low water temperature. However, there is no significant difference in feed efficiency between fish fed with diet treated with protease and those fed with MP diets (dry matter basis).

In the present study, growth of juvenile olive flounder was numerically increased by taurine supplementation resulting in improved FCR and PER, because feed intake was not different among all the groups. Therefore, feed

Table 3 Hematological parameters of olive flounder (*P. olivaceus*) fed the five experimental diets for 10 weeks. The diets were added with graded levels of taurine by 0, 0.25, 0.5, 1.0, and 1.5% (Control, T1, T2, T3, and T4, respectively)

| | Experimental diets | | | | |
	Control	T1	T2	T3	T4
Ht[1]	$15.8 ± 1.58^c$	$20.1 ± 0.56^b$	$20.9 ± 2.47^{ab}$	$24.2 ± 0.51^a$	$22.8 ± 0.95^{ab}$
Hb[2]	$2.81 ± 0.17^d$	$3.54 ± 0.17^{bc}$	$3.00 ± 0.28^{cd}$	$4.66 ± 0.47^a$	$3.81 ± 0.12^b$
Glucose[3]	53.6 ± 5.66	48.9 ± 2.87	47.1 ± 1.81	56.3 ± 6.69	51.1 ± 2.95
Total protein[4]	3.86 ± 0.26	3.86 ± 0.04	4.16 ± 0.26	4.16 ± 0.19	3.97 ± 0.25
Total cholesterol[5]	$82.0 ± 2.53^a$	$68.3 ± 10.3^{ab}$	$71.6 ± 5.76^{ab}$	$67.7 ± 3.04^{ab}$	$67.0 ± 0.76^b$

Values are mean of triplicate groups and presented as mean ± S.D. Values with different superscripts in the same row are significantly different ($P < 0.05$). The *lack of superscript letter* indicates *no significant* differences among treatments
[1]Hematocrit (%)
[2]Hemoglobin (g dL^{-1})
[3]Glucose (mg dL^{-1})
[4]Total protein (g dL^{-1})
[5]Total cholesterol (mg dL^{-1})

Table 4 Non-specific immune response of olive flounder (*P. olivaceus*) fed the five experimental diets for 10 weeks. The diets were added with graded levels of taurine by 0, 0.25, 0.5, 1.0, and 1.5% (control, T1, T2, T3, and T4, respectively)

| | Experimental diets | | | | |
	Control	T1	T2	T3	T4
NBT[1]	0.41 ± 0.01^c	0.51 ± 0.01^b	0.53 ± 0.02^b	0.58 ± 0.02^a	0.52 ± 0.02^b
MPO[2]	1.39 ± 0.08^c	1.80 ± 0.09^b	2.01 ± 0.08^a	2.05 ± 0.06^a	1.64 ± 0.04^b
SOD[3]	70.6 ± 2.50^b	75.0 ± 2.56^{ab}	77.3 ± 2.01^a	76.5 ± 1.55^a	71.7 ± 1.90^{ab}
Ig[4]	20.4 ± 0.35^d	58.4 ± 1.10^c	78.9 ± 1.38^a	74.9 ± 1.83^b	59.4 ± 0.15^c
Lysozyme[5]	15.3 ± 0.70^c	16.9 ± 0.11^c	22.4 ± 1.76^b	27.1 ± 0.79^a	21.8 ± 0.92^b
Antiprotease[6]	13.8 ± 0.63^c	20.6 ± 0.61^b	24.1 ± 0.86^a	23.0 ± 0.81^a	22.6 ± 0.95^{ab}

Values are mean of triplicate groups and presented as mean ± S.D. Values with different superscripts in the same row are significantly different ($P < 0.05$). The lack of superscript letter indicates no significant differences among treatments
[1]Nitro blue tetrazolium activity
[2]Myeloperoxidase level
[3]Superoxide dismutase (% inhibition)
[4]Total immunoglobulin (mg mL^{-1})
[5]Lysozyme activity (µg mL^{-1})
[6]Antiprotease (% inhibition)

efficiency was positively influenced by taurine supplementation. Chatzifotis et al. (2008) reported that in a cold water condition, similar tendency to the current results was observed in common dentex (*Dentex dentex*)-fed diet with taurine supplementation. Cook et al. (2003) investigated on winter syndrome of snapper (*Lutjanus campechanus*) and found similar results by dietary supplementation of β-glucan. Several studies in different fish species have reported similar results. Park et al. (2002) and Kim et al. (2007) reported that dietary taurine can significantly enhance the growth performance of juvenile Japanese flounder. The observations are similar to the results of juvenile yellowtail (Matsunari et al. 2005) and turbot (*Scophthalmus maximus*) (Qi et al. 2012). Growth rate and feed efficiency have been improved with taurine supplementation in European sea bass (*Dicentrarchus labrax*) (Brotons Martinez et al. 2004), rainbow trout (Gaylord et al. 2006), and black tiger shrimp (*Penaeus monodon*) (Shiau and Chou 1994).

Hematological parameters are also improved by the taurine supplementation. Blood parameters indicate the healthiness of fishes (Lemaire et al. 1991; Kader et al. 2010). Results of the present study showed that hematocrit and hemoglobin levels were significantly

increased by taurine supplement. Similar results have been reported by Han et al. (2014) for olive flounder. High levels of cholesterol often indicate physiological disorders of fish (Eslamloo et al. 2012). In the present study, plasma cholesterol levels were high in fish fed the control diet. However, they were significantly reduced by taurine supplement. This might indicate that the fish fed the control diet were relatively under stresses. However, supplemental taurine feeding helped the fish recover at some degree, although such effect was not shown in survival rate. Non-specific immune response was the other major aspect we considered. Since the fish were in high stress conditions or challenging environment like in the suboptimal water temperature, if there was any effect of the dietary supplementation it could appear. Results of this study revealed that all the innate immune parameters of olive flounder could apparently be enhanced by dietary taurine supplement especially in a low water temperature condition. Fish in all groups fed taurine showed significantly higher innate immune responses than fish in the control group. Many studies have determined if taurine has positive effect on immunity of fish. Fang et al. (2002) reported that taurine can directly scavenge free radicals. Higuchi et al. (2012) also showed

Table 5 Whole-body composition of olive flounder (*P. olivaceus*) fed the five experimental diets for 10 weeks (%, dry matter). The diets were added with graded levels of taurine by 0, 0.25, 0.5, 1.0, and 1.5% (control, T1, T2, T3, and T4, respectively)

| | Experimental diets | | | | |
	Control	T1	T2	T3	T4
Dry matter	75.4 ± 0.53	76.1 ± 0.45	75.4 ± 0.68	75.5 ± 0.78	75.9 ± 0.20
Protein	66.4 ± 0.74	69.2 ± 1.80	67.7 ± 1.55	67.3 ± 0.85	67.9 ± 0.45
Lipid	25.7 ± 1.60	23.9 ± 1.92	22.8 ± 2.56	27.7 ± 1.58	25.3 ± 1.82
Ash	14.2 ± 0.06	11.8 ± 2.85	10.2 ± 2.15	12.5 ± 0.83	10.0 ± 2.59

Values are mean of triplicate groups and presented as mean ± S.D. Values with different superscripts in the same row are significantly different ($P < 0.05$). The lack of superscript letter indicates no significant differences among treatments

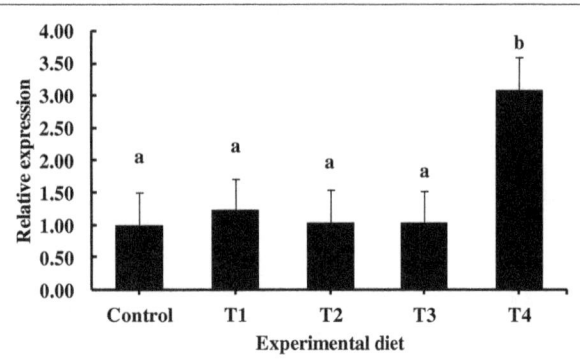

Fig. 1 Relative expression of IGF-1 in livers of olive flounder fed the five experimental diets for 10 weeks. The diets were added with graded levels of taurine by 0, 0.25, 0.5, 1.0, and 1.5% (control, T1, T2, T3, and T4, respectively). Values are mean of triplicate groups and presented as mean ± S.D. Values in the same row having different superscript letters are significantly different (*P* < 0.05)

elevated SOD levels in eels fed a taurine supplemented diet. Similarly, Han et al. (2014) reported higher SOD levels in olive flounder fed diets supplemented with taurine. Interestingly, the present study also showed that SOD and MPO levels were significantly increased in fish fed diets with taurine supplementation. Lysozyme is an important defense enzyme of innate immune system of fishes (Saurabh and Sahoo 2008) and has been used as a key parameter to evaluate non-specific defense ability (Zhou et al. 2006; Ren et al. 2007). Anti-protease is an enzyme which disables enzymes generated by a pathogen inside an organism (Magnadottir et al. 1999). Therefore, an increased level of anti-protease indicates an increase in immunity. In the present study, taurine-fed groups showed significant increment in both lysozyme and anti-protease activities. Similarly, Li et al. (2016) observed significant increase in lysozyme activity of yellow catfish (*Pelteobagrus fulvidraco*). Therefore, these results indicate that taurine not only can improve health and immunity of mammals (Kingston et al. 2004; Gupta et al. 2006), but also can improve the health and innate immunity of fish. IGF-1 mRNA expression was assessed in this study as a new parameter to determine whether taurine supplementation might have effect on gene expression. Interestingly, our results showed that IGF-1 mRNA expression level was significantly increased by 1.5% taurine supplementation. Some studies have been conducted to determine whether taurine supplementation could affect fish at gene expression level (Kingston et al. 2004; Gupta et al. 2006). The function of IGF-1 in fish is similar to that in human or mammals, and IGF-1 can only exert nutritional regulation in the liver of fish (Duan et al. 1994; Duan 1998). Therefore, only liver tissue was analyzed in this study for IGF-1 expression. Niu and Le Bail (1993) reported that IGF-1 level is significantly decreased in fasted rainbow trout. The elevated

peak of IGF-1 in the present study suggests that IGF-1 gene can be up-regulated by the taurine supplementation in diets for olive flounder. Optimum level of taurine supplementation in fish diets has already been elucidated in previous studies for several fish species, such as cobia (*Rachycentron canadum*) (Kousoulaki et al. 2009), Japanese flounder (Kim et al. 2003, 2007, 2008), red seabream (Takagi et al. 2010), and sea bass (Brotons Martinez et al. 2004). According to these studies, the optimum level of taurine supplementation depends on fish species and growth stage of fish. Salze and Davis (2015) have reported that different taurine supplementation levels are needed in different fish species with different growth stages.

Conclusions

This study suggests that taurine supplementation is effective in improving growth performance, feed efficiencies and innate immunity of juvenile olive flounder in low water temperature season. The suggested taurine supplementation level would be between 1.0 and 1.5% (analyzed levels; 0.9–1.3% in diets) in a fish meal-based diet for olive flounder in suboptimal water temperature condition.

Abbreviations

FBW: Final body weight; FCR: Feed conversion ratio; Hb: Hemoglobin; Ht: Hematocrit; Ig: Immunoglobulin; IGF-1: Insulin-like growth factor 1; MPO: Myeloperoxidase; NBT: Nitro blue tetrazolium; PER: Protein efficiency ratio; SGR: Specific growth rate; SOD: Superoxide dismutase; WG: Weight gain

Acknowledgements

This research was supported by the 2017 scientific promotion program funded by Jeju National University.

Funding

This study was funded by Jeju National University under the 2017 scientific promotion program.

Authors' contributions

JMK and GHTM conducted the feeding trial, analysis, and manuscript preparation. GLBEG, CL, and MGK participated in the analyses of samples and supporting the manuscript preparation. BJL and JDK read and revised the manuscript. KJL organized, designed, and completed the manuscript. All authors have read and approved the final manuscript.

Competing interests

The authors declare that they have no competing interests.

Author details

[1]Sajodongaone, 873, Deokpyeong Ro, Sunseong-myeon, Dangjin-Si, Chungcheongnam Do 31759, South Korea. [2]Department of Marine Life

Sciences, Jeju National University, Jeju 63243, South Korea. [3]Aquafeed Research Center, National Institute of Fisheries Science, Pohang 37517, South Korea. [4]College of Animal Life Science, Kangwon National University, Chuncheon 24341, South Korea.

References

Anderson DP, Siwicki AK. Basic hematology and serology for fish health programs. Manila, Philippines: Leetown Science Center; 1995. p. 185–202.

AOAC. Official Methods of Analysis. 18th edition. Association of Official Analytical Chemists: Arlington; 2005.

Brotons Martinez J, Chatzifotis S, Divanach P, Takeuchi T. Effect of dietary taurine supplementation on growth performance and feed selection of sea bass *Dicentrarchus labrax* fry fed with demand-feeders. Fish Sci. 2004;70:74–9.

Brown BA. Routine hematology procedures, in: Brown, B.A. (Ed.), Hematology, Principles and Procedures, Lea and Febiger. Philadelphia; 1980. p.71-112.

Bureau DP, Kaushik SJ, Cho CY. Bioenergetics. In: Fish Nutrition. 3rd ed. USA: Elsevier Science; 2002. p. 2–61.

Carr WES. Chemical stimulation of feeding behavior. In: Hara TJ, editor. Chemoreception in fishes. Amsterdam: Elsevier; 1982. p. 259–73.

Chatzifotis S, Polemitou I, Divanach P, Antonopoulou E. Effect of dietary taurine supplementation on growth performance and bile salt activated lipase activity of common dentex, *Dentex dentex*, fed a fish meal/soy protein concentrate-based diet. Aquaculture. 2008;275:201–8.

Cook MT, Hayball PJ, Hutchinson W, Nowak BF, Hayball JD. Administration of a commercial immunostimulant preparation, EcoActiva™ as a feed supplement enhances macrophage respiratory burst and the growth rate of snapper (*Pagrus auratus*, Sparidae (Bloch and Schneider)) in winter. Fish Shellfish Immunol. 2003;14:333–45.

De la Rosa J, Stipanuk MH. Evidence for the rate-limiting role of cysteinesulfinate decarboxylase activity in taurine biosynthesis *in vivo*. Comp Biochem Physiol. 1985;81:565–71.

Doving KB, Selset R, Thommesen G. Olfactory sensitivity to bile acids in salmonid fishes. Acta Physiol. 1980;108:123–31.

Duan C. Nutritional and developmental regulation of insulin-like growth factors in fish. J Nutr. 1998;128:3056–145.

Duan C, Duguay SJ, Swanson P, Dickhoff WW, Plisetskaya EM. Tissue-specific expression of insulin-like growth factor I messenger ribonucleic acids in salmonids: developmental, hormonal, and nutritional regulation. Perspect Comp Endocrinol. 1994:365–72.

Ellis AE. Serum antiproteases in fish. In: Stolen JS, Fletcher TC, Anderson DP, Roberson BS, editors. Techniques in fish immunology. NJ: SOS Publications; 1990. p. 95–9.

Eslamloo K, Falahatkar B, Yokoyama S. Effects of dietary bovine lactoferrin on growth, physiological performance, iron metabolism and non-specific immune responses of Siberian sturgeon *Acipenser baeri*. Fish Shellfish Immunol. 2012;32:976–85.

Fang YZ, Yang S, Wu GY. Free radicals, antioxidants, and nutrition. Nutrition. 2002; 18:872–9.

Folch J, Lees M, Sloane Stanley GH. A simple method for the isolation and purification of total lipids from animal tissues. J Biol Chem. 1957;226:497–509.

Gaylord TG, Teague AM, Barrows FT. Taurine supplementation of all-plant protein diets for rainbow trout (*Oncorhynchus mykiss*). J World Aquac Soc. 2006;37: 509–17.

Gupta RC, Seki Y, Yosida J. Role of taurine in spinal cord injury. Curr Neurovasc Res. 2006;3:225–35.

Han Y, Koshio S, Jiang Z, Ren T, Ishikawa M, Yokoyama S, Gao J. Interactive effects of dietary taurine and glutamine on growth performance, blood parameters and oxidative status of Japanese flounder *Paralichthys olivaceus*. Aquaculture. 2014;434:348–54.

Hanini I, Sarker MSA, Satoh S, Haga Y, Corneillie S, Ohkuma T, Nakayama H. Effects of Taurine, Phytase and Enzyme Complex Supplementation to Low Fish Meal Diets on Growth of Juvenile Red Sea Bream *Pagrus major*. Aquaculture Sci. 2013;61:367–75.

Hara TJ, Macdonald S, Evans RE, Marui T, Arai S. In: JD MC, Arnold GP, Dodson JJ, Neill WH, editors. Morpholine, bile acids and skin mucus as possible chemical cues in salmonid homing: electrophysiological re-evaluation. New York: Mechanisms of Migration in Fishes; 1984. p. 363–78.

Higuchi M, Celino FT, Shimizu-Yamaguchi S, Miura C, Miura T. Taurine plays an important role in the protection of spermatogonia from oxidative stress. Amino Acids. 2012;43:2359–69.

Hultmark D, Steiner H, Rasmuson T, Boman HG. Insect immunity. Purification and properties of three inducible bactericidal proteins from hemolymph of immunized pupae of *Hyalophora cecropia*. FEBS J. 1980;106:7–16.

Iwata N, Kikuchi K, Honda H, Kiyono M, Kurokura H. Effects of temperature on the growth of Japanese flounder. Fish Sci. 1994;60:527–31.

Kader MA, Koshio S, Ishikawa M, Yokoyama S, Bulbul M. Supplemental effects of some crude ingredients in improving nutritive values of low fishmeal diets for red sea bream, *Pagrus major*. Aquaculture. 2010;308:136–44.

Khaoian P, Ogita H, Watanabe H, Nishioka M, Kanosue F, Nguyen HP, Fukada H, Masumoto T. Effects of Taurine Supplementation to Low Fish Meal Practical Diet on Growth, Tissue Taurine Content and Taste of 1 Year Yellowtail *Seriola quinqueradiata*. Aquaculture Sci. 2014;62:415–23.

Kim KD, Kim KM, Kim KW, Kang YJ, Lee SM. Influence of lipid level and supplemental lecithin in diet on growth, feed utilization and body composition of juvenile flounder (*Paralichthys olivaceus*) in suboptimal water temperatures. Aquaculture. 2006;251:484–90.

Kim SK, Matsunari H, Nomura K, Tanaka H, Yokoyama M, Murata Y, Ishihara K, Takeuchi T. Effect of dietary taurine and lipid contents on conjugated bile acid composition and growth performance of juvenile Japanese flounder *Paralichthys olivaceus*. Fish Sci. 2008;74:875–81.

Kim SK, Matsunari H, Takeuchi T, Yokoyama M, Murata Y, Ishihara K. Effect of different dietary taurine levels on the conjugated bile acid composition and growth performance of juvenile and fingerling Japanese flounder *Paralichthys olivaceus*. Aquaculture. 2007;273:595–601.

Kim SK, Takeuchi T, Yokoyama M, Murata Y. Effect of dietary supplementation with taurine, β-alanine and GABA on the growth of juvenile and fingerling Japanese flounder *Paralichthys olivaceus*. Fish Sci. 2003;69:242–8.

Kim SK, Takeuchi T, Yokoyama M, Murata Y, Kaneniwa M, Sakakura Y. Effect of dietary taurine levels on growth and feeding behavior of juvenile Japanese flounder *Paralichthys olivaceus*. Aquaculture. 2005;250:765–74.

Kingston R, Kelly CJ, Murray P. The therapeutic role of taurine in ischaemia-reperfusion injury. Curr Pharm Des. 2004;10:2401–10.

Kousoulaki K, Albrektsen S, Langmyhr E, Olsen HJ, Campbell P, Aksnes A. The water soluble fraction in fish meal (stickwater) stimulates growth in Atlantic salmon (*Salmo salar* L.) given high plant protein diets. Aquaculture. 2009;289: 74–83.

Lee SM, Cho SH, Kim KD. Effects of dietary protein and energy levels on growth and body composition of juvenile flounder *Paralichthys olivaceus*. J World Aquac Soc. 2000;31:306–15.

Lee SM, Park CS, Bang IC. Dietary protein requirement of young Japanese flounder *Paralichthys olivaceus* fed isocaloric diets. Fish Sci. 2002;68:158–64.

Lemaire P, Drai P, Mathieu A, Lemaire S, Carriere S, Giudicelli J, Lafaurie M. Changes with different diets in plasma enzymes (GOT, GPT, LDH, ALP) and plasma lipids (cholesterol, triglycerides) of sea-bass (*Dicentrarchus labrax*). Aquaculture. 1991;93:63–75.

Li M, Lai H, Li Q, Gong S, Wang R. Effects of dietary taurine on growth, immunity and hyperammonemia in juvenile yellow catfish *Pelteobagrus fulvidraco* fed all-plant protein diets. Aquaculture. 2016;450:349–55.

Lim SJ, Oh DH, Khosravi S, Cha JH, Park SH, Kim KW, Lee KJ. Taurine is an essential nutrient for juvenile parrot fish *Oplegnathus fasciatus*. Aquaculture, 414–415. 2013:274–9.

Magnadottir B, Jonsdottir H, Helgason S, Bjornsson B, Jon T, Pilstroil L. Humoral immune parameters in Atlantic cod (*Gadus morhua* L): I. the effects of environmental temperature. Comp Biochem Physiol. 1999;122:173–80.

Matsunari H, Furuita H, Yamamoto T, Kim SK, Sakakura Y, Takeuchi T. Effect of dietary taurine and cystine on growth performance of juvenile red sea bream *Pagrus major*. Aquaculture. 2008;274:142–7.

Matsunari H, Takeuchi T, Takahashi M, Mushiake K. Effect of dietary taurine supplementation on growth performance of yellowtail juveniles *Seriola quinqueradiata*. Fish Sci. 2005;71:1131–5.

Ministry of Maritime Affairs and Fisheries. Aquaculture Statistic from Ministry of Maritime Affairs and Fisheries of Korea. 2015.

Niu PD, Le Bail PY. Presence of insulin-like growth factor binding protein (IGF-BP) in rainbow trout (*Oncorhynchus mykiss*) serum. J Exp Zool. 1993;265:627–36.

Oja SS, Kontro P. Taurine. In: Metabolism in the Nervous System. US: Springer; 1983. p. 501–33.

Olsen RE, Ringn E. The influence of temperature on the apparent nutrient and fatty acid digestibility of Arctic charr, *Salvelinus alpinus* L. Aquac Res. 1998;29:695–701.

Park GS, Takeuchi T, Yokoyama M, Seikal T. Optimal dietary taurine level for growth of juvenile Japanese flounder *Paralichthys olivaceus*. Fish Sci. 2002;68:824–9.

Peres H, Oliva-Teles A. Influence of temperature on protein utilization in juvenile European sea bass (*Dicentrarchus labrax*). Aquaculture. 1999;170:337–48.

Petrosian AM, Haroutounian JE. Taurine as a universal carrier of lipid soluble vitamins: a hypothesis. Amino Acids. 2000;19:409–21.

Pfaffl MW. A new mathematical model for relative quantification in real-time RT-PCR. Nucleic Acids Res. 2001;29:e45.

Qi G, Ai Q, Mai K, Xu W, Liufu Z, Yun B, Zhou H. Effects of dietary taurine supplementation to a casein-based diet on growth performance and taurine distribution in two sizes of juvenile turbot (*Scophthalmus maximus* L.). Aquaculture. 2012;358:122–8.

Quade MJ, Roth JA. A rapid, direct assay to measure degranulation of bovine neutrophil primary granules. Vet Immunol Immunop. 1997;58:239–48.

Ren T, Koshio S, Ishikawa M, Yokoyama S, Micheal FR, Uyan O, Tung HT. Influence of dietary vitamin C and bovine lactoferrin on blood chemistry and nonspecific immune responses of Japanese eel, *Anguilla japonica*. Aquaculture. 2007;267:31-37.

Salze GP, Davis DA. Taurine: a critical nutrient for future fish feeds. Aquaculture. 2015;437:215–29.

Satoh KI. Effects of supplement with krill extract and krill meal to diet on the growth performance and protein digestibility of yellowtail during low water temperature. Aquaculture Sci. 2003;51:93–9.

Satoh KI, Maita M, Wakatsuki A, Matsuda S. Growth and feed efficiency of adult yellowtail fed extruded pellet diets with two lipid levels and raw-fish diets. Aquaculture Sci. 2003;51:343–8.

Saurabh S, Sahoo PK. Lysozyme: an important defence molecule of fish innate immune system. Aquac Res. 2008;39:223–39.

Schaffer S, Takahashi K, Azuma J. Role of osmoregulation in the actions of taurine. Amino Acids. 2000;19:527–46.

Shiau SY, Chou BS. Grass shrimp, *Penaeus monodon*, growth as influenced by dietary taurine supplementation. Comp Biochem Phys A. 1994;108:137–42.

Siwicki AK, Anderson DP. Nonspecific defense mechanisms assay in fish: II. Potential killing activity of neutrophils and macrophages, lysozyme activity in serum and organs and total immunoglobulin level in serum. In: Disease Diagnosis and Prevention Methods. FAO-project GCP/INT/JPA, IFI Olsztyn, Poland;1993. p.105–112.

Takagi S, Murata H, Goto T, Endo M, Yamashita H, Ukawa M. Taurine is an essential nutrient for yellowtail *Seriola quinqueradiata* fed non-fish meal diets based on soy protein concentrate. Aquaculture. 2008;280:198–205.

Takagi S, Murata H, Goto T, Hatate H, Endo M, Yamashita H, Miyatake H, Ukawa M. Necessity of dietary taurine supplementation for preventing green liver symptom and improving growth performance in yearling red sea bream *Pagrus major* fed nonfishmeal diets based on soy protein concentrate. Fish Sci. 2010;76:119.

Takagi S, Murata H, Goto T, Hatate H, Yamashita H, Takano A, Ukawa M. Long-term Feeding of the Yellowtal *Seriola quinqueradiata* with Soy Protein Concentrate-based Non-fishmeal Diet Supplemented with Taurine. Aquaculture Sci. 2013;61:349–58.

Takeuchi T, Park GS, Seikai T, Yokoyama M. Taurine content in Japanese flounder *Paralichthys olivaceus* and red sea bream *Pagrus major* T. & S. during the period of seed production. Aquac Res. 2001;32:244–8.

Van Waarde A. Biochemistry of non-protein nitrogeneous compounds in fish including the use of amino acids for anaerobic energy production. Comp Biochem Physiol. 1988;91:207–28.

Wright CE, Talan HH, Lin YY, Coaull GE. Taurine: biological update. Annu Rev Biochem. 1986;55:427–53.

Yokoyama M, Takeuchi T, Park GS, Nakazoe J. Hepatic cysteinesulphinate decarboxylase activity in fish. Aquac Res. 2001;32:216–20.

Zhou J, Song XL, Huang J, Wang XH. Effects of dietary supplementation of A3α-peptidoglycan on innate immune responses and defense activity of Japanese flounder (*Paralichthys olivaceus*). Aquaculture. 2006;251:172–81.

Evaluation of visible fluorescent elastomer tags implanted in marine medaka, *Oryzias dancena*

Jae Hyun Im[1], Hyun Woo Gil[2], In-Seok Park[2]*⬥, Cheol Young Choi[2], Tae Ho Lee[2], Kwang Yeol Yoo[3], Chi Hong Kim[4] and Bong Seok Kim[4]

Abstract: The aim of this study was to assess visible implant fluorescent elastomer (VIE) tagging and stress response in marine medaka, *Oryzias dancena*. The experimental fish were anesthetized individually and marked with red, yellow, or green elastomer at each of the following three body locations: (1) the abdomen, (2) the back, and (3) the caudal vasculature. During 12 months, the accumulated survival rates of fish in the experimental treatments were not different among red, yellow, and green elastomers. The experimental fish retained > 85% of the tags injected in the back, > 70% of the tags injected in the caudal vasculature, and > 60% of the tags injected in the abdomen ($P < 0.05$). An important observation was that the abdomen site was associated with poor tag retention. For all injected sites, the red and green tags were able to be detected more easily than the yellow tags when observed under both visible and UV lights. Tag readability was lower for the abdomen site than for the other sites (back and caudal vasculature). Thus, VIE tags were easy to apply to marine medaka (< 1 min per fish) and were readily visible when viewed under UV light.

Keywords: Marine medaka, *Oryzias dancena*, Readability, Visible fluorescent elastomer tag

Background

The marine medaka, *Oryzias dancena*, is nonindigenous to South Korea and is a bony fish with high tolerance to salinity because of its salinity adaptation mechanisms (Inoue and Takei, 2003). In addition to the studies of this euryhaline species, under various salinity conditions, it has been the subject of extensive ecotoxicogenomic research; this should extend the use of the marine medaka as a laboratory model for assessing its responses to salinity changes. Its viability under conditions of maximum tolerable salinity has been measured, and incubation time of fry was assessed by its ability to adapt to various salinity (Cho et al., 2010). This species was recently selected by *i*MLMO (Institute of Marine Living Modified Organisms, Pukyong National University, Busan 608-737, Korea) for use in a project to evaluate living modified organisms. Consistent with this purpose, detailed information on its biology is being collected (Song et al., 2009; Nam et al., 2010), particularly related to its early

gonadogenesis, sexual differentiation, early ontogenesis, embryogenesis, and exceptional capacity for hyperosmoregulation and hypoosmoregulation. In addition, Kim et al. (Song et al., 2009) suggested that this species has a short interval between generations with spawning possible only for 60 days after hatching. A study of the effects of clove oil and lidocaine HCl on the species by Park et al. (2011) has contributed to the safe laboratory handling of this fish, which is required in many studies. The research discussed above has demonstrated that the marine medaka has the ideal characteristics for an experimental animal (Song et al., 2009; Nam et al., 2010; Park et al., 2011).

Identification of individuals is essential in studies of fish growth, migration, and mortality and in stock identification and stock selectivity for tracing particular fish populations (Crossland, 1980). Although short-term tag retention may suffice for some experiments, the effect of a tag on fish survival, behavior, growth, and recognition and the costs of the tagging technique need to be considered. However, traditional external tags (such as spaghetti or dart tags) are commonly lost soon after deployment (Crossland, 1980; Bergman et al., 1992) and can affect growth or survival

---

* Correspondence: ispark@kmou.ac.kr
[2]Division of Marine Bioscience, College of Ocean Science and Technology, Korea Maritime and Ocean University, 727 Taejong-ro, Yeong do-gu, Busan 49112, South Korea
Full list of author information is available at the end of the article

(Crossland, 1976; Tong, 1978; McFarlane and Beamish, 1990; Serafy et al., 1995). Furthermore, these types of tags can only be read by recapturing the fish.

Devices that are located internally but are readable externally, such as acoustic tags, are often limited by short battery life or retention (Ralston and Horn, 1986), and sample sizes are limited by the expense involved. Problems associated with biological compatibility, reliability of identification, fouling of the tag by algae (Jones, 1987; Barrett, 1995), tag retention (Crossland, 1976; Parker, 1990), and external visibility of such devices have reduced confidence in the interpretation of results of in situ studies of reef fish ecology. A less frequently used approach is intrinsic identification, whereby cohorts are identified by size (Jones, 1987; Forrester, 1990) and individuals are recognized by variation in natural markings (Thompson and Jones, 1980; Connell and Jones, 1991) or wound scarring.

The latter approach has cost advantages, so does not influence behavior, but is subject to potentially substantial levels of observer error. Furthermore, many fish species lack unique natural markings and cannot be recognized without an artificial means of verifying identity. Passive integrated tag (PIT) method is the most commonly used. However, the body size of marine medaka is similar to that of PIT chips, so marine medaka is unsuitable for tagging PIT chips. The visible implant fluorescent elastomer (VIE) tag was developed primarily for tagging large batches of small or juvenile fish. The VIE system comprises a viscous liquid elastomer that sets to a pliable solid over a period of hours following application. The elastomer can be injected into transparent or translucent tissues to form a permanent biocompatible mark. When exposed to UV light and viewed through an amber filter, the compound fluoresces brightly. The tag size can easily be varied according to the requirements of the researcher and the size of the fish to be tagged. Thus far, the system has been used for the identification of groups or cohorts of juvenile reef fish (Frederick, 1997) and salmonids, but is also proving potentially effective in controlled laboratory studies of adult blue gills (Dewey and Zigler, 1996). As an externally visible but sub-dermally situated marking system, VIE tags are potentially able to eliminate many of the problems associated with other methods.

Tagging, weighing, measuring standard length, preparing fish for live shipment and transport, injecting vaccines and antibiotics, and collecting blood are causes to increasing stress (Dewey and Zigler, 1996). Stress responses can include physiological changes such as oxygen uptake and transfer, metabolic and hematological changes, mobilization of energy substrates, reallocation of energy away from growth and reproduction, and suppressive effects on immune functions (Schreck et al.,

2001; Redding and Schreck, 1983). These changes can increase disease susceptibility leading to increased mortality and subsequent economic losses (Schreck et al., 2001; Redding and Schreck, 1983). So, analysis of stress response can roughly examine the cause of mortality by tagging. The steroid hormone cortisol is widely accepted as an indicator of stress in fish, generally increasing after exposure to physical stressors (Schreck et al., 2001). Circulating cortisol levels are typically measured to determine the stress status of an individual fish (Redding and Schreck, 1983). Alternatively, whole-body cortisol levels have been used to assess the stress responses of the developing salmonids and flatfish because their blood volumes are insufficient to allow for the measurements of circulating cortisol (Redding and Schreck, 1983). Similarly, whole-body corticosteroids have been measured in smaller adult fish, including the three-spined stickleback, *Gasterosteus aculeatus* (Pottinger et al., 2002), and the zebrafish, *Danio rerio* (Pottinger and Calder, 1995). In this study, we compared various tag colors for readability under visible and UV lights and assessed the likelihood of tag- or handling-related mortality, the retention rates of VIE tags placed, and the stress response in various body sites of marine medaka. So, suitability of VIE tag method in marine medaka was investigated by analysis of readability, mortality, and stress response.

Methods

The fish used in this experiment were adult marine medaka, *O. dancena* (mean body length ± SD 35.1 ± 3.42 mm; mean body weight ± SD 54.4 ± 1.83 mg; age 10 months after hatched). Injection of the VIE tags into the treatment fish, and handling of the control fish, occurred on 16 February 2012. Following to the method of Park et al. (2011), all fish were anesthetized in 800 ppm lidocaine hydrochloride/NaHCO$_3$ at a water temperature of 10 °C. The fish were sedated until they were completely immobile and then individually removed from the anesthetic solution, rinsed in fresh water, and placed on a flat surface for tagging.

Per group of 50, the fish were individually tagged with yellow, red, or green elastomer (Northwest Marine Technology Inc., Shaw Island, Washington, USA) at three body locations (Figs. 1 and 2a): (1) the surface of the abdomen, (2) the inside surface of the back, and (3) the surface of the caudal vasculature, and all experimental groups were triplicated. Control fish (50) were anesthetized but not marked. We used the VIE hand injection Master Kit (Northwest Marine Technology Inc., Shaw Island, Washington, USA) for tagging the fish. Following the kit protocol, the elastomer and curing agent were mixed at a ratio of 10:1 and the prepared elastomer was injected as a liquid (0.3 mL per site). The

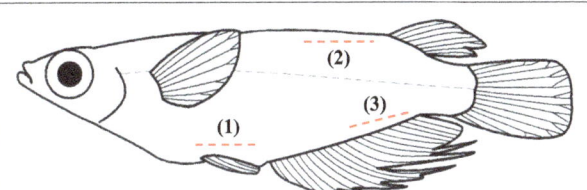

Fig. 1 Elastomer injection locations (red dotted lines) on marine medaka, *Oryzias dancena*: (1) the surface of the abdomen; (2) the inside surface of the back; and (3) the surface of the caudal vasculature

instruments used are shown in Fig. 2b. Tagged fish were divided into different tag colors and injection sites in tanks. The fish were held in 18 flow-through fiber-reinforced plastic tanks (50 × 20 × 20 cm; water volume 2 L) supplied with filtered seawater. The bottom of each tank was fitted with a black sheet to facilitate for obser-vation of the tag. The flow rate was 2 L/min/tank, and the mean water temperature was 26 ± 2.5 °C. A common day–night cycle was established, and all tanks were covered with netting to retain the fish in the tanks. Throughout the 12-month trial, the fish were fed daily to satiation using a dry commercial flounder feed (Agri-brand Furina Korea Co., Korea) that was alternated with a formulated *Artemia* diet. The food was placed on the aquarium floor so that it could be eaten within 2 h.

The survival rate, tag retention, and detection of tags (under visible and UV lights) were recorded at 2-month

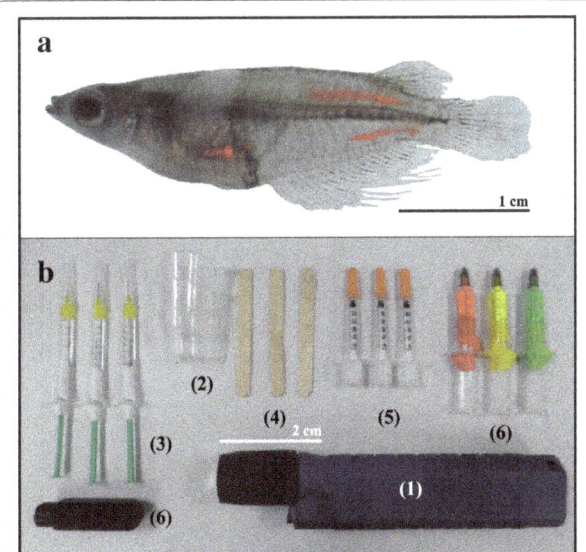

Fig. 2 External morphology of marine medaka, *Oryzias dancena*, tagged with visible implant fluorescent elastomer (VIE), showing the yellow tags at each tagging site under UV light (**a**) and (**b**) the VIE kit (Northwest Marine Technology, Shaw Island, Washington) including the UV lamp (1), mixing beaker (2), silicon (3), mixing stack (4), colored elastomer in a syringe (5), and the injection syringe (6)

intervals. The tanks were checked daily for dead fish, which if present were removed and fixed in 10% neutral formalin solution. The marking with the various colors at the three sites was observed visually from a distance of 30 cm under ambient visible light and UV light and measured the differentiation rate of each group after 1 month. Tag retention data for the dead fish were used to calculate the percent tag retention up to the date that the fish died, but were not used in the subsequent calcu-lations (Zerrenner et al., 1997).

To observe the effects of stress on the whole-body cortisol, glucose, and lactic acid levels of fish under VIE tag injection, we injected the VIE tag in the abdomen, the back, and the caudal vasculature, respectively, and 90 samples were used in each site. The stress responses of the experimental fish were measured at 0, 1, 6, 12, 24, and 48 h. Fifteen samples were used in each measured time. Control fish were not injected VIE tag, but their cortisol levels were measured. For these measurements, 150 fish were used in each experimental group, and no distinction was made between male and female fish. We measured the whole-body cortisol, glucose, and lactic acid levels of the control fish before the experiment. In-dividual fish were blotted onto paper towels to remove excess water, immediately frozen in liquid nitrogen for 10–30 s, and placed in individual 5.0 mL plastic screw cap centrifuge tubes. The samples were stored at – 80 °C until we extracted the cortisol, glucose, and lactic acid. The term "whole-body cortisol" is used to describe the portion of corticosteroid extracted and measured with a cortisol-specific radioimmunoassay (Pottinger et al., 2002). Whole-body glucose concentration was analyzed according to the methodology of Raabo and Terkildsen (1960) (Kit 510, Sigma, St Louis, MO, USA), where the production of H_2O_2 by glucose oxidase in the presence of *o*-dianisidine was evaluated as an absorbance increase at 450 nm. The lactic acid concentrations were analyzed using blood automatic analysis (Boehringer Mannheim Reflotron, Germany).

All measured data were induced by triplicate experi-ments from all experimental samples. Differences in survival rate between control and experimental groups were assessed using the *t* test (Cody and Smith, 1991), and the tag retention rate (%) among tagging sites was assessed using a one-way ANOVA and Duncan's mul-tiple range test (Duncan, 1955). The differences were considered to be significant at a probability of 0.05.

Results

Table 1 shows the retention rate of the VIE tags at each site for marine medaka, *O. dancena*. During the experi-ment, there was no difference in tag retention among the various tag colors for the abdomen site, but the tag retention rate for this site was different from that of the

Table 1 Tagging rate using the naked eye and the UV lamp of visible implant fluorescent elastomer (VIE) tags in each site of marine medaka, *Oryzias dancena*, from 0 to 12 months after VIE tagging

Month	Color	Tagging rate (%)		
		Abdomen	Back	Caudal vasculature
1	Red	57.4 ± 3.26ª	99.3 ± 0.16ᵇ	99.0 ± 0.22ᵇ
	Green	77.6 ± 1.11ª	94.7 ± 0.85ᵇ	99.3 ± 0.10ᶜ
	Yellow	74.3 ± 1.48ª	94.9 ± 0.80ᵇ	95.4 ± 0.67ᵇ
3	Red	60.6 ± 2.32ª	93.5 ± 1.03ᵇ	99.4 ± 0.05ᶜ
	Green	76.4 ± 0.87ª	93.8 ± 0.58ᵇ	95.6 ± 0.57ᶜ
	Yellow	73.2 ± 1.75ª	94.9 ± 0.99ᵇ	92.9 ± 2.10ᵇ
6	Red	57.3 ± 3.74ª	92.0 ± 1.52ᵇ	96.5 ± 0.60ᶜ
	Green	80.7 ± 2.10ª	92.6 ± 0.83ᵇ	94.9 ± 0.81ᵇ
	Yellow	75.1 ± 1.68ª	95.4 ± 1.00ᵇ	92.2 ± 1.88ᵇ
9	Red	64.4 ± 3.19ª	93.4 ± 0.67ᵇ	97.5 ± 0.80ᶜ
	Green	79.0 ± 1.46ª	89.8 ± 1.79ᵇ	99.1 ± 0.09ᶜ
	Yellow	75.9 ± 2.00ª	95.5 ± 0.73ᵇ	94.4 ± 1.73ᵇ
12	Red	67.2 ± 3.05ª	90.5 ± 1.31ᵇ	96.6 ± 0.66ᶜ
	Green	79.5 ± 1.96ª	94.4 ± 0.74ᵇ	97.2 ± 0.38ᶜ
	Yellow	70.4 ± 1.21ª	94.6 ± 0.94ᵇ	95.6 ± 0.86ᵇ

Samples tagged with VIE were investigated. Values are mean ± S.E. ($n = 50$). The experiment was performed in triplicate. The values in each column that do not share a common superscript are significantly different from one another ($P < 0.05$)

Table 2 Differentiation rate using the naked eye of visible implant fluorescent elastomer (VIE) tags in each sites of marine medaka *Oryzias dancena*, from 0 to 12 months after VIE tagging

Month	Color	Differentiation rate (%)		
		Abdomen	Back	Caudal vasculature
1	Red	60.0 ± 4.21ᶜ	100.0ª	100.0ª
	Green	84.0 ± 3.67ª	96.0 ± 2.11ᶜ	100.0ª
	Yellow	73.9 ± 4.55ᵇ	97.8 ± 4.34ᵇ	96.3 ± 3.11ᵇ
3	Red	59.4 ± 3.56ᶜ	97.0 ± 1.55ª	100.0ª
	Green	82.0 ± 5.07ª	93.2 ± 4.74ᶜ	95.6 ± 4.12ᵇ
	Yellow	77.3 ± 3.85ᵇ	96.2 ± 2.65ᵇ	94.3 ± 3.01ᶜ
6	Red	59.8 ± 4.21ᶜ	94.4 ± 2.51ª	97.4 ± 2.33ª
	Green	80.3 ± 6.47ª	92.3 ± 5.78ª	96.9 ± 2.11ᵇ
	Yellow	75.7 ± 4.68ᵇ	97.9 ± 0.91ª	96.3 ± 2.97ᵇ
9	Red	62.4 ± 3.93ᶜ	93.7 ± 2.86ᵇ	98.6 ± 3.12ᵇ
	Green	84.4 ± 5.41ª	93.7 ± 0.74ᵇ	99.5 ± 2.69ª
	Yellow	74.8 ± 5.23ᵇ	98.3 ± 1.99ª	94.3 ± 3.79ᶜ
12	Red	65.2 ± 4.41ᶜ	92.1 ± 4.61ᶜ	98.8 ± 0.72ª
	Green	85.9 ± 2.38ª	94.0 ± 1.68ᵇ	98.1 ± 1.08ª
	Yellow	68.5 ± 5.53ᵇ	97.3 ± 0.23ª	96.3 ± 2.89ᵇ

Samples tagged with VIE were investigated. Both dead samples and eliminated VIE tags were excluded from the analysis. Values are mean ± S.E. ($n = 50$). The experiment was performed in triplicate. The values in each column that do not share a common superscript are significantly different from one another ($P < 0.05$)

back and caudal vasculature sites. For the abdomen site, the retention (%) of the elastomer at 1 month were 57.4 ± 3.26 for red, 77.6 ± 1.11 for green, and 74.3 ± 1.48 for yellow but at 6 months were 57.3 ± 3.74 (red), 80.7 ± 2.10 (green), and 75.1 ± 1.68 (yellow). And in this term, the green and yellow values were significantly higher than the red value. However, the retention rate of the abdomen site was not affected by color. In summary, the retention rate for red was 67.2 ± 3.05, for green was 79.5 ± 1.96, and for yellow was 70.4 ± 1.21. For the back site at 1 month, the retention rate for red was 99.3 ± 0.16 and for green was 94.7 ± 0.85. The retention rate for yellow was 94.9 ± 0.80. After 9 months, the values had declined to 93.4 ± 0.67, 89.8 ± 1.79, and 95.5 ± 0.73 along red, green, and yellow, respectively. These values show that the VIE was removed from the tagged site by the time in each site. In conclusion, at 12 months, the tag retention rates for the back were 90.5 ± 1.31, 94.4 ± 0.74, and 94.6 ± 0.94 for the red, green, and yellow elastomers, respectively. For the caudal vasculature, there were no significant differences ($P < 0.05$) among the elastomer colors (red, 99.0 ± 0.22; green, 99.3 ± 0.10; yellow, 95.4 ± 0.67), but at the end of the experiment, the values were 96.6 ± 0.66, 97.2 ± 0.38, and 95.6 ± 0.86, respectively. Among the three colors, the tag retention rate for the back was the highest ($P < 0.05$), followed by that of the caudal vasculature

Table 3 Differentiation rate using the UV lamp of visible implant fluorescent elastomer (VIE) tags in each site of marine medaka, *Oryzias dancena*, from 0 to 12 months after VIE tagging

Month	Color	Differentiation rate (%)		
		Abdomen	Back	Caudal vasculature
1	Red	88.0 ± 2.61ᵇ	100.0ª	100.0ª
	Green	92.0 ± 1.73ª	100.0ª	100.0ª
	Yellow	86.9 ± 3.11ᶜ	100.0ª	100.0ª
3	Red	89.8 ± 1.73ᵇ	100.0ª	100.0ª
	Green	90.3 ± 2.22ª	99.4 ± 0.03ᵇ	100.0ª
	Yellow	84.7 ± 2.83ᶜ	100.0ª	100.0ª
6	Red	85.9 ± 3.85ᶜ	100.0ª	99.4 ± 0.01ᵇ
	Green	91.8 ± 2.89ª	97.8 ± 1.01ᶜ	98.9 ± 1.07ᶜ
	Yellow	87.5 ± 3.10ᵇ	99.5 ± 0.20ᵇ	100.0ª
9	Red	87.5 ± 2.81ᶜ	100.0ª	99.1 ± 0.04ª
	Green	91.3 ± 2.10ª	97.1 ± 0.78ᶜ	98.2 ± 1.10ᵇ
	Yellow	88.1 ± 1.08ᵇ	98.9 ± 1.00ᵇ	98.5 ± 0.74ᵇ
12	Red	87.2 ± 3.02ᵇ	100.0ª	98.9 ± 0.03ª
	Green	95.0 ± 2.97ª	96.9 ± 1.61ᶜ	97.9 ± 0.69ᵇ
	Yellow	84.0 ± 2.11ᶜ	98.3 ± 0.57ᵇ	98.2 ± 0.42ª

Samples tagged with VIE were investigated. Both dead samples and eliminated VIE tags were excluded from the analysis. Values are mean ± S.E. ($n = 50$). The experiment was performed in triplicate. The values in each column that do not share a common superscript are significantly different from one another ($P < 0.05$)

tags. The abdomen tag retention rate was the lowest among the tagging sites.

We measured the visual differentiation of the tags using two detection methods (visible and UV lights) (Tables 2 and 3). During 1 min, the VIE tags were initially observed using visible light at 30-cm distance from the experimental fish. Table 2 shows the dead fish and those that eliminated the VIE tags were excluded from the analysis. For the abdomen site, the capacity to detect for the red and green tags was significantly greater than that for the yellow tags ($P < 0.05$). At 1 month, the detection rate of the red tags (%) was 60.0 ± 4.21, for the green tags was 84.0 ± 3.67, and for the yellow tags was 73.9 ± 4.55. After 12 months, the values were 65.2 ± 4.41, 85.9 ± 2.38, and 68.5 ± 1.53, respectively, indicating that the red and green colors were more easily detected ($P < 0.05$). For the back site at 1 month, the detection rate (%) of red tags was 100, for green was 96.0 ± 2.11, and for yellow was 97.8 ± 4.34. After 12 months, tag detection were 92.1 ± 4.61 and 94.0 ± 1.68 for the red and green tags, respectively, and for the yellow tags was 97.3 ± 0.23. For the caudal vasculature at 1 month, the detection rates (%) of red and green tags were 100, respectively, but for the yellow tags was 96.3 ± 1.11. At 6 months, the detection rates (%) were 97.4 ± 2.33, 96.9 ± 2.11, and 96.3 ± 1.97, respectively ($P < 0.05$). At the end of the experiment, the detection rates (%) were 98.8 ± 1.72, 98.1 ± 1.08, and 96.3 ± 2.89, respectively. Table 2 shows that abdomen tags were less well detected than back and caudal vasculature tags.

Table 3 shows the results for tag differentiation using the UV light for detection of the VIE tags at each site in the experimental fish. The observation protocol was as described above. For the abdomen site, the detection of the red and green tags was significantly greater than for the yellow tags ($P < 0.05$). At 1 month, the detection rate (%) for the red tags was 88.0 ± 2.61, for green was 92.0 ± 1.73, and for yellow was 86.9 ± 3.11, indicating that the red and green tags were more readily detected than the yellow tags ($P < 0.05$). After 6 months, this had not changed significantly, and at the end of the experiment (12 months), the detection (%) of the red, green, and yellow tags were 87.2 ± 3.02, 95.0 ± 2.97, and 84.0 ± 2.11, respectively, showing that the yellow tags were least detectable when observed by the UV light ($P < 0.05$). For the back site, the tag detection rate was 100% for the three colors, while at 12 months, for the red tags was 100.0, for the green was 96.9 ± 1.61, and for the yellow was 98.3 ± 0.57 ($P < 0.05$). These results indicate that red tags were significantly more readily detected than green and yellow tags ($P < 0.05$). In addition, the back tags were detected more easily relative to those in the abdomen site. On the caudal vasculature, the detection rate (%) of all the color tags were 100%. After the 12 months of the experiment, the detection rate for

the red tags (%) was 98.9 ± 0.03, for the green was 97.9 ± 0.69, and for the yellow was 98.2 ± 0.42, indicating that red and yellow tags were more easily detected than green tags in the caudal vasculature. In conclusion, by UV lamp, the tag readability for the back and caudal vasculature sites were significantly greater than those for the abdomen site ($P < 0.05$; Table 3). Regardless of site, all color tags under UV light were more easily detected than all color tags under visible light.

During experimental period (12 months), accumulated survival rates of back, abdomen, and caudal vasculature groups were not significantly different among red, green, and yellow, respectively (Table 4, $P > 0.05$). However, accumulated survival rates of each color were affected by the tagging site (Table 4, $P < 0.05$). During 12 months, the accumulated survival (%) of the control group was the highest ($P < 0.05$) and the reduction ratio of the accumulated survival in the control group was the most gradual. However, the accumulated survival (%) of the abdomen group in each color were the lowest ($P < 0.05$), and the reduction ratio were the most dramatic in each color (Table 4).

Table 4 Accumulated survival rate using the UV lamp of visible implant fluorescent elastomer (VIE) tags in each site of marine medaka, *Oryzias dancena*, from 0 to 12 months after VIE tagging

Month	Color	Accumulated survival rate (%)			
		Control (no injection)	Abdomen	Back	Caudal vasculature
0	Red	100.0[a]	100.0[a]	100.0[a]	100.0[a]
	Green	100.0[a]	100.0[a]	100.0[a]	100.0[a]
	Yellow	100.0[a]	100.0[a]	100.0[a]	100.0[a]
1	Red	100.0[a]	85.7 ± 0.71[c]	99.3 ± 0.24[a]	97.5 ± 0.41[b]
	Green	100.0[a]	85.0 ± 1.14[c]	99.0 ± 0.59[a]	97.9 ± 0.87[b]
	Yellow	100.0[a]	85.9 ± 1.06[c]	98.8 ± 0.67[a]	96.8 ± 0.91[b]
3	Red	97.5 ± 1.88[a]	80.4 ± 1.73[c]	97.3 ± 1.89[a]	93.6 ± 1.41[b]
	Green	97.5 ± 1.88[a]	79.1 ± 1.92[c]	97.2 ± 1.14[a]	94.1 ± 0.92[b]
	Yellow	97.5 ± 1.88[a]	81.1 ± 1.88[c]	97.8 ± 0.91[a]	93.6 ± 1.10[b]
6	Red	96.1 ± 2.84[a]	77.1 ± 3.24[d]	93.3 ± 1.57[b]	86.4 ± 2.14[c]
	Green	96.1 ± 2.84[a]	76.7 ± 2.88[d]	92.6 ± 1.25[b]	87.0 ± 1.55[c]
	Yellow	96.1 ± 2.84[a]	76.9 ± 3.10[d]	93.4 ± 1.09[b]	86.8 ± 3.81[c]
9	Red	93.8 ± 1.55[a]	71.2 ± 3.44[d]	90.7 ± 2.48[b]	81.1 ± 4.39[c]
	Green	93.8 ± 1.55[a]	70.4 ± 4.05[d]	90.5 ± 3.24[b]	80.1 ± 3.81[c]
	Yellow	93.8 ± 1.55[a]	71.2 ± 2.12[d]	89.9 ± 1.85[b]	81.1 ± 2.58[c]
12	Red	90.4 ± 2.88[a]	64.2 ± 4.32[d]	86.9 ± 3.14[b]	74.1 ± 2.89[c]
	Green	90.4 ± 2.88[a]	63.8 ± 4.75[d]	86.7 ± 2.99[b]	74.6 ± 3.09[c]
	Yellow	90.4 ± 2.88[a]	64.3 ± 4.02[d]	87.4 ± 3.02[b]	73.5 ± 2.77[c]

Samples tagged with VIE were investigated. Values are mean ± S.E. ($n = 50$). The experiment was performed in triplicate. The values in each column that do not share a common superscript are significantly different from one another ($P < 0.05$)

In three colors of VIE tagging groups, the accumulated survival rates of the three sites and the control group were 100% at the initiation of experiment (Table 4). Accumulated survival (%) of the control group declined gradually to 90.4 ± 2.88 during 12 months. In addition, accumulated survival (%) of the back group in three colors declined gradually during 12 months. However, the abdomen group declined drastically to 64.2 ± 4.32 in red, 63.8 ± 4.75 in green, and 64.3 ± 4.02 in yellow during 12 months, respectively. Accumulated survival (%) of the caudal vasculature group in three colors declined gradually to 93.6 ± 1.41 in red, 94.1 ± 0.92 in green, and 93.6 ± 1.10 in yellow until 3 months after injection and declined drastically to 74.1 ± 2.89 in red, 74.6 ± 3.09 in green, and 73.5 ± 2.77 in yellow until 12 months after injection. In summary, the experimental fish of three colors survived > 85% of the tags injected in the back, > 70% of the tags injected in the caudal vasculature, and > 60% of the tags injected in the abdomen (Table 4, $P < 0.05$).

The whole-body cortisol concentration variations of the tagged group during 48 h are shown in Fig. 3. The whole-body cortisol concentration of the control groups was 0.9 μg/dL and has been increased to 1.20 μg/dL in 1 h and became 5.10 μg/dL in 6 h. After 12 h, it rather decreased to 1.26 μg/dL a bit and became 0.90 μg/dL in 24 h and 0.86 μg/dL in 48 h. The whole-body cortisol concentrations of caudal vasculature, abdomen, and back tagged groups were 0.81, 0.92, and 1 μg/dL, respectively, and has been rapidly increased to 14.76, 15.60, and 15.49 μg/dL in 1 h and increased drastically in 6 h ($P < 0.05$). The whole-body cortisol concentrations of the three experimental groups were the highest at 12 h, and became 29.43, 29.80, and 30.43 μg/dL, respectively. In 24 h, the whole-body cortisol concentrations of the three groups decreased rapidly until 48 h ($P < 0.05$). The tagging sites were not affected significantly in whole-body

cortisol concentration ($P > 0.05$), and the change of whole-body cortisol concentration according to exposure was seen compared to that at pre-experiment and the cortisol concentration was the highest at 6 h in the control group. However, the cortisol concentration was the highest at 12 h in the three experimental groups.

The whole-body glucose and lactic acid concentration variations of tagged marine medaka during 48 h are shown in Figs. 4 and 5. The whole-body glucose and lactic acid concentrations of the control groups were 25 mg/dL and 0.8 mmol/L, respectively, and have been rapidly increased to 55 mg/dL and 1.48 mmol/L in 12 h ($P < 0.05$). At 48 h, it rather decreased to 38 mg/dL and 1.0 mmol/L ($P < 0.05$). The whole-body glucose concentrations of three experimental groups were increased rapidly from 12 to 24 h and decreased drastically from 24 to 48 h. The whole-body glucose concentrations of the three experimental groups were the highest at 24 h. The lactic acid concentrations of the three experimental groups were increased rapidly from 24 to 48 h ($P < 0.05$). The lactic acid concentrations of the three experimental groups were the highest at 48 h. The lactic acid concentrations of the three tagged groups were not observed of reduction while at 48 h. Tagging sites were not affected significantly in whole-body glucose and lactic acid concentration ($P > 0.05$). The change of whole-body glucose concentration according to exposure was seen compared to that at pre-experiment, and the whole-body glucose and lactic acid concentrations were the highest at 12 h in the control group. However, the times observed when the highest glucose and lactic acid concentrations of the three groups were delayed were 24 and 48 h.

Discussion

In assessing the tagging sites for fish, it is important to establish the effect of the tag, including the tag retention at

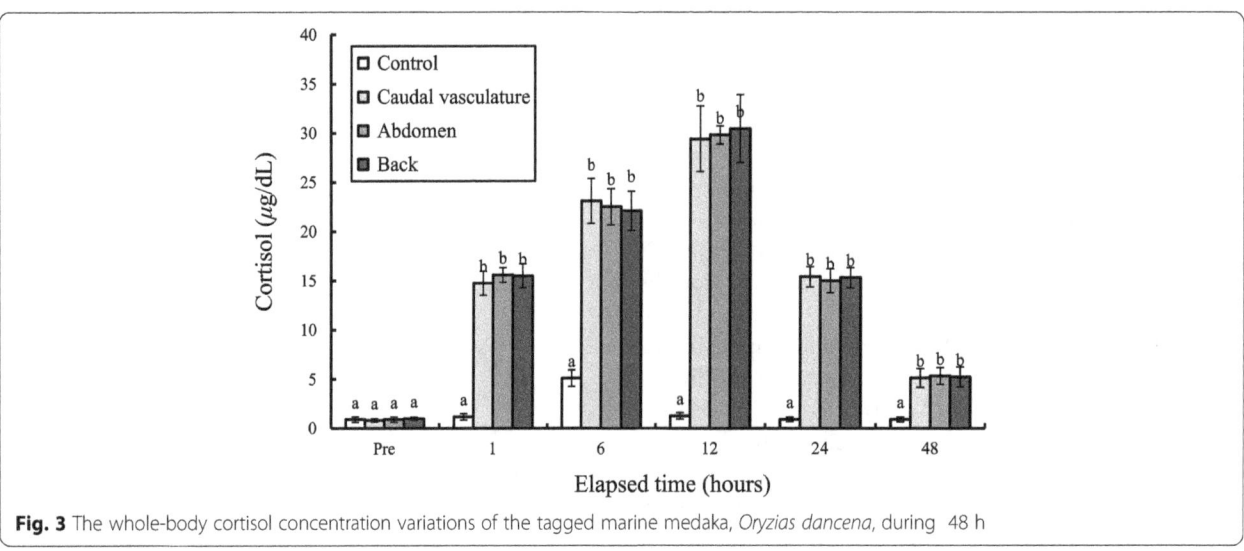

Fig. 3 The whole-body cortisol concentration variations of the tagged marine medaka, *Oryzias dancena*, during 48 h

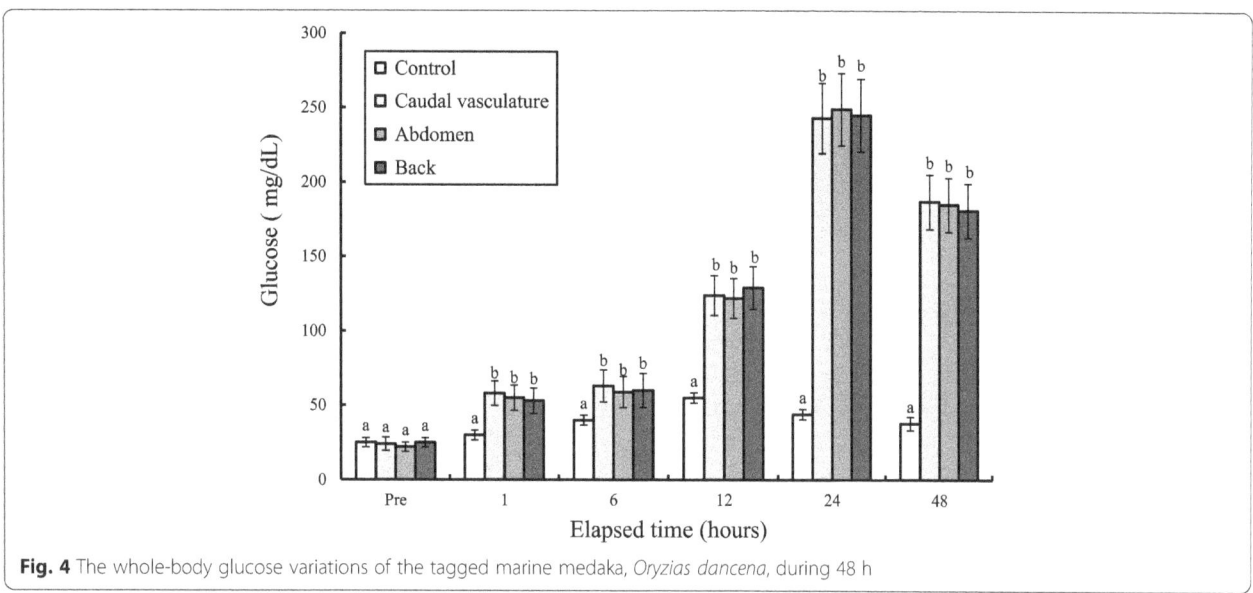

Fig. 4 The whole-body glucose variations of the tagged marine medaka, *Oryzias dancena*, during 48 h

the tagging site, the rate of tag detection following the tagging site, and the survival rate of the tagged fish (Frederick, 1997; Dewey and Zigler, 1996; Park et al., 2013; Willis and Babcock, 1998). Statistically significant differences were found among tag colors and sites, with red and green tags being easier to detect and distinguish under visible and UV lights than yellow tags in marine medaka, *O. dancena*. In contrast, Park et al. (2013) reported that red and orange were easier to detect and identify than green and yellow when viewed under UV light, but green and yellow were easily detected in visible light in a greenling, *Hexagrammos otakii*. However, as in the current study, red tags were more easily detected than green or yellow tags (Willis and Babcock, 1998). In deeper water, where natural light levels

are lower, greater attenuation of red light may occur (Willis and Babcock, 1998). In direct sunlight, red tags were clearly detectable at up to 5 m distant in clear water (Pottinger and Calder, 1995).

As shown in Fig. 3, the VIE tag affected the survival of marine medaka in the laboratory ($P < 0.05$). In conclusion, survival was significantly higher in the control group than in any of the experimental groups. Among the experimental groups, fish tagged in the abdomen site showed the lowest survival. Therefore, skilled injection of the elastomer is crucial for keeping the mortality low, as suggested by the decrease in mortality of marked fish during the laboratory experiment (Frederick, 1997). In previous study, the primary causes of mortality among the tagged

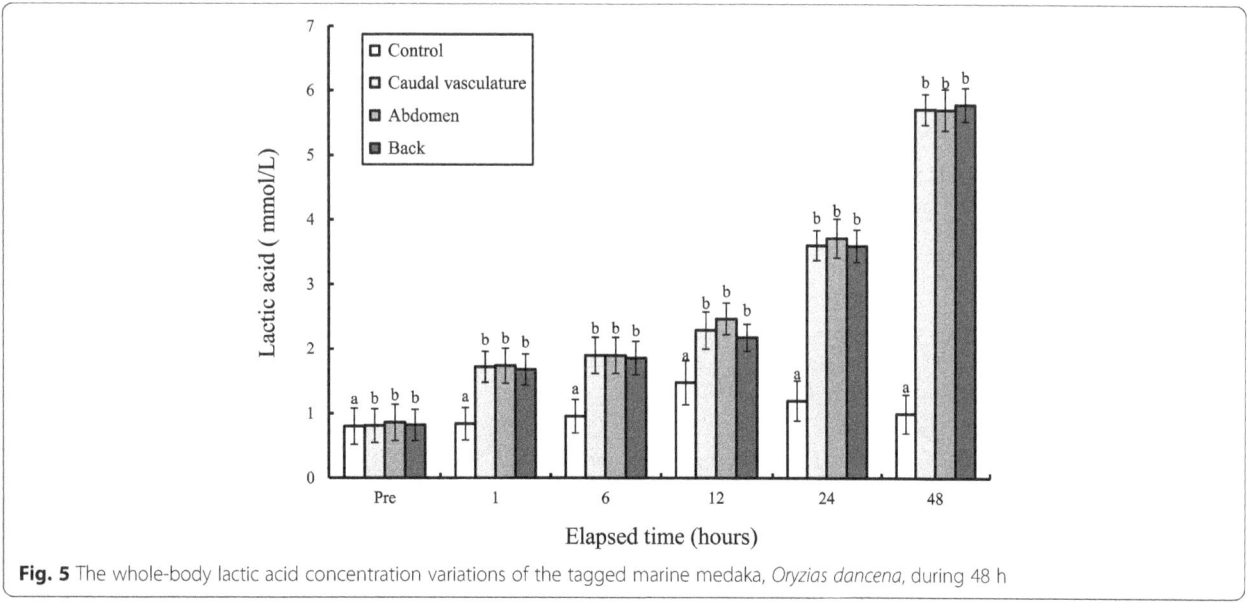

Fig. 5 The whole-body lactic acid concentration variations of the tagged marine medaka, *Oryzias dancena*, during 48 h

samples were internal damage and infection, as a result of gas bladder rupture, and infection from anatomical trauma caused by handling (Willis and Babcock, 1998). The causes of mortality among experimental groups were not determined in this study, and histological observations of post-mortem samples are necessary for investigating the causes of mortality. The results of this experiment are similar to those of the previous studies showing high retention of VIE (Dewey and Zigler, 1996; Willis and Babcock, 1998) in the marine medaka, which had > 90% tag retention for the back tagging site (Table 1), with the caudal vasculature and the abdomen having lower tag retention rates. The various characteristics of this species must be investigated to determine the greater loss of VIE tag retention for the caudal vasculature and the abdomen sites during the experimental period.

The use of VIE tagging in small fish, marine medaka, is advocated as a practical and reliable method for fish identification and monitoring, but it may cause negative effects on growth and mortality. Foreign materials such as tags can lead to stress and may cause changes of the blood reactions of fish. When stress is induced, the fish consume energy, which drives a response of excess secretion of catecholamine and cortisol, and has a considerable influence on the maintenance of homeostasis (Schreck et al., 2001). Plasma cortisol and plasma glucose are recognized as useful indicators of stress in fish (Schreck et al., 2001). In our study, whole-body physiological responses of marine medaka from each tagging region, in the form of high whole-body cortisol, whole-body glucose, and lactic acid values, were generally observed in tagged groups in which a tag had actually been inserted compared to the responses seen in control groups. This finding indicates that the actual insertion of a tag rather than just a pierce injection can result in added stress, and this result shows that tagging sites were not affected significantly in stress response.

The plasma cortisol levels induced by stress appear to increase at various speed and time according to the species of fish (Pickering and Pottinger, 1989). The plasma cortisol concentration after stress is usually reported to increase to a peak value within 1 ~ 3 h and normally recovers within 6 h (Willis and Babcock, 1998). As a whole, the whole-body cortisol values for the tagged group were similar to the values seen in the control group after 48 h. Therefore, the time required for the black rockfish to adapt after the insertion of a tag is approximately 48 h. The whole-body cortisol concentrations showed peak values before 48 h in this study. The trends in cortisol and glucose observed in this experiment indicated generalized stress reactions. Glucose formation was increased simultaneously as the cortisol quantity increased. Elevated cortisol secretion under stress increases the activation of plasma glucose by the

activity of the gluconeogenesis enzyme; also, this increase is the result of a second reaction to the first reaction (response of hormone) to stress (Barton and Iwama, 1991).

Buckley et al. (1994) found that in juvenile reef fish, *Sebastes* spp., the VIE tags could be detected visually in situ for up to 258 days using underwater UV lights. In response to concerns about amphibian declines, Jung et al. (2000) evaluated and validated amphibian monitoring techniques using VIE tags in studies in the Shenandoah and Big Bend national parks, USA. Godin et al. (1995) found that to identify populations of shrimp, *Penaeus vannamei*, individuals could be tagged internally using an externally visible elastomer. Basic considerations in the use of tags in fisheries management or research include the effects of the tags on animal survival, behavior, growth, permanency, and recognition and the cost of the marking technique (McFarlane and Beamish, 1990; Park and Lee, 2001). VIE tags are made of non-toxic medical grade fluorescent elastomer material and have been used successfully to identify fish, amphibians, and decapod crustaceans (Willis and Babcock, 1998; Jerry et al., 2001; Bailey, 2004). The retention rate was 92% for visual implant elastomer (VIE) tags in juvenile crayfish, *Cherax destructor*, and 100% for VIE tags in lobsters, *Homarus gammarus* (Jerry et al., 2001; Uglem et al., 1996). As with VIE tags, passive inductive transponder (PIT) tags are sometimes used in experiments. However, in a study involving injection of small, mid-sized, and large tags into four small Cyprinidae fish species, *Carassius gibelio langsdorfi*, *Hypophthalmichthys molitrix*, *Pseudorasbora parva*, and *Phoxinus phoxinus*, Jang et al. (2007) reported that PIT tags caused high mortality. The larger and heavier PIT tags can affect the swimming ability of small fish, including marine medaka. Thus, Jang et al. (2007) concluded that PIT tags are inappropriate for small individuals. So, the VIE tags are small, light, and made of non-toxic medical grade fluorescent elastomer material and are therefore more appropriate for small individuals and species, including marine medaka, and are considered effective for laboratory experiments and aquaculture facilities. Unfortunately, the relationship among decreasing survival rate, spawning behavior, and VIE tag was not determined by the previous studies. Thus, future investigation will focus on the relationship among reduced survival rates, spawning behavior, and VIE tag.

Conclusions

During 12 months, the accumulated survival rates of marine medaka, *Oryzias dancena*, in the experimental treatments were not different among red, yellow, and green elastomer. The experimental fish retained >85% of the tags injected in the back, >70% of the tags injected in the caudal vasculature, and >60% of the tags injected in the abdomen. For all injected sites the red and green tags were able to be detected more easily than the yellow tags when observed under both visible and UV light. So, the VIE tags are small, light, and made of non-toxic medical grade fluorescent elastomer material and are therefore more appropriate for small individuals and species, including marine medaka, and are considered effective for laboratory experiments and aquaculture facilities.

Abbreviations
PIT: Passive integrated tag; VIE: Visible implant fluorescent elastomer

Acknowledgements
The authors thank the technical staff of the Cheongpyeong Aquaculture Research Center, NIFS, Korea, and the Laboratory for Fishery Genetics and Breeding Sciences at the Korea Maritime and Ocean University, South Korea, for their helpful support, and the anonymous reviewers who greatly improved the quality of this manuscript.

Funding
This work was supported by a grant from the National Institute of Fisheries Science (R2017038) from the Inland Fisheries Research Institute, National Institute of Fisheries Science (NIFS), South Korea.

Authors' contributions
JHI, CYC, KWY, CHK, and BSKS designed the overall plan of the experiment and drafted this manuscript. HYG, ISP, and THL conducted the whole part of the experiment, for example, inserting PIT tags into a sample. All authors read and approved the manuscript.

Competing interests
The authors declare that they have no competing interests.

Author details
[1]Research Cooperation Division, National Institute of Fisheries Science (NIFS), Busan 46083, South Korea. [2]Division of Marine Bioscience, College of Ocean Science and Technology, Korea Maritime and Ocean University, 727 Taejong-ro, Yeong do-gu, Busan 49112, South Korea. [3]Food, Agriculture, Forestry and Fisheries Examination Division, Korean Intellectual Property Office, Daejeon 35208, South Korea. [4]Inland Fisheries Research Institute, National Institute of Fisheries Science (NIFS), Cheongpyeung 12453, South Korea.

References
Bailey LL. Evaluating elastomer marking and photo identification methods for terrestrial salamanders: marking effects and observer bias. Herpetol Rev. 2004;35:38–41.

Barrett NS. Short- and long-term movement patterns of six temperate reef fishes (families Labridae & Monacanthidae). Mar Freshw Res. 1995;46:853–60.

Barton BA, Iwama GK. Physiological changes in fish from stress in aquaculture with emphasis on the response and effects of corticosteroids. Ann Rev Fish Dis. 1991;1:3–26.

Bergman PK, Haw F, Blankenship HL, Buckley RM. Perspectives on design, use, and misuse of fish tags. Fisheries. 1992;17:20–5.

Buckley RM, West JE, Doty DD. Internal micro-tag systems for marking juvenile reef fishes. Bull Mar Sci. 1994;55:850–95.

Cho YS, Lee SY, Kim DS and Nam YK. 2010. Tolerance capacity to salinity change in adult and larva of Oryzias dancena, a euryhaline medaka. [Korean] Kor J Ichthyol 21, 9–16.

Cody RC, Smith JK. Applied statistics and the SAS programming language. 3rd ed. Englewood Cliffs: Prentice Hall; 1991. p. 122–35.

Connell SD, Jones GP. The influence of habitat complexity on postrecruitment processes in a temperate reef fish population. J Exp Mar Biol Ecol. 1991;151:271–94.

Crossland J. Snapper tagging in north-east New Zealand, 1974: analysis of methods, return rates, and movements. N Z J Mar Freshwater Res. 1976;10:675–86.

Crossland J. Population size and exploitation rate of snapper, Chrysophrys auratus, in the Hauraki Gulf from tagging experiments. 1975–76. N Z J Mar Freshwater Res. 1980;14:255–61.

Dewey MR, Zigler SJ. An evaluation of fluorescent elastomer for marking bluegills in experimental studies. Prog Fish Cult. 1996;58:219–20.

Duncan DB. Multiple-range and multiple F tests. Biometrics. 1955;11:1–42.

Forrester GE. Factors influencing the juvenile demography of a coral reef fish. Ecology. 1990;71:1666–81.

Frederick JL. Evaluation of fluorescent elastomer injection method for marking small fish. Bull Mar Sci. 1997;61:399–408.

Godin DM, Carr WH, Hagino G, Segura F, Sweeney JN, Blankenship L. Evaluation of a fluorescent elastomer internal tag in juvenile and adult shrimp, Penaeus vannamei. Aquaculture. 1995;139:243–8.

Inoue K, Takei Y. Asian medaka fishes offer new models for studying mechanisms of seawater adaptation. Comp Biochem Physiol B Biochem Mol Biol. 2003;136:635–45.

Jang MH, Yoon JD, Do YN, Joo GJ. Survival rate on the small cyprinidae by PIT tagging application. [Korean] Kor J Ichthyol. 2007;19:371–7.

Jerry DR, Stewart T, Purvis IW, Piper LR. Evaluation of visible implant elastomer and alphanumeric internal tags as a method to identify juveniles of the freshwater crayfish, Cherax destructor. Aquaculture. 2001;193:149–54.

Jones GP. Competitive interactions among adults and juveniles in a coral reef fish. Ecology. 1987;68:1534–47.

Jung RE, Droege S, Sauer JR, Landy RB. Evaluation of terrestrial and streamside salamander monitoring techniques at Shenandoah National Park. Environ Monitor Assess. 2000;63:65–79.

McFarlane GA, Beamish RJ. Effect of an external tag on growth of sablefish (Anoplopoma fimbria), and consequences to mortality and age at maturity. Can J Fish Aquat Sci. 1990;47:1551–7.

Nam YK, Cho YS, Lee SY, Kim DS. Tolerance capacity to salinity changes in adult and larva of Oryzias dancena, a euryhaline medaka. [Korean] Kor J Ichthyol. 2010;22:9–16.

Park I-S, Kim YJ, Gil HW, Kim D-S. Evaluation of implant fluorescent elastomer tagging greenling, Hexagrammos otakii. Fish Aquat Sci. 2013;16:35–9.

Park I-S, Lee K-K. The effective location of visible implant tags for short-term marking in Nile tilapia (Oreochromis niloticus: Cichlidae). J Fish Sci Tech. 2001;4:159–61.

Park I-S, Park SJ, Gil HW, Nam YK, Kim DS. Anesthetic effects of clove oil and lidocaine-HCl on marine medaka, Oryzias dancena. Lab Animal. 2011;40:45–51.

Parker ROJ. Tagging studies and diver observations of fish populations on live-bottom reefs of the U.S. southeastern coast. Bull Mar Sci. 1990;46:749–60.

Pickering AD, Pottinger TG. Stress responses and disease resistance in salmonid fish: effects of chronic elevation of plasma cortisol. Fish Physiol Biochem. 1989;7:253–8.

Pottinger TG, Calder GM. Physiological stress in fish during toxicological procedures: a potentially confounding factor. Environ Toxicol Water Qual. 1995;10:135–46.

Pottinger TG, Carrick TR, Yeomans WE. The three-spined stickleback as an environmental sentinel: effects if stressors on whole-body physiological indices. J Fish Biol. 2002;61:207–29.

Raabo E, Terkildsen TC. On the enzymatic determination of blood glucose. Scandina J Clinic Lab Invest. 1960;12:402–7.

Ralston SL, Horn MH. High tide movements of the temperate-zone herbivorous fish Cebidichthys violaceus (Girard) as determined by ultrasonic telemetry. J Exp Mar Biol Ecol. 1986;98:35–50.

Redding JM, Schreck CB. Influence of ambient salinity on osmoregulation and cortisol concentration of yearling coho salmon during stress. Trans Amer Fish Soc. 1983;112:800–7.

Song HY, Nam YK, Bang I-C, Kim DS. Early gonadogenesis and sex differentiation of marine medaka, Oryzias dancena (Beloniformes; Teleostei). [Korean] Kor J Ichthyol. 2009;21:141–8.

Serafy JE, Lutz SJ, Capo TR, Ortner PB, Lutz PL. Anchor tags affect swimming performance and growth of juvenile red drum, *Sciaenops ocellatus*. Mar Freshw Behav Physiol. 1995;27:29–35.

Schreck CB, Contreras-Sanchez W, Fitzpatrick MS. Effects of stress on fish reproduction, gamete quality, and progeny. Aquaculture. 2001;197:3–24.

Thompson SM, Jones GP. Social inhibition of maturation in females of the temperate wrasse *Pseudolabrus celidotus* and a comparison with the blennioid *Tripterygion varium*. Mar Biol. 1980;59:247–56.

Tong LJ. Tagging snapper *Chrysophrys auratus* by scuba divers. N Z J Mar Freshwater Res. 1978;12:73–6.

Uglem I, Noess H, Farestveit E, Jorstad KE. Tagging of juvenile lobsters (*Hamarus gammarus* (L.)) with visible implant fluorescent elastomer tags. Aquac Eng. 1996;15:499–501.

Willis TJ, Babcock RC. Retention and in situ detectability of visible implant fluorescent elastomer (VIE) tags in *Pagrus auratus* (Sparidae). N Z J Mar Freshwater Res. 1998;32:247–54.

Zerrenner A, Josephson DC, Krueger CC. Growth, mortality, and mark retention of hatchery brook trout marked with visible implant tags, jaw tags, and adipose fin clips. Prog Fish Cult. 1997;59:241–5.

16

Protective effects of alginate-free residue of sea tangle against hyperlipidemic and oxidant activities in rats

Mi-Jin Yim[1], Grace Choi[1], Jeong Min Lee[1], Soon-Yeong Cho[2] and Dae-Sung Lee[1]*

Abstract: The antihyperlipidemic and antioxidant activities of dietary supplementation of sea tangle from Goseong and the alginate-free residue of sea tangle were investigated in Sprague Dawley rats treated with a high-fat diet, streptozotocin, poloxamer 407, and bromobenzene. The alginate-free residue of Goseong sea tangle induced a significant reduction in triglycerides and total cholesterol levels, as well as a significant increase in high-density lipoprotein cholesterol levels. Alginate-free Goseong sea tangle residue reduced the activities of the phase I enzymes aminopyrine N-demethylase and aniline hydroxylase, which had been increased by intraperitoneal injection of bromobenzene. Pretreatment with Goseong sea tangle residue prevented a bromobenzene-induced decrease in epoxide hydrolase activity. Bromobenzene reduced hepatic glutathione content and increased hepatic lipid peroxide levels. Pretreatment with alginate-free Goseong sea tangle residue prevented lipid peroxidation induced by bromobenzene, but pretreatment with Goseong sea tangle did not. These results suggest that Goseong sea tangle residue exerted antihyperlipidemic and antioxidant activities that were higher than those induced by alginate-containing sea tangle. Therefore, the alginate-free residue may contain physiologically unknown active components, other than alginic acid, which may potentially be used to prevent hyperlipidemic atherosclerosis.

Keywords: *Saccharina japonica*, Sea tangle, Hyperlipidemia, Antioxidant activity

Background

Hyperlipidemia is considered a major risk factor for cardiovascular diseases and events such as atherosclerosis and myocardial infarction (Wald and Law 1995; Talbert 1997). Rates of hyperlipidemia-related diseases are increasing with lifestyle changes. Low-density lipoprotein cholesterol (LDL-C) is regarded as the primary risk factor for atherosclerosis and coronary heart disease (Baigent et al. 2010), and elevated circulating levels of free fatty acids and triglycerides (TG) can lead to these diseases (Pilz et al. 2006; Harchaoui et al. 2009). Therefore, modulating the dysregulation of lipid metabolism and decreasing the levels of serum total cholesterol (TC), TG, and LDL-C are considered beneficial in treating and preventing cardiovascular diseases (Derosa et al. 2006; Zhang et al. 2013). Identifying effective food sources to treat hyperlipidemia would promote this objective (Murata et al. 1999).

The brown alga sea tangle (*Saccharina japonica*) has been used in Korea to promote maternal health (Jin et al. 2004). Sea tangle is also popular in Korea and Japan as a food and has been reported to exhibit hypotensive, antioxidant, antimutagenic, and antibacterial activities (Okai et al. 1993; Han et al. 2002; Wang et al. 2006; Park et al. 2009). Moreover, aqueous extracts of sea tangle and alginate have also been shown to exhibit antioxidant activity and lower hypercholesterolemia (Torsdottir et al. 1991; Lee et al. 2004). However, the alginate-free residue of sea tangle has not been investigated for its biological activities.

In the present study, we evaluated the biological activities of sea tangle residue from which alginate had been removed. The antihyperlipidemic effects of sea tangle residue were assessed in three different experimental rat models, one in which hyperlipidemia was induced by a high-fat diet and two in which hyperlipidemia was induced by streptozotocin and poloxamer 407. In addition, the effects of sea tangle residue on lipid peroxidation and the activities of

* Correspondence: daesung@mabik.re.kr
[1]Department of Applied Research, National Marine Biodiversity Institute of Korea, Seocheon 33662, South Korea
Full list of author information is available at the end of the article

enzymes involved in drug metabolism were examined in the livers of bromobenzene-treated rats.

Methods
Materials
Goseong sea tangle (Goseong, Gangwon-do, South Korea; *S. japonica*) was obtained from a local supplier (Gangneung, Gangwon-do, South Korea) in March 2007. Alginate-free residue from Goseong sea tangle was also used in this study. All samples were powdered after freeze-drying.

Animals and treatments
Male Sprague Dawley rats (Daehan Biolink, Eumsung, South Korea) weighing 190–210 g were housed individually in stainless steel mesh cages in a room maintained at 22 ± 1 °C and 55 ± 3% relative humidity with a normal 12-h light/dark cycle. Rats were fed a commercial standard rat diet (AIN-76). The composition of the experimental diets is shown in Table 1. The high-fat diet-treated rats were orally administered for the last week with a high-fat diet that fed daily for 6 week. Rats were orally administered 100 or 200 mg/kg of body weight of the sea tangle powder in 5% Tween 80 daily for 1 week. During the final 2 days of the oral treatment, rats were injected intraperitoneally (i.p.) with streptozotocin (45 mg/kg in 0.1 M citrate buffer, pH 4.5), poloxamer 407 (300 mg/kg in saline), or bromobenzene (460 mg/kg in 5% Tween 80) four times at 12-h intervals.

All animal experiments procedures were approved by the Committee for Animal Experiments of Kyungsung University.

Sample preparation
At the end of the experimental period and again after 12 h of fasting, the rats were sacrificed by exsanguination under anesthesia with CO_2 and starved for 18 h before sacrifice. Blood was collected from the neck and incubated at room temperature for 30 min. The blood samples were then centrifuged at 3000×g at 4 °C for 10 min, after which the serum was stored at − 70 °C for further biochemical tests.

Liver tissue fat was extracted from the cystic lobe according to the method of Folch et al. (1957). The liver, which had been exhaustively perfused with ice-cold 0.9% NaCl, was homogenized with four volumes of an ice-cold 0.1 M potassium phosphate buffer, pH 7.5. An aliquot of the homogenate was used for the determination of lipid peroxide and glutathione (GSH) contents. The remaining homogenate was centrifuged at 600×g for 10 min, and the resulting supernatant was recentrifuged at 10,000×g for 20 min. The supernatant was centrifuged again at 10,000×g for 60 min to obtain the upper fraction as cytoplasm. The pellet was resuspended in the same volume of 0.1 M potassium phosphate buffer and centrifuged at 10,000×g for 60 min to obtain the microsomal fraction, which was used to measure the activities of aminopyrine N-demethylase (AMND), aniline hydroxylase (ANH), and epoxide hydrolase (EPH).

Glucose analysis
Levels of plasma glucose were determined by the glucose oxidase method using a commercially available enzymatic kit (Embiel Co., Gyeonggi-do, South Korea).

Cholesterol analysis
TG, TC, and high-density lipoprotein cholesterol (HDL-C) levels were determined by enzymatic colorimetric methods using commercial kits (Shinyang Chemical Co., Busan, South Korea).

Lipid peroxide and GSH levels
Hepatic lipid peroxidation was assessed by measuring the concentration of thiobarbituric acid-reactive substances (TBARS) in plasma using the method described by Ohkawa et al. (1979). Hepatic GSH contents were measured by a colorimetric method (Boyne and Ellma 1972).

Enzyme assays
AMND activity in liver microsomes was measured spectrophotometrically by the quantitation of formaldehyde produced from the demethylation of aminopyrine, as described by Nash (1953). ANH activity was assayed by measuring *p*-aminophenol formation from aniline (Bidlack and Lowery 1982). EPH activity was measured spectrophotometrically using the decrease in *trans*-stilbene oxide at 229 nm (Hasegawa and Hammock 1982). Protein contents of the microsome and cytoplasm were determined by the method of Lowry et al. (1951) using bovine serum albumin as a standard.

Table 1 Composition of the experimental diets (g/100 g)

Ingredients	Normal diet	High-fat diet
Casein	20	29
Corn starch	60	10
Sucrose	0	10
Lard	0	35
Corn oil	9	5
α-Cellulose	5.0	5.0
Mineral mixture[a]	3.5	3.5
Vitamin mixture[b]	1.0	1.0
Cholesterol	1.0	1.0
DL-Methionine	0.3	0.3
Choline bitartrate	0.2	0.2

[a]Mineral mixture derived from the rodent diet formula AIN-76
[b]Vitamin mixture derived from the rodent diet formula AIN-76

Statistical analysis

All results are presented as the mean ± SD. Data were evaluated by one-way ANOVA using SPSS (IBM SPSS, Armonk, NY, USA), after which the differences between the mean values were assessed using Duncan's multiple range test. Results were considered statistically significant at $p < 0.05$.

Results

Effects of sea tangle on serum and liver tissue lipid levels in high-fat diet-fed rats

The effects of sea tangle supplementation on serum lipid levels in rats fed a high-fat diet are shown in Table 2. Serum lipid levels were significantly reduced in rats treated with alginate-free Goseong sea tangle residue at doses of 100 and 200 mg/kg, compared with lipid levels in the hyperlipidemia control group. However, administration of Goseong sea tangle did not significantly affect serum lipid levels in rats with hyperlipidemia induced by a high-fat diet.

The effects of dietary supplementation of sea tangle on hepatic lipid levels of rats fed a high-fat diet are shown in Table 3. The rats displayed significantly higher TG and TC levels compared with rats fed a normal diet. Hepatic lipid levels in the alginate-free Goseong sea tangle residue groups were significantly lower than levels in the hyperlipidemia control group.

Effects of sea tangle on blood glucose and lipid levels following streptozotocin administration

Table 4 shows the effects of sea tangle administration on blood glucose and lipid levels in streptozotocin-induced hyperglycemic rats. The group displayed remarkably high serum levels of glucose, TG, and TC compared with normal-diet control rats. Oral administration of all sea tangle types at doses of 200 mg/kg resulted in a significant reduction in TG and TC levels, especially the administration of Goseong sea tangle, compared with the streptozotocin-induced hyperlipidemic control group.

However, blood glucose levels were not elevated by sea tangle treatment in any group.

Effects of sea tangle on serum lipid levels following poloxamer 407 administration

Table 5 shows the effects of sea tangle administration on serum lipid levels in poloxamer 407-induced hyperlipidemic rats. The group displayed significantly high TG and TC serum levels compared with normal-diet control rats. Administration of the alginate-free Goseong sea tangle residue at both 100 and 200 mg/kg doses resulted in a significant, dose-dependent reduction in TG and TC levels, compared with the poloxamer 407-induced hyperlipidemic control group.

Effects of sea tangle on hepatic enzyme activity and lipid peroxidation following bromobenzene administration

Hepatic AMND and ANH activities of bromobenzene-injected rats that had been pretreated with dietary supplementation of sea tangle are shown in Table 6. In comparison with normal-diet control rats, the rats injected with bromobenzene exhibited higher AMND and ANH activities. The increase in AMND activity by bromobenzene was reduced by 8.1 and 12.9% with oral administration of the alginate-free Goseong sea tangle

Table 3 Effects of sea tangle on hepatic tissue lipid levels in high-fat diet-induced rats

	Dose (mg/kg)	TG (mg/g)	TC (mg/g)
Normal diet		24.5 ± 3.38	4.17 ± 0.23
Control		50.8 ± 5.17	32.9 ± 2.48
Goseong sea tangle	100	48.7 ± 3.21	28.1 ± 2.46
	200	46.3 ± 2.55	27.9 ± 2.02
Alginate-free sea tangle residue	100	41.7 ± 3.48	25.9 ± 2.06
	200	35.2 ± 4.22	22.3 ± 1.57

Values are expressed as mean ± SD ($n = 8$) for groups of six replicates.
Different superscripts in each data column indicate significant differences ($p < 0.05$) between groups
TG triglycerides, TC total cholesterol

Table 2 Effects of sea tangle on serum lipid levels in high-fat diet-induced rats

	Dose (mg/kg)	TG (mg/dL)	TC (mg/dL)	HDL-C (mg/dL)
Control		269.0 ± 68.3	93.6 ± 8.53	50.7 ± 2.16
Goseong sea tangle	100	253.8 ± 49.7	100.7 ± 9.09	49.7 ± 1.83
	200	246.0 ± 39.8	98.4 ± 7.33	47.6 ± 1.54
Alginate-free sea tangle residue	100	248.6 ± 18.9	86.3 ± 4.96	55.6 ± 3.11
	200	173.6 ± 24.5	70.8 ± 5.41	58.2 ± 1.37

Values are expressed as mean ± SD ($n = 8$) for groups of six replicates.
Different superscripts in each data column indicate significant differences ($p < 0.05$) between groups
TG triglycerides, TC total cholesterol, HDL-C high-density lipoprotein cholesterol

Table 4 Effects of sea tangle on blood glucose and lipid levels in streptozotocin-induced hyperglycemic rats

	Dose (mg/kg)	Glucose (mg/dL)	TG (mg/dL)	TC (mg/dL)
Normal		92.6 ± 8.97	76.3 ± 7.28	65.3 ± 6.27
Streptozotocin		330.2 ± 23.8	180.7 ± 26.3	147.4 ± 5.19
Goseong sea tangle	100	339.5 ± 33.4	198.3 ± 39.3	138.6 ± 6.34
	200	330.7 ± 24.6	195.3 ± 26.2	120.5 ± 5.44
Alginate-free sea tangle residue	100	295.8 ± 29.6	170.0 ± 23.5	110.5 ± 7.69
	200	287.2 ± 36.5	153.6 ± 15.4	100.8 ± 4.24

Values are expressed as mean ± SD ($n = 8$) for groups of six replicates.
Different superscripts in each data column indicate significant differences ($p < 0.05$) between groups
TG triglycerides, TC total cholesterol

Table 5 Effects of sea tangle on serum lipid levels of poloxamer 407-induced hyperlipidemic rats

	Dose (mg/kg)	TG (mg/dL)	TC (mg/dL)
Normal diet		78.4 ± 7.60	69.4 ± 7.23
Poloxamer 407		1004.1 ± 58.9	780.8 ± 52.7
Goseong sea tangle	100	980.7 ± 46.5	770.3 ± 40.2
	200	916.6 ± 49.2	750.2 ± 51.3
Alginate-free sea tangle residue	100	900.3 ± 33.5	690.4 ± 49.2
	200	719.8 ± 40.0	510.3 ± 29.8

Values are expressed as mean ± SD ($n = 8$) for groups of six replicates.
Different superscripts in each data column indicate significant differences ($p < 0.05$) between groups
TG triglycerides, *TC* total cholesterol

residue at doses of 100 and 200 mg/kg, respectively. The increase in ANH activity by bromobenzene was reduced by 13.1% with oral administration of the alginate-free Goseong sea tangle residue at a dose of 200 mg/kg. However, no such reduction in AMND and ANH activities was observed following oral treatment with other sea tangle preparations.

Hepatic EPH activity in bromobenzene-treated rats was lower than that in normal-diet control rats (Table 7). Pretreatment with alginate-free Goseong sea tangle residue at doses of 100 and 200 mg/kg elevated enzyme activity by 31.5 and 42.6%, respectively.

Hepatic GSH and lipid peroxide contents in bromobenzene-injected rats pretreated with sea tangle are shown in Table 8. Hepatic GSH contents were significantly lower in bromobenzene-injected rats than in normal-diet control rats. No sea tangle type or dose affected GSH contents in bromobenzene-injected rats. Bromobenzene administration resulted in elevation of lipid peroxide contents to 50.0 nmol of TBARS/g from the normal value of 17.8 nmol/g. However, the increase in TBARS content by bromobenzene injection was

Table 6 Hepatic enzyme activities in bromobenzene-injected rats treated with sea tangle

	Dose (mg/kg)	Aminopyrine N-demethylase	Aniline hydroxylase
		Formaldehyde (nmol/mg protein/min)	p-Aminophenol (nmol/mg protein/min)
Normal diet		4.14 ± 0.07	0.71 ± 0.10
Bromobenzene		9.33 ± 0.52	1.34 ± 0.28
Goseong sea tangle	100	8.95 ± 0.44	1.29 ± 0.15
	200	9.28 ± 0.39	1.31 ± 0.20
Alginate-free sea tangle residue	100	8.57 ± 0.66	1.20 ± 0.14
	200	8.13 ± 0.50	1.03 ± 0.09

Values are expressed as mean ± SD ($n = 8$) for groups of six replicates.
Different superscripts in each data column indicate significant differences ($p < 0.05$) between groups

Table 7 Epoxide hydrolase activity of bromobenzene-injected rats pretreated with sea tangle

	Dose (mg/kg)	Epoxide hydrolase
		nmol *trans*-stilbene oxide/mg protein/min
Normal diet		13.6 ± 0.52
Bromobenzene		4.25 ± 0.17
Goseong sea tangle	100	5.10 ± 0.57
	200	4.91 ± 0.45
Alginate-free sea tangle residue	100	6.20 ± 0.30
	200	7.40 ± 0.23

Values are expressed as mean ± SD ($n = 8$) for groups of six replicates.
Different superscripts in each data column indicate significant differences ($p < 0.05$) between groups

inhibited in rats pretreated with alginate-free Goseong sea tangle residue, at doses of both 100 and 200 mg/kg.

Discussion

Hyperlipidemia is a major risk factor in the development of coronary artery disease and the progression of atherosclerotic lesions (McKenney 2001). Developing novel and effective antihyperlipidemia agents warrants increased attention (Sliskovic and White 1991). We investigated the effects of dietary supplementation of alginate-free extracts of sea tangle in rats with hyperlipidemia induced by streptozotocin, poloxamer 407, bromobenzene, or a high-fat diet.

Sea tangle contains alginic acid, carotenoids, xanthophylls, mannitol, and physiologically unknown active components. Aqueous extract of sea tangle has been shown to suppress hyperglycemia and oxidative stress in diabetic rats (Lee et al. 2004). However, the study suggested that dietary supplementation with sea tangle or sodium alginate did not affect plasma glucose and lipid peroxide levels.

When diabetes develops, lipid metabolism is abnormally affected and lipid peroxide and blood lipid levels increase. We found that the alginate-free residue of

Table 8 Hepatic glutathione and lipid peroxide contents of bromobenzene-injected rats treated with sea tangle

Group	Dose (mg/kg)	Glutathione	Lipid peroxide
		GSH (mg/g)	TBARS (nmol/g)
Normal		6.41 ± 0.30	17.8 ± 2.10
Bromobenzene		4.83 ± 0.39	50.0 ± 4.26
Goseong sea tangle	100	5.06 ± 0.19	47.6 ± 3.49
	200	5.18 ± 0.22	45.9 ± 4.66
Alginate-free sea tangle residue	100	5.18 ± 0.27	35.2 ± 1.69
	200	5.21 ± 0.33	31.1 ± 2.47

Values are expressed as mean ± SD ($n = 8$) for groups of six replicates.
Different superscripts in each data column indicate significant differences ($p < 0.05$) between groups
GSH reduced glutathione, *TBARS* thiobarbituric acid-reactive substances

Goseong sea tangle reduced serum, blood, and hepatic lipid levels in hyperlipidemic rats, although Goseong sea tangle did not. This indicates that alginate-free Goseong sea tangle residue could be used to prevent and treat complications from diabetes, in addition to its blood glucose-lowering effect. Thus, we hypothesize that the alginate-free residue of Goseong sea tangle contains components which may exert a protective effect against diabetes.

The present study also evaluated the effects of alginate-free sea tangle residue on several xenobiotic-metabolizing hepatic enzymes in rats injected with bromobenzene. Bromobenzene is a toxic industrial solvent which elicits toxicity predominantly in the liver, where it causes centrilobular necrosis (Park et al. 2005). Although formation of secondary quinone metabolites (Slaughter and Hanzlik 1991; Buben et al. 1988; Narasimhan et al. 1988) and hydrogen peroxide (Wu et al. 1994) has been proposed as a mechanism of action in the toxicity of bromobenzene, much of the chemical's toxicity is known to be associated with cytochrome P450-mediated phase I metabolism to reactive epoxide intermediates (Rogers et al. 2002).

Our results demonstrate that i.p. injection of bromobenzene modulates phase I enzyme activities in the rat liver. The activities of the cytochrome P450-dependent monooxygenases AMND and ANH increased significantly following bromobenzene injection. This increase was suppressed by treatment with Goseong sea tangle residue.

The toxic epoxide intermediate of bromobenzene, produced upon oxidation by cytochrome P450-dependent phase I enzymes, can be detoxified by several pathways, including hydration to 3,4-dihydrodiol catalyzed by EPH (Cohen et al. 1997; Pumford and Halmes 1997). Hepatic EPH activity decreased significantly following bromobenzene injection, but this decrease was inhibited by pretreatment with the Goseong sea tangle residue.

The present study shows that administration of bromobenzene induces oxidative modifications to mitochondrial proteins. Therefore, it is probable that bromobenzene-induced elevations in reactive oxygen species, lipid peroxides, and protein carbonyls may affect the integrity of the mitochondrial membrane, which would lead to mitochondrial dysfunction and, ultimately, to some of the toxic effects observed in this study. However, the alginate-free residue of Goseong sea tangle protected mitochondria against this oxidative damage.

GSH is an important cellular reductant and constitutes the first line of defense against free radicals, peroxides, toxic compounds, and chemically induced hepatotoxicity (Raja et al. 2007). A significant decrease in the level of GSH observed in the rats treated with bromobenzene may be attributable to its increased utilization, which results in increased vulnerability to free radical damage

(Gopi and Setty 2010). However, administration of alginate-free Goseong sea tangle residue increased GSH levels significantly. This may be due to the increase in de novo synthesis and/or GSH regeneration. The alginate-free residue of Goseong sea tangle increased the activity of antioxidant enzymes, counteracting oxidative stress.

Lipid peroxide levels are an index of membrane damage, and increased levels can lead to alterations in membrane structure and function. In this study, elevation in lipid peroxide levels was observed following the administration of bromobenzene and is attributed to the enhanced production of reactive oxygen species (Gopi and Setty 2010). However, administration of Goseong sea tangle residue prevented these changes. The antioxidant effect of the alginate-free residue of Goseong sea tangle may not be due to the GSH-dependent removal of hydroperoxide (Park et al. 2005).

Conclusions

We showed that the alginate-free residue of Goseong sea tangle reduced the perturbation of serum and hepatic lipid levels in hyperlipidemic rats. We also showed the residue's effects on several xenobiotic-metabolizing hepatic enzymes in rats injected with bromobenzene. Altogether, our data suggest that the alginate-free residue of sea tangle contains physiologically unknown compounds which may protect against hyperlipidemic atherosclerosis.

Abbreviations
AMND: Aminopyrine *N*-demethylase; ANH: Aniline hydroxylase; EPH: Epoxide hydrolase; GSH: Glutathione; HDL-C: High-density lipoprotein cholesterol; TBARS: Thiobarbituric acid-reactive substances; TC: Total cholesterol; TG: Triglycerides; TSO: *Trans*-stilbene oxide

Acknowledgements
This work was supported by the National Marine Biodiversity Institute of Korea Research program (2017M01400).

Funding
This work was supported by the National Marine Biodiversity Institute of Korea Research program (2017M01400).

Authors' contributions
MJY carried out the hepatic enzyme analysis and drafted the manuscript. GC carried out the glucose and cholesterol analyses. JML extracted the hepatic fat and prepared the enzyme sources. SYC participated in the design of the study and helped to draft the manuscript. DSL designed the study and completed the manuscript. All authors read and approved the final manuscript.

Competing interests
The authors declare that they have no competing interests.

Author details

[1]Department of Applied Research, National Marine Biodiversity Institute of Korea, Seocheon 33662, South Korea. [2]Department of Food Processing and Distribution, Gangneung-Wonju National University, Gangneung 25457, South Korea.

References

Baigent C, Blackwell L, Emberson J, Holland LE, Reith C, Bhala N, Peto R, Barnes EH, Keech A, Simes J, Collins R, Cholesterol Treatment Trialists' (CTT) Collaborators. Efficacy and safety of more intensive lowering of LDL cholesterol: a meta-analysis of data from 170,000 participants in 26 randomised trials. Lancet. 2010;376:1670–81.

Bidlack WR, Lowery GL. Multiple drug metabolism: p-nitroanisole reversal of acetone enhanced aniline hydroxylation. Biochem Pharmacol. 1982;31:311–7.

Boyne AF, Ellma GL. A methodology for analysis of tissue sulfhydryl components. Anal Biochem. 1972;46:639–53.

Buben JA, Narasimhan N, Hanzlik RP. Effects of chemical and enzymic probes on covalent binding of bromobenzene and derivatives. Evidence of quinones as reactive metabolites. Xenobiotica. 1988;18:501–10.

Cohen SD, Pumford NR, Khairallah EA, Boekelheide K, Pohl LR, Amouzadeh HR, Hinson JA. Selective protein covalent binding and target organ toxicity. Toxicol Appl Pharmacol. 1997;143:1–12.

Derosa G, Salvadeo S, Cicero AF. Prospects for the development of novel anti-hyperlipidemic drugs. Curr Opin Investig Drugs. 2006;7:826–33.

Folch J, Lees M, Sloane Stanley GH. A simple method for the isolation and purification of total lipids from animal tissues. J Biol Chem. 1957;226:497–509.

Gopi S, Setty OH. Protective effect of Phyllanthus fraternus against bromobenzene induced mitochondrial dysfunction in rat liver mitochondria. Food Chem Toxicol. 2010;48:2170–5.

Han J, Kang S, Choue R, Kim H, Leem K, Chung S, Kim C, Chung J. Free radical scavenging effect of Diospyros kaki, Laminaria japonica and Undaria pinnatifida. Fitoterapia. 2002;73:710–2.

Harchaoui KE, Visser ME, Kastelein JJ, Stroes ES, Dallinga-Thie GM. Triglycerides and cardiovascular risk. Curr Cardiol Rev. 2009;5:216–22.

Hasegawa LS, Hammock BD. Spectrophotometric assay for mammalian cytosolic epoxide hydrolase using trans-stilbene oxide as the substrate. Biochem Pharmacol. 1982;31:1979–84.

Jin DQ, Li G, Kim JS, Yong CS, Kim JA, Huh K. Preventive effects of Laminaria japonica aqueous extract on the oxidative stress and xanthine oxidase activity in streptozotocin-induced diabetic rat liver. Biol Pharm Bull. 2004;27:1037–40.

Lee KS, Choi YS, Seo JS. Sea tangle supplementation lowers blood glucose and supports antioxidant systems in streptozotocin-induced diabetic rats. J Med Food. 2004;7:130–5.

Lowry OH, Rosebrough NJ, Farr AL, Randall RJ. Protein measurement with the Folin phenol reagent. J Biol Chem. 1951;193:265–75.

McKenney JM. Pharmacotherapy of dyslipidemia. Cardiovasc Drugs Ther. 2001;15:413–22.

Murata M, Ishihara K, Saito H. Hepatic fatty acid oxidation enzyme activities are stimulated in rats fed the brown seaweed, Undaria pinnatifida (Wakame). J Nutr. 1999;129:146–51.

Narasimhan N, Weller PE, Buben JA, Wiley RA, Hanzlik RP. Microsomal metabolism and covalent binding of [^{13}C/^{14}C]-bromobenzene. Evidence for quinones as reactive metabolites. Xenobiotica. 1988;18:491–9.

Nash T. The colorimetric estimation of formaldehyde by means of the Hantzsch reaction. J Biol Chem. 1953;55:412–6.

Ohkawa H, Ohishi N, Yaki K. Assay for lipid peroxides in animal tissues by thiobabituric acid reaction. Anal Biochem. 1979;95:351–8.

Okai Y, Higashi-okai K, Nakamura S. Identification of heterogenous antimutagenic activities in the extract of edible brown seaweeds, Laminaria japonica (Makonbu) and Undaria pinnatifida (Wakame) by the umu gene expression system in Salmonella typhimurium (TA1535/pSK1002). Mutat Res. 1993;30:63–70.

Park JS, Han WD, Park JR, Choi SH, Choi JW. Change in hepatic drug metabolizing enzymes and lipid peroxdation by methanol extract and major compound of Orostachys japonicas. J Ethnopharmacol. 2005;102:313–8.

Park PJ, Kim EK, Lee SJ, Park SY, Kang DS, Jung BM, Kim KS, Je HY, Ahn CB. Protective effects against H$_2$O$_2$-induced damage by enzymatic hydrolysates of an edible brown seaweed, sea tangle (Laminaria japonica). J Med Chem. 2009;12:159–66.

Pilz S, Scharnagl H, Tiran B, Seelhorst U, Wellnitz B, Boehm BO, Schaefer JR, März W. Free fatty acids are independently associated with all-cause and cardiovascular mortality in subjects with coronary artery disease. J Clin Endocrinol Metab. 2006; 91:2542–7.

Pumford NR, Halmes NC. Protein targets of xenobiotic reactive intermediates. Annu Rev Pharmacol Toxicol. 1997;4:349–59.

Raja KFH, Ahamed N, Kumar V, Mukherjee K, Bandyopadyay A, Mukherjee PK. Antioxidant effect of Cytisus scoparius against carbon tetrachloride treated liver injury in rat. J Ethnopharmacol. 2007;109:41–7.

Rogers JF, Nafziger AN, Bertino JS Jr. Pharmacogenetics affects dosing, efficacy, and toxicity of cytochrome P450-metabolized drugs. Am J Med. 2002;113:746–50.

Slaughter DE, Hanzlik RP. Identification of epoxide- and quinone-derived bromobenzene adducts to protein sulfur nucleophiles. Chem Res Toxicol. 1991;4:349–59.

Sliskovic DR, White AD. Therapeutic potential of ACAT inhibitors as lipid lowering and anti-atherosclerotic agents. Trends Pharmacol Sci. 1991;12:194–9.

Talbert R. Hyperlipidemia, DePiro textbook of therapeutics. Connecticut: Appleton and Lange; 1997. p. 357–86.

Torsdottir I, Alpsen M, Holm G, Sandberg AS, Tolli J. A small dose of soluble alginate-fiber affects postprandial glycemia and gastric emptying in humans with diabetes. J Nutr. 1991;121:795–9.

Wald NJ, Law MR. Serum cholesterol and ischemic heart disease. Atherosclerosis. 1995;118(Suppl):S1–5.

Wang Y, Tang XX, Yang Z, Yu ZM. Effect of alginic acid decomposing bacterium on the growth of Laminaria japonica (Phaephyceae). J Environ Sci (China). 2006;18:543–51.

Wu J, Karlsson K, Danielsson A. Effects of vitamins E, C, and catalase on bromobenzene-and hydrogen peroxide-induced intracellular oxidation and DNA single strand breakage in Hep G2 cells. J Hepatol. 1994;26:669–77.

Zhang X, Wu C, Wu H, Sheng L, Su Y, Zhang X, Luan H, Sun G, Sun X, Tian Y, Ji Y, Guo P, Xu X. Anti-hyperlipidemic effects and potential mechanisms of action of the caffeoylquinic acid-rich Pandanus tectorius fruit extract in hamsters fed a high fat-diet. PLoS One. 2013;8:e61922.

Characterization and expression profiles of aquaporins (AQPs) 1a and 3a in mud loach *Misgurnus mizolepis* after experimental challenges

Sang Yoon Lee[1], Yoon Kwon Nam[1] and Yi Kyung Kim[2]* (iD)

Abstract

Two distinct cDNAs encoding aquaporins (mmAQPs 1a and 3a) were isolated and characterized from mud loach *Misgurnus mizolepis*. The identified mud loach AQP cDNAs encode for polypeptides of 260 and 302 amino acids. Topology predictions confirmed six putative membrane-spanning domains connected by five loops and the N- and C-terminal domains being cytoplasmic. The mud loach AQPs 1a and 3a showed broad distribution in multiple tissues including immune-responsive tissues as well as osmoregulatory tissues. Hence, the diversity of AQP distribution and expression possibly indicated its differential functions in the regulation of fluid movement in response to environmental stimuli. The transcription of mmAQP genes was differentially modulated by immune challenges. In particular, the mmAQP3a expression level in the liver was more responsive to immune challenges than that of mmAQP1a. Taken together, fish stimulation or infection resulted in significant modulation of mud loach AQP genes, suggesting potential functional roles of these proteins in piscine pathophysiological process.

Keywords: *Misgurnus mizolepis*, Aquaporins, Lipopolysaccharide, Polyinosinic: polycytidylic acid, Bacterial infection

Background

Teleostean species possess osmoregulatory system and therefore can overcome salt and water balances along with environmental fluctuations. The epithelia such as gill or intestine are the main site for sensing changes in the salinity, contribute to the alleviation of osmotic stress, and occasionally harbor mechanism to avoid infection (Fiol and Kültz 2007). According to change in the surrounding environment, the macromolecules, such as proteins, trigger complex responses, e.g., cell volume change, changes in cytoskeletal organization or whole tissue remodeling (Fiol and Kültz 2007; Henry et al. 2003).

Aquaporins (AQPs) are important mediators of the movement of water and other small solutes and cell volume regulation. At least 13 aquaporin isoforms have been identified in different organisms ranging from bacteria to humans (Kozono et al. 2003; King et al. 2004) and are categorized into three subfamilies on the basis of their substrate permeability: water-selective aquaporins, aquaglyceroporin (which is permeable to glycerol and certain small, uncharged solutes), and unorthodox subgroup (Ishibashi et al. 2011). The AQPs are ubiquitously expressed in a number of somatic tissues such as the gills, kidney, eye, skeletal muscle, and lung. In addition to their well-established osmoregulatory function, AQPs appear to play important roles in physiological processes including neural signal transduction, skin hydration, brain swelling, and cell migration (Zhu et al. 2011; Verkman 2012). Furthermore, considering its presence in immune-mediated cell, AQPs might also have potential functional role in linkage between unfavorable water flux through its activities and the epithelial barrier immune system (Zhu et al. 2011).

The existence of AQP expression in different tissues clearly indicates its potential functional role in a wide range of biological processes (Gomes et al. 2009;

* Correspondence: yikyung1118@gwnu.ac.kr
[2]Department of Marine Biotechnology, Gangneung-Wonju National University, 7 Jukheon-gil, Gangneung, Gangwon-do 25457, South Korea
Full list of author information is available at the end of the article

Watanabe et al. 2009; Boj et al. 2015). In freshwater prawn *Macrobrachium rosenbergii*, change in salinity has showed a direct or indirect effect on the respiratory metabolism, mortality, growth, and even immune response later (Cheng and Chen 2000; Moshtaghi et al. 2016). To date, extensive studies have been conducted on a number of fish species to investigate tissue-specific functional roles of AQPs under different salinity environments (Cutler and Cramb 2000; Watanabe et al. 2005; Giffard-Mena et al. 2007; Kim et al. 2010; Choi et al. 2013). In addition, many AQP isoforms have been detected in tissues that are not directly involved in osmoregulation, as evidenced by several studies (Watanabe et al. 2009; Kim et al. 2014; Madsen et al. 2014; Boj et al. 2015). However, functional roles of piscine aquaporin genes with regard to immune or bacterial challenges have still remained unexplored. The limited information may hinder a comprehensive understanding on the coordinated role of AQP isoforms in the maintenance of physiological homeostasis.

Mud loach *Misgurnus mizolepis* (Teleostei; Cypriniformes) is a promising candidate for freshwater aquaculture whose market demand is gradually increasing in Korea. In addition to its commercial importance, mud loach has the potential to be used as an experimental model animal for investigating various types of theoretical and practical issues. Some interesting biological features of this species include small adult size, high fecundity, year around spawning capability under controlled conditions, and relatively well-established techniques for its genetic manipulation (Nam et al. 2011; Cho et al. 2012). Considering these facts, mud loach could be an ideal model for investigating the functional roles of the AQP-mediated cellular process in the pathophysiological situation.

Edwardsiella tarda was chosen for this study because it has been demonstrated as a potential pathogen capable of causing disease and stimulating the immune responses in mud loach (Lee et al. 2011; Nam et al. 2011). In the species most commonly associated with *E. tarda* infection, the disease is a generalized septicemia with clinical sign including extensive skin lesions, bloody ascites in abdominal cavity, and damage to internal organs (Plumb 1999).

In line with our long-term goal for comprehensive understanding on the osmophysiology and innate immunity of mud loach, the objective of this study was to examine the expression patterns of AQP genes with regard to immunostimulant or bacterial infection. In the current study, we focused on AQPs 1a and 3a from mud loach, genetically characterized two aquaporin isoforms, and examined their expression patterns in response to immune stimulatory challenges.

Methods
Isolation of mud loach aquaporin 1a and 3a cDNAs
Mud loach *M. mizolepis* expressed sequence tag database (conducted from total RNA) was surveyed to isolate the cDNA sequence for aquaporin. Several partial mud loach AQP clones showing the high homology to previously known vertebrate AQPs were identified using a homology search in NCBI GenBank. Based on the contig assemblies using Sequencher software (Gene Codes Co., Ann Arbor, MI, USA), two distinct isoforms (designated mmAQP1a and mmAQP3a) were identified from various tissues. The full-length cDNA sequence of each aquaporin gene was confirmed from RT-PCR and/or vectorette PCR with an SK vector primer (Stratagene, La Jolla, Ca, USA) and specific primer pair sets (mmAQP1a FW/RV [for *mmAQP1a*] or mmAQP3a FW/RV [for *mmAQP3a*]) (Table 1). The representative

Table 1 List of oligonucleotide primers used in this study

Primer name	Sequence (5'-3')	Thermal cycling conditions	Application
mmAQP1a FW	CAGAAATCTCCACATTACCAGC	35 cycles at 94 °C for 20 s, 58 °C for 20 s, 72 °C for 1.5 min, followed by a final elongation at 72 °C for 3 min	Isolation of mud loach aquaporin cDNAs
mmAQP1a RV	TCCTGAGGTACATACTGATTC		
mmAQP3a FW	ACACACGTTCAAGGGAAAGC		
mmAQP3a RV	CTGGCTGGATTCACAGCATA		
mm18S rRNA RV	GGTTTCCCGTGTTGAGTCAA	Reverse transcription at 37 °C for 60 min	Preparation of normalization control
q-mmAQP1a FW	ATGAGAGTGCTGGTCTCTGG	45 cycles at 94 °C for 20 s, 60 °C for 20 s and 72 °C for 20 s	Real-time PCR assay of mud loach aquaporin mRNAs Real-time PCR assay of 18S rRNA control
q-mmAQP1a RV	AAAGACAGCTTCACAATTGC		
q-mmAQP3a FW	TATGGAGGAGAATGTGAAGC		
q-mmAQP3a RV	AATCTGGAGATGTGCAGCGT		
q-mm18S rRNA FW	AAGCTCGTAGTTGGATCTCG		
q-mm18S rRNA RV	CCTAGCTGCGGTATTCAGGC		

Each PCR amplification reaction was performed with an initial denaturation step at 94 °C for 2 min

cDNA sequences for each AQP isoform were determined with multiple PCR clones ($n \geq 6$) and/or amplified PCR products.

Bioinformatic sequence analysis

Protein-coding sequences for each AQP isoform were obtained using the open reading frame (ORF) finder (https://www.ncbi.nlm.nih.gov/orffinder/) (Wheeler et al. 2003). The molecular mass and theoretical isoelectric point (pI) value were computed using ExPASy ProtParam tool for each isoform (http://web.expasy.org/protparam/) (Gasteiger et al. 2005). We find the corresponding teleostean and human AQPs from BLAST and/or the Ensembl genome browser (http://www.ensembl.org/index.html) to examine their homology. ORFs of mud loach aquaporin were aligned with teleostean and human orthologues using the multiple sequence alignment programs CLUSTAL W or CLUSTAL X 1.81 (Thompson et al. 1994; Chenna et al. 2003). The information about GenBank accession numbers or Ensembl codes for aquaporin sequences are provided in Table 2. Topology prediction for deduced amino acid sequences of AQPs was performed with the software TMHMM (http://www.cbs.dtu.dk/services/TMHMM/) (Krogh et al. 2001).

Preparation of tissue samples for basal expression assay

For the tissue distribution assay of mmAQP transcripts, ten somatic tissues (brain, eye, fin, gill, heart, intestine, kidney, liver, skeletal muscle, and spleen) and two gonad tissues (ovary and testis) were derived from 12 healthy individuals (average body weight $= 9.3 \pm 2.5$ g). Upon surgically removed, biological samples were immediately frozen on dry ice and stored at -80 °C until use.

Immune challenges

To examine whether the expression of mmAQP genes is induced by inflammatory stimuli and immune challenge, lipopolysaccharide (LPS; *Escherichia coli* 0111:B4) or polyinosinic: polycytidylic acid [poly(I:C)] injection and bacterial challenge (*E. tarda*; Gram negative) (Kwon et al. 2005) were performed in vivo. First, LPS (Sigma-Aldrich, St Louis, MO, USA; 5 µg/g body weight [BW]) or poly(I:C) (Sigma-Aldrich, 25 µg/g BW) was injected intraperitoneally into fish individuals (10.5 ± 2.1 g; $n = 8$ for each group). Saline control ($n = 8$) were done with this study. Injection volume was 100 µL. After injection, each group was allocated into each 60-L tanks at 25 °C and no feed was supplied during experimental period. For LPS and poly(I:C) challenges, immune-relevant

Table 2 Amino acid sequence identities of mud loach AQPs 1a and 3a with other orthologues

	Species	Gene name	Accession no.	Identity (%)
mmAQP1a	*Oryzias dancena*	AQP1	AB759557	80
	Osmerus mordax	AQP1a	ACO09149	80
	Danio rerio	AQP1a	NP_996942	87
	Acanthopagrus schlegelii	AQP1a	ABO38816	80
	Dicentrarchus labrax	AQP1a	ABI95464	79
	Fundulus heteroclitus	AQP1a	ACI49538	80
	Danio rerio	AQP1b	NP_001129154	63
	Solea senegalensis	AQP1b	AAV34612	57
	Salmo salar	AQP1b	ACI33306	72
	Sparus aurata	AQP1b	EF011740	61
	Anguilla anguilla	AQP1b	CAD92028	63
	Homo sapiens	AQP1	BAG70089	60
mmAQP3a	*Oryzias dancena*	AQP3	AB759557	70.4
	Danio rerio	AQP3a	EU341833	84.8
	Oreochromis mossambicus	AQP3a	BAD20708	72.2
	Dicentrarchus labrax	AQP3a	ABG36519	69.9
	Fundulus heteroclitus	AQP3a	ACI49539	68.8
	Danio rerio	AQP3b	EU341832	67.9
	Anguilla anguilla	AQP3b	CAC85286	73.6
	Anguilla japonica	AQP3	BAH89253	72
	Astyanax mexicanus	AQP3	XP_007238017	69
	Homo sapiens	AQP3	AB001325	64.7

tissues (kidney, intestine, liver, and spleen) were surgically removed from three individuals in each group at 24 h post challenge.

On other hand, the bacterial challenge was carried out by injecting 1×10^6 cell of *E. tarda* suspended in 100 μL of phosphate-buffered saline (PBS, pH 7.4) intraperitoneally into each individual. Twenty-four individuals (same-size as above) were given *E. tarda* injection, and other 24 individuals were given PBS alone as a control group. Each group (*E. tarda*- or PBS-injected) was transferred to each 100-L tank at 25 °C, and tissue samples were obtained individually from three randomly chosen fish belonging to each tank at 24, 48, and 72 h post injection, respectively. Tissue samples were harvested, frozen, and stored as described above.

RT-PCR analysis of mmAQP transcripts

Total RNA was extracted from dissected tissues using the RNeasy® Plus Micro Kit (Qiagen, Hilden, Germany) including DNase I treatment step. An aliquot of the total RNA (2 μg) was reverse transcribed into cDNA in a reaction volume of 20 μl using the Omniscript® Reverse Transcription Kit (Qiagen). The reverse transcription reaction include an oligo-d(T)$_{20}$ primer (1 μM final concentration) and a mud loach 18S rRNA (0.1 μM) reverse primer, as described by Cho et al. (2012). The use of 18S rRNA as the internal standard can be a valuable alternative for quantifying genes of interest, but it may reduce the variation of expression.

The reaction conditions were performed according to the manufacturer's protocol. RT-PCR reactions were performed in 25-μL reaction volumes including 2-μl cDNA, 2-μl primers, 6.5-μl distilled water, and 12.5-μl 2× iQ SYBR Green Supermix (Bio-Rad, Hercules, CA, USA). Reaction performed on the iCycler iQ Real-time Detection System (Bio-Rad). The plasmid DNAs containing the amplified parts of target mRNAs were prepared as standard samples. The 231-bp (for AQP1a) and 208-bp (for AQP3a) aquaporin gene fragments were amplified with the specific primer pairs, q-mmAQP1a-FW/RV and q-mmAQP3a-FW/RV, respectively (Table 1). Basal expression level of AQP transcripts in tissue types was represented based on ΔCt (Ct of the AQP gene subtracted from the Ct of the 18 s RNA gene). On the other hand, the relative expressions of AQP transcripts in the stimulated groups were expressed as the fold change to non-treated control by using the formula $2^{-\Delta\Delta Ct}$ (Kubista et al. 2006; Schmittgen and Livak 2008). Each reaction was performed at least in triplicates.

Statistical analysis

Numerical data were expressed as means ± S.E.M. Statistical differences were determined by one-way analysis of variance (ANOVA), followed by Duncan's multiple range

tests. All statistical analysis were performed using software SPSS version 10.0 (SAS Inc., Cary, NC, USA), and difference was considered to be significant when $P < 0.05$.

Results

Characteristics of mud loach AQP cDNA 1a and 3a isoforms

We isolated two distinct aquaporin genes from the mud loach. They are 1230 and 1609 bp in the length and contain ORFs of 780 and 906 bp (excluding the termination codon) encoding a polypeptides of 287 and 306 amino acid, respectively. The calculated molecular masses of each isoform are 27.3 and 32.8 kDa, with theoretical pI values of 6.05 and 6.15, respectively. The nucleotide sequences of the two AQP cDNA sequences have been assigned on GenBank under the accession numbers AB971265 (mmAQP1a) and AB971266 (mmAQP3a). The mud loach AQP isoforms showed very lower level of sequence homology (18%) with each other. From multiple sequence alignments with other orthologs, the amino acid sequence of mud loach AQPs shared considerable identities with those from teleostean AQPs (Table 2). The basic features of typical AQP channel structure were observed in mud loach AQP1a and AQP3a isoforms. According to topology prediction for these proteins, those proteins possess six potential transmembrane helices connected by five loops. Both the amino and carboxyl termini are extended into the cytoplasmic side (Fig. 1). Importantly, the two identical asparagine-proline-alanine (NPA) motifs are located in the second and fifth loops, which are the pore-forming signature motifs for AQPs playing a crucial role in the water uptake (Ishibashi et al. 2011). In mmAQP1a, a cysteine residue in position 180, proximal to the C-terminal NPA motif, is possibly pivotal for the inhibition of water permeability by mercurial reagent (Preston et al. 1993).

Basal expression assay of mud loach AQPs 1a and 3a

Under the present RT-PCR conditions, AQP1a and AQP3a transcripts were ubiquitously distributed in all tissue types examined, although the basal levels were significantly different among tissues. In particular, the mRNA levels of AQP1a were higher in non-osmogulatory tissues, brain, eye, heart, and spleen than osmoregulatory tissues, for example, the intestine, kidney, and gill. On the other hand, AQP3a transcripts in non-stimulated fish were detectable markedly in fin, followed by gill, eye and intestine (Fig. 2).

AQP gene expression profiles after immune challenge

The mud loach AQP1a mRNA was rapidly induced by LPS injection in all tissues, and the fold change relative to the expression level in the saline-injected control was

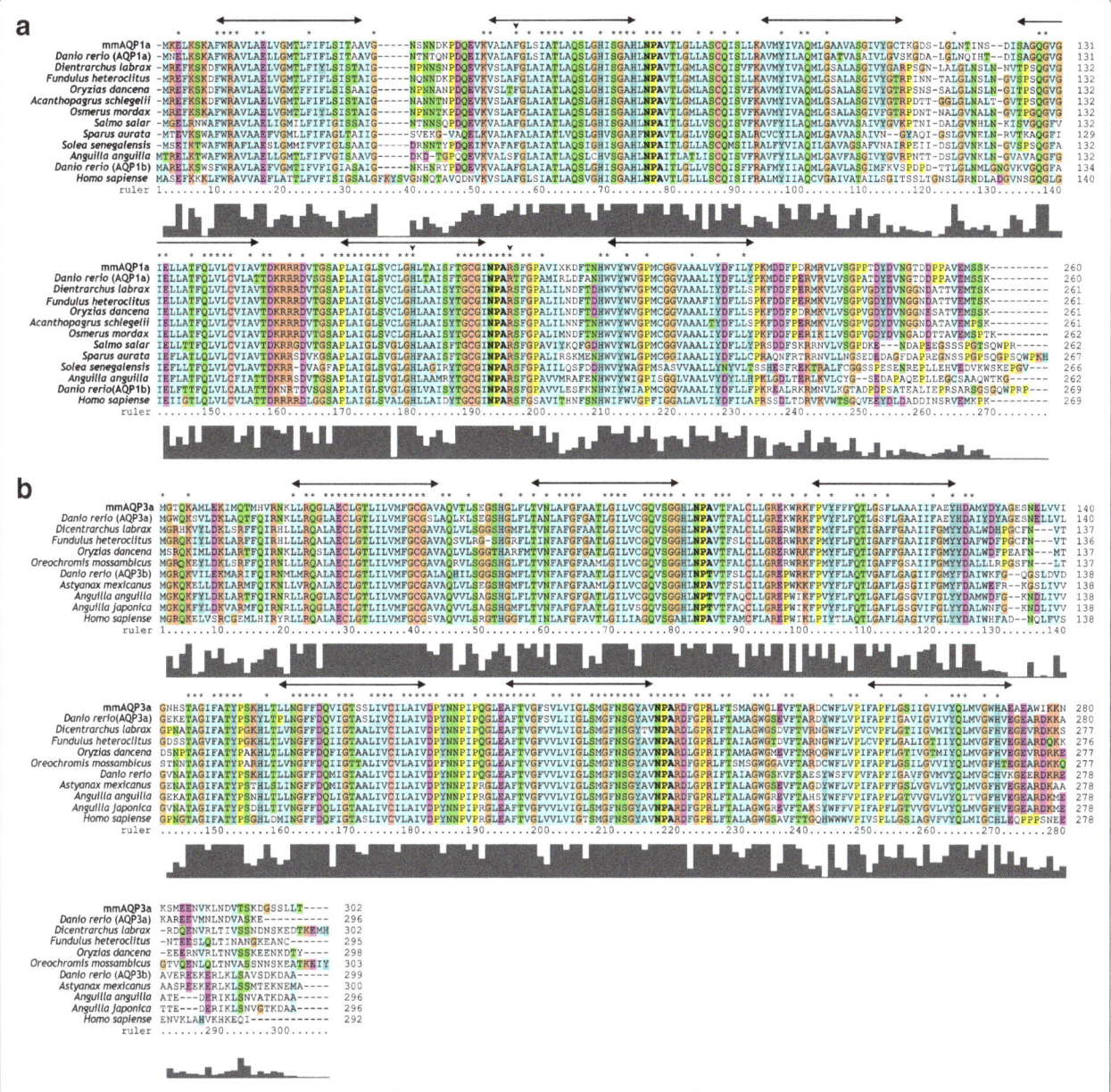

Fig. 1 Multiple amino acid sequence alignments of mud loach *Misgurnus mizolepis* AQP isoforms 1a (**a**) and 3a (**b**) along with those from other selected vertebrate species. Asterisks and hyphens indicate identical residues and gaps introduced for optimum alignments, respectively. The arrowhead above mmAQP1a showed the conserved residues Phe[56], His[180], and Arg[195] (mud loach AQP1a numbering). Two NPA motifs are shown in bold letters. The putative transmembrane locations of membrane-spanning domains are indicated above the alignment. The gray histogram below the ruler depicts sequence conservation between amino acid residues in given AQP isoforms

higher in the liver (1.3-fold) than in the intestine (0.2-fold), kidney (0.5-fold), and spleen (0.4-fold) (Fig. 3a). Also, the transcriptional response of AQP1a to poly(I:C) at 24 h post-injection showed levels highest in the kidney (1.6-fold) compared with other tissues (Fig. 3a). In contrast, the LPS injection significantly stimulated the transcription of AQP3a, which were greatest in the intestine (15.9-fold), followed by the liver (6.14-fold) and spleen (1.05-fold) (Fig. 3b). Moreover, the liver showed a

predominant increase in mmAQP3a transcripts of 23.38-fold induction by poly(I:C) stimulation.

AQP gene profiles after bacterial challenge

An in vivo bacterial injection was conducted with the known pathogen *E. tarda* (Gram negative; FSW910410), which causes edwardsiellosis in mud loach (Fig. 4). The expression patterns of mud loach AQP1a and AQP3a transcripts differed in the intestine, kidney, liver, and

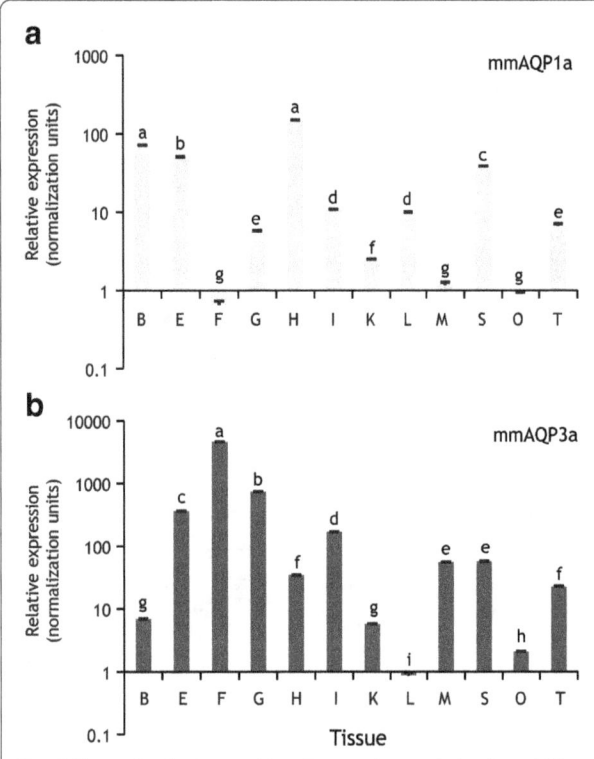

Fig. 2 Tissue distribution and basal expression analysis of mmAQPs 1a (**a**) and 3a (**b**) in adult tissues as assessed by real-time PCR. The mRNA level of AQPs was normalized against the 18S rRNA level in each sample. Abbreviations for tissues are brain (B), eye (E), fin (F), gill (G), heart (H), intestine (I), kidney (K), liver (L), muscle (M), spleen (S), ovary (O), and testis (T)

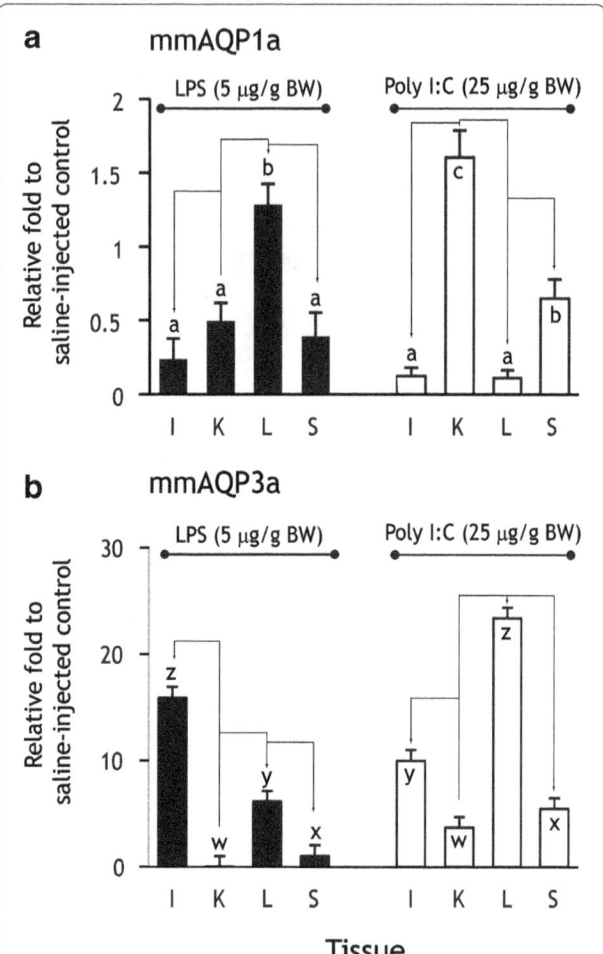

Fig. 3 Transcriptional responses of the mud loach AQPs 1a (**a**) and 3a (**b**) to immunostimulant exposures in differential tissues as assessed by real-time RT-PCR analysis. Levels of AQP isoforms in immunostimulant-exposed groups are expressed as fold changes relative to those in non-exposed control group after normalization against 18S rRNA standard. Mean ± SDs with same letters (a–c for mmAQP1a and x–z for mmAQP3a) are not significantly different based on ANOVA followed by Duncan's multiple range tests

spleen. The transcriptions of mud loach AQP1a and AQP3a in the intestine were significantly decreased by *E. tarda* challenges, and the reduction of intestinal AQP1a and AQP3a expressions was clearly time-dependent. Compared to that of intestinal response, in the kidney, the expression of the two isoform genes was modulated in opposite directions at each time point examined. In the liver, AQP1a isoform was less modulated by bacterial injection; no significant changes were evident during experimental period; instead, the mmAQP3a transcripts were responsive to *E. tarda* challenges, and the maximum induction of AQP3a expression was observed at 72 h after bacterial injection. The spleen showed the differential induction of the two AQP transcripts during *E. tarda* challenge in this study. The splenic mRNA levels of mmAQP1a were highly elevated (up to seven fold relative to the saline-injected controls) at 48 h post injection (hpi) and rapidly deceased at 72 hpi. In contrast, the mmAQP3a transcripts were slightly but significantly induced in time-dependent manner.

Discussion

We cloned two cDNAs encoding AQPs 1a and 3a from mud loach. The identified AQPs possess six

transmembrane domains which are found in known major intrinsic protein (MIP) structures as well as among aquaporin (Borgnia et al. 1999). The traditional NPA sequences have been harbored to form a characteristic pore between the membrane bilayer (Nielsen et al. 1999). These findings suggest that AQPs from mud loach function as water channels that facilitate passage of water and other small solutes through membrane, although we did not investigate the functional role in detail.

In the present study, transcripts encoding AQP1a were detectable in a wide array of tissues of mud loach, which has also been observed in marine medaka *Oryzias dancena*, Japanese medaka *O. latipes*, zebrafish, Japanese eel *Anguilla japonica* and black porgy, *Acanthopagrus*

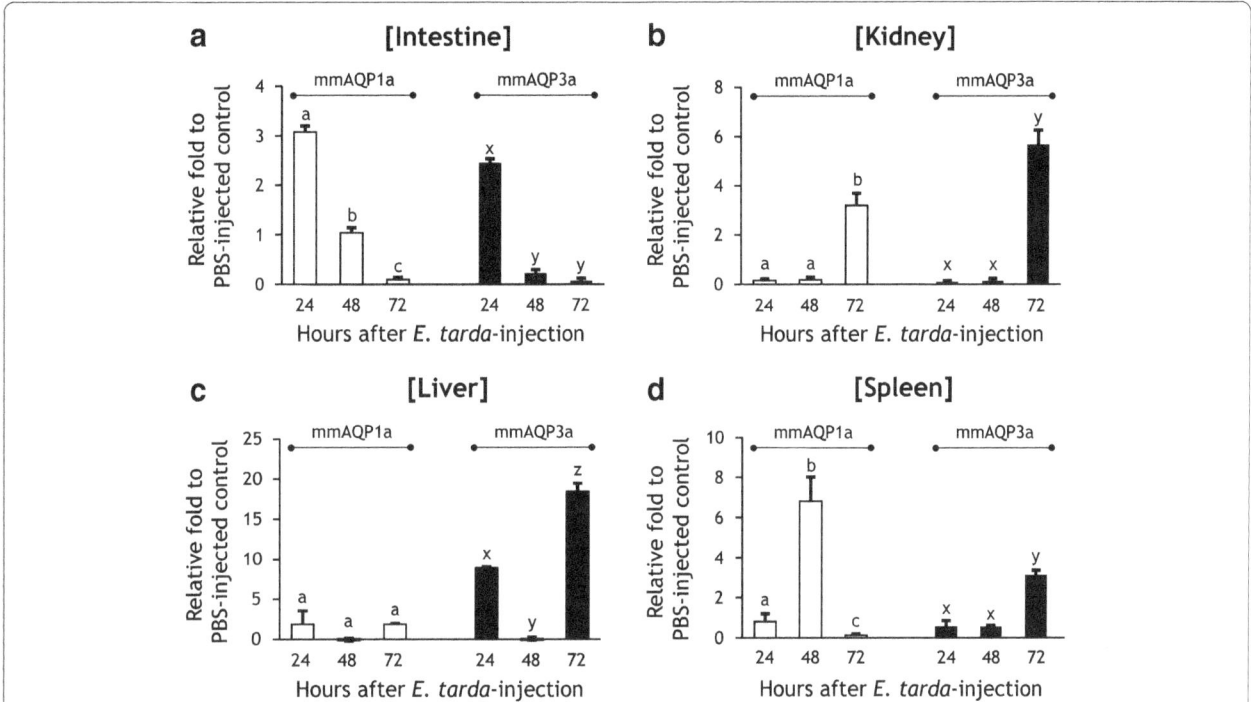

Fig. 4 Differential modulation of mud loach AQP isoforms by *Edwardsiella tarda* challenges (**a–d**). Relative AQP gene expression to reference gene was represented by fold change to control group. Mean ± SDs with same letters are not significantly different based on ANOVA followed by Duncan's multiple range tests

schlegeli (An et al. 2008; Tingaud-Sequeira et al. 2010; Kim et al. 2010, 2014; Madsen et al. 2014). Relatively higher mRNA expression levels were observed in the heart, brain, eye, spleen, and intestine but also present in lower levels in all other tissues examined, which in part may manifest a general expression in erythrocytes and endothelial barriers (Mobasheri and Marples 2004). The higher expression of cardiac mmAQP1a may be a main determinant of its role in myocardial fluid balance, as suggested by Japanese eel and silver sea bream *Sparus sarba* (Aoki et al. 2003; Deane et al. 2011). The splenic AQP has been considered to be involved in the trafficking of hematopoietic cells (Tyagi and Tangevelu 2010). The branchial AQP1a showed lower expression level, compared to those of osmoregulatory tissues. This result is somewhat contradictory to previous report that the gills of freshwater fish are the main pathway for water transport and can account for approximately 90% of the total body water influx (Cutler and Cramb 2000; Deane et al. 2011). Although specific mechanism should be investigated in the further study, the branchial AQP1a might act in concert with other homologs (e.g., AQP3a in this study) to prevent cell swelling. The testicular expression of mmAQP1a was noticeable, while lower expressed in ovary. In gilthead sea bream *Sparus aurata*, the distinct aquaporin paralogs (e.g., AQPs 0a, 1aa, 1ab, 7, 8b, 9b, and 10b) were involved in control of the fluid

balance during spermatogenesis (Boj et al. 2015). Hence, these facts suggest that testicular function in the mud loach is potentially associated with fine-tuned water control by aquaporin channel.

The observation of AQP3a in mud loach was comparable to that described for several teleost (Deane and Woo 2006; Tingaud-Sequeira et al. 2010; Kim et al. 2014). With reference to expression levels, mmAQP3a mRNAs were unequivocally predominant in the fin that is not primarily involved in osmoregulatory systems, which could be due to thin epidermis. The pattern observed in this study is similar to that reported in tilapia and medaka (Watanabe et al. 2005; Kim et al. 2014; Madsen et al. 2014). Another noticeable feature in mmAQP3a mRNA levels was observed in gill and eye, which are externally exposed organs in fish. Relatively higher AQP3a abundance in the gill suggested its involvement in possible osmoreception by mitochondrion-rich chloride cells (Watanabe et al. 2005). In addition, relatively high level was also found in the spleen, in accordance with several teleost species (Watanabe et al. 2005; Kim et al. 2014). However, mmAQP3a mRNA expression was found at very low level in liver, known as major detoxification organ. This expression pattern is similar to that observed previously in other teleostean fish (Watanabe et al. 2005; Tingaud-Sequeira et al. 2010; Madsen et al. 2014). Thus, the mud loach AQPs 1a and

3a, in common with other teleostean fish AQPs, exhibited broad distribution in multiple diverse tissues including immune-responsive and mucosal tissues exposed to external environment. Hence, the variety of AQP distribution and expression presumably suggested its differential functions in the regulation of water movement according to environmental stimuli.

Overall, the RT-PCR quantitative analysis with two AQP transcripts showed the significant response to the challenges. The mmAQP1a mRNAs in the liver and kidney were significantly induced by LPS or poly(I:C) injection. In contrast to expression levels of mmAQP1a transcripts, AQP3a expression was markedly elevated in some but not all tissues examined. In particular, the AQP3a transcript was significantly elevated in the liver or kidney in response to challenge. Furthermore, the significant AQP3a level after poly(I:C) stimulation was striking when compared to basal expression in the liver. Recent study has reported that hepatic AQP3 might be involved in both Kuffer cell migration and proinflammatory cytokine secretion in mammals, reflecting its involvement with immune response (Rodríguez et al. 2011). The expressions of mud loach aquaporin gene under *E. tarda* stimulus are differentially modulated during experimental period, as in the aforementioned results. The intestinal AQP1a and AQP3a transcript levels showed a significant reduction with time. Apart from serving as the site of nutrient uptake and osmoregulation, the intestine functions as a critical defense barrier to the external environment (Grosell 2011). The intestinal fluid balance in the fish as well as mammalian has been indicated to be cooperated by various factors, including hormones, intestinal contents, inflammatory factors, and feeding (Rombout et al. 2011; Zhu et al. 2016). Thus, the bacterial challenge may either directly or indirectly linked to unfavorable change of water balance in the intestine in the pathophysiological stress. In the kidney, the expression of two isoform genes was modulated in opposite directions in response to *E. tarda* challenge. When considering that the role of kidney in freshwater fish is to produce copious quantities of dilute urine, the significant fluctuation in renal AQP expression after bacterial challenge may be associated with the conditions demonstrating severe edema formation due to renal failure, as evidenced by zebrafish (Kramer-Zucker et al. 2005). Usually, the piscine liver performs not only the basically metabolic function including bile formation and excretion but also macrophage contributor (Paulsen et al. 2003; Wolf and Wolfe 2005). Accordingly, modulation of hepatic AQP3 transcript may result in disruption of physiological homeostasis, for example, bile secretory failure (Lehmann et al. 2008). Collectively, these data suggest that the alteration in AQP transcripts expression, especially AQP3a, may be involved in a significant way in the pathophysiology of fish and should be considered in further studies.

Conclusions

In current study, we observed ubiquitous distribution of mud loach AQPs in immune-relevant as well as osmoregulatory tissues. Immune challenge-induced changes in expression pattern of AQP3 indicate that this gene has important role to respond to inflammatory disease or condition. Further investigation is needed to decipher the importance of AQPs in dealing with water homeostasis during infection and inflammation in a finer detail.

Abbreviations
AQP: Aquaporin; EST: Expressed sequence tag; hpi: Hours post injection; LPS: Lipopolysaccharide;; MIP: Major intrinsic protein; NPA: Asparagine-proline-alanine; ORF: Open reading frame; poly(I:C): Polyinosinic: polycytidylic acid

Acknowledgements
We are grateful for the funding by National Research Foundation of Korea.

Funding
This study was supported by the National Research Foundation of Korea (NRF) grant funded by the Korea government (MSIP) (No. 2013R1A1A3013064).

Authors' contributions
YKK and SYL carried out the molecular biology and expression assay. YKK and YKN designed the study and data evaluation and drafted the manuscript. All authors read and approved the final manuscript.

Competing interests
The authors declare that they have no competing interests.

Author details
[1]Department of Marine Bio-Materials and Aquaculture, Pukyong National University, Busan 48513, South Korea. [2]Department of Marine Biotechnology, Gangneung-Wonju National University, 7 Jukheon-gil, Gangneung, Gangwon-do 25457, South Korea.

References
An KW, Kim NN, Choi CY. Cloning and expression of aquaporin 1 and arginine vasotocin receptor mRNA from the black porgy, *Acanthopagrus schlegli*: effect of freshwater acclimation. Fish Physiol Biochem. 2008;34:185–94.

Aoki M, Kaneko T, Katoh F, Hasegawa S, Tsutsui N, Aida K. Intestinal water absorption through aquaporin 1 expressed in the apical membrane of mucosal epithelial cells in seawater-adapted Japanese eel. J Exp Biol. 2003;206:3495–505.

Boj M, Chauvigné F, Zapater C, Cerdà J. Gonadotropin-activated androgen-dependent and independent pathways regulate aquaporin expression during teleost (*Sparus aurata*) spermatogenesis. PLoS One. 2015;10:e0142512.

Borgnia M, Nielsen S, Engel A, Agre P. Cellular and molecular biology of the aquaporin water channels. Annu Rev Biochem. 1999;68:425–58.

Cheng W, Chen JC. Effects of pH, temperature and salinity on immune parameters of the freshwater prawn Macrobrachium rosenbergii. Fish Shellfish Immunol. 2000;10:387–91.

Chenna R, Sugawara H, Koike T, Lopez R, Gibson TJ, Higgins DG, Thompson TD. Multiple sequence alignment with the clustal series of programs. Nucleic Acids Res. 2003;31:3497–500.

Cho YS, Kim BS, Kim DS, Nam YK. Modulation of warm-temperature-acclimation-associated 65-kDa protein genes (Wap65-1 and Wap65-2) in mud loach (Misgurnus mizolepis, Cypriniformes) liver in response to different stimulatory treatments. Fish Shellfish Immunol. 2012;32:662–9.

Choi YJ, Shin HS, Cho SH, Yamamoto Y, Ueda H, Lee J, Choi CY. Expression of aquaporin-3 and -8 mRNAs in the parr and smolt stages of sockeye salmon, Oncorhynchus nerka: effects of cortisol treatment and seawater acclimation. Comp Biochem Physiol A Mol Integr Physiol. 2013;165:228–36.

Cutler CP, Cramb G. Water transport and aquaporin expression in fish. In: Molecular Biology and Physiology of Water and Solute Transport. Springer US; 2000. p. 433-441.

Deane EE, Luk JC, Woo NY. Aquaporin 1a expression in gill, intestine, and kidney of the euryhaline silver sea bream. Front Physiol. 2011;21:2.

Deane EE, Woo NY. Tissue distribution, effects of salinity acclimation, and ontogeny of aquaporin 3 in the marine teleost, silver sea bream (Sparus sarba). Mar Biotechnol (NY). 2006;8:663–71.

Fiol DF, Kültz D. Osmotic stress sensing and signaling in fishes. FEBS J. 2007;274:5790–8.

Gasteiger E GA, Duvaud S, Wilkins MR, Appel RD, Bairoch A. Protein identification and analysis tools on the ExPASy server. In: Walker JM, editor. The Proteomics Protocols Handbook. Humana Press; 2005. p. 571–607.

Giffard-Mena I, Boulo V, Aujoulat F, Fowden H, Castille R, Charmantier G, Cramb G. Aquaporins molecular characterization in the sea-bass (Dicentrarchus labrax): the effect of salinity on AQP1 and AQP3 expression. Comp Biochem Physiol A. 2007;148:430–44.

Gomes D, Agasse A, Thiébaud P, Delrot S, Gerós H, Chaumont F. Aquaporins are multifunctional water and solute transporters highly divergent in living organisms. Biochim Biophys Acta. 2009;1788:1213–28.

Grosell M. The role of the gastrointestinal tract in salt and water balance. In: Grosell M, Farrell AP, Brauner CJ, editors. The multifunctional gut of fish. San Diego: Academic; 2011. p. 136–64.

Henry RP, Gehnrich S, Weihrauch D, Towle DW. Salinity-mediated carbonic anhydrase induction in the gills of the euryhaline green crab, Carcinus maenas. Comp Biochem Physiol A Mol Integr Physiol. 2003;136:243–58.

Ishibashi K, Kondo S, Hara S, Morishita Y. The evolutionary aspects of aquaporin family. Am J Phys. 2011;300:R566–76.

Kim YK, Lee SY, Kim BS, Kim DS, Nam YK. Isolation and mRNA expression analysis of aquaporin isoforms in marine medaka Oryzias dancena, a euryhaline teleost. Comp Biochem Physiol A Mol Integr Physiol. 2014;171:1–8.

Kim YK, Watanabe S, Kaneko T, Huh MD, Park SI. Expression of aquaporins 3, 8 and 10 in the intestines of freshwater- and seawater-acclimated Japanese eels Anguilla japonica. Fish Sci. 2010;76:695–702.

King LS, Kozono D, Agre P. From structure to disease: the evolving tale of aquaporin biology. Nat Rev Mol Cell Biol. 2004;5:687–98.

Kozono D, Ding X, Iwasaki I, Meng X, Kamagata Y, Agre P, Kitagawa Y. Functional expression and characterization of an archaeal aquaporin AqpM from Methanothermobacter marburgenesis. J Biol Chem. 2003;278:10649–56.

Kramer-Zucker AG, Wiessner S, Jensen AM, Drummond IA. Organization of the pronephric filtration apparatus in zebrafish requires Nephrin, Podocin and the FERM domain protein mosaic eyes. Dev Biol. 2005;285:316–29.

Krogh A, Larsson B, Von Heijne G, Sonnhammer EL. Predicting transmembrane protein topology with a hidden Markov model: application to complete genomes. J Mol Biol. 2001;305:567–80.

Kubista M, Andrade JM, Bengtsson M, Forootan A, Jonák J, Lind K, Sindelka R, Sjöback R, Sjögreen B, Strömbom L, Ståhlberg A, Zoric N. The real-time polymerase chain reaction. Mol Asp Med. 2006;27:95–125.

Kwon SR, Nam YK, Kim SK, Kim DS, Kim KH. Generation of Edwardsiella tarda ghosts by bacteriophage PhiX174 lysis gene E. Aquaculture. 2005;250:16–21.

Lee SY, Kim DS, Nam YK. Altered expression of mud loach (Misgurnus mizolepis; Cypriniformes) hepcidin mRNA during experimental challenge with non-pathogenic or pathogenic bacterial species. J Fish Pathol. 2011;24:279–87.

Lehmann GL, Carreras FI, Soria LR, Gradilone SA, Marinelli RA. LPS induces the TNF-α-mediated downregulation of rat liver aquaporin-8: role in sepsis-associated cholestasis. Am J Physiol Gastrointest Liver Physiol. 2008;294:G567–75.

Madsen SS, Bujak J, Tipsmark CK. Aquaporin expression in the Japanese medaka (Oryzias latipes) in freshwater and seawater: challenging the paradigm of intestinal water transport? J Exp Biol. 2014;217:3108–21.

Mobasheri A, Marples D. Expression of the AQP-1 water channel in normal human tissues: a semiquantitative study using tissue microarray technology. Am J Phys. 2004;286:C529–37.

Moshtaghi A, Rahi ML, Nguyen VT, Mather PB, Hurwood DA. A transcriptomic scan for potential candidate genes involved in osmoregulation in an obligate freshwater palaemonid prawn (Macrobrachium australiense). Peer J. 2016;4:e2520.

Nam YK, Cho YS, Lee SY, Kim BS, Kim DS. Molecular characterization of hepcidin gene from mud loach (Misgurnus mizolepis; Cypriniformes). Fish Shellfish Immunol. 2011;31:1251–8.

Nielsen S, Kwon TH, Christensen BM, Promeneur D, Frøkiaer J, Marples D. Physiology and pathophysiology of renal aquaporins. J Am Soc Nephrol. 1999;10:647–63.

Paulsen SM, Lunde H, Engstad RE, Robertsen B. In vivo effects of β-glucan and LPS on regulation of lysozyme activity and mRNA expression in Atlantic salmon (Salmo Salar L.). Fish Shellfish Immunol. 2003;14:39–54.

Plumb JA. Edwardsiella Septicaemias. In: Woo PTK, Bruno DW, editors. Fish diseases and disorders, volume 3: viral, bacterial, and fungal infections, vol. 3. Oxon: CAB International; 1999. p. 479–521.

Preston GM, Jung JS, Guggino WB, Agre P. The mercury-sensitive residue at cysteine 189 in the CHIP28 water channel. J Biol Chem. 1993;268:17–20.

Rodríguez A, Catalán V, Gómez-Ambrosi J, García-Navarro S, Rotellar F, Valentí V, Silva C, Gil MJ, Salvador J, Burrell MA, Calamita G, Malagón MM, Frühbeck G. Insulin-and leptin-mediated control of aquaglyceroporins in human adipocytes and hepatocytes is mediated via the PI3K/Akt/mTOR signaling cascade. J Clin Endocrinol Metab. 2011;96:E586–97.

Rombout JH, Abelli L, Picchietti S, Scapigliati G, Kiron V. Teleost intestinal immunology. Fish Shellfish Immunol. 2011;31:616–26.

Schmittgen TD, Livak KJ. Analyzing real-time PCR data by the comparative C(T) method. Nat Protoc. 2008;3:1101–8.

Thompson JD, Higgins DG, Gibson TJ. Clustal W: improving the sensitivity of progressive multiple sequence alignment through sequence weighting, position specific gap penalties and weight matrix choice. Nucleic Acids Res. 1994;22:4673–80.

Tingaud-Sequeira A, Calusinska M, Finn RN, Chauvigné F, Lozano J, Cerdà J. The zebrafish genome encodes the largest vertebrate repertoire of functional aquaporins with dual paralogy and substrate specificities similar to mammals. BMC Evol Biol. 2010;10:38.

Tyagi MG, Tangevelu P. A possible role of aquaporin water channels in blood cell migration in spleen; interaction with cluster of differentiation molecules. J Exp Sci. 2010;1:41–2.

Verkman AS. Aquaporins in clinical medicine. Annu Rev Med. 2012;63:303–16.

Watanabe S, Hirano T, Grau EG, Kaneko T. Osmosensitivity of prolactin cells is enhanced by the water channel aquaporin-3 in a euryhaline Mozambique tilapia (Oreochromis mossambicus). Am J Phys. 2009;296:R446–53.

Watanabe S, Kaneko T, Aida K. Aquaporin-3 expressed in the basolateral membrane of gill chloride cells in Mozambique tilapia Oreochromis mossambicus adapted to freshwater and seawater. J Exp Biol. 2005;208:2673–82.

Wheeler DL, Church DM, Federhen S, Lash AE, Madden TL, Pontius JU, Schuler GD, Schriml LM, Sequeira E, Tatusova TA, Wagner L. Database resources of the National Center for Biotechnology. Nucleic Acids Res. 2003;31:28–33.

Wolf JC, Wolfe MJ. A brief overview of nonneoplastic hepatic toxicity in fish. Toxico Pathol. 2005;33:75–85.

Zhu C, Chen Z, Jiang Z. Expression, distribution and role of aquaporin water channels in human and animal stomach and intestines. Int J Mol Sci. 2016;17:1399.

Zhu N, Feng X, He C, Gao H, Yang L, Ma Q, Guo L, Qiao Y, Yang H, Ma T. Defective macrophage function in aquaporin-3 deficiency. FASEB J. 2011;25:4233–9.

Effects of different algae in diet on growth and interleukin (IL)-10 production of juvenile sea cucumber *Apostichopus japonicus*

Md Anisuzzaman[1], Jeong U-Cheol[1], Jin Feng[1], Choi Jong-Kuk[1], Kabery Kamrunnahar[1], Lee Da-In[2], Yu Hak Sun[2] and Kang Seok-Joong[1*]

Abstract

The experiment was conducted to investigate the effects of different algae in diet on growth, survival, and interleukin-10 productions of sea cucumber. At first, a 9-week feeding trail was conducted to evaluate the growth performance and survival of the sea cucumber fed one of the six experimental diets containing ST (*Sargassum thunbergii*), UL (*Ulva lactuca*), UP (*Undaria pinnatifida*), LJ (*Laminaria japonica*), SS (*Schizochytrium* sp.), and NO (*Nannochloropsis oculata*) in a recirculating aquaculture system. The result showed that survival was not significantly different among the dietary treatments, and the specific growth rate (SGR) of sea cucumber fed the UL diet (1.58% d^{-1}) was significantly higher than that of sea cucumber fed the other diets ($P < 0.05$), except for the LJ and NO diets. Secondly, interleukin (IL)-10 gene expression was determined where mice splenocytes were stimulated with 10 μg ml^{-1} of sea cucumber extracts for 2 h. The result showed that IL-10 gene expression levels were significantly increased in UL, LJ, and NO diets fed sea cucumber extracts compared to other experimental diets. The results suggest that dietary inclusion with *Ulva lactuca*, *Laminaria japonica*, and *Nannochloropsis oculata* algae may improve the growth of juvenile sea cucumber and could upregulate IL-10 gene expression in mice splenocytes. Such detailed information could be helpful in further development of more appropriate diets for sea cucumber culture.

Keywords: Sea cucumber (*Apostichopus japonicus*), Algae, Growth, Interleukin (IL)-10

Background

The sea cucumber, *Apostichopus japonicus*, has become an important mariculture species in Russia, China, Japan, and South Korea because of its relatively high economic value (Sloan 1984). Market demand for this species increased because of its aphrodisiac and curative properties (Liao 1997). However, the production of sea cucumbers obtained from the natural environment has declined due to overexploitation and pollution (Conand 2004). Depletion of wild production together with high commercial value has encouraged the people to develop aquaculture methods for holothurians, especially *A. japonicus* (Conand 2004; Yuan et al. 2006).

Successful culture of juvenile sea cucumbers requires proper knowledge about feed intake behavior and dietary requirements (Slater et al. 2009). However, little is known regarding which artificial diets are capable of inducing rapid growth and healthy conditions of commercially important sea cucumbers (Slater et al. 2009; Yuan et al. 2006; Zhou et al. 2006).

Sea cucumbers are deposit feeders that ingest sediment containing organic matter, including bacteria, protozoa, diatoms, and detritus of plants or animals (Yingst 1976; Moriarty 1982; Zhang et al. 1995; Feng et al. 2016a, 2016b). *A. japonicus* preferentially inhabits the sea bottom in flourishing large algae, rich detritus of which provide sea cucumber with its main organic nutrient (Li et al. 1994; Zhang et al. 1995). Traditionally, sea cucumbers are cultured in earthen ponds without artificial feed. But recently, farmers have started to feed the sea

* Correspondence: sjkang@gnu.ac.kr
[1]Department of Seafood and Aquaculture Science, Gyeongsang National University, Tongyeong 53064, Republic of Korea
Full list of author information is available at the end of the article

cucumbers with formulated diets to increase production (Shi et al. 2013). Formulated diets for sea cucumbers are commonly made of macroalgal powder and sea mud. Among macroalgae, the brown algal *Sargassum thunbergii* is widely distributed over shallow coastal areas in Korea, Japan, and China and commonly used as a feed ingredient in sea cucumber culture (Sui 1989; Battaglene et al. 1999). However, it is difficult to satisfy demand for sea cucumber culture because this algal species is not produced commercially and its use as feed ingredients is also expensive (Lobban and Harrison 1994). In addition, more and more *S. thunbergii* have been harvested in recent years with the rapid expansion of sea cucumber farming scale, which results in severe damage to *S. thunbergii* resource (Yuan 2005; Wang et al. 2006).

Meanwhile, by feeding with commercial feed which mostly used *S. thunbergii*, sea cucumber have a high level of n-6 fatty acids and low n-3 fatty acids and the balance of the n-3/n-6 ratio is not good (Feng et al. 2016a, 2016b). But n-3 fatty acids and a good balance of the n-3/n-6 ratio is very important to protect from allergic and inflammatory diseases like asthma. So, reducing the *S. thunbergii* content of sea cucumber feed will be one strategy to increase the sustainability of the sea cucumber culture.

Therefore, it is critical to find good substitutes for *S. thunbergii* to relieve the pressure on natural *S. thunbergii* resource and produce good quality sea cucumber. Several researchers reported that juvenile sea cucumbers fed commercially available dried powdered macroalgae (*Ulva lactuca, Laminaria japonica, Sargassum thunbergii, Sargassum polycystum*) and sea mud exhibited significant growth (Battaglene et al. 1999; Liu et al. 2010; Zhu et al. 2007). In our study, we used *Ulva lactuca, Undaria pinnatifida, Laminaria japonica, Nannochloropsis oculata,* and *Schizochytrium* sp. as a partial alternative source of *Sargassum thunbergii* to produce significant growth and good quality sea cucumber. *Ulva lactuca, Undaria pinnatifida,* and *Laminaria japonica* are popular and cheaper algae in Korea and widely used in the culture of sea urchins and abalone (Agatsuma 2000, Qi et al. 2010). *Nannochloropsis oculata* and *Schizochytrium* sp. are considered promising algae for aquaculture and offer high levels of polyunsaturated fatty acids (PUFAs), especially eicosapentaenoic acid (EPA, 20:5n-3) and docosahexaenoic acid (DHA, 22:6n-3) respectively (Kandilian et al. 2013 Yue et al. 2004).

Sea cucumbers have many therapeutic effects against various diseases (Bordbar et al. 2011; Guo et al. 2015). Moreover, sea cucumber extracts have potent biological effects and have antiviral, anticancer, antibacterial, antioxidant, and anti-inflammation effects (Esmat et al. 2013; Farshadpour et al. 2014; Kiani et al. 2014; Wijesinghe et al. 2013). In China and Malaysia, sea cucumbers have been traditionally used for the remedy of different inflammatory diseases like asthma. Asthma is a chronic inflammatory disease and a major public health problem. Interleukin (IL)-10 is a potent anti-inflammatory cytokine that downregulates the synthesis of Th1 (T helper 1)- and Th2 (T helper 2)-associated cytokines, chemokines, and inflammatory enzymes. It plays a vital role for the mitigation of allergic responses. But till now, there are no reports demonstrating the effect of different algae in sea cucumber on IL-10 production.

In this study, the effects of different algae in diet on growth, survival, and anti-inflammatory cytokine (IL-10) production of the juvenile sea cucumber were examined.

Methods
Experimental diets
Six experimental diets designed as *Sargassum thunbergii* (ST), *Ulva lactuca* (UL), *Undaria pinnatifida* (UP), *Laminaria japonica* (LJ), *Schizochytrium* sp. (SS), and *Nannochloropsis oculata* (NO) were prepared. Ingredients and proximate compositions of experimental diets are presented in Table 1. Most sea cucumbers are deposit feeders that ingest sediment with organic matter. Several studies showed that juvenile sea cucumbers fed different algae and sea mud exhibited significant growth (Battaglene et al. 1999; Liu et al. 2009; Hai-Bo et al. 2015. ST diet was used as the control diet where 15% *Sargassum thunbergii* and 15% wheat flour were used. For diets UL, UP, LJ, SS, and NO, wheat flour was replaced by 15% UL, UP, LJ, SS, and NO respectively. All ingredients were ground into fine powder through a 200-µm mesh, thoroughly mixed, and stored at −20 °C.

Experimental animal and feeding trail
The experiment was carried out for 9 weeks in the laboratory of Marine Biology and Aquaculture, Gyeongsang National University, Republic of Korea. Sea cucumbers used in this experiment were collected from the Goseong Sea cucumber farm. Prior to the experiment, sea cucumbers were transferred to the laboratory in fiberglass aquaria and acclimated for 2 weeks at 18 °C.

After 2 days starvation, 240 sea cucumbers with initial wet body weights of 2.98 ± 0.06 g (mean ± SE) were randomly selected from acclimatized sea cucumbers and placed in equal number into 24 fiberglass aquaria ($45 \times 60 \times 50$ cm^3) to form six groups in tetraplicate. The six groups were fed with different experimental diets such as ST, UL, UP, LJ, SS, and NO respectively. A complete randomized block design was used to arrange the 24 aquaria of six treatment groups.

Table 1 Ingredients and composition of experimental diets for *Apostichopus japonicus* (% dry matter basis)

Ingredients	ST (control)	UL	UP	LJ	SS	NO
Ulva lactuca powder	0	15	0	0	0	0
Undaria pinnatifida powder	0	0	15	0	0	0
Laminaria japonica powder	0	0	0	15	0	0
Schizochytrium sp. powder	0	0	0	0	15	0
Nannochloropsis oculata powder	0	0	0	0	0	15
Wheat flour	15	0	0	0	0	0
Seaweed powder	15	15	15	15	15	15
Soybean meal	8	8	8	8	8	8
Shellfish powder	8	8	8	8	8	8
Shell powder	2	2	2	2	2	2
Calcium phosphate	2	2	2	2	2	2
Yeast protein	5	5	5	5	5	5
Soyabean lecithin	4	4	4	4	4	4
Mineral premix[a]	0.5	0.5	0.5	0.5	0.5	0.5
Vitamin premix[b]	0.5	0.5	0.5	0.5	0.5	0.5
Sea mud	40	40	40	40	40	40
Proximate composition (%)						
Crude protein	17.74	17.03	17.48	16.87	17.59	19.64
Crude lipid	3.39	3.72	3.11	3.32	7.44	5.64
Ash	41.20	44.80	45.55	45.85	41.5	42.10

[a]Mineral premix (g kg^{-1} premix): MgSO$_4$·7H$_2$O, 80.0; NaH$_2$PO$_4$·2H$_2$O, 200.0; KH$_2$PO$_4$, 130.0; Ferric citrate, 40.0; ZnSO$_4$·7H$_2$O, 10; Ca-lactate, 25.5; CuCl, 0.2; AlCl$_3$·6H$_2$O, 0.15; KIO$_3$, 0.15; Na$_2$Se$_2$O$_3$, 0.01; MnSO$_4$·H$_2$O, 2.0; CoCl$_2$·6H$_2$O, 1.0
[b]Vitamin premix (g kg^{-1} premix): ascorbic acid, 92.7; α-tocopheryl acetate, 14.5; thiamine hydrochloride, 7; riboflavin, 7.0; pyridoxine hydrochloride, 1.4; niacin, 27.8; Ca-D-pantothenate, 9.7; myo-inositol, 139.1; D-biotin, 0.5; folic acid, 0.5; p-amino benzoic acid, 13.9; menadione, 4.2; retinyl acetate, 0.65; cholecalciferol, 0.8; cyanocobalamin, 0.004

During the experiment, aeration was provided continuously and to ensure water quality two thirds volume of the water in each aquarium was exchanged every day. Seawater temperature was controlled at 18 ± 1.0 °C. Dissolved oxygen was maintained above 5.0–7.0 mg l^{-1}, and the levels of ammonia in the water of the aquaria were less than 0.25 mg l^{-1}. Other conditions were salinity 32 ± 1 ppt, pH 7.7–8.3, and photoperiod 24-h dark.

Procedure and sample collection

Twenty-four sea cucumbers were collected as an initial sample before starting the experiment. During the experiment, sea cucumbers were fed once per day (at about 1700 hours). Uneaten feed were collected by siphon after 24 h and then dried at 65 °C to constant weight. Sea cucumber feces were also collected by siphon once per day (1600 hours). The feces were dried at 65 °C to constant weight, and those from each aquarium were pooled for further analysis. At the end of the 9-week experiment, all the experimental sea cucumbers were deprived of food to clear their guts for 2 days, weighed, and then dried at 65 °C until constant weight was achieved.

Data calculation

Survival rate (SR), specific growth rate (SGR), ingestion rate (IR), feces production rate (FPR), and food conversion efficiency (FCE) were calculated as follows:

$$\text{SR} (\%) = 100 \times (N_2/N_1)$$
$$\text{SGR} (\%\text{d}^{-1}) = 100 \, (\ln W_2 - \ln W_1)/T$$
$$\text{IR} (\text{g g}^{-1}\text{d}^{-1}) = I/[T \, (W_2 + W_1)/2]$$
$$\text{FPR} (\text{g g}^{-1}\text{d}^{-1}) = F/[T \, (W_2 + W_1)/2]$$
$$\text{FCE} (\%) = 100 \, (W_2 - W_1)/I$$

where N_1 is the number of individuals alive at start of the experiment and N_2 is the number of individuals alive at end of the experiment; W_1 and W_2 are initial and final combined dry weights, respectively, of all 10 sea cucumbers in each aquarium; T is the experimental period; I is the dry weight of the total feed ingested; and F is the dry weight of feces.

Preparation of sea cucumber extract

At first, our experimental sea cucumbers were cleaned and the visceral organs removed. After that, the sea cucumbers were cut into small pieces and homogenized. One hundred fifty grams of samples were boiled in 300 ml distilled water for 20 min. After removing solid materials from the water, the boiled water was vaporized using a microwave until the mixture was reduced by 50%. After centrifugation of the extracts at 500×g for 10 min, a fivefold volume of 100% ethyl alcohol was added to the supernatant and incubated at 20 °C for 24 h. After that, the supernatant was discarded. The extract pellet was washed with 70% ethyl alcohol and centrifuged under the same conditions. The supernatant was discarded, and the pellet was evaporated under a vacuum. The final extracts were prepared by re-suspending the pellet in 20 ml distilled water (Lee et al. 2016).

IL-10 gene expression

In order to analyze the IL-10 gene expression, mice splenocytes were stimulated with 10 μg ml^{-1} of each experimental diet-fed sea cucumber extract for 2 h. The total RNAs were isolated by Qiazol reagent (Qiagen Science, USA) according to the manufacturer's protocols. Two micrograms of total RNAs were transcribed using M-MLV reverse transcriptase (Promega, USA), according to the manufacturer's protocols. IL-10 mRNA expression levels were synthesized by real-time PCR using the iCyclerTM (Bio-Rad Laboratories, Hercules, CA, USA). GAPDH was used for the reference gene. IL-10 and GAPDH primer sequence are previously described (Lee et al. 2016).

Statistical analysis

Statistical analysis was performed by the software SPSS 18.0 with possible differences among diet treatments being tested by using one-way ANOVA. Tukey multiple comparison tests were used to analyze the differences among treatments. Differences were considered significant at a probability level of 0.05.

Results

Growth and survival

The growth performance and survival of sea cucumber are shown in Table 2. All sea cucumbers were alive at the end of the 9-week feeding trial. The growth performance of the sea cucumbers differed significantly among treatments. Final wet and dry body weights of sea cucumbers showed the highest value for the UL diet group and the lowest value for the ST diet group ($P < 0.05$).

The highest SGR (1.58% d^{-1}) was observed in sea cucumber fed the UL diet. SGR of sea cucumbers fed the ST diet was significantly ($P < 0.05$) lower than UL, LJ, and NO diets, but not significantly ($P > 0.05$) different from that fed the UP and SS diets (Fig. 1).

Ingestion rate and feces production rate

Ingestion rates (Fig. 2) and feces production rates (Fig. 3) of the sea cucumbers showed significant differences among different diet treatments. Sea cucumbers fed with diets LJ and UL showed significantly higher IR (0.62 and 0.59 g g^{-1} d^{-1} respectively) and FPR (0.53 and 0.52 g g^{-1} d^{-1} respectively) than those fed other diets ($P < 0.05$). Sea cucumbers fed with diet ST showed the lowest IR (0.32 g g^{-1} d^{-1}) ($P < 0.05$) and FPR (0.26 g g^{-1} d^{-1}) ($P < 0.05$) among all treatments.

Food conversion efficiency

Food conversion efficiency (%) was significantly different among different diet treatments (Fig. 4). FCE of the sea cucumbers fed with diet NO was 3.74%, which was significantly higher than those fed with other diets ($P < 0.05$). Sea cucumbers fed the diet UP showed the lowest FCE (2.21%).

Anti-inflammatory cytokine, IL-10 expression level

In order to establish proper algae for sea cucumber diet, we synthesize IL-10 gene expression levels. Splenocytes were stimulated with each experimental diet-fed sea cucumber extract for 2 h. Result showed that IL-10 gene expression levels were significantly increased in the UL, LJ, and NO diets compared to the other experimental diets (Fig. 5). The highest IL-10 gene expression levels were found when the sea cucumbers were fed the *Ulva lactuca* algae diet. However, IL-10 gene expression levels were not increased by ST, UP, and SS diets and have no significant differences. These results suggest that UL, LJ, and NO algae could upregulate IL-10 gene expression.

Discussion

In all treatments, no sea cucumber died and survival rates of sea cucumbers were excellent (100%) and were

Table 2 Initial and final wet weight (WW) and dry weight (DW) of *Apostichopus japonicus* fed different test diets (mean ± SE)

Experimental diets	Initial WW (g)	Initial DW (g)	Final WW (g)	Final DW (g)	Survival (%)
ST	3.04 ± 0.09	0.28	5.07 ± 0.92	0.47	100
UL	2.93 ± 0.12	0.27	7.65 ± 0.85	0.71	100
UP	2.94 ± 0.11	0.27	5.94 ± 0.62	0.55	100
LJ	3.03 ± 0.07	0.28	7.27 ± 0.45	0.67	100
SS	2.92 ± 0.10	0.27	6.16 ± 0.15	0.57	100
NO	3.02 ± 0.06	0.28	7.05 ± 0.18	0.65	100

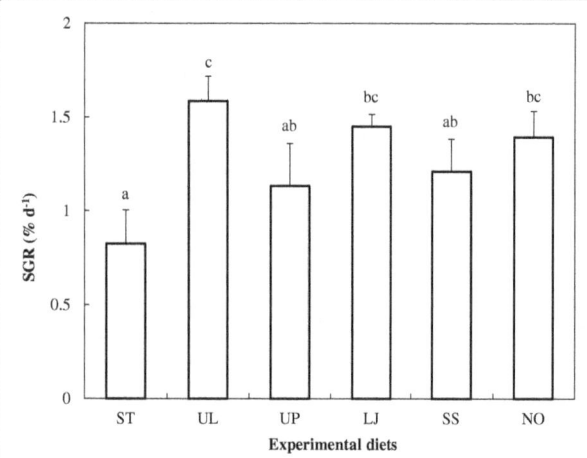

Fig. 1 Specific growth rate of *Apostichopus japonicus* fed different test diets. Different letters indicate significant differences (*P* < 0.05) between treatments within the same group, and bars represent standard errors

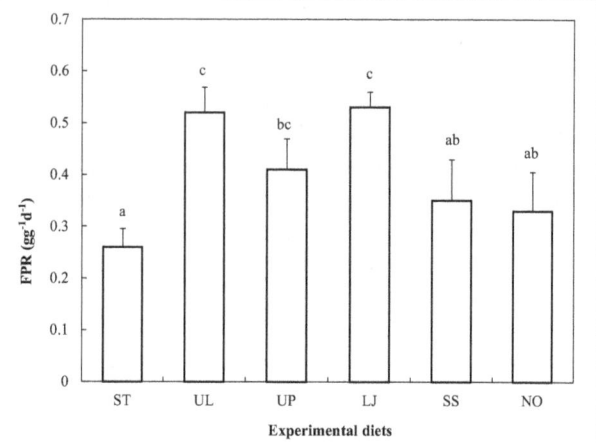

Fig. 3 Feces production rate (FPR) of *Apostichopus japonicus* fed different test diets. Different letters indicate significant differences (*P* < 0.05) between treatments within the same group, and bars represent standard errors

higher than the rates reported in previous similar studies (Hai-Bo et al. 2015; Slater and Carton 2007; Zhou et al. 2006). The results showed that sea cucumbers might have the ability to tolerate the different algae such as UL, UP, LJ, NO, and SS in diet.

Systematic studies on the use of algae as an ingredient in commercial feeds for sea cucumber culture are very rare. Many researchers have used different types of algae such as *Sargassum polycystum*, *Sargassum thunbergii*, *Laminaria japonica*, *Spirulina platensis*, *Ulva lactuca*, *Undaria pinnatifida*, and *Pyropia spheroplasts* to study sea cucumber nutrition requirements (Li et al. 2009; Liu et al. 2009; Seo and Lee 2011; Seo et al. 2011; Slater et al. 2009; Yuan et al. 2006, Shahabuddin et al. 2017). Most researchers have used *S. thunbergii* as a main feed

ingredient in land-based intensive culture systems (Battaglene et al. 1999; Slater et al. 2009). However, in our study of various experimental diets, the SGR was much higher in sea cucumbers fed UL, LJ, or NO diets compared to UP, SS, and ST diets (Fig. 1). Seo et al. (2011) reported that sea cucumbers fed 20% *L. japonica* and 20% *S. thunbergii* containing diet grew much better than those only eating *S. thunbergii* (40%) diet. Zhu et al. (2007) also reported that the SGR was increased significantly when sea cucumbers were fed *U. lactuca* compared with *S. thunbergii* and the SGR of sea cucumbers decreased when they were fed *L. japonica*. They used much smaller sized (0.49 g) sea cucumbers compared to ours (2.98 g). Different size of sea cucumber may have different diet choice and nutrition requirement (Yanagisawa 1998).

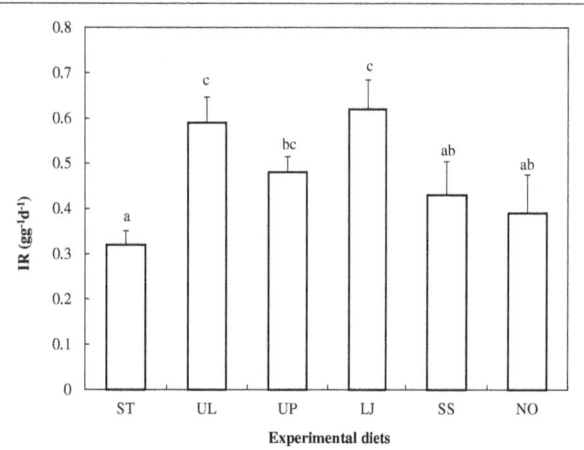

Fig. 2 Ingestion rate (IR) of *Apostichopus japonicus* fed different test diets. Different letters indicate significant differences (*P* < 0.05) between treatments within the same group, and bars represent standard errors

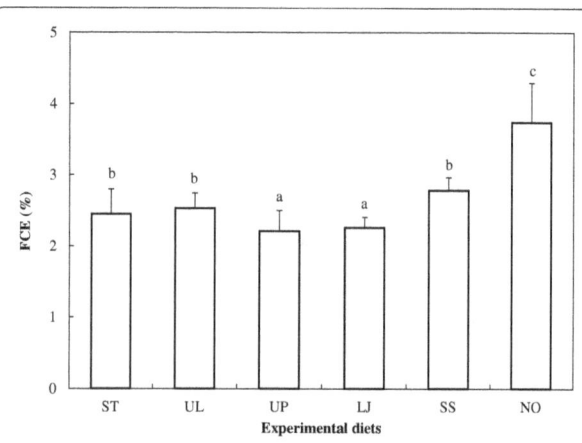

Fig. 4 Food conversion efficiency (FCE) of *Apostichopus japonicus* fed different test diets. Different letters indicate significant differences (*P* < 0.05) between treatments within the same group, and bars represent standard errors

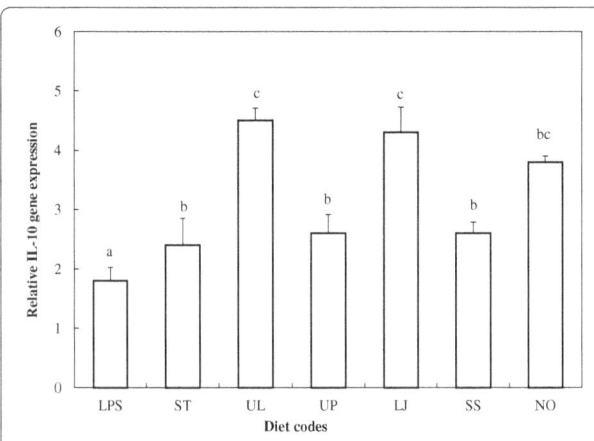

Fig. 5 IL-10 gene expression. IL-10 gene expressions were significantly increased by administration of the UL, LJ, and NO diets

Algae are an important food source for sea cucumbers. Zhu et al. (2007) reported that *U. lactuca* as a feed ingredient for the sea cucumbers was better than *S. thunbergii* in growth performance. In the present studies, NO, ST, SS, and UP diets of which the protein contents were higher than those of the UL and LJ diets were selected to test the effects on growth of the sea cucumbers (Table 2). The results showed that SGR of the sea cucumbers fed the UL and LJ diets was significantly higher than that of those fed the ST, SS, and UP diets. The results indicate that some other factors were responsible for the nutrient effects of the algae besides the protein, lipid, and energy contents. One of the three mechanisms such as acid hydrolysis, enzymatic digestion, or mechanical trituration is necessary to break the cell wall of microalgal cell contents (Bitterlich 1985). For instance, many fishes like tilapia are the species which rely on acid hydrolysis. They are able to disrupt cyanobacterium cell walls because of low pH values of their stomach fluid (Caulton 1976; Payne 1978). But for holothurians like the sea cucumber, the structure and environment of the digestive tract of the sea cucumber are quite different from tilapia. Sea cucumbers have no specialized organ for grinding or gland for chemical digestion (Massin 1982), and digestive enzyme activities are very low and have very little cellulose activity (Wang et al. 2007). Therefore, sea cucumbers are able to assimilate a specific amount of cellulose content.

In our studies, the higher SGR of the sea cucumber was observed in the treatments fed with the UL and LJ diets though the protein and lipid contents of those diets were comparatively low ($P < 0.05$). UL algae are two cells thick, soft and translucent, and LJ is multicellular, filamentous. Their cell walls are easy to break and have comparatively lower cellulose content (Burrows 1991; Miyai et al. 2008). So, sea cucumbers could easily digest and take full advantage of the nutrients in the UL and LJ algae.

The algae UP and SS contain a significant amount of cellulose and need specific enzyme activities to break the cellulose content (Jurkovic et al. 1995). Sea cucumbers have lower specific enzyme activities and may not digest macroalgae such as SS and UP efficiently (Wang et al. 2007). Moreover, *Schizochytrium* algae have a higher amount of lipids (Menghe et al. 2009). But sea cucumbers have low tolerance to lipids and do not require a high dietary lipid (Seo and Lee 2011). Therefore, in our studies, sea cucumbers fed the UP and SS diets had lower SGR among all the treatments with the exception of the ST diets.

Ingestion rates (IRs) of sea cucumbers were significantly affected by different experimental diets. There was a negative relationship between IR and the protein level. In the natural ecosystem, low nutritional value of sediment consumed by deposit feeders means those animals need to consume large amounts of sediment in order to obtain a net input of energy (Santos et al. 1994; Hudson et al. 2004). Vice versa, when food quality becomes better, internal appetite regulation would work actively to decrease food ingestion. In this study, the ingestion rate of sea cucumbers decreased when the protein content of the diets increased. The same phenomenon was also found in other echinoderms. McBride et al. (1998) reported that for sea urchin (*Strongylocentrotus franciscanus*), prepared diets of different protein levels resulted in different ingestion rate. Otero-Villanueva et al. (2004) also found in *Psammechinus miliaris* that the lowest ingestion rate was related to a high energetic diet.

Regulatory T cells (Treg cells), known as suppressor T cells, are a subpopulation of T cells and modulate the immune systems (Kikodze et al. 2016). IL-10 is one of the Treg cells and known as a key regulator of immunity to many infections or inflammatory diseases (Gutierrez-Murgas et al. 2016). For instance, high levels of IL-10 have protective effect against asthma disease (Raeiszadeh Jahromi et al. 2014). Conversely, lack of IL-10 promotes cell apoptosis during virus infection in the small intestine (Pan et al. 2014). In a previous study, we already investigated that administration of sea cumber total extract can upregulate IL-10 and ameliorate asthma disease (Lee et al. 2016). Here, we suggest that UL, LJ, and NO algae increase interleukin (IL)-10 gene expression.

Conclusions

In conclusion, the results of this experiment suggest that dietary inclusion with *Ulva lactuca*, *Laminaria japonica*, and *Nannochloropsis oculata* algae may improve growth of juvenile sea cucumbers and could upregulate IL-10 gene expression. Such detailed information could be helpful in further development of more appropriate diets for the culture of sea cucumber.

Abbreviations

FCE: Food conversion efficiency; FPR: Feces production rate; IL: Interleukin; IR: Ingestion rate; SGR: Specific growth rate; SR: Survival rate

Acknowledgements

This research was a part of the project titled "Feed development for the production of sea cucumber contains a substance improving asthma ameliorated material" funded by the Ministry of Oceans and Fisheries, Korea.

Funding

The design of the study; collection, analysis, and interpretation of the data; and writing of the manuscript were funded by a grant from the Ministry of Oceans and Fisheries, Korea (20150297).

Authors' contributions

AM, UCJ, and SJK designed the study. AM wrote the article. AM, UCJ, FJ, JKC, and KK manufactured the experimental feed, conducted the feeding trial, and performed the analyses. DIL and HSY performed the interleukin-10 experiment. SJK conceived, coordinated, and revised the manuscript. All authors read and approved the final manuscript.

Competing interests

The authors declare that they have no competing interests.

Author details

[1]Department of Seafood and Aquaculture Science, Gyeongsang National University, Tongyeong 53064, Republic of Korea. [2]Department of Parasitology, School of Medicine, Pusan National University, Yangsan-si, Gyeongsangnam-do 626-870, Republic of Korea.

References

Agatsuma Y. Food consumption and growth of the juvenile sea urchin *Strongylocentrotus intermedius*. Fish Sci. 2000;66:467–72.

Battaglene SC, Seymour EJ, Ramofafia C. Survival and growth of cultured juvenile sea cucumbers *Holothuria scabra*. Aquaculture. 1999;178:293–322.

Bitterlich G. Digestive process in silver carp *Hypophthalmichthys molitrix* studied in vitro. Aquaculture. 1985;50:123–31.

Bordbar S, Anwar F, Saari N. High-value components and bioactives from sea cucumbers for functional foods: a review. Marine Drugs. 2011;9:1761–805.

Burrows EM. Seaweeds of the British Isles, vol. 2. London: Natural History Museum; 1991.

Caulton MS. The importance of pre-digestive food preparation to *Tilapia rendalli* Boulanger when feeding on aquatic macrophytes. Trans Rhod Sci Assoc. 1976;57:22–8.

Conand C. Present status of world sea cucumber resources and utilization: an international overview. Advances in sea cucumber aquaculture and management (Lovatelli, A., Conand,C., Purcell, S., Uthicke, S., Hamel, J.F. & Mercier, A. eds), FAO, Rome, Italy; 2004. p.13–23.

Esmat AY, Said MM, Soliman AA, El-Masry KS, Badiea EA. Bioactive compounds, antioxidant potential, and hepatoprotective activity of sea cucumber (*Holothuria atra*) against thioacetamide intoxication in rats. Nutrition. 2013;29:258–67.

Farshadpour F, Gharibi S, Taherzadeh M, Amirinejad R, Taherkhani R, Habibian A, Zandi K. Antiviral activity of Holothuria sp. a sea cucumber against herpes simplex virus type 1 (HSV-1). Eur Rev Med Pharmacol Sci. 2014;18:333–7.

Feng J, Jong-Kuk C, U-Cheol J, Anisuzzaman M, Chung-Ho R, Byeong-dae C, Seok-Joong K. Effects of fermented fecal solid diets on growth of the sea cucumber *Apostichopus japonicus*. Korean J Fish Aqua Sci. 2016b;49:161–7.

Feng J, Anisuzzaman M, U-Cheol J, Jong-Kuk C, Hak-Sun Y, Seung-Wan K, Seok-Joong K. Comparison of fatty acid composition of wild and cultured sea cucumber *Apostichopus japonicus*. Korean J Fish Aqua Sci. 2016a;49:474–85.

Guo Y, Ding Y, Xu F, Liu B, Kou Z, Xiao W, Zhu J. Systems pharmacology-based drug discovery for marine resources: an example using sea cucumber (holothurians). J Ethnopharmacol. 2015;165:61–72.

Gutierrez-Murgas YM, Gwenn S, Danielle R, Matthew B, Jessica N. IL-10 plays an important role in the control of inflammation but not in the bacterial burden in S. epidermidis CNS catheter infection. J Neuroinflammation. 2016;13(1):271.

Hai-Bo Y, Qin-Feng G, Shuang-Lin D, Bin W. Changes in fatty acid profiles of sea cucumber *Apostichopus japonicus* (Selenka) induced by terrestrial plants in diets. Aquaculture. 2015;442:119–24.

Hudson IR, Wigham BD, Tyler PA. The feeding behavior of a deep-sea holothurian, *Stichopus tremulus* (Gunnerus) based on in situ observation and experiments using a remotely operated vehicle. J Exp Mar Biol Ecol. 2004;301:75–91.

Jurkovic N, Kolb N, Colic I. Nutritive value of marine algae *Laminaria japonica* and *Undaria pinnatifida*. Mol Nutr Food Res. 1995;39:63–6.

Kandilian R, Lee E, Pilon L. Radiation and optical properties of *Nannochloropsis oculata* grown under different irradiances and spectra. Bioresour Technol. 2013;137:63–73.

Kiani N, Heidari B, Rassa M, Kadkhodazadeh M, Heidari B. Antibacterial activity of the body wall extracts of sea cucumber (Invertebrata; Echinodermata) on infectious oral streptococci. J Bas Clin Physiol Pharmacol. 2014;1:1–7.

Kikodze N, Pantsulaia I, Chikovani T. The role of T regulatory and Th17 cells in the pathogenesis of rheumatoid arthritis (review). Georgian Med News. 2016;261:62–8.

Lee DI, Park MK, Kang SA, Choi JH, Kang SJ, Lee JY, Yu HS. Preventive intra oral treatment of sea cucumber ameliorate OVA-induced allergic airway inflammation. Am J Chin Med. 2016;44:1663–74.

Li Y, Wang Y, Wang P. Habitat environment and water area selection of stock increment of *Apostichopus japonicus*. Trans Oceano Limno. 1994;4:42–7.

Li B, Zhu W, Feng Z, Li L, Hu Y. Affection of diet phospholipid content on growth and body composition of sea cucumber, *Apostichopus japonicus*. Mar Sci. 2009;33:25–8.

Liao. Fauna Sinica: Phylum Echiondermta Class Holthuroidea. Beijing: Science Press; 1997. p. 21–37.

Liu Y, Dong S, Tian X, Wang F, Gao Q. Effects of dietary sea mud and yellow soil on growth and energy budget of the sea cucumber *Apostichopus japonicus* (Selenka). Aquaculture. 2009;286:266–70.

Liu Y, Dong SL, Tian XL, Wang F, Gao QF. The effect of different macroalgae on the growth of sea cucumbers (*Apostichopus japonicus* Selenka). Aquac Res. 2010;2:1–5.

Lobban CS, Harrison PJ. Seaweed ecology and physiology. New York: Cambridge University Press; 1994.

Massin C. Food and feeding mechanisms: Holothuroidea. In: Lawrence JM, editor. Echinoderm Jangoux, M. Rotterdam: A. A. Balkema Publishers; 1982. p. 115–20.

McBride SC, Lawrence JM, Lawrence AL, Mulligan TJ. The effects of protein concentration in prepared diets on growth, feeding rate, total organic absorption, and gross assimilation efficiency of the sea urchin *Strongylocentrotus franciscanus*. J Shellfish Res. 1998;17:1562–70.

Menghe HL, Edwin HR, Craig ST, Lester K. Effects of dried algae *Schizochytrium* sp. , a rich source of docosahexaenoic acid, on growth, fatty acid composition, and sensory quality of channel catfish *Ictalurus punctatus*. Aquaculture. 2009; 292:232–6.

Miyai K, Tokushige T, Kondo M. Suppression of thyroid function during ingestion of seaweed "Kombu" (*Laminaria japonoca*) in normal Japanese adults. Endocr J. 2008;55:1103–8.

Moriarty DJW. Feeding of *Holothuria atra* and *Stichopus chloronotus* on bacteria, organic carbon and organic nitrogen in sediments of the Great Barrier Reef. Aus J Mar Fresh Res. 1982;33:255–63.

Otero-Villanueva MM, Kelly MS, Burnell G. How diets influence energy partitioning in the regular echinoid *Psammechinus miliaris*; constructing an energy budget. J Exp Mar Biol Ecol. 2004;304:159–81.

Pan D, Kenway-Lynch CS, Lala W, Veazey RS, Lackner AA, Das A, Pahar B. Lack of interleukin-10-mediated anti-inflammatory signals and upregulated interferon gamma production are linked to increased intestinal epithelial cell apoptosis in pathogenic simian immunodeficiency virus infection. J Virol. 2014;88: 13015–28.

Payne AI. Gut pH and digestive strategies in estuarine grey mullet Mugilidae and tilapia Cichlidae. J Fish Res Board Can. 1978;13:627–9.

Qi ZH, Liu HM, Li B, Mao YZ, Jiang ZJ, Zhang JH, Fang JG. Suitability of two seaweeds, Gracilaria lemaneiformis and Sargassum pallidum, as feed for the abalone *Haliotis discus hannai ino*. Aquaculture. 2010;300:189–93.

Raeiszadeh Jahromi S, Mahesh PA, Jayaraj BS, Madhunapantula SR, Holla AD, Vishweswaraiah S, Ramachandra NB. Serum levels of IL-10, IL-17F and IL-33 in patients with asthma: a case-control study. J Asthma. 2014;51:1004–13.

Santos V, Billett DSM, Rice AL, Wolff GA. Organic matter in deep-sea sediments from the porcupine abyssal plain in the north-east Atlantic Ocean: 1. Lipids. Deep-Sea Res. 1994;41:789–819.

Seo JY, Lee SM. Optimum dietary protein and lipid levels for growth of juvenile sea cucumber *Apostichopus japonicaus*. Aquac Nutr. 2011;17:56–61.

Seo JY, Shin IS, Lee SM. Effect of dietary inclusion of various plant ingredients as an alternative for *Sargassum thunbergii* on growth and body composition of juvenile sea cucumber *Apostichopus japonicaus*. Aquac Nutr. 2011;17:549–56.

Shahabuddin AM, Khan NMD, Koji M, Araki T, Yoshimatsu T. Dietary supplementation of red alga *Pyropia spheroplasts* on growth, feed utilization and body composition of sea cucumber, *Apostichopus japonicus* (Selenka). Aquac Res. 2017;1:1–10.

Shi C, Dong SL, Pei SR, Wang F, Tian XL, Gao QF. Effects of diatom concentrationin prepared feeds on growth and energy budget of the sea cucumber *Apostichopus japonicus* (Selenka). Aquac Res. 2013;46:609–17.

Slater MJ, Carton AG. Survivorship and growth of the sea cucumber *Australostichopus (Stichopus) mollis* (Hutton 1872) in polyculture trials with green-lipped mussel farms. Aquaculture. 2007;272:389–98.

Slater MJ, Je AG, Carton AG. The use of the waste from green-lipped mussels as a food source for juvenile sea cucumber, *Australostichopus mollis*. Aquaculture. 2009;292: 219–24.

Sloan NA. Echinorderm fisheries of the world: a review. Echinodermata (Proceedings of the Fifth International Echinoderm Conference). Rotterdam: A. A. Balkema Publishers; 1984. p. 109–24.

Sui X. The main factors influencing the larval development and survival rate of the sea cucumber *Apostichopus japonicus*. Oceanolo et Limnolo Sinica. 1989;20:314–21.

Wang FJ, Sun XT, Li F. Studies on sexual reproduction and seedling-rearing of *Sargassum thunbergii*. Mar Fish Res. 2006;27:1–6.

Wang JQ, Tang L, Xu C, Cheng JC. Histological observation of alimentary tract and annual changes of four digestive enzymes in sea cucumber (*Apostichopus japonicus*). Fish Sci. 2007;26:481–4.

Wijesinghe WA, Jeon YJ, Ramasamy P, Wahid ME, Vairappan CS. Anticancer activity and mediation of apoptosis in human HL-60 leukaemia cells by edible sea cucumber (*Holothuria edulis*) extract. Food Chem. 2013;139:326–31.

Yanagisawa T. Aspects of the biology and culture of the sea cucumber. In: De Silva S, editor. Tropical Mariculture. London: Academic; 1998. p. 291–308.

Yingst JY. The utilization of organic matter in shallow marine sediments by an epibenthic deposit-feeding holothurian. J Exp Mar Biol Ecol. 1976;23:55–69.

Yuan CY. Current status and development of feed in sea cucumber. Fish Sci. 2005;24:54–6.

Yuan X, Yang H, Zhou Y, Mao Y, Zhang T, Liu Y. The influence of diets containing dried bivalve feces and/or powdered algae on growth and energy distribution in sea cucumber *Apostichopus japonicus* (Selenka) (Echinodermata: Holothuroidea). Aquaculture. 2006;256:457–67.

Yue J, King-Wai F, Raymond TYW, Feng C. Fatty acid composition and squalene content of the marine microalga *Schizochytrium mangrovei*. J Agri Food Chem. 2004;52:1196–200.

Zhang BL, Sun DL, Wu YQ. Preliminary analysis on the feeding habit of *Apostichopus japonicus* in the rocky coast waters off Lingshan Island. Mar Sci. 1995;3:11–3.

Zhou Y, Yang H, Liu S, Yuan X, Mao Y, Zhang T, Liu Y, Zhang F. Feeding on biodeposits of bivalves by the sea cucumber *Stichopus japonicus* Selenka (Echinidermata: Holothuroidea) and a suspension coculture of filter-feeding bivalves with deposit feeders in lantern nets from longlines. Aquaculture. 2006;256:510–20.

Zhu J, Liu H, Leng K, Qu K, Wang S, Xue Z, Sun Y. Studies on the effects of some common diets on the growth of *Apostichopus japonicaus*. Mar Fish Res. 2007;25:48–53.

19

New record of an economic marine alga, *Ahnfeltiopsis concinna*, in Korea

Pil Joon Kang and Ki Wan Nam[*] (ID)

Abstract

An economic marine alga, which is considered to be an important source of carrageenan, was collected from Jindo of the southern coast of Korea. This species shares the vegetative and female reproductive features of *Ahnfeltiopis* and is characterized mostly by its small size (up to 8 cm), terete to subterete thalli at the lower portion, cartilaginous in texture, dichotomous branches, rarely produced proliferations, and an absence of hypha-like filaments in the medulla. It is distinguished from other Korean species within the genus by the thallus feature. In a phylogenetic tree based on the molecular data, this alga nests in the same clade with *A. concinna* from Japan but forms a sister clade to *A. concinna* from Mexico and Hawaii (type locality). However, the genetic distance among those sequences was calculated as 0.1–1. 3% for *rbc*L and 1.1% for COI sequences, considered to be intraspecific variation within the genus. Based on the morphology and molecular analysis, this alga is identified as *A. concinna* originally described from Hawaii. This is the first record of the species in the Korean marine algal flora.

Keywords: *Ahnfeltiopsis concinna*, Korea, Economic marine alga, Molecular analysis, *rbc*L, COI, Morphology, First record

Background

Ahnfeltiopsis P.C. Silva et DeCew belongs to Gigartinales F. Schmitz, which is considered to be one of the economic marine algal taxa as an important source of carrageenan (Craigie 1990; Donald et al. 1993). Particularly, it has been known that this alga can be used as a potential commercial material of antioxidant compounds in the medicine, food, pharmaceutical, and cosmetic industries in Hawaii (Kelman et al. 2012).

This genus was first proposed by Silva and DeCew in Silva (1979), but was invalid. Later, the generic name, *Ahnfeltiopsis*, was validly published. It was characterized by internal cystocarps and a heteromorphic type of life history in which upright unisexual gametophytes alternate with a crustose tetrasporophyte (Silva and DeCew 1992; Masuda 1993). However, it has been reported that *Ahnfeltiopsis* is polyphyletic in molecular phylogeny (Fredericq and Lopez-Bautista 2002; Maggs et al. 2013; Calderon and Boo 2016; Calderon et al. 2016; the present study). This suggests that the generic features should be revised for delimitation of *Ahnfeltiopsis*.

Ahnfeltiopsis involves 33 species distributed from temperate to tropical waters (Dawson 1954; Masuda 1993; Silva et al. 1996; Guiry and Guiry 2017). Among these, three species, *Ahnfeltiopsis catenata* (Yendo) Masuda, *A. paradoxa* (Suringar) Masuda, and *A. flabelliformis* (Harvey) Masuda, had been reported in Korea (Kim et al. 2013). However, recently, the former two species have been transferred to *Besa* Setchell based on molecular and morphological examination (Calderon et al. 2016). Accordingly, only *A. flabelliformis* in this genus is currently recorded in the Korean marine algal flora. A gigartinalean species was collected from Jindo of the southern coast of Korea during a survey of marine algal flora. Based on the morphology and molecular data, this species was identified as *Ahnfeltiopsis concinna*, which was established from Hawaii (Dawson 1961), and is newly recorded in Korea in the present study.

Methods

Specimens for this study were collected from Jindo located in southern coast of Korea. Taxonomic data were obtained from fresh, liquid-preserved, and herbarium specimens. Liquid-preserved material was stored in a 10% solution of formalin/seawater. For anatomical

* Correspondence: kwnam@pknu.ac.kr
Department of Marine Biology, Pukyong National University, Busan 48513, South Korea

observations, the material was cleared in 5–10% NaOH in distilled water for 2–7 days and then rinsed in distilled water. Blades dissected from the cleared materials were hand sectioned, transferred to a slide with a drop of distilled water, and mounted in pure glycerin. In some instances, a smearing method for microscopic examination was employed. Measurements are given as width and length. For photographs, the sections were stained with 0.5–1.0% aqueous methylene blue, aniline blue, or hematoxylin. For permanent slides, the glycerin was exchanged with 10–20% corn syrup.

Total genomic DNA was extracted from silica-gel-preserved sample using the DNeasy Plant Mini Kit (Qiagen, Hilden, Germany) according to the manufacturer's protocol. Before extraction, dried material was crushed with liquid nitrogen using a mortar and pestle. Concentrations of extracted DNA were assessed by using gel electrophoresis on a 1% agarose gel. Extracted DNA was used for amplification of ribulose-1, 5-bisphosphate carboxylase/oxygenase large subunit (*rbc*L) regions and cytochrome oxidase I (COI). PCR amplifications were performed in a TaKaRa PCR Thermal Cycler Dice (TaKaRa Bio Inc., Otsu, Japan). The PCR products were moved to Macrogen Sequencing Service for sequencing (Macrogen, Seoul, Korea). The sequences of PCR primers for amplification are as follows: *rbc*L (forward: 5′ GGAG GATTAGGGTCCGATTCC 3′, reverse: 5′ CTTCCGTCA ATTCCTTTAAG 3′), COI (forward: 5′ GCTGCGTTCT TCATCGATGC 3′, reverse: 5′ TCCTCCGCTTATTGA TATGC 3′) (Lin et al. 2001).

Sequences for the *rbc*L region were aligned using BioEdit (Hall 1999). Phylogenetic analyses were performed using neighbor joining, maximum-likelihood, and maximum parsimony methods with Mega 6 program (Tamura et al. 2013). Bootstrap values were calculated with 1000 replications. RbcL and COI sequences of other species were obtained from GenBank. *Ahnfeltia plicata* was used as an outgroup.

Results and discussion

Ahnfeltiopsis concinna (J. Agardh) P.C. Silva & DeCew 1992: 577

Type locality: Hawaii (Dawson 1961)

Korean name: Go-un-bu-chaet-sal nom. nov. (신칭: 고운부챗살)

Specimens examined: NIRBAL0000146348, PKNU 0000127011 - 0000127015, PKNU 0000127025 (Jindo: 13.ii.2014)

Habitat: Growing on rock near upper to lower intertidal

Morphology: Thalli 5–8 cm high, terete to subterete at the lower portion, somewhat compressed at the upper portion, fan-shaped, brown to yellow in color, cartilaginous in texture, attached to substratum by discoid holdfast (Fig. 1 and Fig. 2a); main axes issuing dichotomous

Fig. 1 *Ahnfeltiopsis concinna* (J. Agardh) P.C. Silva & DeCew. Habit of vegetative plant. *Bar* 1 cm

branches at short intervals; branches divided dichotomously to subdichotomously, with rounded or blunt apex, 1–2 mm wide, 200–300 µm thick; proliferations rare, arranged pinnately to irregularly; multiaxial; cortex consisted with small and pigmented cells (Fig. 2c, d), five to eight round cell layers thick, 2–3 × 4–5 µm; pseudoparenchymatous medulla compact (Fig. 2a), ellipsoid in transverse section (Fig. 2b), without hypha-like filaments, 200–300 × 80–100 µm; gonimoblast filaments developing inwardly (Fig. 3a, b), carposporophytes producing masses of carposporangia; carposporangia round (Fig. 3c), 10–12 µm in diam.; cystocarps formed at middle portion of branches, internally immersed in medulla, surrounded by some layers of secondary medullary cells, with carpostomes (Fig. 3c, d). Male and tetrasporangial plants were not collected during the present study.

Ahnfeltiopsis was established to accommodate several species, which had been previously assigned to *Ahnfeltia* E.M. Fries and *Gymnogongrus* C. Martius (Silva and DeCew 1992) and which have common internal cystocarps with specialized pores (carpostomes) and crustose tetrasporic life history (Silva and DeCew 1992; Masuda 1993). However, since the genus is known to be polyphyletic based on molecular data (Fredericq and Lopez-Bautista 2002; Maggs et al. 2013; Calderon and Boo 2016; Calderon et al. 2016; the present study), the generic delimitation cannot be used for *Ahnfeltiopsis*. Recently, some combined features of female structures have been adopted for this genus (Calderon and Boo 2016; Calderon et al. 2016). The vegetative feature of multiaxial thalli with a compact and pseudoparenchymatous medulla is also common in the genus (Masuda 1993). Even though male and tetrasporangial plants were not observed, our gigartinalean species collected from Jindo, Korea, during this study can be referred to *Ahnfeltiopsis* based on these vegetative and female features in addition to the

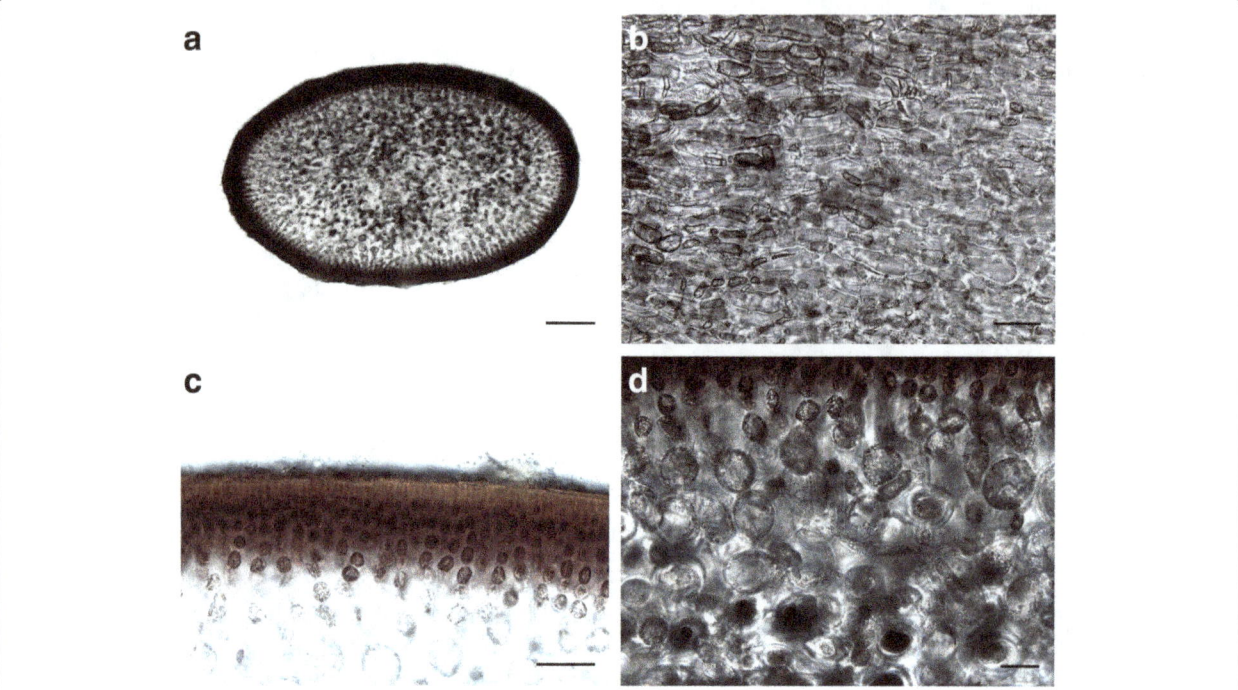

Fig. 2 *Ahnfeltiopsis concinna* (J. Agardh) P.C. Silva & DeCew. **a** Compact and pseudoparenchymatous medulla in transverse section of branches. **b** Ellipsoid medullary cells in longitudinal section. **c** Cortical cell layers. **d** Round inner cortical cells. *Bars* in **a** 200 µm; **b** 70 µm; **c** 50 µm; **d** 30 µm

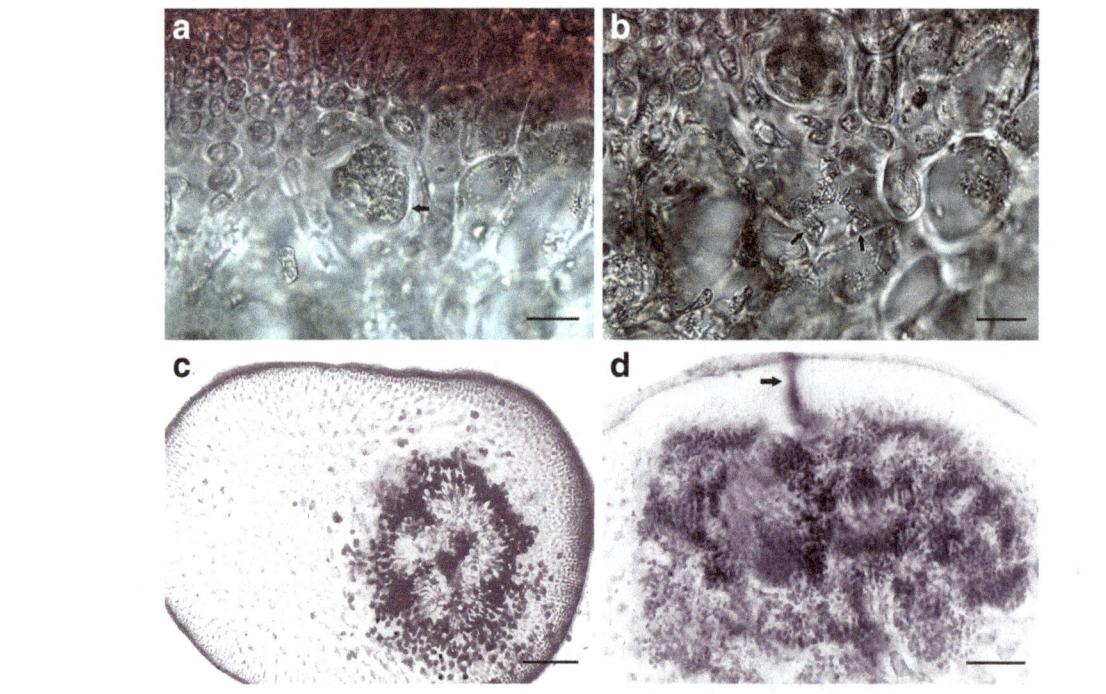

Fig. 3 *Ahnfeltiopsis concinna* (J. Agardh) P.C. Silva & DeCew. **a** Large auxiliary cell (arrow). **b** Initials (arrows) of gonimoblast filaments developing inwardly. **c** Cystocarp internally immersed in medulla of middle portion of branch. **d** Fully developed cystocarp with a carpostome (arrow). *Bars* in **a**, **b** 20 µm; **c** 300 µm; **d** 800 µm

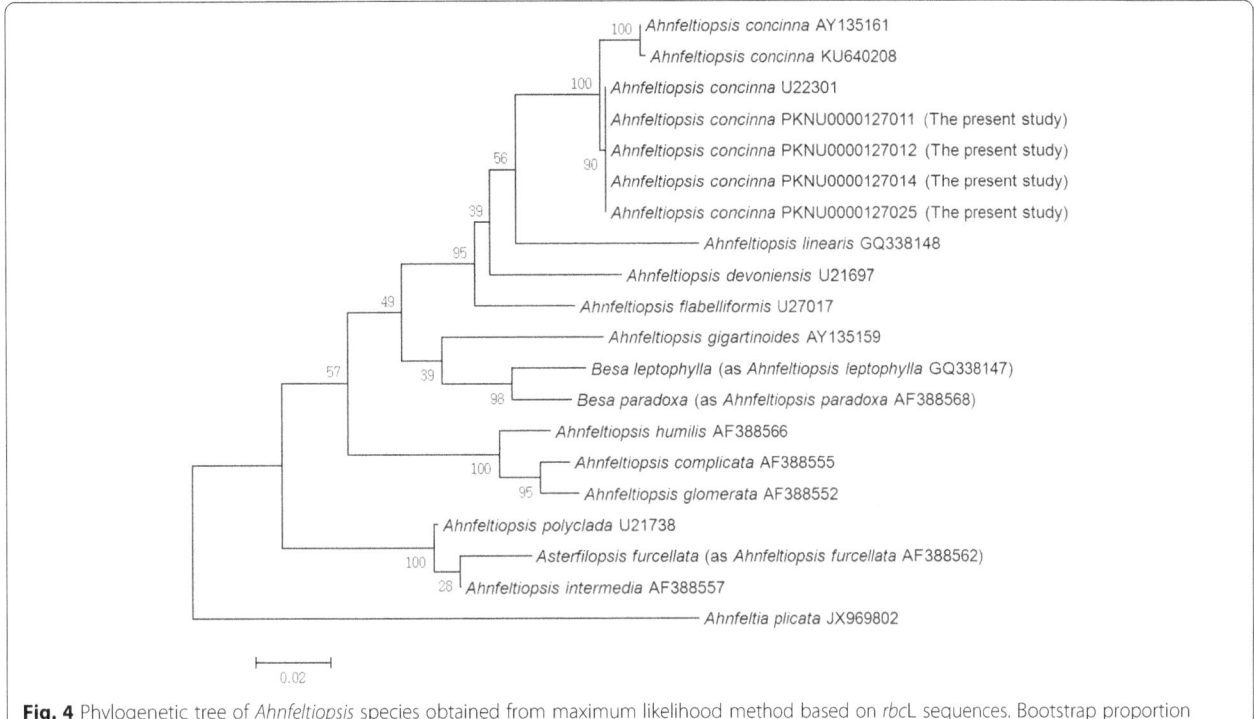

Fig. 4 Phylogenetic tree of *Ahnfeltiopsis* species obtained from maximum likelihood method based on *rbc*L sequences. Bootstrap proportion values (1000 replicates samples) are shown above branches. *Bar* indicates 0.02 substitutions/site

gross morphology. According to the original and other publications (Silva and DeCew 1992; Masuda et al. 1994; Braune and Guiry 2011), *Ahnfeltiopsis concinna* is distinct from similar species within the genus by its small size (up to 8 cm) thalli, terete to subterete thalli, cartilaginous in texture, dichotomous branches, rarely produced proliferations, and an absence of hypha-like filament in the medulla. This Korean alga shares these characteristics and is distinguished from *A. flabelliformis* which is currently recorded in Korea (Kang 1966, 1968; Lee and Kang 1986; Lee and Kang 2002; Boo and Ko 2012; Kim

et al. 2013), by the thallus feature. It has terete to subterete thalli particularly at the lower portion, while *A. flabelliformis* shows compressed thalli (Masuda 1987, 1993; Masuda et al. 1994; Yoshida 1998; Lee 2008).

In a phylogenetic tree based on *rbc*L sequence data (Fig. 4), the Korean alga nests in the same clade with *A. concinna* from Japan but forms a sister clade to *A. concinna* from Mexico and Hawaii (type locality). However, the genetic distance among those sequences was calculated as 0.1–1.3%, considered to be intraspecific variation within the genus. In general, the value of

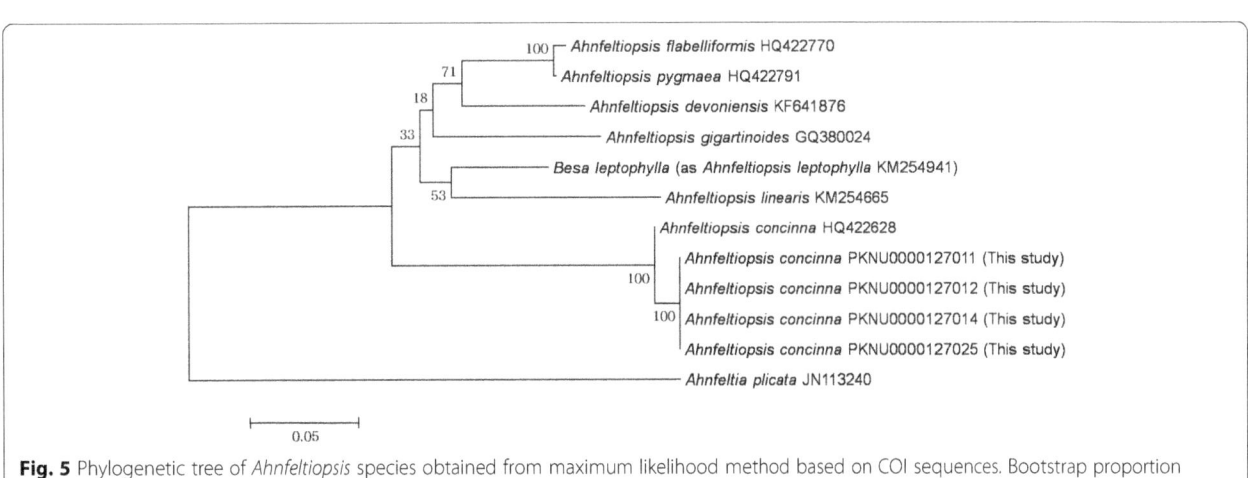

Fig. 5 Phylogenetic tree of *Ahnfeltiopsis* species obtained from maximum likelihood method based on COI sequences. Bootstrap proportion values (1000 replicates samples) are shown above branches. *Bar* indicates 0.05 substitutions/site

interspecific divergence in the Gigartinaceae and Peyssonneliaceae within the Gigartinales varies from 2.8 to 16.5% (Hommersand et al. 1994; Kato et al. 2009). Moreover, the Korean alga also nests in the same clade with *A. concinna* from the type locality based on the COI sequence data (Fig. 5), with the genetic distance of 1.1% between the two sequences.

This morphological and molecular evidence leads to the conclusion that our specimens from Jindo, Korea, should be identified as *Ahnfeltiopsis concinna* originally described from Hawaii. This species is newly recorded in Korea in the present study.

Conclusions

A gigartinalean species was collected from Jindo, Korea, during a survey of marine algal flora. This species was identified as *Ahnfeltiopsis concinna*, which was originally described from Hawaii, based on morphological and molecular data. This is the first record of the species in the Korean marine algal flora.

Abbreviations

COI: Cytochrome C oxidase subunit 1; *rbc*L: ribulose-1, 5-bisphosphate carboxylase large subunit

Funding

This work was supported by a grant from the National Institute of Biological Resources funded by the Ministry of Environment (MOE) of the Republic of Korea (NIBR201701204), and by a grant from Marine Biotechnology Program (20170431) funded by the Ministry of Oceans and Fisheries of the Korean Government.

Authors' contributions

PJK conducted the research, analyzed the materials, and prepared the draft manuscript. KWN designed and directed the study and finalized the manuscript. Both authors read and approved the final manuscript.

Competing interests

The authors declare that they have no competing interests.

References

Boo SM, Ko YD. Marine plants from Korea. Seoul: Marine & Extreme Genome Research Centre Program; 2012.

Braune W, Guiry MD. Seaweeds. A color guide to common benthic green, brown and red algae of the world's oceans. Ruggell: A.R.G. Gantner Verlag K.G; 2011.

Calderon MS, Boo SM. Phylogeny of Phyllophoraceae (Rhodophyta, Gigartinales) reveals *Asterfilopsis* gen. nov. from the Southern Hemisphere. Phycologia. 2016;55(5):543–54.

Calderon MS, Miller KA, Seo TH, Boo SM. Transfer of selected *Ahnfeltiopsis* (Phyllophoraceae, Rhodophyta) species to the genus *Besa* and description of *Schottera koreana* sp. nov. Eur J Phycol. 2016;51(4):431–43.

Craigie JS. Cell walls. In: Cole KM, Sheath RG, editors. Biology of the red algae. New York: Cambridge University Press; 1990. p. 221–58.

Dawson EY. Marine plants in the vicinity of the Institute Oc'eanographique de Nha Trang. Vietnam Pac Sci. 1954;8:372–469.

Dawson EY. Marine red algae of Pacific Mexico. Part 4. Gigartinales Pac Nat. 1961;2:191–343.

Donald FK, Dutcher JA, Bird KT, Capecchi MF. Nuclear genome characterization and carrageenan analysis of *Gymnogongrus griffithsiae* (Rhodophyta) from North Carolina. J Appl Phycol. 1993;5:99–107.

Fredericq S, Lopez-Bautista JM. Characterization and phylogenetic position of the red alga *Besa papillaeformis* Setchell: An example of progenetic heterochrony? Constancea. 2002;83(9):1–12.

Guiry MD, Guiry GM. AlgaeBase. World-wide electronic publication, National University of Ireland, Galway. 2017. http://www.algaebase.org. Accessed 6 Mar 2017.

Hall TA. BioEdit: a user-friendly biological sequence alignment editor and analysis program for Windows 95/98/NT. Nucleic Acids Symp Ser. 1999;41:95–8.

Hommersand MH, Fredericq S, Freshwater DW. Phylogenetic systematics and biogeography of the Gigartinaceae (Gigartinales, Rhodophyta) based on sequence analysis of *rbc*L. Bot Mar. 1994;37:193–203.

Kang JW. On the geographical distribution of marine algae in Korea. Bull Pusan Fish Coll. 1966;7:1–125.

Kang JW. Illustrated encyclopedia of fauna and flora of Korea, Vol.8. Marine algae. Seoul: Samhwapress; 1968.

Kato A, Guimarães SMPB, Kawai H, Masuda M. Characterization of the crustose red alga *Peyssonnelia japonica* (Rhodophyta, Gigartinales) and its taxonomic relationship with *Peyssonnelia boudouresquei* based on morphological and molecular data. Phycol Res. 2009;57:74–86.

Kelman D, Posner EK, McDermid KJ, Tabandera NK, Wright PR, Wright AD. Antioxidant activity of Hawaiian marine algae. Mar Drugs. 2012;10:403–16.

Kim HS, Boo SM, Lee IK, Sohn CH. National List of Species of Korea 「Marine Algae」. Seoul: Jeonghaengsa; 2013.

Lee Y. Marine algae of Jeju. Seoul: Academy Publication; 2008.

Lee IK, Kang JW. A check list of marine algae in Korea. Kor J Phycol. 1986;1:311–25.

Lee Y, Kang SY. A catalogue of the seaweeds in Korea. Jeju: Jeju National University Press; 2002.

Lin S-M, Frederucq S, Hommersand MH. Systematics of the Delesseriaceae (Ceramiales, Rhodophyta) based on large subunit rDNA and rbcL sequences, including the Phycodryoideae, subfam. nov. J Phycol. 2001;37:881–99.

Maggs CA, Le Gall L, Mineur F, Provan J, Saunders GW. *Fredericqia deveauniensis*, gen. et sp. nov. (Phyllophoraceae, Rhodophyta), a new cryptogenic species. Cryptogam Algol. 2013;34:273–96.

Masuda M. Taxonomic notes on the Japanese species of *Gymnogongrus* (Phyllophoraceae, Rhodophyta). J Fac Sci Hokkaido Univ, Series V (Botany). 1987;14:39–72.

Masuda M. *Ahnfeltiopsis* (Gigartinales, Rhodophyta) in the western Pacific. Jap J Phycol. 1993;41:1–6.

Masuda M, Zhang JF, Xia BM. *Ahnfeltiopsis* from the western Pacific: key, description and distribution of the species. In: Abbott IA, editor. Taxonomy of economic seaweeds, vol. 4. California: A Publication of the California Sea Grant College; 1994. p. 159–83.

Silva PC. The benthic algal flora of central San Francisco Bay. In: Conomos TJ, editor. San Francisco Bay: The urbanized estuary. San Francisco: Pac Div Amer Ass Adv Sci; 1979. p. 287–345.

Silva PC, Basson PW, Moe RL. Catalogue of the benthic marine algae of the Indian Ocean. Univ Calif Publ Bot. 1996;79:1–1259.

Silva PC, DeCew TC. *Ahnfeltiopsis*, a new genus in the Phyllophoraceae (Gigartinales, Rhodophyceae). Phycologia. 1992;31:576–80.

Tamura K, Stecher G, Peterson D, Filipski A, Kumar S. MEGA6: Molecular Evolutionary Genetics Analysis version 6.0. Mol Biol Evol. 2013;30:2725–9.

Yoshida T. Marine algae of Japan. Tokyo: Uchida Rokakuho Publishing; 1998.

20

Comparison of different ploidy detection methods in *Oncorhynchus mykiss*, the rainbow trout

Hong Seab Kim[1†], Ki-Hwa Chung[2†] and Jung-Ho Son[1*]

Abstract

The objective of this study was to determine a simple and reliable ploidy identification protocol for the rainbow trout (RT), *Oncorhynchus mykiss*, in the field condition. To evaluate the ploidy level and compare different detection protocols, triploid RT and gynogenesis were induced by UV irradiation and/or heat shock. The hatching rate at day 30 was 85.2% and the survival rate at day 90 was 69.4% (fingerling). The sex ratio of female RT was 93.75% in the gynogenesis group, illustrating that the UV irradiation inactivated the sperm DNA. The hatching rate and survival rate were 82.0 and 74.7%, respectively, in the triploid-induced group. The triploid induction rate by heat shock procedure was 73.9%. Cytogenetic protocols for ploidy identification such as chromosome counting, erythrocyte nuclear size comparison, and analysis of nucleolar organizing regions (NORs) by silver staining were compared. Silver nitrate staining showed the greatest success rate (22/23 and 32/32 for the triploid-induced group and gynogenesis group, respectively), followed by erythrocyte nuclear size comparison (16/23 and 19/32 for the triploid-induced group and gynogenesis group, respectively) and, lastly, chromosome preparation (2/23 and 6/32 for the triploid-induced group and gynogenesis group, respectively) with the lowest success rate. Based on our findings, silver staining for RT ploidy identification is speculated to be highly applicable in a wide range of research conditions, due to its cost-effectiveness and simplicity compared to other numerous ploidy detection protocols.

Keywords: Rainbow trout, Nucleolar organizing regions, Silver staining, Triploid, Gynogenesis

Background

It has been 50 years since the domestication of rainbow trout (RT) in South Korea, reaching a production of more than 3000 tons per year (Ministry of Ocean and Fisheries 2016). However, the lack of a systematic control of brood stock, recessive growth due to inbreeding, and the increased male ratio are causing the overall productivity of RT to slump (Hwang 2012). In the global aquaculture industry, the induction of numerous artificial triploid fish species is already an important subject of study (Felip et al. 1997; Gjedrem et al. 2012; Maxime 2008). The usage of triploid fish for industrial purposes has numerous advantages as it contains three sets of chromosomes and is genetically sterile. Above all, these types of fish have reduced gonadal development

(Cal et al. 2006; FAO 2005), meaning that instead of sexual maturation, the energy is directed towards the development of flesh quality and somatic growth (Felip et al. 2001; Kizak et al. 2013; Piferrer et al. 2009). These characteristics have drawn the attention of people for the preference of triploid fish over diploid.

Although diploid and triploid fish are morphologically equal throughout their life cycle, they are cytologically different. Hence, there are many ways, direct or indirect, to identify the ploidy of a fish (Maxime 2008; Tiwary et al. 2004). Among those are the measurement of nuclear and cellular size (Alcantar-Vazquez 2016; Thomas and Morrison 1995), electrophoresis of proteins (Liu et al. 1978; Shimizu et al. 1993), nuclear and cell size measurement of erythrocyte (Olele and Tiguiri 2013; Pradeep et al. 2011), chromosome counting (Thogaard 1983; Tiwary et al. 1997), DNA content determination with flow cytometry (Alcantar-Vazquez et al. 2008; Lamatsch et al. 2000), and staining of nucleoli with silver nitrate (Howell and Black 1980; Porto-Foresti et al.

* Correspondence: jhson@noahbio.com
†Equal contributors
[1]Noah Biotech Inc., Cheonan 31035, South Korea
Full list of author information is available at the end of the article

2002). Yet, regardless of type, it is believed that an easy, simple, and inexpensive method for ploidy identification is most advantageous and productive. The silver staining method for nucleoli identification, nuclear and cell size measurement of erythrocyte, and chromosome counting meet the criterion mentioned above since they are functional and have the capacity for a hasty identification of ploidy level, whereas most other methods require specific equipments and expensive materials (Carman et al. 1992). In this study with RT, we have preferentially focused on silver staining over chromosome counting and erythrocyte nuclear size comparison because of two main reasons: first is randomness. Chromosome preparation is known to be very random (Deng et al. 2003). There are too many factors to consider such as relative humidity (Spurbeck et al. 1996), cell dropping height (Barch et al. 1997; Hlics et al. 1997), and *flame vs. air* drying method of the slide (Karami et al. 2015). Still, even taking into account all these aspects, getting a well-spread metaphase is overly time consuming and not always rewarding. This is not an exception with the erythrocyte nuclear size comparison method. Although it is widely used, as mentioned in reports by Felip et al. (2001) and Caterina et al. (2014), the nuclear size of red blood cells is not always ~ 1.5 times bigger and it depends on the type of anticoagulant used while collecting the blood samples, as well as the preservation time of samples and slide preparation conditions. Second is inconsistency in chromosome numbers. Due to the Robertsonian translocation in RT chromosome (Inokuchi et al. 1994; Jankun et al. 2007), the change in number is unavoidable. The numbers range from $2n = 56$ (Kenanoglu et al. 2013), $2n = 56$ to 68 (Oliveira et al. 1995), and $2n = 58$ to 63 (Colihueque et al. 2001), making chromosome preparation less reliable. To the contrary, considering there is a direct relationship between the numbers of nucleolar organizing regions (NORs) per chromosome pair (Jankun et al. 2007; Phillips et al. 1986) in RT, silver staining is a more reliable method of ploidy identification.

In order to identify polyploidy of the samples (gynogenetic diploid females and presumed triploid RT), three different ploidy detection methods were compared. Furthermore, hatching rate, survival rates, sex ratio determination, and triploid induction rate were also measured.

Methods
Fish
RT were randomly selected from Dong Gang Aquaculture located in Pyeongchang. Males ($n = 5$, length 63.4 ± 2.3 cm; body weight 3415 ± 576.8 g) and females ($n = 19$, length 58.6 ± 4.2 cm; body weight 3519 ± 835.7 g) were anesthetized by MS-222 (Tricaine methane sulfonate, 25 mg/l) in a 50-l container. All eggs and milt used in this experiment were obtained by abdominal massage. Egg quality was evaluated by visual inspection. By calculating the average mass of an RT egg (~ 0.6 g), the total number of eggs collected was

calculated based on the mass of the container (35.7 l), giving a total of approximately 59,500 eggs. A total of about 47 ml of milt was collected from five males and divided into two for the treatment of gynogenesis and triploid production.

Gynogenesis and triploid production
Milt stripped from males was diluted (1:10) with saline solution and transferred to Petri dishes, 10 cm in diameter, forming a thin layer of sperm. The Petri dishes were exposed to UV (Phillips 6 W UV lamp) for 15 min on ice for the inactivation of sperm DNA (Fernandez-Diez et al. 2016). The eggs were divided into two groups, and each group was treated with normal intact milt (triploid-induced group) and UV-irradiated milt (gynogenesis group) for 2 min and stirred with a feather. For every ~ 3000 eggs, 1 ml of milt was used. After 10 min of fertilization, eggs were exposed to heat shock at 28 °C for 20 min to prevent extrusion of the second polar body. Hatching rate was calculated 30 days post fertilization, and survival rate was determined as the fish reached 90 days post fertilization. To further confirm ploidy by means of erythrocyte nuclear size, chromosome counting, and NOR identification, RT fingerlings ($n = 23$, age 3 months old; body weight 1.5–2 g) were randomly selected and kept alive while being transported to the lab in a *1-gal dispensing bag* connected to an air pump.

In addition, gonadal tissue slices obtained from the pool of gynogenesis group fingerlings ($n = 32$) were set onto a slide and gently squashed using a cover glass for sex ratio determination by histological examination under a microscope.

Detection of NORs by silver staining
Small pieces of fin tissue were obtained without sacrificing the samples (triploid-induced group), then sheared on a pre-cleaned slide with few drops of 50% acetic acid and finally let dry in air at room temperature. Samples were stained with silver nitrate following the procedures proposed by Howell and Black (1980) with a modification to remove silver residue precipitation. The first solution, solution A (Sol A), was made with 0.5 g of gelatin, 25 ml of double-distilled water, and 0.25 ml of formic acid containing formaldehyde (2% final concentration). An aqueous solution, solution B (Sol B), was a mix of 5 g of silver nitrate and 10 ml of double-distilled water. Both Sol A and Sol B were covered with aluminum foil and stored in the dark to avoid photoreaction. As for the staining of the slide, 50 µl of Sol A and 100 µl of Sol B were dropped on the slide and the solutions were gently mixed using the side of a pre-cleaned 3-ml disposable pipette. Next, the slide was placed on a hot plate (60 °C) that was covered well to provide as much darkness as possible for the stain to take place. As the solution became golden brown, the slide was removed from the hot plate, gently washed under running double-distilled water, and let dry in air.

Chromosome preparation

Fingerling samples were prepared as described by Kligerman and Bloom (1977) but modified to suit our experiment. To intercept cell division by interrupting the polymerization of microtubules, the fish were transferred into a 2-l glass beaker and then treated with 0.005% colchicine for 3 h. After colchicine treatment, the fish were sacrificed, and fins and gills were collected and placed in individual 1.5-ml Eppendorf tubes. Immediately after, samples were treated with 0.075 M potassium chloride (KCl) hypotonic solution for 20 min at room temperature twice. Samples were centrifuged at 3000 rpm for 2 min, supernatant was removed, and Carnoy's fixative solution (3:1 methanol/acetic acid) was added twice, each lasting 20 min. At the end of the last fixation procedure, samples were stored at 4 °C until assay. Each sample was placed on a slide with two to three drops of 50% acetic acid. Tissues were gently minced into tiny pieces using a 14-gauge needle attached to a 1-ml syringe under a dissecting microscope. Afterwards, 7 µl of the minced solution was pipetted and dropped onto a pre-cleaned slide at a height of 30~40 cm and air-dried. The slide was then stained with 5% Giemsa for 20 min at room temperature, washed with running double-distilled water, and let dry in air before observing under the microscope.

Erythrocyte nuclear size comparison

Due to the difficulty of blood withdrawal from fingerling (3 months old), fish were sacrificed and blood samples were aspirated using a 14-gauge needle in a 1-ml syringe coated with EDTA solution, while preparing the samples for chromosome preparation. On a pre-cleaned slide, 20 µl of blood was placed and smeared using a cover glass. The smeared blood was then stained with 0.22% Coomassie blue stain (composed of 220 mg Coomassie blue in 50 ml methanol, 10 ml acetic acid, and 40 ml double-distilled water) for 3 min, washed with double-distilled water, and let dry in air.

Microscope and camera equipment

All slides were observed using a Zeiss Axiovert 200 inverted microscope with a magnification of × 600, × 900, and × 1000, and photographs were taken using a Canon PowerShot G9 digital camera connected to the microscope via a Soligor adapter tube.

Results and discussion

Hatching rate, survival rate, and sex ratio determination

The average hatching and survival rates were calculated from 250 randomly selected samples of each group. The hatching rate of triploid-induced group and gynogenesis group was 85.2% ($n = 212$) and 82.0% ($n = 205$), respectively. The survival rate for each group was 69.6% ($n = 174$, triploid-induced group) and 74.4% ($n = 186$, gynogenesis group) at 90 days post fertilization (Table 1).

Based on the gonadal tissue examination (Fig. 1), the female sex ratio of gynogenesis group was 93.75% (30:32), indicating a fairly high induction of female. The histological section of female gonadal tissue showed corrugated structural morphology with signs of immature oocytes (Fig. 1a). On the other hand, the male testis showed an overall silky surface with immature spermatogonial development (Fig. 1b).

Triploid induction rate

The triploid induction rate measured by silver staining was 73.9% (17/23, Fig. 2). Throughout our experiments, we encountered samples with four NORs (Fig. 2e, f), which show similar patterns to the previous results reported by Flajshans et al. (1992) on the existence of four NORs in the course of triploid fish production.

Ploidy identification

The success rate for ploidy identification of each method was recorded. Chromosome preparation, erythrocyte nuclear size comparison, and silver nitrate staining methods were performed in all samples (gynogenesis group and triploid-induced group). The results of each method are shown in Table 2.

Chromosome preparation showed a very poor success rate of 6/32 and 2/23 for the gynogenesis group and triploid-induced group, respectively (Table 2). Attaining a clear image for chromosomal count was very random (Fig. 3a, b). After many experimental attempts, in which we tried our best to maintain a uniform working condition, we were occasionally able to obtain a justifiable spread of chromosomes. An approximate of 60 chromosomes, a characteristic of a diploid cell, was observed (Fig. 3b). Incomplete spread of metaphase chromosome, disturbing the viewer while performing chromosomal count, is shown in Fig. 3a.

Erythrocytes of gynogenesis group and triploid-induced group are shown in Fig. 4. The difference of nuclear length of triploid samples from those of diploid was at the major axis as mentioned by Jankun et al. (2007). However, the majority of the samples had the tendency to display a minor length difference showing

Table 1 Hatching and survival rate of induced gynogenesis and triploid rainbow trout

Group	Hatching rate	Survival rate
	Days[a]	
	30	90
Gynogenesis group (%)	82.0 (205/250)[b]	74.4 (186/250)
Triploid-induced group (%)	85.2 (212/250)	69.6 (174/250)

[a]Days after fertilization
[b]Counted no. of samples/total no. of samples ($n = 250$)

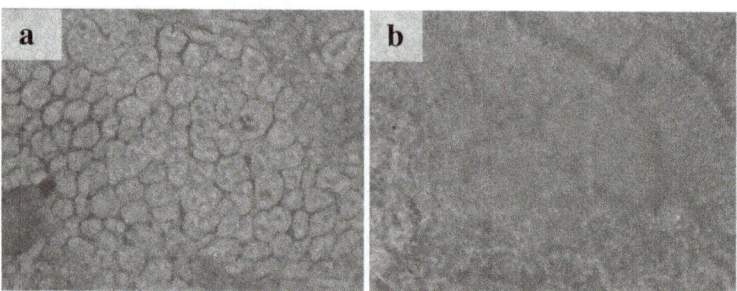

Fig. 1 Histological examination of gonadal tissue from rainbow trout. **a** Section of squeezed female gonadal structure showing signs of primordial oocytes (× 600). **b** Section of squeezed male gonadal structure showing signs of primordial spermatogonia (× 600)

difficulties in ploidy detection. Additionally, although an anticoagulant (EDTA) was used to prevent aggregation of erythrocytes, some samples showed signs of coagulation while others displayed signs of hemorrhage (data not shown). Overall, the success rate for ploidy detection in erythrocyte nuclear size comparison method was of 19/32 and 16/23 for the gynogenesis group and triploid-induced group, respectively (Table 2).

Phillips and Ihssen (1985) and Phillips et al. (1986) reported that *Oncorhynchus* species have only one NOR per chromosome pair. Therefore, if the samples from the triploid-induced group were triploids, the cells would be expected to have a maximum of three NORs. Ploidy detection using silver nitrate was the most successful (Table 2) compared to the other two methods. The results were 32/32 in the gynogenesis group and

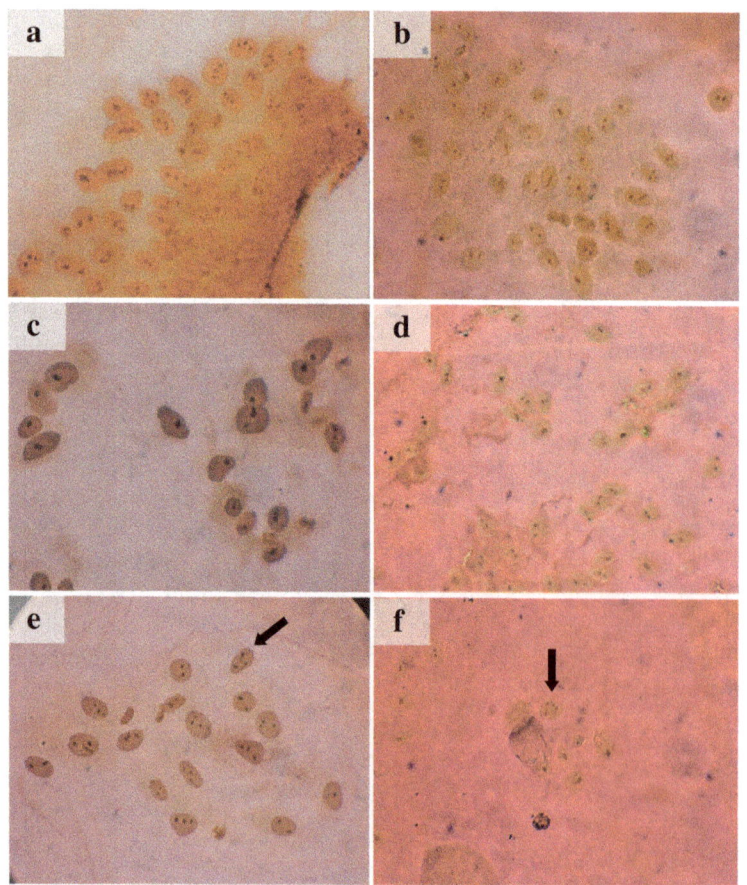

Fig. 2 Interphase nuclei from rainbow trout stained with silver nitrate. **a** Triploid *O. mykiss* sample containing up to three NORs. The excess number of cells but identifiable (× 600). **b** Ideal triploid sample with a maximum of three NORs (× 600). **c, d** Diploid *O. mykiss* sample containing one and two NORs, respectively (× 600). **e, f** Samples with a maximum of four NORs shown in arrows (× 900 and × 600, respectively)

Table 2 Number of successful ploidy detection experiments from three different ploidy detection methods

Ploidy detection method[a]	Group	
	Gynogenesis (n = 32)	Triploid-induced (n = 23)
Chromosome preparation	6/32[b]	2/23
Erythrocyte nuclear size comparison	19/32	16/23
Silver nitrate staining	32/32	22/23

[a]All ploidy detection methods were tested under identical samples
[b]Ploidy-identified sample/total no. of samples

22/23 in the triploid-induced group. Moreover, as previously mentioned by Kavalco and Pazza (2004), silver debris precipitation in the conventional silver nitrate staining procedure is responsible for false positive results, giving difficulties to the viewer when identifying the ploidy of a sample. Nonetheless, our results show clearer stains with few or no silver debris reason being the filtration (0.45 μm) of the staining solution before usage. A difference in coloration of stain can be seen in Fig. 2, which is due to the amount of time exposed to the silver nitrate stain. As reported by Howell and Black (1980), within 30 s, the stain turns yellow, and within 2 min, it turns golden brown. Because the time taken for the stain to transform into golden brown was not always the same, avoiding the stain to become too dark was critical. Through our study, we recommend that the optimal staining time should be less than 90 s, because longer exposure to the stain would negatively affect the imaging of the sample.

Through our study, we have compared and demonstrated three different but easily approachable methods for ploidy detection in RT and, hereinabove, presented the results (Table 2).

It can be denoted from our results that all three methods have their advantages. However, the most field-applicable, easy, and rapid method of ploidy identification funnels down to silver nitrate staining for NOR identification. Although chromosome counting is precise and excels to identify different ploidy levels, chromosome analysis requires technically sophisticated skill. Furthermore, our study shows that the erythrocyte nuclear size comparison method is, in fact, faster when compared to chromosome counting with higher success rate. Yet, we speculated that nuclear size comparison from fish blood cells was, to some degree, subjective and an inaccurate ploidy detection protocol since it depended on numerous factors such as the anticoagulant used, sample preservation time, and preparation conditions (Felip et al. 2001; Caterina et al. 2014).

Despite the fact that ploidy identification using silver nitrate in fish specimen is not as widely used as in animals, plants, and insects, silver staining for ploidy identification is fast and, at the same time, easy and very reliable since neither special skills nor expensive equipments are necessary. There are also several advantages when identifying the ploidy in RT; for instance, in place of sacrificing the specimens, samples could be obtained by cutting small pieces of fin from different yearlings and applying the staining method directly in the field without the inconvenience of returning to the laboratory. Moreover, this method could be applied in the early embryonic stage and therefore obviate the high raising cost and waste of time until being fully grown for ploidy identification. According to Phillips et al. (1986), the majority of these rapidly dividing embryonic cells are composed of their maximum number of nucleoli, thus making silver staining possible for the identification of triploids in the early developmental stage of fish.

Furthermore, the trial to induce triploid RT from our study [diploid 26.1%; triploid (including those with four NORs) 73.9%] is somewhat different from the previously reported studies (Hwang 2012). This may be due to the contributed experimental condition's discrepancies, such as temperature applied to eggs and the prevention timing of the second polar body extrusion.

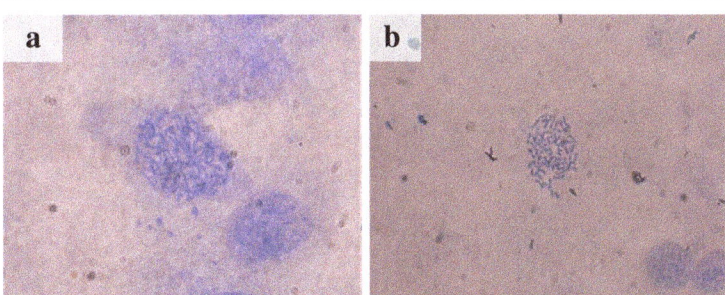

Fig. 3 Metaphase chromosome from rainbow trout stained with Giemsa. **a, b** Triploid and diploid chromosomes stained with 5% Giemsa stain under ×900 and ×600, respectively

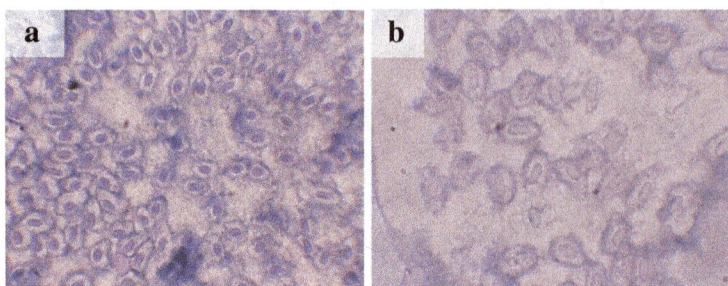

Fig. 4 Erythrocyte size comparison. Samples stained with 0.22% Coomassie blue. **a** Diploid sample (× 1000). **b** Triploid sample (× 1000)

Conclusions

According to the data obtained in this study, it is speculated that silver staining is a suitable ploidy detection method in RT not only for technically unsophisticated farms but also to fish research personnel. We hope that this silver staining method is useful to those who seek to produce an all-female and/or triploid brood in RT and/or other fish species.

Abbreviations
NORs: Nucleolar organizing regions; RT: Rainbow trout

Acknowledgements
Our deepest gratitude to Mr. Hyeong-Kook Kim for his technical assistance and helpful suggestions while performing this research and to the staff members in Dong Kang Fishery for their kindest allowance in the usage of their facility and invaluable assistance.

Funding
This work was supported by the Gyeongnam National University of Science and Technology Grant 2016.

Authors' contributions
HSK and JHS carried out the collection of the samples and interpreted the results altogether with KHC. HSK, JHS, and KHC designed the experiment. HSK and JHS performed the experiment and prepared the manuscript. All authors have read and approved the final manuscript.

Competing interests
The authors declare that they have no competing interests.

Author details
[1]Noah Biotech Inc., Cheonan 31035, South Korea. [2]Department of Animal Resources and Technology, Gyeongnam National University of Science and Technology, Jinju 52725, South Korea.

References
Alcantar-Vazquez JP. Fisiologia de los peces triploides. Lat Am J Aquat Res. 2016; 44:1–15.

Alcantar-Vazquez JP, Dumas S, Puente-Carreon E, Pliego-Cortés HS, Peña R. Induction of triploidy in spotted sand bass (*Paralabrax maculatofasciatus* Steindachner, 1868) by cold shock. Aquac Res. 2008;39:59–63.

Barch MJ, Knusten T, Spurbeck JL. The AGT cytogenetic laboratory manual. 3rd ed. New York: Raven; 1997. p. 25–50.

Cal RM, Vidal S, Gómez C, Álvarez-Blázquez B, Martínez P, Piferrer F. Growth and gonadal development in diploid and triploid turbot (Scophthalmus maximus). Aquaculture. 2006;251:99–108.

Carman O, Oshiro T, Takashima F. Variation in the maximum number of nucleoli in diploid and triploid common carp. Nippon Suisan Gakkai Shi. 1992;58:2303–9.

Caterina F, Francesca A, Giuseppe P, Alessandro Z, Francesco F. Effect of three different anticoagulants and storage time on haematological parameters of *Mugil cephalus* (Linneaus, 1758). Turk J Fish Aquat Sci. 2014;14:615–21.

Colihueque N, Ittura P, Estay F, Diaz NF. Diploid chromosome number variations and sex chromosome polymorphism in five cultured stains of rainbow trout Oncorhynchus mykiss. Aquaculture. 2001;198:63–77.

Deng W, Tsao SW, Lucas JN, Leung CS, Cheung ALM. A new method for improving metaphase chromosome spreading. Cytometry A. 2003;51A:46–51.

Felip A, Piferrer F, Zanuy S, Carrillo M. Comparative growth performance of diploid and triploid European sea bass over the first four spawning seasons. J Fish Biol. 2001;58:76–88.

Felip A, Zanuy S, Carrillo M, Martinez G, Ramos J, Piferrer F. Optimal conditions for the induction of triploidy in the sea bass (Dicentrarchus labrax L.). Aquaculture. 1997;152:287–98.

Fernandez-Diez C, Gonzalez-Rojo S, Lombo M, Herraez MP. Reproduction. 2016; 152:57–67.

Flajshans M, Rab P, Dobosz S. Frequency analyses of active NORs in nuclei of artificially induced triploid fishes. Theor Appl Genet. 1992;85:68–72.

Food and Agriculture Organization (FAO). Bartley D. Fisheries and aquaculture topics. Genetic biotechnologies. Topics fact sheets. Rome: FAO Fisheries and Aquaculture Department [online]; 2005.

Gjedrem T, Robinson N, Rye M. The importance of selective breeding in aquaculture to meet future demands for animal protein: a review. Aquaculture. 2012;350–353:117–29.

Hlics R, Muhlig P, Claussen U. The spreading of metaphases is a slow process which leads to a stretching of chromosomes. Cytogenet Cell Genet. 1997;76:167–71.

Howell WM, Black DA. Controlled silver staining of nucleolus organizer with a protective colloidal developer: a 1-step method. Experientia. 1980;36:1014–45.

Hwang GD. Inland fisheries research institute yearly research report. Korea: Chungcheongbuk-do; 2012.

Inokuchi T, Abe S, Yamaha E, Yamazaki F, Yoshida MC. BrdU replication banding studies on the chromosomes in early embryos of salmonid fishes. Hereditas. 1994;121:255–65.

Jankun M, Kuzminski H, Furgala-Selezniow G. Cytologic ploidy determination in fish—an example of two salmonid species. Environ Biotechnol. 2007;3:52–6.

Karami A, Araghi PE, Syed MA, Wilson SP. Chromosome preparation in fish: effects of fish species and larval age. Int Aquat Res. 2015;7:201–10.

Kavalco KF, Pazza R. A rapid alternative technique for obtaining silver-positive patterns in chromosoms. Genet Mol Biol. 2004;27(2):196–8.

Kenanoglu ON, Yilmaz S, Ergun S, Aki C. A preliminary study on the determination of triploidy by chromosome analysis at the different stages of development in rainbow trout, Oncorhychus mykiss. Mar Sci Tech Bull. 2013; 2:17–21.

Kizak V, Güner Y, Türel M, Kayim M. Comparison of growth performance, gonadal structure and erythrocyte size in triploid and diploid brown trout (Salmo trutta fario L., 1758). Turk J Fish Aquat Sci. 2013;13:571–80.

Kligerman AD, Bloom SE. Rapid chromosome preparation from solid tissues of fishes. J Fish Res Board Can. 1977;34:266–9.

Lamatsch DK, Steinlein C, Schmid M, Schartl M. Non-invasive determination of genome size and ploidy level in fishes by flow cytometry detection of triploid Poecilia formosa. Cytometry. 2000;39:91–5.

Liu S, Sezakki D, Hashimoto K, Kobayashi H, Nakamura M. Simplified techniques for determination of polyploidy in ginbuna, Carassius auratus langsdorfi. Bull Jpn Soc Sci Fish. 1978;44:601–6.

Maxime V. The physiology of triploid fish: current knowledge and comparisons with diploid fish. Fish Fish. 2008;9:67–78.

Ministry of Ocean and Fisheries. Fishery production survey, statistics by type of fishery and species (inland water fisheries). Korea; 2016. p. 205–9.

Olele NF, Tiguiri OH. Optimization of triploidy induction and growth performance of Claria anguillarias (African catfish) using cold shock. Acad J Interdisc Stud. 2013;2:189–96.

Oliveira C, Foresti F, Rigolino MG, Tabata YA. Synaptonemal complex analysis in spermatocytes and oocytes of rainbow trout, Oncorhynchus mykiss (Pisces, Salmonidae): the process of autosome and sex chromosome synapsis. Chromosom Res. 1995;3:182–90.

Phillips RB, Ihssen PE. Chromosome banding in salmonid fishes: nucleolar organizer regions in Salmo and Salvelinus. Can J Genet Cytol. 1985;27:433–40.

Phillips RB, Zajicek KD, Ihssen PE, Johnson O. Application of silver staining to the identification of triploid fish cells. Aquaculture. 1986;54:313–9.

Piferrer F, Beaumont B, Falguiere JC, Flajshans DM, Haffray PE, Colombo L. Polyploy fish and shellfish: production, biology and applications to aquaculture for performance improvement and genetic containment. Aquaculture. 2009;293:125–56.

Porto-Foresti F, Oliveira C, Tabata YA, Rigolino MG, Foresti F. NORs inheritance analysis in crossings including individuals from two stocks of rainbow trout (Oncorhynchus mykiss). Hereditas. 2002;136:227–30.

Pradeep PJ, Srijaya TC, Jose D, Papini A, Hassan A, Chatterji AK. Identification of diploid and triploid red tilapia by using erythrocyte indices. Caryologia. 2011; 64:485–92.

Shimizu Y, Oshiro T, Sakaizumi M. Electrophoretic studies od diploid, triploid and tetraploid forms of the Japanese silver crucian carp, Carassius auratus langsdorfi. Jpn J Ichthyol. 1993;40:65–75.

Spurbeck JL, Zinsmeister AR, Meyer KJ, Jalal SM. Dynamics of chromosome spreading. Am J Med Genet. 1996;61:387–93.

Thogaard GH. Chromosome set manipulation and sex control in fish. In: Hoar WS, Randall DJ, Donaldson EM, editors. Fish physiology. New York: Academic; 1983. p. 405–34.

Thomas P, Morrison R. A method to assess triploidy in swim-up rainbow trout. Austasia Aquacult. 1995;9:62–3.

Tiwary BK, Kirubagaran R, Ray AK. Induction of triploid by cold shock in catfish, Heteropneustes fossilis (Bloch). Asian Fish Sci. 1997;10:123–9.

Tiwary BK, Kirubagaran R, Ray AK. The biology of triploid fish. Rev Fish Biol Fish. 2004;14:391–402.

21

Effects of water physico-chemical parameters on tilapia (*Oreochromis niloticus*) growth in earthen ponds in Teso North Sub-County, Busia County

Agano J. Makori[1], Paul O. Abuom[1], Raphael Kapiyo[1], Douglas N. Anyona[1] and Gabriel O. Dida[2*]

Abstract

Small-scale fish farmers in developing countries are faced with challenges owing to their limited information on aquaculture management. Nile tilapia farmers in Teso North Sub-County recorded lower yields than expected in 2009 despite having been provided with required inputs. Water quality was suspected to be the key factor responsible for the low yields. This study sought to assess the effects of earthen pond water physico-chemical parameters on the growth of Nile tilapia in six earthen fish ponds under semi-intensive culture system in Teso North Sub-County. The study was longitudinal in nature with pond water and fish being the units of analysis. Systematic sampling was used to select five ponds while a control pond was purposively selected based on its previously high harvest. Four ponds were fed by surface flow and two by underground water. Each pond was fertilized and stocked with 900 fry of averagely 1.4 g and 4.4 cm. Physico-chemical parameters were measured in-situ using a multi-parameter probe. Sixty fish samples were randomly obtained from each pond fortnightly for four months using a 10 mm mesh size and measured, weighed and returned into the pond. Mean range of physico-chemical parameters were: dissolved oxygen (DO) 4.86–10.53 mg/l, temperature 24-26 °C, pH 6.1–8.3, conductivity 35–87 μS/cm and ammonia 0.01–0.3 mg/l. Temperature ($p = 0.012$) and conductivity ($p = 0.0001$) levels varied significantly between ponds. Overall Specific Growth Rate ranged between 1.8% (0.1692 g/day) and 3.8% (1.9 g/day). Ammonia, DO and pH in the ponds were within the optimal levels for growth of tilapia, while temperature and conductivity were below optimal levels. As temperature and DO increased, growth rate of tilapia increased. However, increase in conductivity, pH and ammonia decreased fish growth rate. Temperature and DO ranging between 27 and 30 °C and 5–23 mg/l, respectively, and SGR of 3.8%/day and above are recommended for higher productivity.

Keywords: Earthen fish ponds, Culture, Growth rate, Nile tilapia, Physico-chemical parameters

Background

Tilapia culture has been practised in more than 100 countries around the globe, yet most tilapia farmers, farm owners, farm managers, researchers, and graduate students in developing countries have little or lack accurate and critical information on tilapia culture (Abdel-Fattah, 2006). Lack of information on the basic requirements of an effective aquaculture system by small-scale fish farmers has handicapped the orderly, rapid development and high yield of the aquaculture industry in developing countries (Machena and Moehl, 2001). While the Sub-Saharan Africa region has numerous attributes such as underutilized land and water resources, cheap labor, high demand for fish and a favorable climate all year round, aquaculture production is still not at its maximum (Machena and Moehl, 2001). Optimal production of fish in culture systems has frequently been curtailed by several factors among them limited information on aquaculture set-up and poor information on pond water quality requirements for optimal fish production (Machena and Moehl, 2001).

Growth of fish is dependent on a wide range of positive or negative impacting factors. Studies show that

* Correspondence: gdidah@gmail.com
[2]School of Public Health, Maseno University, P. O. Box 333-40105, Maseno, Kenya
Full list of author information is available at the end of the article

growth of fish in aquaculture mainly depends on feed consumption and quality (Slawski et al., 2011); stocking density (Ma et al., 2006); biotic factors such as sex and age (Imsland and Jonassen, 2003); genetic variance; and abiotic factors such as water chemistry, temperature (Imsland et al., 2007), photoperiod (Imsland and Jonassen, 2003), and oxygen level (Bhatnagar and Devi, 2013). Therefore, successful management of fish ponds requires an understanding of water quality, which is determined by abiotic factors such as temperature, dissolved oxygen (DO), transparency, turbidity, water color, carbon dioxide, pH, alkalinity, hardness, unionized ammonia, nitrite, nitrate, primary productivity, biological oxygen demand (BOD), plankton population among others (Bhatnagar and Devi, 2013). A study by Bryan et al. (2011) conducted in Pennsylvania in 1998 on 557 pond owners established that 10% of the respondents had experienced water quality problems in their ponds ranging from muddy water to toxicity leading to fish kills.

Concerns about pond water quality are directly related to its production and therefore water quality parameters of greatest concern to fish farming are important to consider in fish culture (Bryan et al., 2011). Therefore, when evaluating and selecting sites for earthen fish pond siting, the source of water and its quality are some of the main factors to consider while ensuring that the water source has a high concentration of dissolved oxygen and optimal temperatures which should be kept at the right levels throughout the culture period among other critical factors (Ngugi et al., 2007).

Studies show that a special set of water chemistry requirements, and optimal water quality is essential to a healthy, balanced, and functioning aquaculture system (DeLong et al., 2009). The growth of different fish species is also influenced by a different range of factors, among them water quality parameters. Fish growth is generally greater in ponds with optimal levels of DO, temperature among other parameters (Bartholomew, 2010), though different fish species have ideal levels of water quality parameters within which they grow optimally (Kausar and Salim, 2006). Nile tilapia (*Oreochromis niloticus*) is ideal for culture due to its high growth rates, adaptability to a wide range of environmental conditions, ability to grow and reproduce in captivity and ability to feed at low trophic levels (Abdel-Fattah, 2006). The most preferred temperature range for optimal growth of tilapia is 25 to 27 °C, while the ideal pH ranges between 6 and 9 (DeWalle et al., 2011).

In 2009, the government of Kenya through the ministry of fisheries development introduced Fish Farming Enterprise & Productivity Programme (FFE & PP) in 140 constituencies across the country. The principal aim was to enhance food security, generate income among the rural community and stimulate rural enterprise development.

In Teso North Sub-County, where the current study focused, 100 farmers benefited from the program at the time. Despite having been sensitized and supplied with adequate nutritious supplementary fish feeds and other crucial inputs essential for aquaculture, fish farmers in the selected constituencies still realized low yields from the semi-intensive aquaculture system. According to Ngugi et al. (2007), the expected yields from each pond at the stocking density of 3 fish/m^2 for the mono-sex tilapia was 240 kg (8000 kg/ha/year).

However, 99 FFE & PP ponds out of the 100 ponds within Teso North Sub-County recorded harvests below 150 kg with the lowest recording as low as 70 kg, while the highest recorded 200 kg. Some of the reasons attributed to low production include the quality of water that was reportedly not taken into consideration during the initial set-up stages (Abdel-Fattah, 2006). Such vital knowledge alongside other important information on culture conditions of the pond water, fish feeds, growth rates, stocking density, and expected yields was missing among small scale tilapia farmers in Teso North Sub-County, Busia County, Kenya. This study thus sought to generate accurate information on water-quality parameters required for optimal production of Nile tilapia by small scale farmers in Teso North Sub-County, Busia County, Kenya.

Methods

Study area

This study was carried out in Teso North Sub-County (Fig. 1) which covers 236.8 Km^2 in surface area with an estimated population of 117,947 (KNBS 2010). Teso North Sub-County has its main town located at Amagoro which lies at Latitude 0° 37′ 40.335′ N and Longitude 34° 19′ 57.2736 E at 1200 and 1500 m asl. The Sub-County is characterized by undulating terrain with highlands intersected by numerous valleys and two rivers namely: Malakisi and Malaba (Jaetzold et al., 2007). Most parts of the Sub-County receive mean annual precipitation of between 800 and 1700 mm (MoPND, 2008). Mean maximum temperatures range between 26 and 30 °C while the mean minimum temperature ranges between 14 and 22 °C. A total of 184 households were engaged in fish farming at the time of the study (RoK, 2010). About 95% of the farmers cultured tilapia in earthen fish ponds.

Research design

A longitudinal research design was adopted for this study in which measurements were done fortnightly for four months. A total of six earthen fish ponds were selected, five of them through systematic random sampling and one selected purposively to act as a control based on its high yields recorded during the previous harvesting

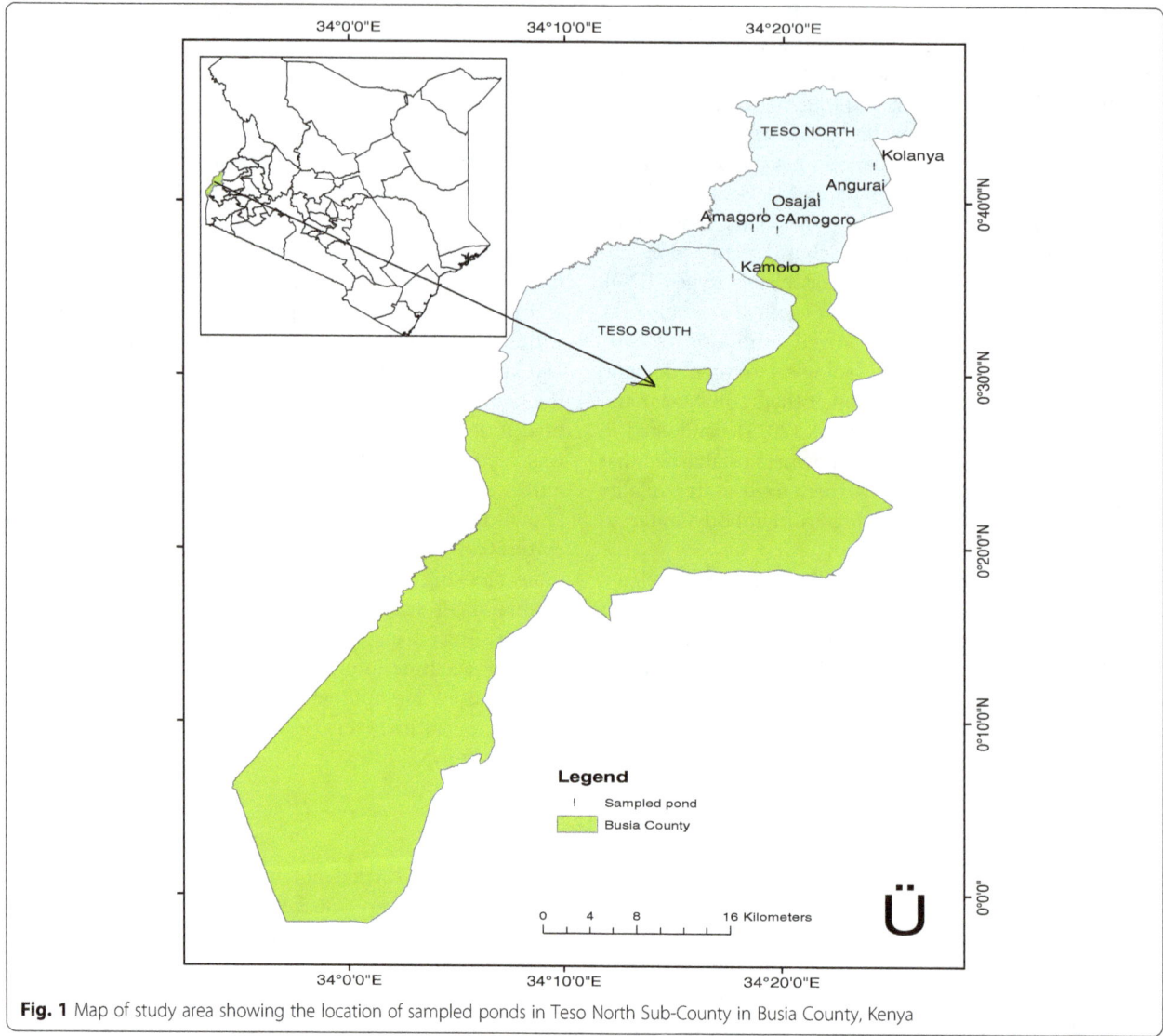

Fig. 1 Map of study area showing the location of sampled ponds in Teso North Sub-County in Busia County, Kenya

season. The fish ponds were classified into two classes based on the sources of their water supplies, i.e., well-surface flow ponds which comprised of Kamolo, Amagoro, and Osajai and underground spring-fed ponds comprising of Kolanya and Angurai. Each pond had an area of 300 m² and an average depth of 1 m.

Ponds fertilization, stocking and feeding
Each of the six ponds was initially fertilized 7 days prior to stocking, using organic manures from farm animals at a rate of 5 kg/100 m²/week (Ngugi et al., 2007). Male Nile tilapia fingerlings each weighing about 5 g and measuring about 0.4 cm total length were stocked at a stocking density of three fish per m² in the six earthen ponds. Each pond was stocked with a total of 900 fingerlings. The culture system was semi-intensive and expected production from each pond based on the preferred stocking density was 240 kg

(Ngugi et al., 2007). The fish were fed twice a day (at 10:00 and 16:00) with the amount of feed being proportional to 10% body weight.

Sampling
Six sampling sites (two sites near the inlet, two in the middle, and two near the outlet) were identified within each fish pond. Fish growth (mean length and width) was conducted fortnightly alongside measurement of physico-chemical parameters for a period of 4 months. Sixty fish were randomly sampled from each earthen pond using a seine net of 10 mm mesh size. The fish samples were anesthetized using AQUI-S (a sedative and/or anesthetic that provides control in animal husbandry, fish transportation and research operations) at a dose of 2.5 mL/100 L of water in a bucket prior to taking measurements to avoid stressing them. Their total lengths (TL) in centimeters and weights in grams were

taken immediately using a fish measuring board and an electronic weighing scale, respectively. After the measurements, fish were transferred into a bucket containing clean water for stabilization before releasing them back to the pond. Specific Growth Rate (SGR) was calculated fortnightly for a period of 112 days. Specific growth rate refers to percentage increase in body dimensions per time and the results are given in percentage increase per day (Hopkins, 1992). The formula for calculating the SGR is as follows:

$$\text{SGR} \, (\%/\text{day}) = \frac{\log(w_t) - \log(w_i)}{t} \times 100$$

Where: wi is initial weight/length,
wt is final weight/length,
t is time in days.

A YSI multi-parameter probe (HI 9828 - YSI Incorporation, Yellow Spring, USA) was used to measure dissolved oxygen, temperature, pH and conductivity in situ. Total ammonia nitrogen (TAN) was also measured in situ using ammonia test kit (Model HI28049, Hach, USA). Each of the six ponds had recordings taken at the bottom, mid-water, and near the surface at 08:00, 10:00, 12:00, 14:00, 16:00 and 18:00 on different sampling days.

Data analysis

Descriptive statistics was used to outline the basic features of the data in the study by giving simple summaries like the mean and standard deviation of weight and length of fish and other physico-chemical parameters. Analysis of variance (ANOVA) was used to test the study hypotheses. The relationships between fish growth and physico-chemical parameters were determined using correlation and multiple regression analyses.

Results and discussions

Physico-chemical parameters

Dissolved oxygen levels measured in the six earthen fish ponds showed the highest (10.6 ± 8.4 mg/L) levels to be in the control earthen pond and lowest (4.9 ± 2.8 mg/L) levels at Osajai earthen pond, though the difference was not statistically significant ($F_{(5,53)}$ = 1.72, p = 0.1483) (Table 1).

The average dissolved oxygen (DO) levels in the six ponds ranged from 4.86 mg/L to 10.53 mg/L during the entire study period of 112 days with a mean of 7.066 mg/L (Table 2). According to Riche and Garling (2003), the preferred DO for optimum growth of tilapia is above 5 mg/L. Other researchers have however proved that tilapia can tolerate condition of high oxygen super saturation of up to 40 mg/L (Tsadik and Kutty, 1987). On the lower limit, Ross (2002) noted that DO concentration of 3 mg/L should be the minimum for optimum

growth of tilapia. Generally, fish growth and yields are greater in ponds with higher DO concentration (Bartholomew, 2010). In the current study, all ponds recorded minimum DO levels of less than 3 mg/L at certain times during sampling save for Angurai pond where the minimum DO concentration was 6.04 mg/L. Dissolved oxygen levels were significantly different per growth period (days) (F = 2.02, p = 0.044), with highest (9.29 ± 4.22 mg/L) DO levels recorded on day 0 and lowest (5.33 ± 3.83 mg/L) on day 56. The reduction in DO on day 56 could have been as a result of increased uptake by microorganisms during breakdown of accumulated organic matter in the pond. As regards diurnal oxygen variations at different sampling times, the findings showed different trends in each pond in relation to time of sampling, though the lowest DO levels were recorded at dawn across all the ponds. Consistent with the current study findings, Boyd (2010) also noted that in pond aquaculture, the lowest level of DO concentration is likely to occur at night.

Conductivity levels varied significantly between the six earthen ponds ($F_{(5,53)}$ = 6.31, p = 0.0001) with further analysis showing significantly lower conductivity levels at Amagoro compared to all the other ponds. The average conductivity in each of the six ponds ranged between 34.67 µS/cm and 86.67 µS/cm. Diurnal mean conductivity between 08.00 and 16.00 at different ponds varied ranging from 24.32 to 99.42 µS/cm (Table 1). Conductivity levels also varied significantly at different durations (days) during sampling ($F_{(8,311)}$, =9.92, p = 0.0001), with highest conductivity (83.81 µS/cm) recorded on day 70 and lowest (42.58 µS/cm) on day 112 (Table 2).

Given that conductivity in aquatic ecosystems is mostly driven by soil composition or the bedrock on which a river flows (Russell et al., 2011), the varying conductivity levels observed in the six ponds could be attributed to the bedrock material on which the ponds were sited but could also be a result of human activities. Crane (2006) noted that conductivity values greater than 100 µS/cm were indicative of human activity. According to Russell et al. (2011), water conductivity of between 150 and 500 µS/cm is ideal for fish culture (Russell et al., 2011). Stone et al. (2013), however, put the desirable range of conductivity for fish ponds at between 100 and 2000 µS/cm. It is therefore important to consider the rock type and soil composition of a site before construction of a fish pond.

pH levels showed a narrow range of variation between different fish ponds, with the highest (6.76 ± 0.31) mean pH recorded at Kamolo and the lowest (6.32 ± 0.15) at Amagoro. There were, however, no significant difference in pH between the six earthen ponds (p = 0.091) (Table 1). pH level differed significantly between sampling days

Table 1 Mean physico-chemical parameters in the six earthen fish ponds

Earthen Ponds	Dissolved oxygen (mg/L) (Mean ± SD)	pH (Mean ± SD)	Temperature (°C) (Mean ± SD)	Conductivity (µS/cm) (Mean ± SD)	Ammonia (mg/L) (Mean ± SD)
Control	10.6 ± 8.4	6.73 ± 0.29	26.49 ± 2.41[A]	61.74 ± 21.66[BC]	0.03 ± 0.04
Amagoro	5.44 ± 2.61	6.32 ± 0.15	23.87 ± 1.38[B]	34.52 ± 21.1[D]	0.07 ± 0.16
Angurai	8.34 ± 3.22	6.71 ± 0.28	24.24 ± 1.12[B]	60.63 ± 29.5[BC]	0.02 ± 0.02
Kamolo	6.45 ± 4.86	6.69 ± 0.31	23.66 ± 0.88[B]	57.98 ± 14.8[C]	0.01 ± 0.01
Kolanyo	6.61 ± 4.47	6.76 ± 0.67	23.48 ± 2.70[B]	81.52 ± 10.1[AB]	0.56 ± 0.07
Osajai	4.86 ± 2.81	6.50 ± 0.25	24.55 ± 1.69[B]	86.66 ± 29.60[A]	0.03 ± 0.02
F – value	1.72	2.03	3.31	6.31	0.75
P – value	0.1483	0.091	0.012	0.0001	0.059

Means with different supercripts in the same column are significantly different at $P<0.05$ (Data analyzed by Duncan's Multiple Range Test)

$(F_{(8311)} = 4.86, p = 0.0001)$ with the lowest pH (6.35) recorded on day 42 and highest (6.87) on day 70 (Table 2), while the mean daily pH between 08.00 and 16.00 fluctuated between 6.07 and 6.94 with only three ponds (Kolanya, Angurai, and Kamolo) having their pH ranging between 6.5 and 9; which is the ideal range for tilapia culture as reported by Bolorunduro and Abba (1996). Boyd (1998) noted that the daily fluctuations in pH result from changes in the rate of photosynthesis in response to daily photoperiod. As carbon dioxide accumulates in the water during the night, the pH falls. The process could be responsible for the low pH levels recorded in some of the ponds in this study. BEAR (1992) reported a pH range of between 6.5 and 9.0 as optimum for growth of tilapia. Crane (2006) reported that highly acidic water with pH less than 5.5 limited fish growth and reproduction, noting that the ideal pH range for freshwater aquaculture should range between 6.5 and 7.0, though a pH range of 6.1 to 8.0 is also considered satisfactory for the survival and reproduction of fish. Bryan et al. (2011) concurs that most fish would do

better in ponds with a pH near 7.0 and that ponds with a pH less than 6.0 may result in stunting or reduced fish production.

Temperature levels differed significantly across the six earthen ponds $(F_{(5,53)} = 3.31, p = 0.012)$, with further analysis (DMRT) indicating a significantly higher mean temperature (26.5 ± 2.4 °C) at the control pond than all the other earthen ponds, whose levels ranged between 23.5 and 24.5 °C (Table 1). There were significant differences in temperature in relation to duration of study (days) $(F_{(8311)} = 8.71, p = 0.0001)$ whereby highest temperature (25.67 ± 2.53 °C), was recorded on day 98, while the lowest (22.99 ± 1.63 °C) was recorded on day 42 (Table 2). Temperatures between 20 and 36 °C have been reported by various researchers as being suitable for tilapia culture. According to Kausar and Salim (2006), for instance, the preferred temperature range for optimum tilapia growth in ponds is between 25 and 27 °C. FAO (2011) reported the preferred temperature ranges of between 31 and 36 °C, while Ngugi et al. (2007) gave a range of between 20 and 35 °C as ideal for tilapia culture. These previous studies are

Table 2 Mean physico-chemical parameters against duration of study (days)

Period of study	Dissolved oxygen (mg/L) (Mean ± SD)	Temperature (°C) (Mean ± SD)	pH (Mean ± SD)	Conductivity (µS/cm) (Mean ± SD)	Ammonia (mg/L) (Mean ± SD)
Day 0	9.29 ± 4.22[A]	23.7 ± 0.75[B]	6.79 ± 0.38[A]	72.7 ± 18.8[A]	0.02 ± 0.01[CD]
Day 14	7.69 ± 7.15[AB]	25.1 ± 2.93[B]	6.62 ± 0.33[A]	44.3 ± 34.2[A]	0.01 ± 0.01[B]
Day 28	6.30 ± 2.97[AB]	23.3 ± 0.61[B]	6.46 ± 0.40[BC]	76.4 ± 28.1[BC]	0.04 ± 0.06[D]
Day 42	6.93 ± 5.34[AB]	23.2 ± 2.40[B]	6.66 ± 0.23[B]	57.3 ± 26.1[B]	0.01 ± 0.01[BC]
Day 56	6.99 ± 4.30[B]	22.9 ± 1.63[A]	6.35 ± 0.26[A]	59.2 ± 25.1[A]	0.031 ± 0.04[CD]
Day 70	5.33 ± 3.83[B]	25.0 ± 3.01[A]	6.53 ± 0.46[A]	73.3 ± 12.4[A]	0.01 ± 0.02[A]
Day 84	5.44 ± 5.79[AB]	25.6 ± 2.43[A]	6.87 ± 0.81[BC]	83.8 ± 29.5[BC]	0.09 ± 0.06[CD]
Day 98	6.37 ± 6.03[A]	25.5 ± 2.91[A]	6.70 ± 0.52[D]	57.5 ± 20.1[D]	0.02 ± 0.02[CD]
Day 112	8.73 ± 8.09[AB]	25.7 ± 2.53[A]	6.68 ± 0.21[CD]	42.6 ± 36.9[CD]	0.01 ± 0.05[D]
F statistic	2.02	8.71	4.86	9.92	18.0
P value	0.0437	0.0001	0.0001	0.0001	0.0001

Means with different supercripts in the same column are significantly different at $P<0.05$ (Data analyzed by Duncan's Multiple Range Test)

consistent with the current study findings in which the highest SGR of 4.4%/day was recorded at a water temperature of 35 °C in the control pond.

The mean ammonia levels in the six ponds ranged from 0.01–0.4 mg/L though the differences were not statistically significant ($p = 0.59$) (Table 1). Ammonia levels were significantly different with respect to the period of study (days) ($F_{(8311)} = 18$, $p = 0.0001$), with the lowest (0.008 mg/L) levels recorded on day 28 and highest (0.086 mg/L) on day 70 (Table 2). According to TNAU, (2008), the optimal range of un-ionized ammonia is 0.02–0.05 mg/L in fish ponds. Consistent with this study, BFAR (1992) also reported ammonia levels of between 0.02–0.05 mg/L as the optimum for tilapia growth. Emerson et al. (1975), however, noted that a concentration of 0.6 mg/L of un-ionized ammonia, is capable of killing fish even if exposed briefly while chronic exposure to levels as low as 0.06 mg/L can cause gill and kidney damage and reduction in growth.

Mean fish length and weight

There was no significant difference in mean length and weight of fish obtained from the six earthen ponds ($p = 0.858$ and $p = 0.42$, respectively). Nevertheless, fish obtained from the control earthen pond registered the highest mean length (9.82 ± 4.24 cm) and weight (31.03 ± 28.5 g) while Kolanya fish pond recorded the lowest mean length (7.74 ± 2.9 cm) and Kamolo fish pond recorded the lowest mean weight (17.1 ± 11.95 g) (Table 3). The growth of fish in total length was almost uniform increasing steadily from below 5 cm to over 12 cm on day 112 (Fig. 2). However, the highest mean total length was recorded in the control earthen pond and the least in Kolanya earthen pond (Table 3).

The control earthen fish pond recorded the highest mean fish weight (31.03 ± 28.5 g) while Kamolo earthen fish pond recorded the lowest mean fish weight (17.1 ± 11.95 g) (Table 3). The difference in weight gain betweeen the control pond and Kamolo fish pond could be attributed to ideal conditions in the control fish pond,

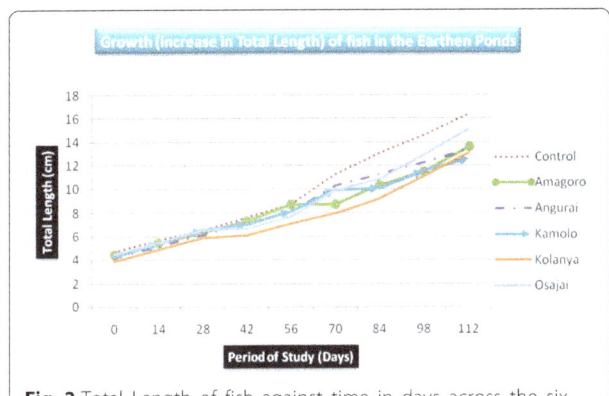

Fig. 2 Total Length of fish against time in days across the six earthen ponds

quality of water in the pond, feed availability, and stocking density, all of which may have favored the development of tilapia in the control pond (Ngugi et al., 2007). Further analysis of weight data collected on the 28th and 42nd day in Kolanya pond indicated a constant specific growth rate of 2.3% over the 14 days period, implying stunted growth of fish in this particular pond (Fig. 3).

Specific growth rate (%/day) and physico-chemical parameters

The SGR attained at the end of the study (after 112 days), ranged from 3.7–4.4%/day. The highest SGR of 4.4%/day was attained in the control pond under the following water quality parameters: DO was 23.2 mg/l, pH was 6.94, temperature was 30.25 °C, ammonia was 0.01 mg/l and conductivity was 23.5 µS/cm. Given that the SGR was highest in the control pond corresponding to a DO of 23.2 mg/l (Fig. 4), it is clear that high DO levels influenced the growth of fish positively. Osajai pond recorded 4.0%/day SGR, while the other four earthen ponds all recorded the same growth rate of 3.7%/day SGR (Table 4).

Dissolved oxygen affects the growth, survival, distribution, behavior, and physiology of fish and other aquatic organisms, and therefore oxygen depletion in water leads to poor feeding of fish, starvation, reduced growth, and more

Table 3 Mean length and weight of fish recorded in six ponds during the study period (112 days)

Earthen ponds	Mean length (cm)	Mean weight (g)
Control	9.82 ± 4.24	31.03 ± 28.47
Amagoro	8.46 ± 2.99	17.21 ± 11.84
Angurai	8.70 ± 3.21	18.05 ± 12.79
Kamolo	8.37 ± 2.79	17.13 ± 11.95
Kolanyo	7.74 ± 2.91	15.80 ± 12.55
Osajai	8.85 ± 3.57	20.98 ± 17.10
F value	0.38	1.01
P value	0.858	0.423

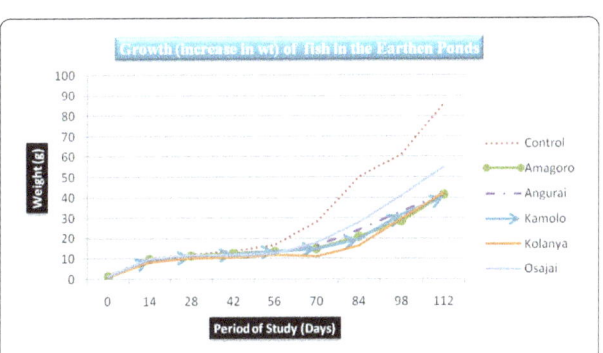

Fig. 3 A line graph showing the increase in weight with time across the six earthen ponds

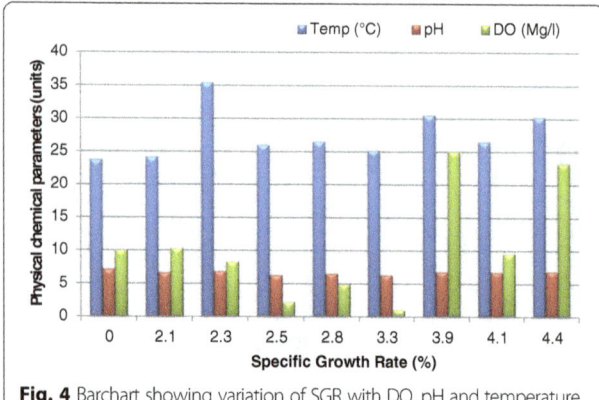

Fig. 4 Barchart showing variation of SGR with DO, pH and temperature in the control pond

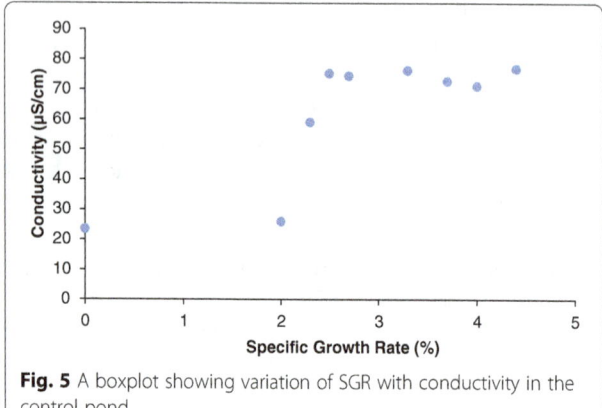

Fig. 5 A boxplot showing variation of SGR with conductivity in the control pond

fish mortality, either directly or indirectly (Bhatnagar and Garg, 2000).

According to Bhatnagar and Singh (2010) and Bhatnagar et al. (2004) DO level > 5 ppm is essential to support good fish production. Bhatnagar et al. (2004) also suggests that DO levels of 1–3 ppm has sub lethal effect on growth and feed utilization, while DO levels of 0.3–0.8 ppm is lethal to fishes. Ekubo and Abowei (2011) also cautioned that fish are likely to die if exposed to less than 0.3 mg L-1 of DO for a long period of time. Avoiding excessive application of fertilizer or organic manure can help manage the levels of DO in fish ponds. Control of aquatic weeds and phyto-plankton, as well as recycling of water and use of aerators can also be used by small-scale fish farmers to manage DO levels in earthen pond (Bhatnagar and Devi, 2013).

An increase in SGR was noted with a corresponding increase in temperature before falling sharply at a temperature of 35.53 °C. The highest SGR of 4.4%/day was recorded at a temperature of 30.25 °C (Fig. 4). In the current study, the highest SGR was also obtained at a pH of 6.94 (near neutral) (Fig. 4), conductivity of 77.0 μS/cm (Fig. 5), and ammonia of 0.01 mg/l in the control pond. Consistent with the current study findings, Santhosh and Singh (2007) reported the suitable pH range for fish culture ranges between 6.7 and 9.5. Fish

have an average blood pH of 7.4, and therefore a little deviation from this value, generally between 7.0 to 8.5 is more optimum and conducive to fish life. However, fish can become stressed in water with a pH ranging from 4.0 to 6.5 and 9.0 to 11.0 and death is almost certain at a pH of less than 4.0 or greater than 11.0 (Ekubo and Abowei, 2011). High pH levels can be reduced by addition of gypsum or organic matter, while low pH levels can be rectified bu. application of quick lime (Bhatnagar and Devi, 2013).

Optimum conductivity for high fish production differs from one species to another. Stone and Thomforde (2004) recommended the desirable range 100–2000 μS/cm and acceptable range 30–5000 μS/cm for pond fish culture. The conductivity level of 77.0 μS/cm recorded in the control pond in the current study and which cor-responded to the highest SGR was however slightly lower than the desirable range described by Stone and Thomforde (2004). Studies indicate that ammonia is a by-product from the metabolism of proteins excreted by fish and bacterial decomposition of organic matter like wasted food, feces, dead plankton among others (Bhat-nagar and Devi, 2013). It has been observed that ammo-nia in the range > 0.1 mg/L tends to cause gill damage in fish, destroy mucous producing membranes, and cause sub-lethal effects like reduced growth, poor feed conversion, and reduced disease resistance (Bhatnagar and Devi, 2013).

Maximum limit of ammonia concentration for aquatic organisms is 0.1 mg/L (Santhosh and Singh, 2007), while Bhatnagar and Singh (2010) recommended that ammo-nia levels of < 0.2 mg/l are suitable for pond fishery. Consistent with these scholars, it was noted in the current study that the highest SGR for the control pond was registered at an ammonia level of 0.01 mg/l, which was within the recommended < 0.2 mg/l ammonia for fish culture in ponds. Increasing pond aeration, regular water change, addition of quicklime are some of the ways that farmers can use in managing ammonia within

Table 4 Percentage SGR of fish in six earthen ponds over the study period (112 days)

Days	Control	Amagoro	Angurai	Kamolo	Kolanya	Osajai
14	2.1	12.0	1.8	2.0	1.8	2.0
28	2.3	2.3	2.4	2.3	2.3	2.3
42	2.5	2.5	2.5	2.4	2.3	2.4
56	2.8	2.6	2.6	2.6	2.5	2.5
70	3.3	2.7	2.8	2.7	2.4	2.9
84	3.9	3	3.2	3	2.8	3.3
98	4.1	3.3	3.5	3.4	3.4	3.7
112	4.4	3.7	3.7	3.7	3.7	4

their fish ponds (Bhatnagar and Devi, 2013). Given the specificity of certain parameters to the growth of fish, it is imperative to ensure that the various physico-chemical parameters are at their optimal levels at all levels during the entire process of fish culture in ponds.

Correlation between fish growth (weight and length) and physico-chemical parameters

A strong positive correlation was observed between mean fish weight and: mean length (r = 0.92949, p = 0.0001), temperature (r = 0.57488, p = 0.0001), and dissolved oxygen (r = 0.30620, p = 0.0243), while mean fish weight showed a negative correlation with conductivity (r = –0.37 724, p = 0.0049) (Table 5).

A positive correlation was also observed between mean fish length and temperature (r = 0.54232, p = 0.0001) and a negative correlation between mean fish length and conductivity (r = –0.34323, p = 0.0111). A fish pond with good water quality is likely to produce more and larger fish than a pond with poor water quality (Boyd, 1998). For instance, analysis of weight data collected on the 28th and 42nd day in Kolanya pond indicated a constant specific growth rate of 2.3% over the 14 days period, which was an indication of stunting of fish over that duration probably due to the effect of poor water quality.

An analysis of physico-chemical parameters during that period revealed that the mean dissolved oxygen and temperature levels during the same period ranged from 3.11–3.88 mg/l and 20.77–20.94 °C, respectively. The levels of these two parameters could have been low for the optimal growth of tilapia in the Kolanyo earthen pond. In a different study, Abo-State et al. (2009), reported a SGR of between 3.308 and 3.513%/day in tilapia fish cultured within a 70 day period under the following average values of water quality parameters: DO was 7.5 mg/L, pH was 7.6, temperature was 27.5 °C, total ammonia nitrogen (TAN) was 0.040 mg/L (Abo-State et al., 2009). Mbugua (2008) reported that at stocking

density of 2 fish/ m^2, male tilapia can attain 200 g and above in 4 to 5 months. However in this study in which the stocking density was 3 fish/m^2, the male tilapia in the six ponds attained mean weights ranging between 42.4 ± 0.9 g and 86.1 ± 1.3 g.

According to USDA (1996), water quality in ponds changes continuously and this often affects the optimal levels of physical and biological characteristics. As was the case in the current study, high dissolved oxygen levels and high temperatures tend to favor the growth of fish in earthen ponds. Studies concur that most biological and chemical processes within fish ponds are influenced by temperature, hence the need to ensure that the temperatures are maintained at optimal levels and that the same temperature favor the optimal growth if the fish species reared.

Regression between fish growth and physico-chemical parameters

In the control pond, 99.9% of the variation in fish weight was explained by the five physico-chemical parameters, while that of the other five fish ponds ranged between 66.8 and 99.5%. The multiple regression analysis generated different R^2 values for each of the six earthen ponds as indicated in Table 6.

Partial regression coefficient (B value) defines the direction and the magnitude of the slope of a regression line. In the current study, the B values associated with pH (–4.56) and conductivity (–0.154) bore negative signs, implying that for every increase in one unit of pH and conductivity, the regression equation predicted a decrease of 4.56 g and 0.154 g of fish weight (dependent variable), respectively. The B values associated with DO, temperature and ammonia bore positive signs, implying that for every increase of one unit of these parameters, there was a corresponding increase in fish weight by a certain unit. For instance, the regression equation

Table 5 Correlation matrix of fish length, weight and physic-chemical parameters

	Dissolved oxygen	Temp	pH	Cond.	Ammonia	Mean length	Mean weight
Dissolved oxygen	1.0000						
Temp	0.52790 0.0001	1.0000					
pH	0.43224 0.0011	0.45172 0.0006	1.0000				
Cond.	– 0.28925 0.0339	– 0.20314 0.1407	0.02350 0.8661	1.0000			
Ammonia	– 0.15896 0.2509	0.17194 0.2138	– 0.07659 0.5820	0.17190 0.2139	1.0000		
Mean length	0.12424 0.3708	0.54232 0.0001	0.08711 0.5311	– 0.34323 0.0111	– 0.04147 0.7659	1.0000	
Mean weight	0.30620 0.0243	0.57488 0.0001	0.08139 0.5589	– 0.37724 0.0049	– 0.11036 0.4269	0.92949 0.0001	1.0000

Table 6 Coefficient of determination of the six ponds showing amount of variation in fish growth explained by physico-chemical parameters

Ponds	R	R^2	Adjusted R Square	Std. Error of the Estimate	Change Statistics				
					R^2	F Change	df1	df2	Sig. F Change
Control	.999[a]	.999	.997	1.54593	.999	539.117	5	3	.000
Amagoro	.817[a]	.668	.115	11.0730	.668	1.208	5	3	.476
Angurai	.943[a]	.889	.705	6.9885	.889	4.823	5	3	.113
Kamolo	.865[a]	.749	.330	9.80052	.749	1.788	5	3	.335
Kolanya	.893[a]	.798	.461	9.23804	.798	2.366	5	3	.255
Osajai	.998[a]	.995	.987	1.93360	.995	124.704	5	3	.001

[a]Predictors: (constant), ammonia, conductivity, pH, temperature, dissolved oxygen

predicted an increase of 2.409 g of fish weight with every increase of one unit of temperature.

Conclusions

In conclusion, dissolved oxygen, temperature and ammonia in the earthen ponds were within the optimum range for growth of tilapia, while conductivity and pH were not. The water used for culture of tilapia was slightly acidic. The control pond registered the highest growth of tilapia while the same differed in other fish ponds with stunted growth experienced in Kolanya fish pond between day 28 and 42. The stunting was attributed to unfavorable temperatures that did not support the optimal growth of fish. Based on the findings therefore, it can be concluded that the low yields from the ESP fish ponds in 2010, was directly attributed to poor water quality in the earthen ponds, either because of lack of the right information or ignorance among the fish farmers.

Any changes to their environment add stress to the fish, and the larger and faster the changes, the greater the stress. It is therefore recommended that water of good quality as well as maintenance of all the other factors is very essential for ensuring maximum yield in a fish pond. This can only be achieved through sensitization and educating the fish farmers on comprehensive management of aquaculture systems.

Good water quality refers to that with adequate oxygen, proper temperature, transparency, limited levels of metabolites, and optimum levels of other environmental factors affecting fish culture. This information would enable the farmers to take better care of their fish ponds by frequently monitoring the conditions of the ponds, fish behavior, and water color for any abnormal changes.

Abbreviations

ANOVA: Analysis of variance; DO: Dissolved oxygen; GOK: Government of Kenya; NACOSTI: National Commission for Science, Technology and Innovation; SGR: Specific growth rate; TAN: total ammonia nitrogen

Acknowledgements

We wish to acknowledge the fish farmers from Teso North Sub-County, within Busia County for allowing us the use of their fish ponds for this study. The authors also thank the research assistants for their assistance throughout the study.

Funding

This study was financially supported by the National Commission for Science, Technology and Innovation (NACOSTI), Kenya, through Maseno University.

Authors' contributions

PA, GD, RK and AM conceived and designed the study. DA, AM, and PA organized the field work. AM, PA, and DA collected the data. GD and DA performed the statistical analysis. AM, GD, and DA drafted the manuscript. All authors read and approved the final manuscript.

Competing interests

The authors declare that they have no competing interests.

Author details

[1]School of Environment and Earth Science, Maseno University, P. O. Box 333-40105, Maseno, Kenya. [2]School of Public Health, Maseno University, P. O. Box 333-40105, Maseno, Kenya.

References

Abdel-Fattah ME. Tilapia Culture. 2006. (Available at: https://www.goodreads.com/book/show/661835.Tilapia_Culture.) Accessed 22 July 2017.

Abo-State HA, Tahoun AM, Hammouda YA. Effects of Replacement of Soybean meal by DDGS combined with Commercial Phytase on Nile tilapia (Oreochromisniloticus) Fingerlings Growth Performance and Feed Utilization. 2009; Available at. http://www.idosi.org/aejaes/jaes5(4)/4.pdf.

Bartholomew WG. Effect of channel catfish stocking rate on yield and water quality in an intensive, mixed suspended-growth production system. 2010.

Available at: http://afs.tandfonline.com/doi/abs/10.1577/A09-020.1#. WfMAxxLg97k.

BFAR. Basic biology of Tilapia. 1992. https://www.bfar.da.gov.ph/bfar/download/ nfftc/BasicBiologyofTilapia.pdf. Accessed 21 June 2017.

Bhatnagar A, Devi P. Water quality guidelines for the management of pond fish culture. Int J Environ Sci. 2013;3(6):1980–2009. https://doi.org/10.6088/ijes. 2013030600019.

Bhatnagar A, Garg SK. Causative factors of fish mortality in still water fish ponds under sub-tropical conditions. Aquaculture. 2000;1(2):91–6.

Bhatnagar A, Jana SN, Garg SK, Patra BC, Singh G, Barman UK. Water quality management in aquaculture. In: Course manual of summerschool on development of sustainable aquaculture technology in fresh and saline waters, CCS Haryana agricultural, Hisar (India); 2004. p. 203–10.

Bhatnagar A, Singh G. Culture fisheries in village ponds: a multi-location study in Haryana, India. Agric Biol J N Am. 2010;1(5):961–8.

Bolorunduro PI, Abba YA. Water quality Management in Fish Culture 1996.

Boyd CE. Water quality for pond aquaculture. Res Develop. 1998;43:1–11.

Boyd CE. Dissolved oxygen concentration in pond aquaculture. 2010. Available at https://www.researchgate.net/publication/281309202_Dissolvedoxygen_ concentration_in_pond_aquaculture. Accessed 11 Aug 2017.

Bryan R, Soderberg W, Blanchet H, Sharpe WE. Management of Fish Ponds in Pennsylvania. 2011. Available at http://www.water-rresearch.net/Waterlibrary/ Lake/waterqualityponds.pdf. Accessed 28 Nov 2016.

Crane B. Results of Water quality Measurements in Messer Pond. 2006. Available at http://www.messerpond.org/Ecology/WaterSamplingSummary. pdf. Accessed 23 Aug 2017.

DeLong DP, Losordo, TM, Rakocy JE. Tank culture of tilapia. Online; SRAC Publication No. 282. Southern Regional Aquaculture Center. 2009. Available at https://thefishsite.com/articles/tank-culture-of-tilapia

Dewalle DR, Swistock BR, Sharpe WE. Episodic flow – duration analysis: assessing toxic exposure of brook trout (Salvenius fontinalis) to episodic increases in aluminium. 2011. Available at: http://www.nrcresearchpress.com/doi/abs/10. 1139/f95-081#.WfMIZBLg97k. Accessed 20 Aug 2017.

Ekubo AA, Abowei JFN. Review of some water quality management principles in culture fisheries. Res J Appl Sci Eng Technol. 2011;3(2):1342–57.

Emerson K, Russo RC, Lund RE, Thurston RV. Aqueous ammonia equilibrium calculations: effect of pH and temperature. J Fish Res Board Can. 1975; 32:2379–83.

FAO. Fisheries and Aquaculture Department. 2011. Available at http://www.fao. org/fishery/culturedspecies/Oreochromis_niloticus/en Accessed 29.08.2017.

Hopkins KD. Reporting fish growth, a review of the basics. Journal of World Aquaculture Society. 1992;23:173–9.

Imsland AK, Jonassen TM. Growth and age at first maturity in turbot and halibut reared under different photoperiods. Aquac Int. 2003;11:463–75.

Imsland AK, Schram E, Roth B, Schelvis-Smit R, Kloet K. Growth of juvenile turbot Scophthalmus maximus (Rafinesque) under a constant and switched temperature regime. Aquac Int. 2007;15:403–7.

Jaetzold R, Schmidt H, Hornetz B, Shisanya C. Farm Management Handbook of Kenya. Natural Conditions and Farm Management Information. Western Kenya. 2007;2(A) Available at: http://www.fao.org/fileadmin/user_upload/drought/ docs/FMHB%20Nyanza%20Province.pdf. Accessed 14 Dec 2016.

Kausar R, Salim M. Effect of Water Temperature on Growth Performance and Feed Conversion Ratio of Labeo rohita. 2006. http://www.oalib.com/paper/ 2147234#.WfMKSRLg97k. Accessed 23 Apr 2017.

Kenya National Bureau of Statistics (KNBS) (2010). Population and Housing Census Results 2009, Minister of State for Planning, National Development and Vision 2030, Kenya.

Ma A, Chen C, Lei J, Chen S, Zhuang Z, Wang Y. Turbot Scophthalmus maximus: stocking density on growth, pigmentation and feed conversion. Chin J Oceanol Limnol. 2006;24:307–12.

Machena C, Moehl J. Sub-Saharan African aquaculture: regional summary. In Subasinghe RP, Bueno P, Phillips MJ, Hough C, McGladdery SE, Arthur JR. eds. Aquaculture in the Third Millennium. Technical Proceedings of the Conference on Aquaculture in the Third Millennium, Bangkok, Thailand, 2001; 341–355. NACA, Bangkok and FAO, Rome.

Mbugua HM. A Comparative Economic Evaluation of Fish Farming of three Important Aquaculture Species in Kenya. 2008. Available at: http://www. unuftp.is/static/fellows/document/henry07prfi.pdf. 23.07.2017.

Ministry of Planning and National Development (MoPND). Teso North District Development Plan (2008–2012). 2008.

Ngugi CC, James RB, Bethuel OO. A New Guide to Fish Farming in Kenya, Oregon State University, USA. 2007.

Republic of Kenya (2010). Fish Farming in Teso District, Ministry of Fisheries Development.

Riche M, Garling D. Fish: Feed and Nutrition. Feeding Tilapia in Intensive Recirculating Systems. 2003. Available at; http://www.hatcheryfeed.com/ hf-articles/141/. Accessed 13 May 2017.

Ross LL. Environmental physiology and energetic. Fish and Fisheries Series. 2002; 25:89–128. Accessed at http://link.springer.com/chapter/10.1007%2F978-94- 011-4008-9_4 Accessed on 14 Sept 2013.

Russell M, Shuke R, Samantha S. Effects of Conductivity on Survivorship and Weight of Goldfish (Carassius auratus). 2011. Available at http://departments. juniata.edu/biology/eco/documents/Russell_et al.pdf. 23 Apr 2017.

Santhosh B, Singh NP. Guidelines for water quality management for fish culture in Tripura, ICAR Research Complex for NEH Region, Tripura Center, Publication no.29. 2007.

Slawski H, Adem H, Tressel RP, Wysujack K, Kotzamanis Y, Schulz C. Austausch von Fischmehl durch Rapsproteinkonzentrat in Futtermitteln für Steinbutt (Psetta maxima L). Züchtungskunde. 2011;83:451–60.

Stone N, Shelton JL, Haggard BE, Thomforde HK. Interpretation of Water Analysis Reports for Fish Culture. Southern Regional Aquaculture Center (SRAC) Publication No. 4606. 12 pg. 2013.

Stone NM, and Thomforde HK. Understanding Your Fish Pond Water Analysis Report. Cooperative Extension Program, University of Arkansas at Pine Bluff Aquaculture / Fisheries. 2004.

Tamil Nadu Agricultural University (TNAU). Water quality Management. Accessed at http://www.agritech.tnau.ac.in/fishery/fish_water.html. 2008. Accessed 19 Aug 2017.

Tsadik GG, Kutty MN. Influence of ambient temperature and dissolved oxygen on feeding and growth of the tilapia (Oreochromis niloticus) 1987. (www.fao.org/ docrep/field/003/AC168E/AC168E00.htm).

USDA US. Department of Agriculture. Aquaculture outlook. 1996;4:26 – 28.

Gene structure and expression characteristics of liver-expressed antimicrobial peptide-2 isoforms in mud loach (*Misgurnus mizolepis*, Cypriniformes)

Sang Yoon Lee and Yoon Kwon Nam[*]

Abstract

Background: Liver-expressed antimicrobial peptide-2 (LEAP-2) is an important component of innate immune system in teleosts. In order to understand isoform-specific involvement and regulation of LEAP-2 genes in mud loach (*Misgurnus mizolepis*, Cypriniformes), a commercially important food fish, this study was aimed to characterize gene structure and expression characteristics of two paralog LEAP-2 isoforms.

Results: Mud loach LEAP-2 isoforms (LEAP-2A and LEAP-2B) showed conserved features in the core structure of mature peptides characterized by four Cys residues to form two disulfide bonds. The two paralog isoforms represented a tripartite genomic organization, known as a common structure of vertebrate LEAP-2 genes. Bioinformatic analysis predicted various transcription factor binding motifs in the 5′-flanking regions of mud loach LEAP-2 genes with regard to development and immune response. Mud loach LEAP-2A and LEAP-2B isoforms exhibited different tissue expression patterns and were developmentally regulated. Both isoforms are rapidly modulated toward upregulation during bacterial challenge in an isoform and/or tissue-dependent fashion.

Conclusion: Both LEAP-2 isoforms play protective roles not only in embryonic and larval development but also in early immune response to bacterial invasion in mud loach. The regulation pattern of the two isoform genes under basal and stimulated conditions would be isoform-specific, suggestive of a certain degree of functional divergence between isoforms in innate immune system in this species.

Keywords: Mud loach *Misgurnus mizolepis*, LEAP-2 isoforms, Development, Bacterial challenge

Background

Antimicrobial peptide (AMP) is the vital component of the innate immune system of fish as a central player in the first defense line against bacterial invasion (Magnadóttir 2006; Hancock et al. 2016). AMPs disrupt the physical integrity of microbial membranes and also they function as a modulatory effector in the innate immunity of fish (Townes et al. 2009; Li et al. 2012). Liver-expressed antimicrobial peptide-2 (LEAP-2) is the second blood-derived antimicrobial peptide, which was first identified in human (Krause et al. 2003). Similar with LEAP-1 (also termed hepcidin), LEAP-2 is cysteine-rich and predominantly expressed in the liver. LEAP-2s have been reported to exhibit selective antimicrobial activity against various microbes, and they may have potential ability to induce the hydrolysis of bacterial DNA, suggesting that LEAP-2 would be important in the modulation of fish innate immunity (Henriques et al. 2010; Li et al. 2015). In addition, most teleost species represent external fertilization and development without any close interconnection with parents, leading proposed needs of fish embryos and early larvae to protect themselves from microbial invasions (Nam et al. 2010). Potential involvements of LEAP-2s in the host protection during embryonic and early larval developments have been claimed in a few fish species (Liu et al. 2010; Liang et al. 2013).

* Correspondence: yoonknam@pknu.ac.kr
Department of Marine Bio-Materials and Aquaculture, Pukyong National University, Busan 48513, South Korea

Insofar, LEAP-2 sequences have been isolated and characterized from a considerable number of teleost species belonging to a wide range of taxonomic positions. Most of previously reported fish LEAP-2s (including LEAP-2-like sequences available in public GenBank database) seem to share a conserved structural homology particularly regarding the core structure with two disulfide bonds in their predicted mature peptides (Chen et al. 2016). However, in spite of their structural homology, regulation of LEAP-2 gene expression under both non-stimulated (i.e. basal expression in healthy fish) and stimulated (i.e. modulation of expression in response to immune/stress challenges) conditions have been largely variable or even contradictory among fish species (Liang et al. 2013; Zhang et al. 2004; Bao et al. 2006). Moreover, unlike mammals, many fish species are thought to multiple LEAP-2 isoforms (two or three isoforms depending on species). Usually, such paralog isoforms of host defense proteins have often been reported to exhibit the certain subfunctionalization in their physiological roles related with host defense. However, comparative information on isoform-dependent or isoform-specific regulations of LEAP-2 paralog genes within a given species has been available only in a limited number of fish species (Zhang et al. 2004; Li et al. 2014; Yang et al. 2014).

Mud loach (*Misgurnus mizolepis*, Cypriniformes) is a commercially important, aquaculture-relevant fish in Korea with a gradually increasing market demand as a food fish (Nam et al. 2001). Mud loach, as its name indicates, usually inhabits muddy bottom of ponds, paddy fields, and creaks. Muddy bottom of their habitats may often contain abundant and diverse microbial populations, and hence, the AMP-mediated protective system should be a fundamental requirement for this fish species. Previously, we have reported the multivalent involvements of LEAP-1 (hepcidin) as a central orchestrator to modulate immunity and iron homeostasis in this species (Nam et al. 2011). In line with our goal to understand coordinated regulation of LEAP-1 and LEAP-2 in mud loach, we reported here the characterization of two mud loach LEAP-2 isoforms (LEAP-2A and LEAP-2B). In this study, we isolated complementary DNA (cDNA) and genomic genes including their 5′-flanking regions, examined tissue and developmental expression patterns, and scrutinized transcriptional responses of the two isoforms to bacterial challenge.

Methods
Isolation of mud loach LEAP-2A and LEAP-2B genes
From the mud loach liver transcriptome next generation sequencing (NGS) database (unpublished data), NGS clones showing the significant homology to previously known vertebrate LEAP-2s were collected and subjected to the contig assembly using Sequencher® software

(Gene Codes, Ann Arbor, MI, USA). Mud loach liver cDNA template was prepared by reverse transcription (RT) with oligo-d(T) primer and liver total RNA (2 μg) using the Omniscript® Reverse Transcription Kit (Qiagen, Hilden, Germany) according to manufacturer's instruction. Based on the contig sequences, two putative mud loach LEAP-2 cDNAs were isolated by reverse transcription-polymerase chain reaction (RT-PCR) using the liver cDNA template. Amplified RT-PCR products were cloned into pGEM-T easy vectors (Promega, Madison, WI, USA), and recombinant clones (*n* = 12) carrying correct insert size were subjected to sequencing. The representative cDNA sequence for each LEAP-2 isoform was confirmed again by direct sequencing of RT-PCR product amplified from mud loach liver. Oligonucleotide primers used in this study are listed in the Additional file 1: Table S1.

Based on the cDNA sequence, genomic fragment corresponding to each isoform was PCR-isolated, TA-cloned, and sequenced as above. In order to get the 5′- and/or 3′-flanking regions of LEAP-2 isoforms, genome walking was carried out by using Universal Genome Walker® Kit (Clontech Laboratories Inc., Mountain View, CA, USA). Preparation of library and primer walking were carried out according to manufacturer's instruction. For each LEAP-2 isoform, amplified fragments were TA-cloned, sequenced, and assembled into a contig. Finally, the continuous version of genomic fragment spanning from 5′-flanking region to 3′-untranslated region (UTR) was PCR-isolated. Representative genomic sequence of each isoform was determined by direct sequencing of amplified products with primer walking method.

Bioinformatic sequence analysis
Sequence homology with orthologs was examined using NCBI BLASTx (http://blast.ncbi.nlm.nih.gov/Blast.cgi). Open reading frame (ORF) of mud loach LEAP-2 isoforms were predicted with the ORF Finder program (https://www.ncbi.nlm.nih.gov/orffinder/). Amino acid sequence of each isoform was deduced from the corresponding cDNA using the same program. Parameter scores for primary structure of each LEAP-2 isoform were estimated using ExPASy ProtParam tool (http://web.expasy.org/protparam/). ClustalW program (http://www.genome.jp/tools-bin/clustalw) was used to carry out multiple sequence alignment of LEAP-2 isoforms along with their orthologs. Putative cleavage sites for signal peptide and propeptide were predicted with SignalP 4.1 Server (http://www.cbs.dtu.dk/services/SignalP/) and ProP 1.0 Server (http://www.cbs.dtu.dk/services/ProP/), respectively. Phylogenetic relationship among LEAP-2 isoforms (whole protein region) in the teleost lineage was inferred with Molecular Evolutionary Genetics

Analysis tool (MEGA ver. 7.0; http://www.megasoftware.net/). Putative transcription factor (TF) binding motifs in the abalone LEAP-2A and LEAP-2B promoters were predicted with TRANSFAC® software (http://genexplain.com/transfac; GeneXplain GmbH, Wolfenbüttel, Germany).

Tissue distribution assay of LEAP-2A and LEAP-2B transcripts

From 12 heathy female and 12 male adults (average total body weight (BW) = 27.1 ± 4.6 g), tissues including the brain, eye, fin, gill, heart, intestine, kidney, liver, muscle, spleen, skin, ovary, and testis were surgically removed. Tissue samples were immediately frozen on dry ice upon sampling and stored at − 80 °C untiled used. Total RNA was extracted from each tissue type using RNeasy plus Mini Kit (Qiagen, Hilden, Germany) according to manufacturer's recommendations including the DNase I treatment step. Purified total RNA was reverse transcribed to cDNA for quantitative reverse transcription PCR (RT-qPCR) assay to examine the distribution pattern and basal expression levels of each LEAP-2 isoform in adult tissues.

Expression assay of LEAP-2A and LEAP-2B transcripts in developing embryos and early larvae

Expression patterns of LEAP-2 transcripts during embryonic development and early larval period were examined by RT-qPCR assay. Induced spawning was performed using carp pituitary extracts according to the method described previously (Kim et al. 1994). Pooled eggs from three females were inseminated with pooled sperm collected from three males. Fertilized eggs were incubated at 25 ± 1 °C until hatch using constant aeration (dissolved oxygen = 7 ± 1 °C ppm). During development, embryos (approximately 200–300 embryos) were sampled at 0 h (just fertilized), 2 h (32–64 cells), 4 h (early blastula), 6 h (early gastrulation), 8 h (late gastrulation), 12 h (3–4 myotomes stage with the formation of optic vesicles), 16 h (12–14 myotomes stage), 20 h (20–22 myotomes stages with the beginning of eye lens formation), 24 h (tail-beating stage, almost close to hatching), and 28 h (hatch-out) post fertilization (HPF). After hatching, hatchlings were transferred to a new 25 °C tank (60 L). Larvae (approximately 100 larvae) were further sampled at day 1 (D1) and day 2 (D2, yolk sac absorption). Two independent induced spawning trials were made. From each spawning trial, sampling of embryos and larvae at each time point was carried out in triplicates (i.e., six biological replications for each developmental stage). Total RNAs were extracted from sampled embryos and larvae as above in order to perform RT-qPCR analysis of LEAP-2 transcripts.

In vivo bacterial challenge

In order to examine the potential modulation of each LEAP-2 isoform in response to in vivo immune stimulatory treatment, mud loaches were experimentally challenged with *Edwardsiella tarda*, a causative agent for edwardsiellosis in this fish species. Freshly grown *E. tarda* (strain FSW910410; (Nam et al. 2011; Cho et al. 2009)) was washed twice with phosphate buffered saline (PBS, pH 6.8) and suspended in PBS. Individuals (average BW = 12.1 ± 3.4 g; n = 40 in total) were intraperitoneally injected with *E. tarda* (1×10^6 cells/g body weight) (Nam et al. 2011). Injection volume was 200 μL. The same volume of PBS was also injected to 40 individuals in order to prepare the non-challenged control group. After injection, fish belonging to each group (challenged or non-challenged group) were transferred to one of two 200 L tanks at 25 °C. Six individuals were randomly selected from both groups at 6, 12, 24, 48, 96, and 192 h post injection (HPI). From each individual, the liver, kidney, and spleen were surgically sampled for RT-qPCR assay of LEAP-2 isoforms.

RT-qPCR assay and statistics

Reverse transcription reaction was carried out with Omniscript® Reverse Transcription Kit (Qiagen, Germany) with an inclusion of mud loach 18S rRNA reverse primer in order to prepare a normalization control in RT product. An aliquot (2 μL) of fourfold diluted cDNA template was subjected to qPCR amplification. Quantitative PCR amplification was performed with LightCyler480® Real-Time PCR System (Roche Applied Science, Mannheim, Germany) and LightCycler® DNA Master SYBR Green I (Roche Applied Science, Germany). Specific amplifications of both LEAP-2 genes and 18S rRNA genes were verified with ethidium bromide-stained gel electrophoresis of amplified fragments and also confirmed with the melting curve analysis following the qPCR amplification. Each primer pair for target (LEAP-2 isoforms) and normalization control (18S rRNA) genes was confirmed to show the PCR efficiency higher than 94% based on the standard curves prepared using a serial dilution of cDNA samples. For each cDNA sample, triplicate assays were made.

Expression levels of LEAP-2 isoforms under non-stimulated conditions (i.e., tissue and developmental expression assays) were addressed as ΔCt method relative to 18S rRNA control level (Nam et al. 2011; Schmittgen and Livak 2008). Differential expression of LEAP-2 isoforms in response to stimulatory treatments (i.e., bacterial challenge) was presented as the fold difference relative to the non-treated control group using the $2^{-\Delta\Delta Ct}$ method (Nam et al. 2011; Schmittgen and Livak 2008). Significant differences in expression levels between or among groups were tested using student's t

test or one-way ANOVA (followed by Duncan's multiple ranged tests) at $P = 0.05$ level.

Results and discussion

Mud loach LEAP-2 cDNA and amino acid sequences

Mud loach LEAP-2A (designated based on molecular phylogeny, see below) cDNA exhibited 268-bp 5′-UTR, 282-bp ORF encoding a polypeptide comprising 93 amino acids (aa) and a long, 966-bp 3′-UTR including 19-bp poly(A+) tail. A putative polyadenylation signal (AATAAA) was found at 24 bp upstream from the poly(A+) (GenBank accession no. KX372543). The whole protein of LEAP-2A revealed 10.54 KDa of calculated molecular mass and 9.23 of theoretical pI value. As previously known in other LEAP-2 orthologs, mud loach LEAP-2A showed potential cleavage sites for signal peptide (between [28]Cys and [29]Ser) and propeptide (between [52]Arg and [53]Met), respectively. As a result, the mature peptide of mud loach LEAP-2A (44 aa) was estimated to have 4.62 kDa of molecular weight and 8.88 of pI value (Fig. 1). On the other hand, mud loach LEAP-2B cDNA was comprised of 33-bp 5′-UTR, 279-bp ORF, and 546-bp 3′-UTR including 25-bp poly(A+) tail. A polyadenylation signal (AATAAA) was found at − 16 bp from the poly(A+) tail (KX372544). LEAP-2B encoded a 91-aa polypeptide (10.36 KDa and pI = 8.38). Signal peptide cleavage was predicted at the site between [27]Ser and [28]Val, while propeptide cleavage site was between [50]Arg and [51]Met. Resultant 41-aa mature peptide (LEAP-2B) showed the 4.66 KDa (Mw) and 6.68 (pI) (Fig. 1). Mud loach LEAP-2A and LEAP-2B shared only a low sequence identity (38.7%) of each other at amino acid level through the entire protein region. However, when only mature peptides considered, the sequence identity between the two isoforms was 61.0%. Both mud loach LEAP-2 isoforms conserved the four Cys residues

predicted to form two disulfide bonds ([69]Cys-[80]Cys and [75]Cys-[85]Cys for LEAP-2A and [67]Cys-[78]Cys and [73]Cys-[83]Cys for LEAP-2B) (Fig. 1), which is a typical feature of almost previously known LEAP-2 orthologs (Henriques et al. 2010; Li et al. 2015).

Mud loach LEAP-2 isoforms (mature peptide region) were further aligned with sequences of representative teleostean LEAP-2 orthologs (Additional file 2: Figure S1A). In the alignment, all the LEAP-2s except one isoform sequence from common carp *Cyprinus carpio* (XP_018919135, assumed to be a carp LEAP-2C variant) reveal the four Cys residues involved in two disulfide bonds at clearly conserved positions. In overall, high degree of sequence homology among orthologs was observed within a given isoform type (LEAP-2A, LEAP-2B, or LEAP-2C). LEAP-2A mature peptides were consisted of either 41 aa or 46 aa, while mature LEAP-2Bs were uniform to be 41 aa, except an ortholog (40 aa) from northern pike *Esox lucius* (Esociformes). On the other hand, lengths of LEAP-2C isoforms were variable depending on species, ranging from 34 aa to 40 aa. A closer look into LEAP-2 isoforms from fish species belonging to Cypriniformes indicated that cypriniform LEAP-2s would display isoform-dependent pI ranges (8.48–8.88 for LEAP-2A, 6.78–7.69 for LEAP-2B, and 9.06–9.20 for LEAP-2C) (Additional file 2: Figure S1B). In general, mature peptides of AMPs represent positively charged, cationic characteristics because they should bind to anionic microbial membranes (Hancock et al. 2016; Townes et al. 2009). With this context, the non-cationic charge of cypriniform LEAP-2Bs might suggest the possibility of potential divergences with regard to their roles in the immune system. From our survey of pI values of other teleostean LEAP-2B isoforms (mature peptide region), such non-cationic pI values were hardly seen in orthologs from other teleost taxa (data not

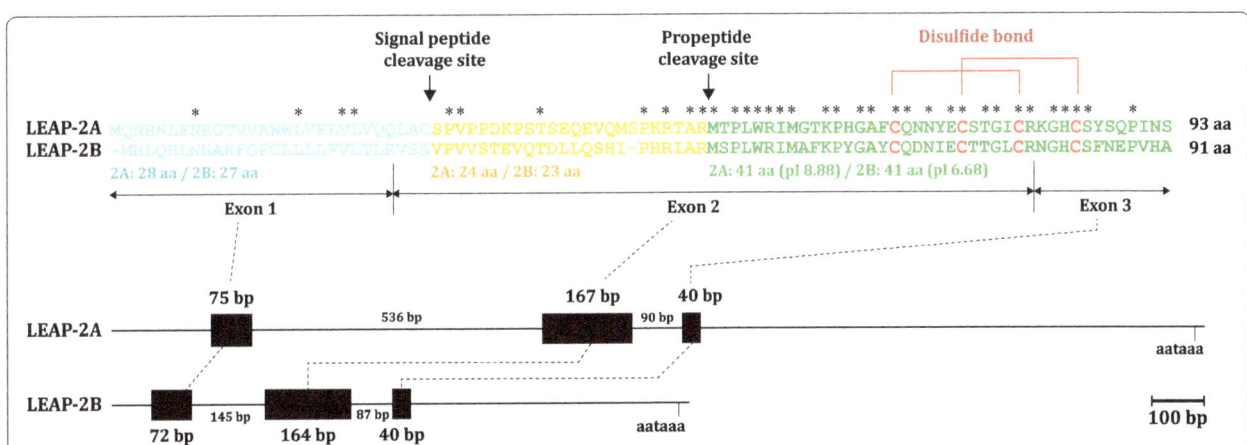

Fig. 1 Primary polypeptide structures of mud loach LEAP-2A and LEAP-2B isoforms. In the pairwise alignment (upper), the identical amino acid residues are indicated by asterisks. Tripartite gene structures of the two LEAP-2 isoforms are also provided (lower). In the presentation of gene structure, exons are indicated by solid boxes while introns by horizontal lines

shown). Within the cypriniform group, the low pI values of LEAP-2Bs would be caused by the replacement of positively charged ^{13}His in LEAP-2A with uncharged Tyr in LEAP-2B as well as substitutions of uncharged ^{19}Asn and ^{37}Gln (in LEAP-2A), respectively, to negatively charged Asp and Glu (in LEAP-2B). Hence, further structural and functional studies are needed for gaining deeper insight into the consequences of these changes.

Molecular phylogenetic analysis of teleosts LEAP-2 isoforms (with entire protein region) resulted that LEAP-2 sequences were clustered into main clades according to isoform types (i.e., LEAP-2A, LEAP-2B, and LEAP-2C) (Fig. 2). In overall tree topology, LEAP-2A and LEAP-2B were more closely affiliated, while LEAP-2C exhibited an independent group separately. From the phylogenetic tree, teleostean LEAP-2C isoforms formed a single, main clade supported by the high bootstrap confidence value, whereas both LEAP-2As and LEAP-2Bs were found to be non-monophyletic. LEAP-2As from species belonging to Neoteleostei formed an independent group (labeled LEAP-2A-(2)) separated from another group consisting of other LEAP-2As (i.e., orthologs from Ostariophysi, Protacanthopterygii, Stomiatii, and an anguilliform species; LEAP-2A-(1)). On the other hand, LEAP-2Bs were assigned into two groups; one was protacanthopterygian group comprising of species belonging to Esociformes or Salmoniformes (labeled LEAP-2B-(1) in Fig. 2), and the other was ostariophysian group consisting of species belonging to Cypriniformes or Characiformes (labeled LEAP-2B-(2)). Taken together, our molecular phylogenetic results suggest that evolutionary divergences among LEAP-2 isoforms might be lineage-dependent in this infraclass group Teleostei. Based on this molecular phylogeny, the two LEAP-2 paralogs from the mud loach should be designated as members of ostariophysian LEAP-2A and LEAP-2B, respectively. Within an isoform group, mud loach LEAP-2A and LEAP-2B isoforms were closely affiliated respectively with orthologs from other cypriniform species.

The number of LEAP-2 isoforms in a given species has been reported to be species-specific, although majority of fish species are likely to possess two functional LEAP-2 isoforms (usually LEAP-2A/LEAP-2B or LEAP-2A/LEAP-2C). For example, only one LEAP-2 isoform was reported as a single copy gene in channel catfish, *Ictalurus punctatus* (Bao et al. 2006), whereas three isoforms (LEAP-2A, LEAP-2B, and LEAP-2C) were identified in Salmoniformes (rainbow trout *Oncorhychus mykiss* and Atlantic salmon *Salmo salar*) (Zhang et al. 2004). From the molecular phylogeny inferred in the present study, common carp *C. carpio* and zebrafish *Danio rerio* also seemed to possess three LEAP-2 isoforms. Furthermore, our data suggest that several fish species display multiple subisoform copies within a given

LEAP-2 isoform type, as particularly exemplified in common carp and salmoniform species (rainbow trout and Atlantic salmon). Those species are known to have experienced additional whole genome duplication (WGD; 50–80 million years ago (mya) for common carp and 5.6–11.3 mya for Salmoniformes) after a WGD event for the occurrence of divergent teleost fishes (320–350 mya) (Glasuer and Neuhauss 2014). Another example for multiple subisoforms of LEAP-2 could be observed in LEAP-2A and LEAP-2C isoforms from large yellow croaker *Larimichthys crocea* belonging to Eupercaria (Neoteleostei). Because this taxonomic group is not thought to have undergone additional WGDs, the presence of multiple LEAP-2 subisoforms in this fish species might be due to the gene-specific duplication(s) during its adaptive evolution. Duplication and/or amplification of AMP genes in certain Eupercaria fish groups have been previously reported with a proposed explanation based on the positive Darwinian selection (i.e., an adaptive evolutionary process directed by pathogens when the host fish is exposed to new environments) (Padhi and Verghese 2007), as highlighted by diversified hepcidin (LEAP-1) isoforms and subisoforms in those fish species (Cho et al. 2009; Yang et al. 2007; Lee and Nam 2011). Currently, it is unclear whether or not mud loach possesses additional LEAP-2 isoform (i.e., LEAP-2C) and/or subisoforms. However, we have not yet found LEAP-2C-like sequence from our several rounds of NGS analyses for mud loach transcriptomes.

Gene structure and promoter characteristics

Both mud loach LEAP-2A (KX372541) and LEAP-2B (KX372542) genes showed a tripartite structure (three exons) that is also the common organization of previously reported vertebrate LEAP-2 genes (Fig. 1) (Li et al. 2014). For mud loach LEAP-2A gene, three exons (75-bp exon-1, 167-bp exon-2, and 40-bp exon-3) were interrupted by two introns (536-bp intron-1 and 90-bp intron-2). Mud loach LEAP-2B exhibited similar lengths for its three exons (72, 164, and 40 bp for exon-1, exon-2, and exon-3). However, the intron-1 (145 bp) of mud loach LEAP-2B gene was significantly shorter than that of LEAP-2A gene. For each of mud loach LEAP-2 isoform genes, exon sequences clearly matched the coding region of its corresponding cDNA counterpart. The GT-AG exon-intron junction rule was consistently upheld for each boundary region. Bioinformatic analysis predicted various TF binding motifs on 5′-flanking regions of mud loach LEAP-2 isoform genes (Additional file 3: Table S2). They included sites targeted by aryl hydrocarbon receptor (AhR), activator protein-1 (AP-1), activating transcription factor-2 (ATF-2), CCAAT-enhancer binding protein (C/EBP), cyclic AMP-responsive element binding protein (CREBP), hypoxia inducible factor-1 (HIF-

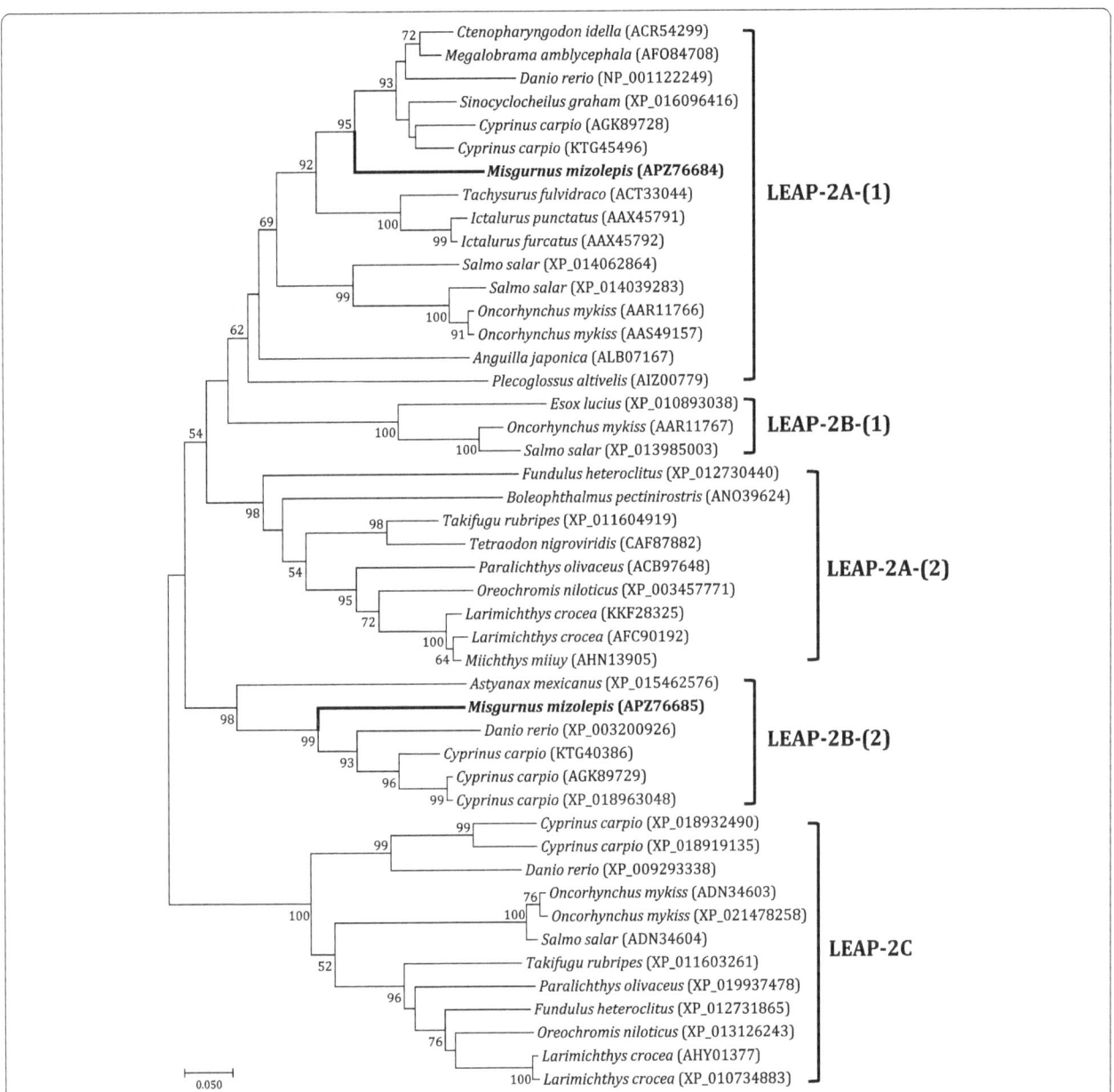

Fig. 2 Molecular phylogenetic relationships among teleostean LEAP-2s. Neighbor-joining tree was drawn with complete amino acid sequences of LEAP-2s from representative teleosts using MEGA7 software (ver. 7.0.26). Tree topology was tested by bootstrap tests (1000 replicates), and only bootstrap values higher than 50% are shown

1), interferon regulatory factor (IRF), nuclear factor of activated T cells (NF-AT), and signal transducer and activated transcription factor (STAT). Although functional recruitments of these transcription factors to the regulatory regions of mud loach LEAP-2 genes should be explored in future, all of them have been already known to be closely involved in immune modulation and stress response (Cho et al. 2009; Truksa et al. 2009). They have been reported to be frequently found in the 5′-flanking regions of acute-phase battery of genes (Nam et al. 2011). However, a clear-cut distinction between mud loach LEAP-2A and LEAP-2B isoforms based on the predicted TF binding profiles is almost impossible at this moment, suggesting the need of further empirical analyses and functional typing. Besides the immune/stress-related TFs above, both mud loach LEAP-2 isoforms also exhibited diverse TF binding motifs related with cell proliferation, organ development, and reproduction such as caudal type homeobox (CdxA), distal-less homeobox (Dlx) group, Krüppel-like factor 6 (KLF6), similar to mothers against decapentaplegic (SMAD) factors, and sex determining region Y box (SOX) factors (Beck and Stringer 2010; Panganiban and Rubenstein 2002; Matsumoto et al. 2006; Budi et al. 2017; Boweles et al. 2000). It is suggestive of that LEAP-2s may play roles in ontogenic development and possibly also in gonad development of this species. Potential involvement of LEAP-2 in the developmental process including the prediction of similar TF binding motifs has been proposed in several fish species (Liu et al. 2010; Bao et al. 2006).

Expression pattern in adult tissues

Based on RT-qPCR assay, mud loach LEAP-2A transcripts were detectable in a wide range of tissue; however, basal expression levels significantly varied across tissue types (Fig. 3a). The organ showing the most robust expression of LEAP-2A transcripts was not the liver; rather, the highest expression level was found in the ovary (more than twofold relative in the liver). This highest expression in the ovary was followed by those in the liver, testis, intestine, and skin. The muscle displayed

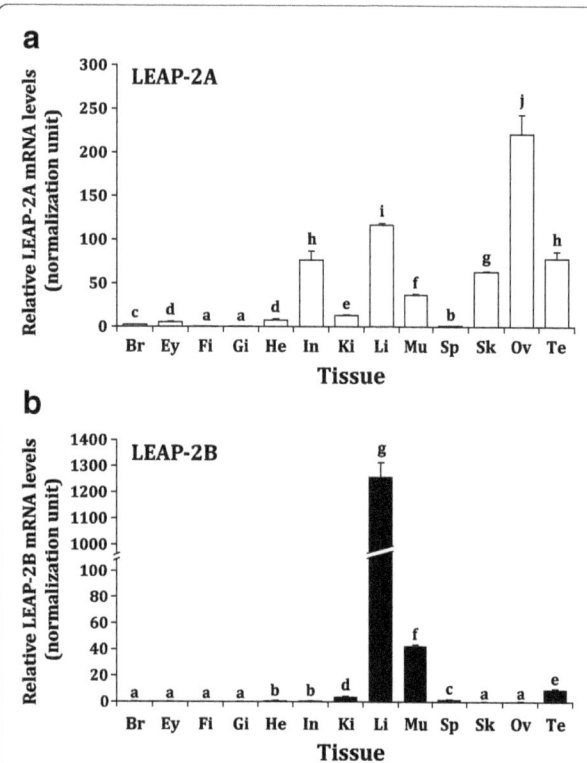

Fig. 3 Tissue distribution patterns and basal expression levels of mud loach LEAP-2A (**a**) and LEAP-2B (**b**) transcripts, as determined by RT-qPCR assay. Expression levels were normalized against 18S rRNA gene. Tissue abbreviations are brain (Br), eye (Ey), fin (Fi), gill (Gi), heart (He), intestine (In), kidney (Ki), liver (Li), muscles (Mu), spleen (Sp), skin (Sk), ovary (Ov), and testis (Te). Statistically different means (± s.d.) are indicated by different letters (a–j in (**a**) and a–g in (**b**)) based on ANOVA followed by Duncan's multiple ranged tests ($P < 0.05$). T bars indicate standard deviations

a moderate level of LEAP-2A transcripts. The remaining other tissues showed only a weak expression of LEAP-2A. On the other hand, LEAP-2B displayed the apparently different pattern of tissue expression compared to LEAP-2A (Fig. 3b). LEAP-2B transcripts were predominantly expressed in the liver, and the hepatic mRNA level of LEAP-2B was more than 10-fold relative to that of LEAP-2A. Except a modest expression level in the muscle, the mRNA expression of LEAP-2B was readily low or minute in all other non-liver tissues. This finding suggests that two paralog LEAP-2A isoforms have undergone certain functional differentiation in their tissue-dependent roles. For example, the strong expression of LEAP-2A in the ovary suggests its presumed roles in female reproductive immunity to protect the ovary against bacterial invasion during ovarian development and maturation. Similarly, even not as much as in the ovary, LEAP-2A showed a fairly high expression level in the mud loach testis, which is suggestive of certain protective roles in male reproduction in this species. Insofar, basal expression data of LEAP-2s in fish gonads have been limited. Our finding on the strong expression of LEAP-2A in the mud loach ovary was similar with the observation from the grass carp *Ctenopharyngodon idella* (Liu et al. 2010), however, apparently different from findings in common carp *C. carpio* (Yang et al. 2014) and blunt snout bream *Megalobrama amblycephala* (Liang et al. 2013) to represent very low or negligible expression of LEAP-2 in gonadic tissues. Meanwhile, the protective role of a hepcidin (LEAP-1) isoform in male reproductive immunity has been reported in the mudskipper *Boleophthalmus pectinirostris* (Li et al. 2016).

On the other hand, the liver-predominant expression with abundant amounts of LEAP-2B may indicate that its main playground would be focused on the liver rather than other systematic or mucosal lymphoid tissues. Tissue expression pattern of LEAP-2 has been reported to be species-specific. Certain teleost species have been reported to express LEAP-2 exclusively in the liver (e.g., rainbow trout *O. mykiss*) (Zhang et al. 2004), while other fish species showed a wide distribution of LEAP-2 transcripts in various tissues including the liver (Bao et al. 2006; Ren et al. 2014). Furthermore, the liver has not been always the main organ showing the highest or predominant expression of LEAP-2s. Channel catfish (*I. punctatus*) (Bao et al. 2006) and yellow catfish (*Pelteobagrus fulvidraco*) (Ren et al. 2014) exhibited very low expression of LEAP-2 in their liver. Blunt snout bream (*M. amblycephala*) (Liang et al. 2013) and large yellow croaker (*Larimichthys crocea*) (Li et al. 2014) showed the highest expression of LEAP-2 in midgut and intestine, although there was also robust expression of LEAP-2 in the liver. Other species such as miiuy croaker

(*Miichthys miiuy*) (Liu et al. 2014) and mudskipper (*B. pectinirostris*) (Chen et al. 2016) displayed the strongest expression of LEAP-2 in the liver.

Expression pattern in developing embryos and early larvae

Expression of mud loach LEAP-2A was hardly detectable in early stages of embryonic development until 12 HPF (Fig. 4a). Active transcription for LEAP-2A began to be observed from 16 HPF, gradually elevated with development until 24 HPF and sharply increased to reach the peak at hatch-out stage (28 HPF, day 0). Afterward, the expression level of LEAP-2A during the yolk sac absorption period was decreased down until 2 days after hatching (day 2). The expression of LEAP-2B was also not active in early developmental stages. Clear sign for its transcription began to be found at 8 HPF, and a small peak was formed at 12 HPF. After that, the expression level of LEAP-2B was kept to be constantly low until hatching. However, after hatching, the transcription of

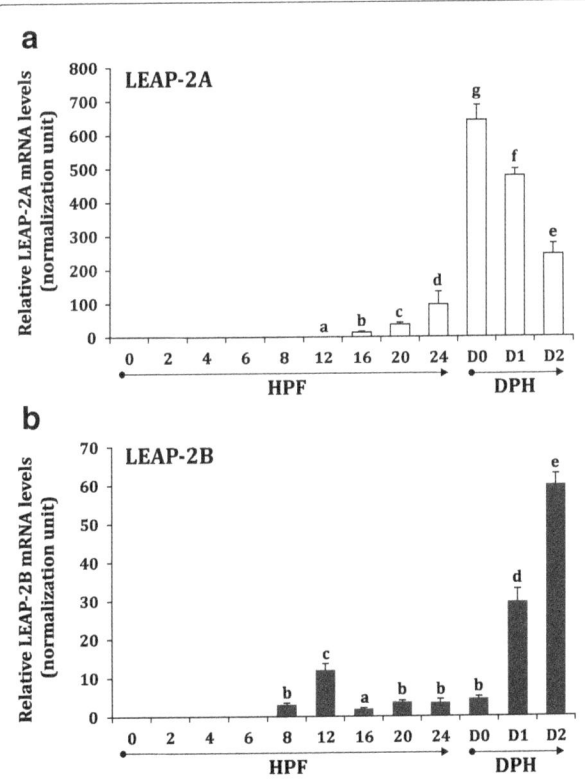

Fig. 4 Altered mRNA levels of mud loach LEAP-2A (**a**) and LEAP-2B (**b**) during embryonic development (0 to 28 h post fertilization, HPF) and early larval development up to 2 days post hatching (DPH) at 25 °C, as judged by RT-qPCR assay. Day 0 (D0) is the time of just hatching corresponding to 28 HPF. Expression levels were normalized against 18S rRNA gene. LEAP-2 transcripts were not detected at early stages of development. Statistically different means (± s.d., as T bars) are indicated by different letters (a–g in LEAP-2A and a–e in LEAP-2B) based on ANOVA followed by Duncan's multiple ranged tests (*P* < 0.05)

LEAP-2B began to be rapidly stimulated with a remarkable increase of its transcript level until day 2 (Fig. 4b).

Our data indicate that transcripts of both LEAP-2 isoforms would be little detected in the developmental period from early cleavages to blastula stage, suggesting that there was no significant contribution of maternally transmitted LEAP-2 copies to these early embryos. This finding is in accordance with the observations made with channel catfish, in which mature LEAP-2 mRNAs would be detectable only after hatching (Bao et al. 2006). However, in contrast, our finding on developmental expression is largely different from those of other previous studies to report active expression of LEAP-2 transcripts in early embryos, as exemplified by grass carp (early detection at 16-cell stage) (Liu et al. 2010) and blunt snout bream (expression level peaked at mid-gastrula) (Liang et al. 2013). Collectively, the developmental regulation of LEAP-2 genes in early embryos might be species-specific, although the information on the developmental expression of LEAP-2 genes in teleosts has been still limited to only couples of species (Liu et al. 2010; Liang et al. 2013; Bao et al. 2006). Additionally, in channel catfish, LEAP-2 gene has been reported to be regulated at the level of splicing where the primary transcripts would remain unspliced until 6 days after hatching (Bao et al. 2006). However, we have not yet found differentially spliced transcripts of LEAP-2 in mud loach.

On the other hand, the significant upregulation of mud loach LEAP-2 isoforms around the hatching event (from the prophase of hatching to day 1 for LEAP-2A, and from the post hatching to day 2 for LEAP-2B) is broadly congruent with findings from most of previous studies, essentially including findings in grass carp (Liu et al. 2010) and blunt snout bream (Liang et al. 2013). More specifically, the pattern of mud loach LEAP-2A is obviously similar with that of blunt snout bream LEAP-2 in terms of that the peak of expression level at hatched larvae was declined with the progress of early larval development (Liang et al. 2013). On the other hand, expression pattern of mud loach LEAP-2B is similar with that of grass carp in the sense that no decrease of expression in post hatched larvae (Liu et al. 2010), suggesting that developmental modulation of LEAP-2 in teleosts may be an isoform-dependent as well as specific-specific.

Rapid upregulation of AMPs including LEAP-2 isoforms at hatching phase is generally explained by the preparation of antimicrobial and/or immune modulatory function for hatched larvae that are no longer protected from the egg membrane (Liang et al. 2013). Besides LEAPs, active expression of multiple beta-defensin isoforms and cathelicidin in early larval stages have also been characterized in olive flounder *Paralichthys olivaceus* (Nam et al. 2010) and Atlantic cod *Gadus morhua* (Broekman et al. 2011). In this loach species, the main

LEAP-2 isoform to confer defensive function on newly hatched larvae might be LEAP-2A based on its much higher expression level than LEAP-2B counterpart. However, with progress of ontogenic development of post hatched larvae, the roles of LEAP-2B seem to become important, which is also in agreement with the period when the liver begins to be developed in loach species belonging to genus *Misgurnus* (Fujimoto et al. 2006; Kim et al. 1987). LEAP-2B may also have a certain role in onset of myogenic development as inferred by its upregulation at the 3–4 myotome stages; however, the mechanism behind this finding should be clarified in future.

Differential expression in response to bacterial challenge

Under present challenge conditions, mortality was found in neither *E. tarda*-challenged group nor PBS-injected group, although some pathological symptoms could be observable in a few *E. tarda*-injected individuals at 96 and 192 HPI (photos not shown).

Upon *E. tarda* challenge, LEAP-2A was rapidly upregulated in all the three tissues (the liver, kidney, and spleen) examined. Induced fold change relative to non-challenged fish was the highest in the liver (up to 20-fold at 12 HPI), and this highest induction was followed by that in the kidney (more than eightfold at 12 HPI) (Fig. 5a). In the spleen, the upregulation of LEAP-2A was not significant (only 1.5-fold) compared to fold inductions observed in the liver and kidney. Considering

the time course expression pattern, the induction of LEAP-2A is an early response during bacterial invasion, as evidenced by the apparent upregulation of LEAP-2A as early as 6 HPI. Increased level of LEAP-2A transcripts in early phase was decreased down to control levels at late phases (i.e., 48, 96, and 192 HPI). This expression pattern was broadly in agreement with the previous observations made in yellow catfish (challenged with *E. tarda*) (Ren et al. 2014) and large yellow croaker (challenged with *Vibrio alginolyticus*) (Li et al. 2014).

Mud loach LEAP-2B also followed, in general, the early response pattern in terms of the rapid induction at 6–12 HPI in all the three tissues and subsequent decreases in late phases (Fig. 5b). However, even down-regulated in later phases, LEAP-2B showed a tendency of more persistent expression pattern, compared to LEAP-2A. Unlike LEAP-2A showing the rapid drop of its transcripts soon after early induction (i.e., recovery to control level at 48 HPI in the liver and kidney), the mRNA expression levels of LEAP-2B were still higher than control level until 96 HPI in all the three tissues, although there was a rebound of expression (i.e., at 48 HPI in the kidney and at 24 HPI in the spleen). Isoform-dependent difference in the time course modulation of LEAP-2 genes in this study is similar with the previous finding on the expression pattern of large yellow croaker LEAP-2 isoforms (Li et al. 2014). During the challenge using *V. alginolyticus* in large yellow croaker, one

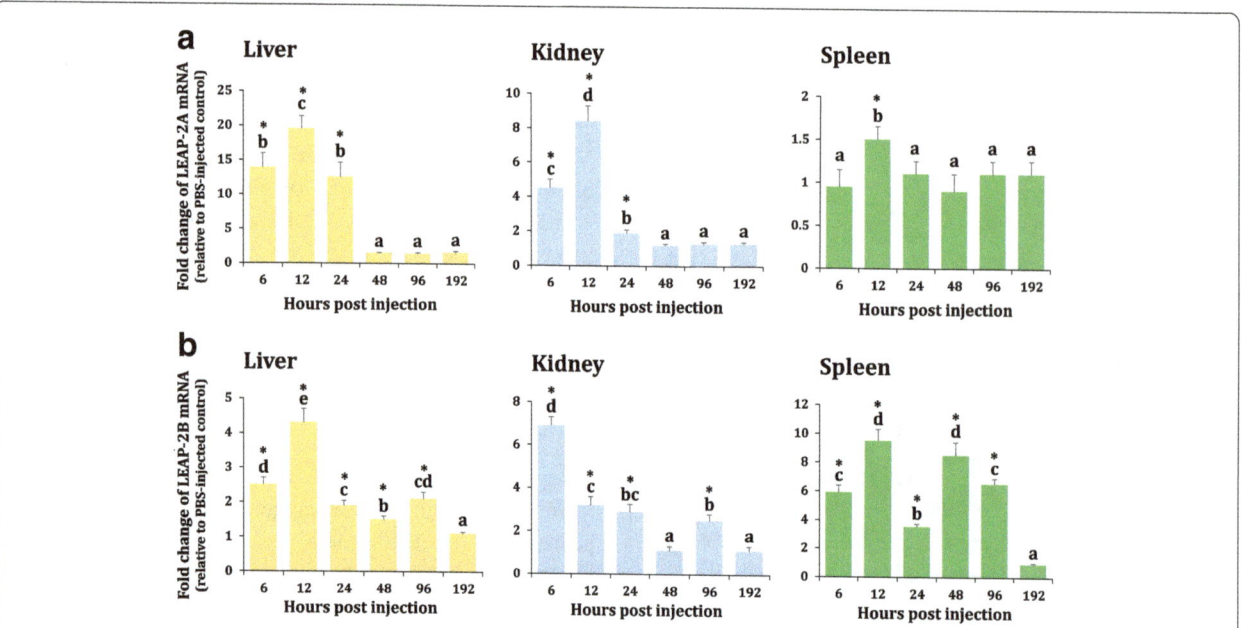

Fig. 5 Differential expression patterns of mud loach LEAP-2A (**a**) and LEAP-2B (**b**) in the liver, kidney, and spleen during experimental challenge with *Edwardsiella tarda*. Expression levels of *E. tarda*-challenged group are presented as fold change relative to PBS-injected control based on RT-qPCR assay. Statistically different means (± s.d.) are indicated by different letters based on ANOVA followed by Duncan's multiple ranged tests (*P* < 0.05). Asterisks indicate statistically different expression levels (*P* < 0.05) of *E. tarda*-injected group when compared to expression levels of PBS-injected control group based on student's *t* test

isoform (LEAP-2A) is rapidly upregulated in early phase and afterward declined, while the other isoform (LEAP-2C) is gradually upregulated with the time after challenge (Li et al. 2014). In addition, expression pattern between mud loach LEAP-2A and LEAP-2B isoforms was apparently different in the spleen. In contrast to only the modest increase of LEAP-2A in the spleen, LEAP-2B was significantly upregulated by bacterial challenge (up to 10-fold relative to non-challenged control). It suggests that isoform-specific involvement in innate immune pathways may differ depending on tissue types. Previously, channel catfish LEAP-2 has been reported to be moderately upregulated in the spleen during *E. ictaluri* challenge, but not differentially modulated in the kidney and liver (Bao et al. 2006). Transcriptional responses of fish LEAP-2 genes to bacterial challenge have been known to be variable among tissues and species. Although bacterial challenges have been reported usually to induce the transcription of LEAP-2 genes in diverse tissues of fish essentially including the liver, there have been also some exceptional or contradictory results in several fish species. For example, rainbow trout was found to display induced expression of LEAP-2 genes only in non-liver tissues such as intestine and skin (Zhang et al. 2004). One isoform of common carp LEAP-2 (LEAP-2B) was significantly downregulated in the liver during *V. anguillarum* challenge unlike its paralog counterpart LEAP-2A showing the highest induction of the expression during the same bacterial challenge (Yang et al. 2014).

Overall, data form bacterial challenge experiment in this study suggest that both mud loach LEAP-2 isoforms are potentially associated with the early response to bacterial invasion, in which their involvements may be tissue- and isoform-dependent (i.e., induction of LEAP-2A mainly in the liver and LEAP-2B in the spleen). However, in order to better hypothesize subfunctionalization(s) between the two LEAP-2 isoforms as antimicrobial components of acute-phase battery in innate immunity, further study should be needed particularly with regard to the evaluation of actual bactericidal activity of each isoform (Hancock et al. 2016; Li et al. 2012). In addition, further study on the possible divergence of the two LEAP-2 isoforms as immune modulatory effectors (i.e., effects of LEAP-2 isoforms on the modulation of expression for other immune-related genes) (Katzenback 2015) would be valuable to get a deeper insight into the genetic pathways interconnected with LEAP-2s in this species.

Conclusions

Two novel isoforms for liver-expressed antimicrobial peptide-2 (LEAP-2A and LEAP-2B) were isolated from a mud loach (*Misgurnus mizolepis*, Cypriniformes). Based on molecular characterization of gene structure and bioinformatic analysis, the two mud loach LEAP-2 genes share conserved characteristics with their orthologs in the teleost lineage. Our findings from expression analyses suggest that both LEAP-2 isoforms play host defense roles not only in early larval developments but also in acute immune response to invading bacteria in adults. Expression assay also indicates the two paralog genes exhibit isoform-specific regulations particularly in a tissue-dependent manner. Data from this study could be a fundamental basis to understand functional involvements of LEAP-2s in the innate immune system of mud loach.

Abbreviations
AMP: Antimicrobial peptide; DPH: Days post hatching; HPF: Hours post fertilization; HPI: Hours post injection; LEAP-2: Liver-expressed antimicrobial peptide-2; ORF: Open reading frame; RT: Reverse transcription

Acknowledgements
Not applicable

Funding
This study was supported by the grant from the Korea Institute of Marine Science & Technology Promotion (KIMST, project no. 20130369).

Authors' contributions
SYL carried out the molecular cloning and gene expression analyses. YKN designed the study, performed the bioinformatic analyses, and drafted the manuscript. Both authors read and approved the final manuscript.

Competing interests
The authors declare that they have no competing interests.

References
Bao B, Peatman E, Xu P, Li P, Zeng H, He C, Liu Z. The catfish liver-expressed antimicrobial peptide 2 (LEAP-2) gene is expressed in a wide range of tissues and developmentally regulated. Mol Immunol. 2006;43:367–77.

Beck F, Stringer E. The role of Cdx genes in the gut and in axial development. Biochm Soc Trans. 2010;38:353–7.

Bowles J, Schepers G, Koopman P. Phylogeny of the SOX family of developmental transcription factors based on sequence and structural indicators. Dev Biol. 2000;227:239–55.

Broekman DC, Frei DM, Gylfason GA, Steinarsson A, Jörnvall H, Agerberth B, Gudmundsson GH, Maier VH. Cod cathelicidin: isolation of mature peptide, cleavage site characterization and developmental expression. Dev Comp Immunol. 2011;35:296–303.

Budi EH, Duan D, Derynck R. Transforming growth factor-β receptors and smads: regulatory complexity and functional versatility. Trends Cell Biol. 2017;27:658–72.

Chen J, Chen Q, Lu XJ, Chen J. The protection effect of LEAP-2 on the mudskipper (*Boleophthalmus pectinirostris*) against *Edwardsiella tarda* infection is associated with its immunomodulatory activity on monocytes/macrophages. Fish Shellfish Immunol. 2016;59:66–76.

Cho YS, Lee SY, Kim KH, Kim SK, Kim DS, Nam YK. Gene structure and differential modulation of multiple rockbream (*Oplegnathus fasciatus*) hepcidin isoforms resulting from different biological stimulations. Dev Comp Immunol. 2009;33:46–58.

Fujimoto T, Kataoka T, Sakao S, Saito T, Yamaha E, Arai K. Developmental stages and germ cell lineage of the loach (*Misgurnus anguillicaudatus*). Zool Sci. 2006;23:977–89.

Glasuer SMK, Neuhauss SCF. Whole-genome duplication in teleost fishes and its evolutionary consequences. Mol Gen Genomics. 2014;289:1045–60.

Hancock RE, Haney EF, Gill EE. The immunology of host defence peptides: beyond antimicrobial activity. Nat Rev Immunol. 2016;16:321–34.

Henriques ST, Tan CC, Craik DJ, Clark RJ. Structural and functional analysis of human liver-expressed antimicrobial peptide 2. Chembiochem. 2010;11:2148–57.

Katzenback BA. Antimicrobial peptides as mediators of innate immunity in teleosts. Biology. 2015;4:607–39.

Kim DS, Jo JY, Lee TY. Induction of triploidy in mud loach (*Misgurnus mizolepis*) and its effect on gonad development and growth. Aquaculture. 1994;120:263–70.

Kim YU, Park YS, Kim DS. Development of eggs, larvae and juveniles of loach *Misgurnus mizolepis* Günther. Bull Kor Fish Soc. 1987;20:16–23.

Krause A, Sillard R, Kleemeier B, Klüver E, Maronde E, Conejo-García JR, et al. Isolation and biochemical characterization of LEAP-2, a novel blood peptide expressed in the liver. Protein Sci. 2003;12:143–52.

Lee SY, Nam YK. Isolation of novel hepcidin isoforms from the rockbream *Oplegnathus fasciatus* (Perciformes). Fish Aquat Sci. 2011;14:31-42.

Li HX, Lu XJ, Li CH, Chen J. Molecular characterization and functional analysis of two distinct liver-expressed antimicrobial peptide 2 (LEAP-2) genes in large yellow croaker (*Larimichthys crocea*). Fish Shellfish Immunol. 2014;38:330–9.

Li HX, Lu XJ, Li CH, Chen J. Molecular characterization of the liver-expressed antimicrobial peptide 2 (LEAP-2) in a teleost fish, *Plecoglossus altivelis*: antimicrobial activity and molecular mechanism. Mol Immunol. 2015;65:406–15.

Li Y, Xiang Q, Zhang Q, Huang Y, Su Z. Overview on the recent study of antimicrobial peptides: origins, functions, relative mechanisms and application. Peptides. 2012;37:207–15.

Li Z, Hong WS, Qiu HT, Zhang YT, Yang MS, You XX, Chen SX. Cloning and expression of two hepcidin genes in the mudskipper (*Boleophthalmus pectinirostris*) provides insights into their roles in male reproductive immunity. Fish Shellfish Immunol. 2016;56:239–47.

Liang T, Ji W, Zhang GR, Wei KJ, Feng K, Wang WM, Zou GW. Molecular cloning and expression analysis of liver-expressed antimicrobial peptide 1 (LEAP-1) and LEAP-2 genes in the blunt snout bream (*Megalobrama amblycephala*). Fish Shellfish Immunol. 2013;35:553–63.

Liu F, Li JL, Yue GH, Fu JJ, Zhou ZF. Molecular cloning and expression analysis of the liver-expressed antimicrobial peptide 2 (LEAP-2) gene in grass carp. Vet Immunol Immunopathol. 2010;133:133–43.

Liu T, Gao Y, Wang R, Xu T. Characterization, evolution and functional analysis of the liver-expressed antimicrobial peptide 2 (LEAP-2) gene in miiuy croaker. Fish Shellfish Immunol. 2014;41:191–9.

Magnadóttir B. Innate immunity of fish (overview). Fish Shellfish Immunol. 2006;20:137–51.

Matsumoto N, Jubo A, Liu H, Akita K, Laub F, Ramirez F, Keller G, Friedman SL. Developmental regulation of yolk sac hematopoiesis by Krüppel-like factor 6. Blood. 2006;107:1357–65.

Nam BH, Moon JY, Kim YO, Kong HJ, Kim WJ, Lee SJ, Kim KK. Multiple β-defensin isoforms identified in early developmental stages of the teleost *Paralichthys olivaceus*. Fish Shellfish Immunol. 2010;28:267–74.

Nam YK, Cho YS, Lee SY, Kim BS, Kim DS. Molecular characterization of hepcidin gene from mud loach (*Misgurnus mizolepis*; Cypriniformes). Fish Shellfish Immunol. 2011;31:1251–8.

Nam YK, Noh JK, Cho YS, Cho HJ, Cho KN, Kim CG, Kim DS. Dramatically accelerated growth and extraordinary gigantism of transgenic mud loach *Misgurnus mizolepis*. Transgenic Res. 2001;10:353–62.

Padhi A, Verghese B. Evidence for positive Darwinian selection on the hepcidin gene of perciform and pleuronectiform fishes. Mol Divers. 2007;11:119–30.

Panganiban G, Rubenstein JLR. Developmental function of the Distal-less/Dlx homeobox genes. Development. 2002;129:4371–86.

Ren G, Shen WY, Li WF, Zhu YR. Molecular characterization and expression pattern of a liver-expressed antimicrobial peptide 2 (LEAP-2) gene in yellow catfish (*Pelteobagrus fulvidraco*). J Aquacult Res Dev. 2014;5:229.

Schmittgen TD, Livak KJ. Analyzing real-time PCR data by the comparative CT method. Nat Protoc. 2008;3:1101–8.

Townes CL, Michailidis G, Hall J. The interaction of the antimicrobial peptide cLEAP-2 and the bacterial membrane. Biochem Biophys Res Commun. 2009;387:500–3.

Truksa J, Lee P, Beutler E. Two BMP responsive elements, STAT, and bZIP/HNF4/COUP motifs of the hepcidin promoter are critical for BMP, SMAD1, and HJV responsiveness. Blood. 2009;113:688–95.

Yang G, Guo H, Li H, Shan S, Zhang X, Rombout JHWM, An L. Molecular characterization of LEAP-2 cDNA in common carp (*Cyprinus carpio* L.) and the differential expression upon a *Vibrio anguillarum* stimulus: indications for a significant immune role in skin. Fish Shellfish Immunol. 2014;37:22–9.

Yang M, Wang KJ, Chen JH, Qu HD, Li SJ. Genomic organization and tissue specific expression analysis of hepcidin-like genes from black porgy (*Acanthopagrus schlegelii* B). Fish Shellfish Immunol. 2007;23:1060-71.

Zhang YA, Zou J, Chang CI, Secombes CJ. Discovery and characterization of two types of liver-expressed antimicrobial peptide 2 (LEAP-2) genes in rainbow trout. Vet Immunol Immunopathol. 2004;101:259–69.

The effect of feeding frequency, water temperature, and stocking density on the growth of river puffer *Takifugu obscurus* reared in a zero-exchange water system

Gwang-Yeol Yoo[1] and Jeong-Yeol Lee[2*]

Abstract

The effects of daily feeding frequency (Exp I), water temperature (Exp II), and stocking density (Exp III) on the growth of river puffer, *Takifugu obscurus*, juvenile fish of 10 and 40 g in body weight were examined to develop effective techniques to produce river puffer in a non-exchange water system. In Exp I, fish were fed commercial floating feed with 45 % protein one to five times per day to apparent satiation each by hand daily for 8 weeks at 25 °C. In both the 10- and 40-g size groups, the final body weight, daily feed consumption, and weight gain of fish fed one meal per day were significantly lower than those of fish fed five meals per day ($P < 0.05$). However, there were no significant differences in the final body weight, daily feed consumption, and weight gain among fish fed two, three, and five meals per day. Feed efficiency showed decreasing tendency with increasing size of fish. In Exp II, fish of 10 and 40 g in initial body weight were reared with the commercial feed at 15–30 °C for 8 weeks. The weight gain of fish increased with raising water temperature up to 25 °C and decreased drastically at 30 °C for both sizes. The Q10 of specific growth rate was decreased with raising water temperature from 5.04 (temperature interval, 15–20 °C) to 0.66 (25–30 °C) for the 10-g fish and from 4.98 to 0.31 for the 40-g fish. In Exp III, the effect of stocking density on growth was examined with fish of 10 and 40 g in initial body weight. The final body weight for initial stocking densities of 4, 8, and 12 kg/m^3 was significantly higher than that of 20 kg/m^3 for the 10-g fish, and the final stocking density reached 10.1, 19.2, 28.7, and 39.9 kg/m^3, respectively. For the 40-g fish, the final body weight for initial stocking densities of 3 and 6 kg/m^3 was significantly higher than that of 9 and 15 kg/m^3 and the final stocking density reached 7.38, 13.5, 17.1, and 27.5 kg/m^3, respectively ($P < 0.05$). In both groups, weight gain tended to decrease with increasing stocking density; however, survival showed no significant difference.

Keywords: *Takifugu obscurus*, Feeding frequency, Stocking density, Water temperature, Q10

Background

There are over 100 different species of puffers in the world (Bai et al. 1999). Among them, tiger puffer, *Takifugu rubripes*, and river puffer, *Takifugu obscurus*, are the main culture species. River puffer is distributed in the south-western and the south-eastern seawaters of Korea and China, respectively (Kato et al. 2005). River puffer moves from sea to river for reproduction during the spawning season of May to June in Korea. Therefore,

it has typical anadromous life history (Yang and Chen 2004). River puffer is one of the promising cultured fish species in South Korea because of its desirable taste and high market price (Ministry of Oceans and Fisheries 2013).

The current commercial culture practice involves the growing of fingerlings with a size of 2–5 g using floating net cages along the coast or indoor recirculating aquaculture system (RAS). River puffer takes 30 months to grow to marketable size, 400 g. One of the most serious problems in the aquaculture of river puffer is the extremely low productivity that is linked with high mortalities from outbreaks of parasitic diseases as shown in tiger puffer (Kikuchi et al. 2006). Survival through the

* Correspondence: yjeong@kunsan.ac.kr
[2]Department of Aquaculture and Aquatic Science, Kunsan National University, Gunsan 54150, South Korea
Full list of author information is available at the end of the article

production period is estimated to be less than 50 %, and prospects for developing methods for the prevention and treatment of diseases remain as a major challenge faced by aquaculture enterprise. Nevertheless, the river puffer is a promising candidate for aquaculture production because of its high market price being more than four times that of olive flounder and rock fish (Ministry of Oceans and Fisheries 2013).

In recent years, a new production technology, based on mixed biofloc communities, has been implemented in shrimp and tilapia production because they feature high production yield, water quality control, and feed protein recycling simultaneously in the same culture unit (Avnimelech 2006; Crab et al. 2007; Little et al. 2008). Such units are known as intensive zero-exchange systems (BFT), and they utilize direct carbon inputs (sugar or molasses) to promote the growth of biofloc communities (McIntosh 2001; Burford et al. 2004). Mixed biofloc communities remove ammonia-nitrogen from the water and assimilate it directly into bacterial biomass. Containing as much as 34 % crude protein (Tacon et al. 2002), this bacterial biomass has been shown to contribute substantially to fish nutrition, thereby potentially deducing feed costs, lowering production costs, and improving the overall economics of the system (Moss 2002; Samocha et al. 2004; Azim and Little 2008). This also reduces equipment and energy costs for the production system by eliminating the need for external components such as biofiltration, disinfection through ultraviolet light or ozonation, and possibly solid filtration.

Key biological parameters, such as nitrogenous waste levels, effects of water temperature, and impacts of stocking density on growth, and proper feed composition need to be determined for a specific species in order to develop a profitable production system (Kikuchi et al. 2006). Previous investigations on the river puffer, however, have focused mainly on the development of larvae and juvenile, reproductive physiology, and seed production (Jang et al. 1996; Park et al. 1997; Son et al. 2001; Yang and Chen 2004; Kato et al. 2005; Yang and Chen 2006). We confirmed that BFT tank (217 %) showed a higher weight gain than the running seawater tank (164 %) in a preliminary study where 5-g-size river puffers were carried out in BFT tank and running seawater tank for 8 weeks (unpublished data). However, information for developing a cost-effective, zero-exchange system for this species is still scarce. Therefore, the objective of the present study was to evaluate the effects of feeding frequency, water temperature, and stocking density on the growth of river puffer in non-exchange water mixed biofloc community systems.

Methods

River puffers produced by Chungcheongnam-do Fishery Research Institute were reared with commercial diet (Woosong, Daejeon, South Korea) at 20 °C until the start of the experiments. The experiments were conducted in six circular concrete tanks (two tanks have an inner diameter of 6 m and height of 1 m, and four tanks have an inner diameter of 3 m and height of 0.8 m) illuminated by 100-lx fluorescent light during the day (07:00–19:00). Each tank was filled with 20-psu seawater and was aerated and agitated continuously using pipe diffusers connected to an air pump. Fish (average weight of 13 ± 4.8 g) were stocked in each tank with an initial stocking density of approximately 2 kg/m^3. Fish were grown without intentional water exchange until the end of feeding trials and were fed twice daily at 09:00 and 17:00 with floating commercial feed containing 45 % crude protein (Woosong, Daejeon, South Korea) at 2 % of total fish weight per day. Molasses (44 % C) was added in each tank as an external carbon source to be the estimated C:N ratio of 15 (Avnimelech 2007). The pH was maintained between 7 and 8, and alkalinity was kept between 100 and 150 mg/L CaCO$_3$ by periodic addition of NaHCO$_3$. Dissolved oxygen concentration was maintained greater than 4.0 mg/L. After 50 days of rearing experiments, feeding trials were conducted by installing plastic cages in each rearing tank. A tight net of 10 cm was lined inside the plastic cages to prevent feed loss through the wall and bottom of the cages.

The effect of temperature on the rate of the physiological processes is generally expressed as Q10, where

$$Q10 = (R1/R2)^{10/(T1-T2)}$$

R1 and R2 are specific growth rate (SGR) at a higher and lower temperature, respectively, and T1 and T2 are the higher and lower temperatures, respectively.

Throughout the experimental period, water temperature, salinity, dissolved oxygen (DO), and pH were measured daily at 08:00–10:00 h using YSI 556 (YSI, Inc., Yellow Springs, USA). Water samples (50 mL) were collected twice a week at 14:00 h from each tank. Half of the water sample was analyzed spectrophotometrically for total ammonia-nitrogen (TAN) and nitrite nitrogen (NO$_2$–N) following the *Standard methods for the examination of water and wastewater* (APHA 1998); the remaining half was filtered under vacuum pressure through predried and preweighed GF/C filter paper. The filter paper containing suspended materials was dried at 105 °C in an oven until constant weight, and the dried sample was weighed to 0.01 mg. The weight difference and the total suspended solids (TSS) were calculated. Alkalinity was determined twice a week by titration (Hach digital titrator, Hach, Loveland, CO, USA).

All data were subjected to one-way ANOVA using SPSS for Windows (release 14.0). When a significant treatment effect was observed, Duncan's multiple range test was used to compare means. Treatment effects were considered significant at $P \leq 0.05$.

Feeding frequency

The effect of feeding frequency on the growth of river puffer was examined with fish of 10 and 40 g in initial body weight. Thirty and 20 fish with average body weight of 10 and 40 g, respectively, were stocked in plastic cages. Plastic cages ($0.5 \times 0.4 \times 0.3$ m) were placed in a tank (inner diameter of 6 m, height of 1 m). Fish were fed floating commercial pellet feed one, two, three, and five meals per day to apparent satiation by hand. Each feeding group was triplicated, and the experiment lasted for a period of 8 weeks. Any uneaten food was collected, and feed intake was calculated for each meal as feed provided minus feed waste. Feeding was conducted between 07:30 and 08:00 for the once daily feeding; 07:30 and 08:00 and 17:30 and 18:00 for the twice daily feeding; 07:30 and 08:00, 12:30 and 13:00, and 17:30 and 18:00 for the three times feeding; and 07:30 and 08:00, 10:00 and 10:30, 12:30 and 13:00, 15:00 and 15:30, and 17:30 and 18:00 for the five times feeding. Water temperature was maintained at 25 ± 1 °C.

Water temperature

The effect of water temperature on the growth of river puffer was examined with fish of 10 and 40 g in initial body weight. Fish were stocked in plastic cages ($0.5 \times 0.4 \times 0.3$ m) in four rearing experiment tanks (inner diameter of 3 m, height of 0.8 m), and water temperature of rearing tanks was controlled using a heat pump at 15 ± 1, 20 ± 1, 25 ± 1, and 30 ± 1 °C. Thirty fish of 10-g size and 20 fish of 40-g size were distributed randomly to each plastic cage and were fed floating commercial pellet feed in triplicate groups at twice a day to apparent satiation each by hand for 8 weeks. Prior to the rearing experiment, fish were acclimated to each experimental temperature during a week.

Stocking density

The effect of stocking density on the growth of river puffer was studied using two size groups, 10 and 40 g. Fish averaging 10 g were randomly distributed to floating plastic cages $0.4 \times 0.17 \times 0.3$ m in a group of 8, 16, 24, and 40 at a density of 4, 8, 12, and 20 kg/m^3, respectively. Fish in the second size group with average weight of 40 g were distributed to floating plastic cages of $0.6 \times 0.45 \times 0.3$ m in a group of 6, 12, 18, and 30 fish at a density of 3, 6, 9, and 15 kg/m^3, respectively. The plastic cages were randomly assigned to triplicate groups of each stocking density in the experimental tank (inner diameter of 6 m, height of 1 m). Fish in both size groups were fed floating commercial feed twice a day to apparent satiation for 8 and 12 weeks, respectively. Water temperature was kept at 25 ± 1 °C.

Results

Water quality parameters of different experimental groups are shown in Table 1. The biofloc development was observed in terms of total suspended solids (TSS). The TSS gradually increased during the experiment. Bioflocs were observed as brown in color and were composed of suspended organic particles in the form of flocculated aggregates, which were colonized by a number of heterotrophic bacteria, microalgae, and protozoa. All the water quality parameters in the six experimental tanks were found within suitable ranges for river puffer culture throughout the experimental period.

Feeding frequency

The growth of river puffer reared at different daily feeding frequencies is shown in Table 2. In both the 10- and 40-g size groups, the final body weight, daily feed consumption, and weight gain of fish fed one meal per day were significantly lower than those of fish fed five meals per day ($P < 0.05$). However, there were no significant differences in the final body weight, daily feed

Table 1 Water quality parameters of different experimental groups

| | Feeding frequency | Stocking density | Water temperature | | | |
			15 °C	20 °C	25 °C	30 °C
Temperature (°C)	25 ± 1	25 ± 1	15 ± 1	20 ± 1	25 ± 1	30 ± 1
Alkalinity (mg CaCO₃)	120 ± 31.2	132 ± 24.7	112 ± 18.8	121 ± 23.4	129 ± 13.9	127 ± 26.3
pH	7.7 ± 0.34	7.5 ± 0.27	7.4 ± 0.26	7.6 ± 0.19	7.6 ± 0.26	7.5 ± 0.31
Dissolved oxygen (mg/L)	5.57 ± 0.26	5.69 ± 0.42	6.39 ± 0.34	5.86 ± 0.44	5.49 ± 0.32	5.43 ± 0.28
Total ammonia-N (mg/L)	0.42 ± 1.0	0.51 ± 0.08	0.36 ± 0.06	0.47 ± 0.06	0.52 ± 0.08	0.53 ± 0.11
Nitrite-N (mg/L)	1.47 ± 0.14	0.83 ± 0.11	0.49 ± 0.04	0.53 ± 0.04	0.76 ± 0.07	0.87 ± 0.08
Nitrate-N (mg/L)	2.89 ± 0.13	2.23 ± 0.12	1.47 ± 0.07	1.86 ± 0.06	2.47 ± 0.11	2.56 ± 0.14
Total suspended solids (mg/L)	184 ± 19.4	167 ± 12.7	143 ± 13.6	157 ± 17.5	197 ± 19.4	201 ± 24.1

Means and standard deviations of triplicate for each experimental group

Table 2 The effect of feeding frequency on the growth of river puffer

	Treatment	Body weight (g)		DFC[1] (%)	Weight gain (%)	Feed efficiency[2] (%)	Survival (%)
		Initial	Final				
10-g size	1 meal per day	8.9 ± 0.06	19.3 ± 2.2	1.29 ± 0.18^b	114 ± 16.3^b	89.7 ± 1.54^b	97.8 ± 1.92^a
	2 meals per day	8.8 ± 0.03	23.0 ± 1.4	1.60 ± 0.08^{ab}	159 ± 9.55^{ab}	93.6 ± 2.13^a	100 ± 0.00^a
	3 meals per day	8.8 ± 0.05	21.6 ± 2.5	1.50 ± 0.15^{ab}	156 ± 17.3^{ab}	92.8 ± 1.27^{ab}	98.9 ± 1.92^a
	5 meals per day	8.8 ± 0.07	24.2 ± 2.8	1.65 ± 0.19^a	174 ± 35.8^a	94.4 ± 2.43^a	100 ± 0.00^a
40-g size	1 meal per day	38.4 ± 0.3	61.6 ± 5.02	0.82 ± 0.13^b	60.4 ± 11.9^b	75.5 ± 3.22^a	96.7 ± 2.89^a
	2 meals per day	38.7 ± 0.2	70.7 ± 5.16	1.04 ± 0.12^a	82.8 ± 14.2^a	80.6 ± 2.75^a	100 ± 0.00^a
	3 meals per day	38.5 ± 0.2	72.7 ± 2.01	1.10 ± 0.04^a	88.9 ± 5.06^a	81.3 ± 1.86^a	100 ± 0.00^a
	5 meals per day	38.7 ± 0.2	71.3 ± 1.94	1.06 ± 0.04^a	84.3 ± 4.98^a	81.0 ± 3.44^a	98.3 ± 2.89^a

Means and standard deviations of triplicate for each experimental group. For each body weight group, values in the same column having the same superscript are not significantly different ($P < 0.05$)
[1]DFC = daily feed consumption: (final body weight − initial body weight)/[(final body weight + initial body weight)/2 × feeding days] × 100
[2]Feed efficiency: (weight gain/feed intake) × 100

consumption, and weight gain among fish fed two, three, and five meals per day. Feed efficiency of the 10-g-size fish fed one meal per day was significantly lower than that of similar-sized fish fed two, three, and five meals per day ($P < 0.05$). However, there was no significant difference in feed efficiency among treatments for the 40-g size group. There was no significant difference in survival among treatments for both the 10- and 40-g size groups.

Table 3 shows the average feed consumption for the 10-g-size river puffer in response to the number of daily feedings during the 8-week feeding experiment. Total feed consumption increased with increasing daily feeding frequency. However, it did not differ among the groups that were fed two, three, and five meals per day. Feed consumption at the first feeding (07:30 to 08:00) for groups that were fed three and five meals per day was almost equal and was much smaller than that for the one and two meals per day feeding groups. For the groups that were fed three and five meals per day, feed

consumption on and after the second feeding was much smaller than that of the first and was almost constant at each of the other feeding times, about 125 g for the group fed three meals per day and 78 g for the group fed five meals per day. Similar trends were observed for the remainder of the experimental period for the 10-g size group as well as the 40-g size group.

Water temperature

The effects of water temperature on the growth of river puffer are shown in Table 4. The final body weight, weight gain, specific growth rate, and feed efficiency increased with raising water temperature up to 25 °C and drastically decreased at 30 °C for both fish sizes. Survival rate was observed to be lower in fish reared at 20 and 30 °C. However, the mortalities were apparently not the result of parasites or of a bacterial disease. The observed Q10 of SGR decreased with raising water temperature from 5.04 (temperature interval, 15–20 °C) to 0.66 (25–30 °C) for the 10-g-size fish and from 4.98 to 0.31 for the 40-g-size fish.

Stocking density

The effects of stocking density on the growth of river puffer are shown in Table 5. The final body weight for initial stocking densities of 4, 8, and 12 kg/m^3 was significantly higher than that of 20 kg/m^3 for the 10-g-size fish, and the final stocking density reached 10.1, 19.2, 28.7, and 39.9 kg/m^3, respectively. For the 40-g-size fish, the final body weight for initial stocking densities of 3 and 6 kg/m^3 was significantly higher than that of 9 and 15 kg/m^3 and the final stocking density reached 7.38, 13.5, 17.1, and 27.5 kg/m^3, respectively ($P < 0.05$). In both groups, weight gain tended to decrease with increasing stocking density and survival showed no significant difference among groups.

Table 3 Feed consumption of 10-g-size river puffer in response to the number of daily feedings (g)

	1 meal per day	2 meals per day	3 meals per day	5 meals per day
07:30–08:00	317 ± 6.76	229 ± 7.75^a	182 ± 10.7^a	159 ± 7.24^a
10:00–10:30				78.2 ± 4.96^b
12:30–13:00			125 ± 6.14^b	79.4 ± 7.14^b
15:00–15:30				75.6 ± 5.36^b
17:30–18:00		188 ± 9.17^b	134 ± 9.28^b	80.4 ± 4.04^b
Total	317	417	441	473

Data are the total feed consumption of each cage during the 8-week feeding experiment. Means and standard deviations of triplicate for each experimental group. Values in the same column having the same superscript are not significantly different ($P < 0.05$)

Table 4 The effect of water temperature on the growth of river puffer

	WT(°C)	Body weight (g)		DFC (%)	Weight gain (%)	SGR[1] (%)	Q10	Feed efficiency[2] (%)	Survival (%)
		Initial	Final						
10-g size	15	10.2 ± 0.48	14.8 ± 4.35	0.65 ± 0.06[c]	44.4 ± 5.09[c]	0.66 ± 0.06[c]		61.4 ± 3.72[c]	94.4 ± 1.92[b]
	20	10.5 ± 0.09	23.9 ± 4.73	1.39 ± 0.06[b]	128 ± 8.45[b]	1.47 ± 0.07[b]	5.04	81.7 ± 5.64[b]	100 ± 0.00[a]
	25	10.6 ± 0.06	27.4 ± 6.04	1.57 ± 0.09[a]	157 ± 11.3[a]	1.69 ± 0.08[a]	1.32	94.8 ± 5.17[a]	100 ± 0.00[a]
	30	10.6 ± 0.11	22.8 ± 5.17	1.30 ± 0.07[b]	115 ± 12.8[b]	1.37 ± 0.11[b]	0.66	78.3 ± 3.99[b]	90.0 ± 3.33[b]
40-g size	15	41.5 ± 0.32	50.1 ± 7.31	0.34 ± 0.06[c]	20.8 ± 2.25[d]	0.34 ± 0.03[c]		47.2 ± 3.24[c]	96.7 ± 2.89[ab]
	20	40.9 ± 0.73	64.4 ± 11.6	0.80 ± 0.11[a]	57.3 ± 6.58[b]	0.81 ± 0.08[a]	4.98	62.7 ± 4.76[b]	98.3 ± 2.89[ab]
	25	41.0 ± 0.65	69.1 ± 9.39	0.91 ± 0.08[a]	68.4 ± 5.46[a]	0.93 ± 0.05[a]	1.73	77.6 ± 4.17[a]	100 ± 0.00[a]
	30	41.4 ± 0.05	56.4 ± 8.01	0.55 ± 0.08[b]	36.2 ± 4.89[c]	0.55 ± 0.06[b]	0.31	53.5 ± 3.67[c]	90.0 ± 5.00[b]

Means and standard deviations of triplicate for each experimental group. For each body weight group, values in the same column having the same superscript are not significantly different ($P < 0.05$)
[1]Specific growth rate: [(ln final wt. − ln initial wt.)/days] × 100
[2]Refer to Table 1

Discussion

The results of this study indicated that the growth rate and feed consumption of the river puffer, *T. obscurus* (between 10 and 40 g), fed daily multiple diets seem to be higher than those of fish provided with a single meal. Numerous experiments have been carried out on feeding frequency in several kinds of aquaculture species (Kikuchi et al. 2006; Flood et al. 2012; Zhao et al. 2014; Kousoulaki et al. 2015), but it is difficult to make general conclusions on optimal feeding frequencies as each study has different conditions, species, and life stages of fish (Kousoulaki et al. 2015). High feeding frequency has been reported to obtain high growth by increasing feed consumption in most species (Schnaittacher et al. 2005; Seo and Lee 2008; Zhao et al. 2014). The present study was also consistent with this opinion. Different results have also reported that the highest feeding frequency did not gain the most weight in yellowtail flounder, a species with small stomachs and relatively long intestines, due to the short interval between meals (Dwyer et al. 2002). When the interval between meals is short, the food passes through the digestive tract more quickly, resulting in a less effective digestion (Zhao et al. 2014).

The effects of water temperature on the growth of fish have been examined with several kinds of fish such as the olive flounder (Fonds et al. 1995), channel catfish (Buentello et al. 2000), sea bass (Ruyet et al. 2004), parrot fish (Kim et al. 2008), and turbot (Ham et al. 2003). Generally, the growth of fish increases with raising water temperature up to a certain level and then decreases at higher temperatures as was observed in the current study and reported for other fishes (Jobling 1994a). It has also been reported that the optimum temperature for growth tended to decrease with the age of the fish (Björnsson et al. 2001; Kikuchi et al. 2006). In the current study, the optimum temperature for the growth of river puffer less than 50 g in body weight is around 25 °C. Further research will be required with larger size fish in order to determine the effective temperature regimen needed for the production of river puffer because the commercial size of river puffer at harvest is more than 400 g in Korea.

Table 5 The effect of stocking density on the growth of river puffer

	Initial		Final		DFC (%)	Weight gain (%)	Feed efficiency[1] (%)	Survival (%)
	Body weight (g)	Stocking density (kg/m³)	Body weight (g)	Stocking density (kg/m³)				
10-g size	10.1 ± 0.05	4	25.7 ± 0.99[a]	10.1 ± 0.39	1.55 ± 0.06[a]	154 ± 11.0[a]	96.3 ± 3.93[a]	100 ± 0.00[a]
	10.0 ± 0.02	8	24.4 ± 2.14[a]	19.2 ± 1.68	1.49 ± 0.13[a]	143 ± 20.9[ab]	87.9 ± 5.70[ab]	100 ± 0.00[a]
	10.0 ± 0.01	12	24.4 ± 2.33[a]	28.7 ± 2.74	1.47 ± 0.19[ab]	134 ± 16.7[ab]	85.4 ± 5.15[b]	98.6 ± 2.41[a]
	10.2 ± 0.06	20	20.4 ± 2.43[b]	39.9 ± 4.77	1.21 ± 0.20[b]	118 ± 6.31[b]	80.7 ± 2.05[b]	98.3 ± 2.89[a]
40-g size	40.5 ± 0.32	3	98.3 ± 3.01[a]	7.38 ± 0.23	0.99 ± 0.05[a]	143 ± 11.7[a]	82.8 ± 3.40[a]	100 ± 0.00[a]
	40.2 ± 0.26	6	89.9 ± 8.64[a]	13.5 ± 1.30	0.91 ± 0.09[a]	124 ± 20.3[a]	80.9 ± 4.22[a]	100 ± 0.00[a]
	40.0 ± 0.31	9	75.9 ± 3.37[b]	17.1 ± 0.76	0.74 ± 0.04[b]	90.1 ± 7.42[b]	78.7 ± 3.13[a]	98.1 ± 3.21[a]
	40.0 ± 0.28	15	73.2 ± 5.45[b]	27.5 ± 2.04	0.69 ± 0.09[b]	83.6 ± 15.2[b]	77.1 ± 5.89[a]	96.7 ± 3.33[a]

For each body weight group, values in the same column having the same superscript are not significantly different ($P < 0.05$)
[1]Refer to Table 1

The Q10 of SGR obtained in the current study tended to decrease with raising water temperature and was consistent with what was reported by Jobling (1994b). The values in the same temperature range were similar for fish between 10- and 40-g initial body weight and were not largely affected by the fish size, although SGR was considerably different.

In the effects of stocking density, the numbers of fish were held constant for the duration of the experiment, and so stocking density was allowed to increase with time because of the growth of fish. Judging from the final biomass per cage, the maximum stocking densities of river puffer with BFT tanks are estimated to be more than 28 kg/m^3 for fish of 24-g body weight and decreased to about 13 kg/m^3 for the 90-g-size fish. The growth of river puffer in our study was affected by stocking density. Similar results have been European sea bass (Papoutsoglou et al. 1998), rock fish (Oh et al. 2013), and Amur sturgeon (Ni et al. 2014). Kikuchi et al. (2006) reported that the relationship between stocking density and growth of fish may not be uniformly positively or negatively linear for a given species. In our study, mortality showed no significant difference among different stocking densities in both the 10- and 40-g size groups. Similarly, many studies have not found any significant effects of density on survival (Gomes et al. 2006; Rafatnezhad et al. 2008; Ni et al. 2014). Kikuchi et al. (2006) reported that injuring of caudal fin was observed at high stocking densities of 8, 20, and 10 kg/m^3 for 8-, 13-, and 100-g-size tiger puffers, respectively. In our study, no injury was observed in both size groups. It may be for the reason that tiger puffer has more interspecific aggression than river puffer. Meanwhile, previous studies have reported that stocking density affected skeletal anomalies (Boglione et al. 2009), metamorphosis (Hosfeld et al. 2009), sex ratios (Saillant et al. 2003), gonad development Claudia et al. 2004, egg production (Peck and Holste 2006), skin pigmentation (Doolan et al. 2008), and survival (Tagawa et al. 2004) on various fishes.

Conclusions

In the present study, the effects of feeding frequency, water temperature, and stocking density on the growth of river puffer were examined by the use of fingerlings of 10 and 40 g with the objective of developing effective techniques for producing river puffer in zero-exchange mixed biofloc community systems. The results of the current study showed that keeping water temperature around 25 °C, stocking density less than 15 kg/m^3, and feeding frequency of three meals per day are suitable conditions for the production of river puffer when starting with fish that are between 10 and 90 g in body weight. These parameters, however, still require additional work as the suitable market size for river puffer

in Korea is about 400 g, and the results obtained in this study did not cover the full size range for commercial-scale production. The results of the current study are a major first step in defining the culture parameters that are needed to establish rearing conditions of river puffer in zero-exchange mixed biofloc community systems. Further research will be necessary to understand the influence of growth on water quality control and other size classes of river puffer.

Acknowledgements

Both authors are grateful to the anonymous reviewers for their valuable advice and suggestions which helped improve this paper. This paper was also partly supported by research funds of Kunsan National University, South Korea, for the data analysis.

Authors' contributions

GYY analyzed the chemical composition and prepared the draft paper and manufactured the feed and conducted the feeding trial. JYL designed this study, the feeding system, and the revised paper. Both authors read and approved the final manuscript.

Competing interests

The authors declare that they have no competing interests.

Author details

[1]Chungcheongnam-do Fisheries Research Institute, Boryeong 33508, South Korea. [2]Department of Aquaculture and Aquatic Science, Kunsan National University, Gunsan 54150, South Korea.

References

APHA. Standard methods for the examination of the water and wastewater (22nd edn). Washington: American Public Health Association; 1998.

Avnimelech Y. Bio-filters: the need for an new comprehensive approach. Aquac Eng. 2006;34:172–8.

Avnimelech Y. Feeding with microbial flocs by tilapia in minimal discharge bioflocs technology ponds. Aquac Eng. 2007;34:171–8.

Azim ME, Little DC. The biofloc technology (BFT) in indoor tanks: water quality, biofloc composition, and growth and welfare of Nile tilapia (Oreochromis niloticus). Aquaculture. 2008;283:29–35.

Bai SC, Wang XJ, Cho ES. Optimum dietary protein level for maximum growth of juvenile yellow puffer. Fish sci. 1999;65:380–3.

Björnsson B, Steinarsson A, Oddgeirsson M. Optimal temperature for growth and feed conversion of immature cod (Gadus morhua L.). J Mar Sci. 2001;58:29–38.

Boglione C, Marino G, Giganti M, Longobardi A, Marzi PD, Cataudella S. Skeletal anomalies in dusky grouper Epinephelus marginatus (Lowe 1834) juveniles reared with different methodologies and larval densities. Aquaculture. 2009;291:48–61.

Buentello JA, Gatlin III DM, Neill WH. Effects of water temperature and dissolved oxygen on daily feed consumption, feed utilization and growth of channel catfish (Ictalurus punctatus). Aquaculture. 2000;182:339–52.

Burford MA, Thompson PJ, Mcintosh RP, Bauman RH, Pearson DC. The contribution of flocculated material to shrimp (Litopenaeus vannamei) nutrition in a high-intensity, zero-exchange system. Aquaculture. 2004;232:525–37.

Claudia CO, Miguel RS, Miguel AON, Gutiérrez-Yurrita PJ. Effect of density and sex ratio on gonad development and spawning in the crayfish Procambarus llamasi. Aquaculture. 2004;236:331–9.

Crab R, Avnimelech Y, Defoirdt T, Bossier P, Verstraete W. Nitrogen removal techniques in aquaculture for a sustainable production. Aquaculture. 2007;270:1–14.

Doolan BJ, Allan GL, Booth MA, Jones PL. Effects of cage netting colour and density on the skin pigmentation and stress response of Australian snapper Pagrus auratus (Bloch & Schneider, 1801). Aquacult Res. 2008;39:1360–8.

Dwyer KS, Brown JA, Parrish C, Lall SP. Feeding frequency affects food consumption, feeding pattern and growth of juvenile yellowtail flounder (*Limanda ferruginea*). Aquaculture. 2002;213:279–92.

Flood MJ, Purser GJ, Carter CG. The effects of changing feeding frequency simultaneously with seawater transfer in Atlantic salmon *Salmo salar* L. smolt. Aquaculture Int. 2012;20:29–40.

Fonds M, Tanaka M, Van der Veer HW. Feeding and growth of juvenile Japanese flounder *Paralichthys olivaceus* in relation to temperature and food supply. Netherlands J Sea Res. 1995;34:111–8.

Gomes LC, Chagas EC, Martins-Junior H, Roubach R, Ono EA, Lourenço JNP. Cage culture of tambaqui (*Colossoma macropomum*) in a central Amazon floodplain lake. Aquaculture. 2006;253:374–84.

Ham EHV, Berntssen MHG, Imsland AK, Parpour AC, Bonga SEW, Stefansson SO. The influence of temperature and ration on growth, feed conversion, body composition and nutrient retention of juvenile turbot (*Scophthalmus maximus*). Aquaculture. 2003;217:547–58.

Hosfeld CD, Hammer J, Handeland SO, Fivelstad S, Stefansson SO. Effects of fish density on growth and smoltification in intensive production of Atlantic salmon (*Salmo salar* L.). Aquaculture. 2009;294:236–41.

Jang SI, Kang HW, Han HK. Embryonic, larval, and juvenile stages in yellow puffer Takifugu obscurus. Korean J Aquaculture. 1996;9:11–8 (in Korean with English abstract).

Jobling M. Environmental factors and growth. Pages 155–168 in Fish bioenergetics. London: Chapman & Hall; 1994a.

Jobling M. Respiration and metabolism. Pages 121–142 in Fish bioenergetics. London: Chapman & Hall; 1994b.

Kato A, Doi H, Nakada T, Sakai H, Hirose S. *Takifugu obscurus* is a euryhaline fugu species very close to *Takifugu rubripes* and suitable for studying osmoregulation. BMC Physiology. 2005;5:18.

Kikuchi K, Iwata N, Kawabata T, Yanagawa T. Effect of feeding frequency, water temperature, and stocking density on the growth of tiger puffer, *Takifugu rubripes*. J World Aquaculture Soc. 2006;37:12–20.

Kim KM, Lee JU, Kim JW, Han SJ, Kim KD, Jo JY. Daily feeding rates of parrot fish Oplegnathus fasciatus fed extruded pellet at the different water temperatures. Korean J Aquaculture. 2008;21:294–8 (in Korean with English abstract).

Kousoulaki K, Sæther BS, Albrektsen S, Noble C. Review on European sea bass (*Dicentrarchus labrax*, Linnaeus, 1758) nutrition and feed management: a practical guide for optimizing feed formulation and farming protocols. Aquacult Nutr. 2015;21:129–51.

Little DC, Murray FJ, Azim E, Leschen W, Boyd K, Watterson A, Young JA. Option for producing a warm water fish in the UK: limit to "green growth"? Trends Food Sci Technol. 2008;19:255–64.

McIntosh RP. High rate bacterial systems for culturing shrimp. Pages 117–129 in Summerfelt S.T. In: Watten BJ, Timmons MB, editors. Proceedings from the Aquacultural Engineering Society's 2001 Issues Forum. Shepherdstown: Aquaculture Engineering; 2001.

Ministry of Oceans and Fisheries. *Statistical* Yearbook of Maritime Affairs & Fisheries, Sejong, Korea: Ministry of Oceans and Fisheries; 2013.

Moss SM. Marine shrimp farming in the Western Hemisphere: past problems, present solutions, and future visions. Rev Fish Sci. 2002;10:601–20.

Ni M, Haishen W, Jifang L, Meili C, Yan B, Yuanyuan R, Mo Z, Zhifei S, Houmeng D. Effects of stocking density on mortality, growth and physiology of juvenile Amur sturgeon (*Acipenser schrenckii*). Aquaculture Research. 2014; doi:10.1111/are.12620.

Oh DH, Song JW, Kim MG, Lee BJ, Kim KW, Han HS, Lee KJ. Effect of food particle size, stocking density and feeding frequency on the growth performance of juvenile Korean rockfish *Sebastes schlegelii*. Korean J Aquaculture. 2013;46:407–12 (in Korean with English abstract).

Papoutsoglou SE, Tziha G, Vrettos X, Athanasiou A. Effects of stocking density on behavior and growth rate of European sea bass (*Dicentrarchus labrax*) juveniles reared in a closed circulated system. Aquaculture Eng. 1998;18:135–44.

Park IS, Kim HS, Kim ES, Kim JH, Park CW. Cytogenetic analysis of river puffer, *Takifugu obscurus* (Teleostomi : Tetraodontiformes). J Korean Fishery Soc. 1997;30:408–12 (in Korean with English abstract).

Peck MA, Holste L. Effects of salinity, photoperiod and adult stocking density on egg production and egg hatching success in *Acartia tonsa* (Calanoida: Copepoda): optimizing intensive cultures. Aquaculture. 2006;255:341–50.

Rafatnezhad S, Falahatkar B, Gilani MHT. Effects of stocking density on haematological parameters, growth and fin erosion of great sturgeon juveniles. Aquacult Res. 2008;14:1506–13.

Ruyet PLJ, Mahé K, Le Bayon N, Le Delliou H. Effects of temperature on growth and metabolism in a Mediterranean population of European sea bass Dicentrarchus labrax. Aquaculture. 2004;237:269–80.

Saillant E, Fostier A, Haffray P, Menu B, Laureau S, Thimonier J, Chatain B. Effects of rearing density, size grading and parental factors on sex ratios of the sea bass (*Dicentrarchus labrax* L.) in intensive aquaculture. Aquaculture. 2003;221:183–206.

Samocha TM, Lawrence AL, Collins CA, Castille FL, Bray WA, Davies CJ, Lee PG, Wood GF. Production of the Pacific white shrimp, *Litopenaeus vannamei*, in high-density greenhouse-enclosed raceways using low salinity groundwater. J Appl Aquac. 2004;15:1–19.

Schnaittacher G, King WV, Berlinsky DL. The effects of feeding frequency on growth of juvenile Atlantic halibut, *Hippoglossus hippoglossus* L. Aquacult Res. 2005;36:370–7.

Seo JY, Lee SM. Effects of dietary macronutrient level and feeding frequency on growth and body composition of juvenile rockfish (*Sebastes schlegeli*). Aquac Int. 2008;16:551–60.

Son KH, Han KN, Chang CS. The changes of digestive enzyme in early stage of the river puffer, *Takifugu obscurus*. Korean J Fisheries Aquatic Sci. 2001;34:577–83.

Tacon AGJ, Cody JJ, Conquest LD, Divakaran S, Forster IP, Decamp OE. Effect of culture system on the nutrition and growth performance of Pacific white shrimp, *Litopenaues vannamei*, (Boone) fed different diets. Aquacult Nutr. 2002;8:121–37.

Tagawa M, Kaji T, Kinoshita M, Tanaka M. Effect of stocking density and addition of proteins on larval survival in Japanese flounder *Paralichthys olivaceus*. Aquaculture. 2004;230:517–25.

Yang Z, Chen YF. Induced ovulation in obscure puffer Takifugu obscurus by injections of LHRH-α. Aquac Int. 2004;12:215–23.

Yang Z, Chen YF. Salinity tolerance of embryos of obscure puffer Takifugu obscurus. Aquaculture. 2006;253:393–7.

Zhao S, Han D, Zhu X, Jin J, Yang Y, Xie S. Effects of feeding frequency and dietary protein levels on juvenile allogynogenetic gibel carp (*Carassius auratus gibelio*) var. CAS III: growth, feed utilization and serum free essential amino acids dynamics. Aquaculture Research. 2014; doi:10.1111/are.1241.

24

Osteological development of wild-captured larvae and a juvenile *Sebastes koreanus* (Pisces, Scorpaenoidei) from the Yellow Sea

Hyo Jae Yu and Jin-Koo Kim[*] ⓘ

Abstract

The osteological development in *Sebastes koreanus* is described and illustrated on the basis of 32 larvae [6.11–11.10 mm body length (BL)] and a single juvenile (18.60 mm BL) collected from the Yellow Sea. The first-ossified skeletal elements, which are related to feeding, swimming, and respiration, appear in larvae of 6.27 mm BL; these include the jaw bones, palatine, opercular, hyoid arch, and pectoral girdle. All skeletal elements are fully ossified in the juvenile observed in the study. Ossification of the neurocranium started in the frontal, pterotic, and parietal regions at 6.27 mm BL, and then in the parasphenoid and basioccipital regions at 8.17 mm BL. The vertebrae had started to ossify at ~7.17 mm BL, and their ossification was nearly complete at 11.10 mm BL. In the juvenile, although ossification of the pectoral girdle was fully complete, the fusion of the scapula and uppermost radial had not yet occurred. Thus, the scapula and uppermost radial fuse during or after the juvenile stage. The five hypurals in the caudal skeleton were also fused to form three hypural elements. The osteological results are discussed from a functional viewpoint and in terms of the comparative osteological development in related species.

Keywords: *Sebastes koreanus*, Korean fish, Larvae, Juvenile, Osteological development

Background

Osteological development in teleost fishes involves a sequence of remarkable morphological and functional changes, occurring in different developmental stages (Löffler et al. 2008; Kang et al. 2012; Ott et al. 2012). These ontogenetic changes strongly influence the feeding, breathing, and swimming behaviors of both larvae and juveniles, and are therefore useful in functional and ecological analyses and as a basis for phylogenetic inferences about relationships among teleost taxa (Omori et al. 1996; Faustino and Power 1999; Koumoundouros et al. 2000, 2001a, b; Liu 2001; Lima et al. 2013; Voskoboinikova and Kudryavtseva 2014). Practically speaking, an accurate knowledge of skeletal development is essential for the detection and elimination of skeletal deformities appearing during artificial seedling production and to promote effective aquacultural and resource management (Koumoundouros et al. 1997a, b).

Sebastes koreanus Kim and Lee 1994, in the family Scorpaenidae (or Sebastidae *sensu* Nakabo and Kai 2013), is smaller than its congeneric species and is regarded as endemic to the Yellow Sea (Kim and Lee 1994; Kim et al. 2005; Choi and Yang 2008). The species may be a good model fish with which to understand the phylogenetic relationships within the suborder Scorpaenoidei, because the species is specifically adapted to the unique marine environment of the Yellow Sea. Comparisons of *S. koreanus* with other *Sebastes* species have shown that *S. koreanus* collected in the wild exhibit wide ontogenetic variations in their pigmentation patterns and in their head–spine development (Yu et al. 2015). In addition, the restricted distribution of the species makes populations susceptible to collapse as a consequence of environmental pollution and/or the influence of climate change on the Yellow Sea. In this respect, artificial seedling production presents a viable approach to the conservation of susceptible species. However, no studies have yet been conducted on the details of skeletal development in *S. koreanus*, except for studies of morphological development and parturition season (Yu et al.

* Correspondence: taengko@hanmail.net
Department of Marine Biology, Pukyong National University, 45, Yongso-roNam-gu, Busan 608-737, South Korea

2015) and a small study of osteological development of reared larvae and juveniles in the Korea Strait (Park et al. 2015). Also, we confirmed some differences in osteological development of *S. koreanus* when compared with reared larvae (Park et al. 2015) and wild-captured larvae from the Yellow Sea. Therefore, in this study, we describe in detail the early skeletal development of *S. koreanus* in the context of functional changes, based on wild-captured larvae and juvenile. We also compare the osteological development of *S. koreanus* with that of congeneric species.

Methods

All individuals were collected from the eastern margin of the Yellow Sea. Larvae of *S. koreanus* [6.11–11.10 mm body length (BL), *n* = 32] were collected off the Taean Peninsula in May 2011 using a bongo net (0.6 m mouth opening, with 330 and 500 μm mesh size; bottom depth 15–24 m). The juvenile of *S. koreanus* (18.60 mm BL, *n* = 1) was collected off Gang-hwa-do in July 2012 using a stow net. The individuals were preserved in 5 % formalin immediately after collection. The specimens fixed in formalin were washed with distilled water and then preserved in 99 % ethanol. Before staining, each sample was identified according to the morphological characteristics of Yu et al. (2015), which were measured to the nearest 0.01 mm with the stereomicroscope (Olympus SZX16, Japan). The methods of measurement followed Leis and Carson-Ewart (2000). The measured body parts included BL and total length (TL). The anatomical terminology relating to skeletal structures follows Russell (1976), and the terminology of developmental stages follows Kim et al. (2011). The skeletal staining technique was derived from the double staining protocol of Darias et al. (2010). After staining, the specimens were examined on their right and dorsal sides with a stereomicroscope and photographs taken with a camera lucida (Olympus SZX-DA, Japan) attached to the microscope. Drawings of the different skeletal parts were prepared from the photographs. We compared the skeletal structures of the larvae and the juvenile with those of adult *S. koreanus* specimens, to observe the precise locations and shapes of the skeletal elements. We also compared stained specimens with a stained *Sebastes inermis* complex juvenile (17.06 mm BL, *n* = 1) collected in the wild. The stained specimens were preserved in 100 % glycerin in glass bottles and were deposited at Pukyong National University (PKU).

Materials examined

The examined materials included preflexion larvae (6.11–6.27 mm BL, *n* = 2; PKUI 367–368), flexion larvae (6.43–8.40 mm BL, *n* = 19; PKUI 369–387), postflexion larvae (8.44–11.10 mm BL, *n* = 11; PKUI 388–398), and a juvenile (18.60 mm BL, *n* = 1; PKUI 21).

Results

The osteological development of *S. koreanus* at various developmental stages was described in the following skeletal regions: neurocranium, jaw bones, palate series, opercular series, hyoid arch, pectoral girdle, infraorbital bone, caudal skeleton, and vertebrae. The development results are summarized in Table 1.

Neurocranium

The development and ossification of the neurocranium for individuals at different developmental stages are illustrated in Figs. 1 and 2. In the smallest larvae (6.11 mm BL; preflexion stage), no skeletal structures of the neurocranium were visible. Ossification of the neurocranium started at 6.27 mm BL, with ossification of the parietal, frontal, and pterotic bones (Fig. 1a); ossification of these elements appeared to begin at the tips of the spines. In the 7.11 mm BL larva, the skeletal elements that had appeared in earlier stages continued to ossify, but no ossification of additional elements was observed (Fig. 1b). At 8.11 mm BL, the posterior of the parasphenoid and the anterior of the basioccipital had started to ossify, and the two elements were joined. At the same time, the exoccipital began to ossify along its posterior margin (Fig. 1c). At 9.06 mm BL, ossification of the frontal had extended to the dorsal area of the neurocranium, and then the frontal boundary line joined the parietal. In addition, the supraoccipital, sphenotic, and prootic elements had started to ossify along their margins at this stage (Fig. 1d). At 10.20 mm BL, the lateral ethmoid had started to ossify along its dorsal margin, and the parietal, pterotic, parasphenoid, and basioccipital were almost fully ossified (Fig. 2a). At 11.10 mm BL, ossification of the frontal had extended to most regions and ossification of the sphenotic and supraoccipital was complete. The epiotic, which appeared relatively late compared with the other neurocranial elements, had started to ossify at this stage (Fig. 2b), and the vomer and medial ethmoid appeared simultaneously along their anterior margin. At this stage, although all the elements of the neurocranium had started to ossify, some elements continued to ossify. In the juvenile stage (18.60 mm BL), the ossification of the neurocranium was fully complete (Fig. 2c).

Jaw bones, palatine, and opercular series

The development and ossification of the jaw bones, palatine, and opercular series for individuals at different developmental stages are illustrated in Fig. 3. No skeletal structures were visible in the smallest preflexion larva (6.11 mm BL). At 6.27 mm BL, the maxillary

Table 1 Developmental sequence of ossification in *Sebastes koreanus*

Elements	Preflexion larvae		Flexion larvae		Postflexion larvae				Juvenile
	6.11	6.27	7.17	8.17	9.06	10.20	11.10	···	18.60
Neurocranium									
Parasphenoid				▶	—	—	—	—	—
Exoccipital				▶	—	—	—	—	—
Basioccipital				▶	—	—	—	—	—
Supraoccipital					▶	—	—	—	—
Epiotic							▶	—	—
Prootic					▶	—	—	—	—
Frontal		▶	—	—	—	—	—	—	—
Sphenotic					▶	—	—	—	—
Pterotic		▶	—	—	—	—	—	—	—
Vomer							▶	—	—
Lateral ethmoid						▶	—	—	—
Medial ethmoid							▶	—	—
Parietal		▶	—	—	—	—	—	—	—
Jaw bone									
Maxillary		▶	—	—	—	—	—	—	—
Premaxillary		▶	—	—	—	—	—	—	—
Dentary		▶	—	—	—	—	—	—	—
Articular					▶	—	—	—	—
Angular						▶	—	—	—
Hyoid arch									
Ceratohyal				▶	—	—	—	—	—
Epihyal							▶	—	—
Hypohyal							▶	—	—
Branchiostegal ray		▶	—	—	—	—	—	—	—
Interhyal							▶	—	—
Palate									
Palatine						▶	—	—	—
Metapterygoid						▶	—	—	—
Ectopterygoid					▶	—	—	—	—
Endopterygoid					▶	—	—	—	—
Hyomandibular		▶	—	—	—	—	—	—	—
Synplectic			▶	—	—	—	—	—	—
Quadrate			▶	—	—	—	—	—	—
Opercular									
Opercle		▶	—	—	—	—	—	—	—
Subopercle				▶	—	—	—	—	—
Preopercle		▶	—	—	—	—	—	—	—
Interopercle				▶	—	—	—	—	—
Pectoral girdle									
Clavicle		▶	—	—	—	—	—	—	—
Upper clavicle			▶	—	—	—	—	—	—
Actimost								▶	—
Coracoid								▶	—
Scapula								▶	—
Scapula foramen								▶	—
Lower postclavicle					▶	—	—	—	—
Upper postclavicle					▶	—	—	—	—
Supratemporal				▶	—	—	—	—	—
Posttemporal				▶	—	—	—	—	—
Caudal skeleton									
Equral								▶	—
Hypural								▶	—
Parhypral								▶	—
Urostyle					▶	—	—	—	—
Parapophysis								▶	—
Caudal bony plate								▶	—
Infraorbital bone									
Preorbital					▶	—	—	—	—
Suborbital						▶	—	—	—
Vertebrae									
Neural spine			▶	—	—	—	—	—	—
Hemal spine					▶	—	—	—	—
Parapophysis					▶	—	—	—	—
Centrum				▶	—	—	—	—	—

black arrowhead, initial ossification; black bar, ossified state

and premaxillary had both begun to ossify at their anterior and ventral margins, respectively (Fig. 3a). The dentary also started to ossify along its V-shaped anterior margin. The hyomandibular started to ossify at opposite medial margins (Fig. 3a). At the same time, the strongest three preopercular spines on the preopercle began to ossify, and the opercle had simultaneously ossified at its anterior margin (Fig. 3a). At 7.17 mm BL, the quadrate and the symplectic started to ossify in the region in which the two elements join (Fig. 3b). At 8.17 mm BL, the interopercle and preopercle had begun to ossify at their margins (Fig. 3c). The premaxillary, maxillary, and dentary also continued to ossify, and then the premaxillary had formed the ascending process and articular process. At 9.06 mm BL, the angular had ossified (Fig. 3d), and the endopterygoid and ectopterygoid had started to ossify along their adjacent margins. In particular, the upper part of the hyomandibular had quickly and fully ossified, and the opercle had extended to the strongest first spine. At 10.20 mm BL, the articular had started to ossify, and the maxillary and premaxillary had fully ossified and assumed their adult forms (Fig. 3e). The palatine started to ossify along its anterior margin, but the degree of ossification was small. At 11.10 mm BL, the ossification of the jaw bones was complete, and the opercular series was almost fully ossified at this stage, except for small parts of the subopercle and interopercle (Fig. 3f). At the juvenile stage (18.60 mm BL), the ossification of the jaw bones, palatine, and opercular series was complete (Fig. 3g).

Hyoid arch and pectoral girdle

The development and ossification of the hyoid arch and pectoral girdle in individuals at all stages of development are illustrated in Figs. 4 and 5. At 6.27 mm BL, development of the hyoid arch and pectoral girdle had begun, with the ossification of the branchiostegal ray and clavicle, respectively (Figs. 4a and 5a). The fifth branchiostegal ray, which was the first branchiostegal ray to begin ossification, started to ossify in its middle region, and the clavicle was fully ossified as a long needle-shape. At 7.17 mm BL, the fourth branchiostegal ray and the upper clavicle had started to ossify (Figs. 4b and 5b). At 8.17 mm BL, all of the branchiostegal rays, except the first ray, had started to ossify, and the ceratohyal had started to ossify along its dorsal and ventral margins (Fig. 4c). The posttemporal and supratemporal had also started to ossify and were connected to the upper clavicle (Fig. 5c). At 9.06 mm BL, the first branchiostegal ray had started to ossify, and the other branchiostegal rays were fully ossified (Fig. 4d). The upper postclavicle and lower postclavicle of the pectoral girdle had also started to ossify and were

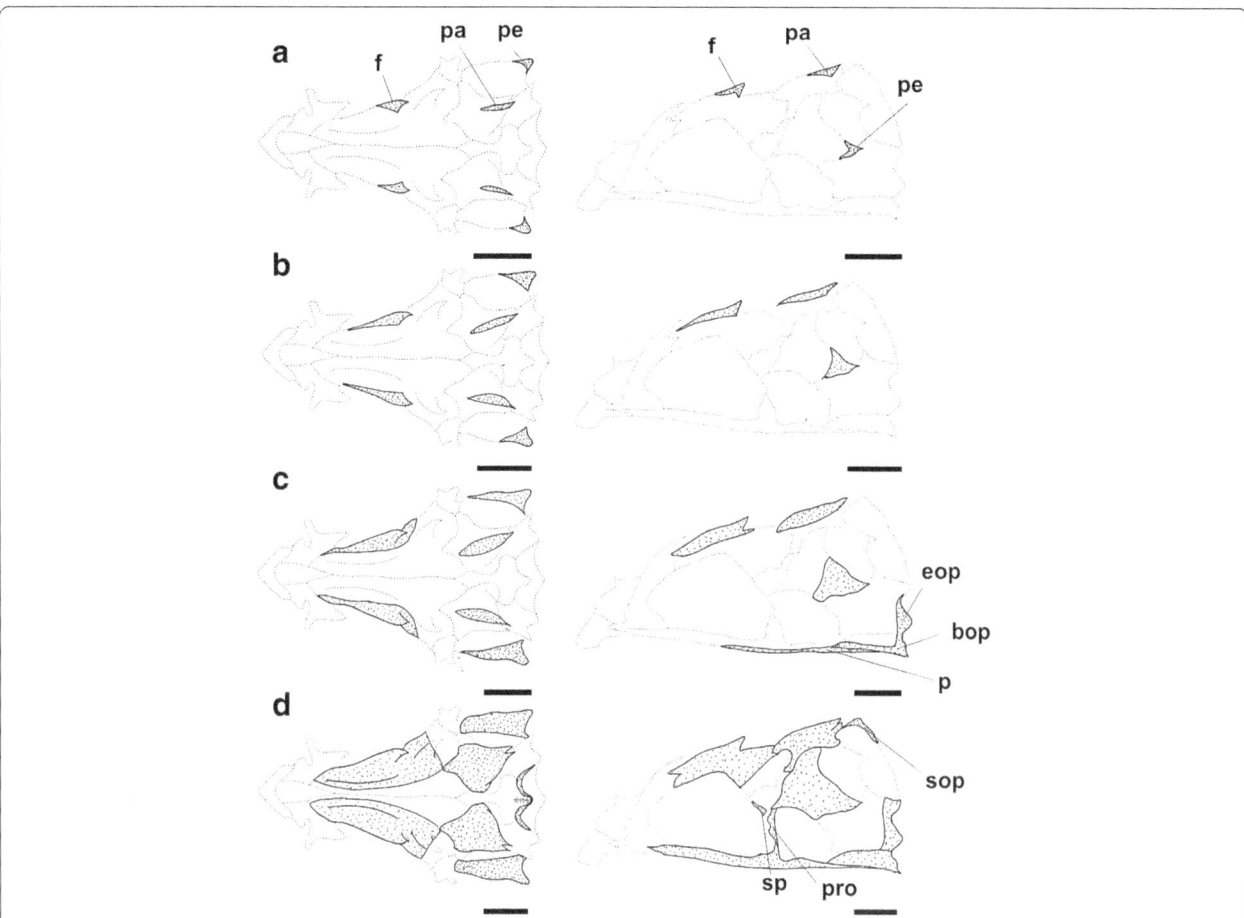

Fig. 1 Developmental sequence of the neurocranium of *Sebastes koreanus*, showing dorsal (*left*) and lateral (*right*) views of preflexion to postflexion larvae. **a** Preflexion larva; 6.27 mm BL. **b** Flexion larva; 7.17 mm BL. **c** Flexion larva; 8.17 mm BL. **d** Postflexion larva; 9.06 mm BL. *Dotted lines* show the outlines of skeletal structures in the adult. *Open areas* show nonskeletal structures. *Solid lines* show the boundaries of ossified areas. *Dotted areas* show ossified elements. bop, basioccipital; eop, exoccipital; f, frontal; p, parasphenoid; pa, parietal; pe, pterotic; pro, prootic; sop, supraoccipital; sp, sphenotic. Bars 0.5 mm

connected to one other (Fig. 5d). At 10.20 mm BL, the ceratohyal had enlarged anteriorly and the posttemporal was fully ossified (Figs. 4e and 5e). At 11.10 mm BL, the anterior parts of the ceratohyal and clavicle were fully ossified, and the seven pairs of branchiostegal rays had approached their adult number and shape (Figs. 4f and 5f). In the juvenile stage (18.60 mm BL), the hyoid arch and pectoral girdle were fully ossified (Figs. 4f and 5f). However, although the scapula and uppermost radial of the pectoral girdle were nearly joined, they had not fused (Fig. 5f).

Infraorbital bone

The development and ossification of the infraorbital bone in individuals at all stages of development are illustrated in Fig. 6. At 9.06 mm BL, the infraorbital bone elements on the preorbital had started to ossify (Fig. 6a). At 10.20 mm BL, the area of ossification of the preorbital had increased, but no additional elements

were visible (Fig. 6b). At 11.10 mm BL, the first and second suborbital bones had started to ossify along their dorsal margins (Fig. 6c). In the juvenile stage (18.60 mm BL), the infraorbital bone was fully ossified (Fig. 6d).

Vertebrae and caudal skeleton

The development and ossification of the vertebrae and caudal skeleton in individuals at all stages of development are illustrated in Fig. 7. The skeletal elements of the vertebrae were first apparent at 7.17 mm BL (Fig. 7a). The first visible ossified elements of the vertebrae were the neural spines; no ossification of the centra was observed at this stage. The centra first started to ossify in the dorsal regions at 8.17 mm BL (Fig. 7b). After a centrum had formed, the neural spines appeared to elongate dorsally. The first hemal spines were observed at 9.06 mm BL, at which time ossification was visible in 10 centra, 14 neural spines, and three hemal spines

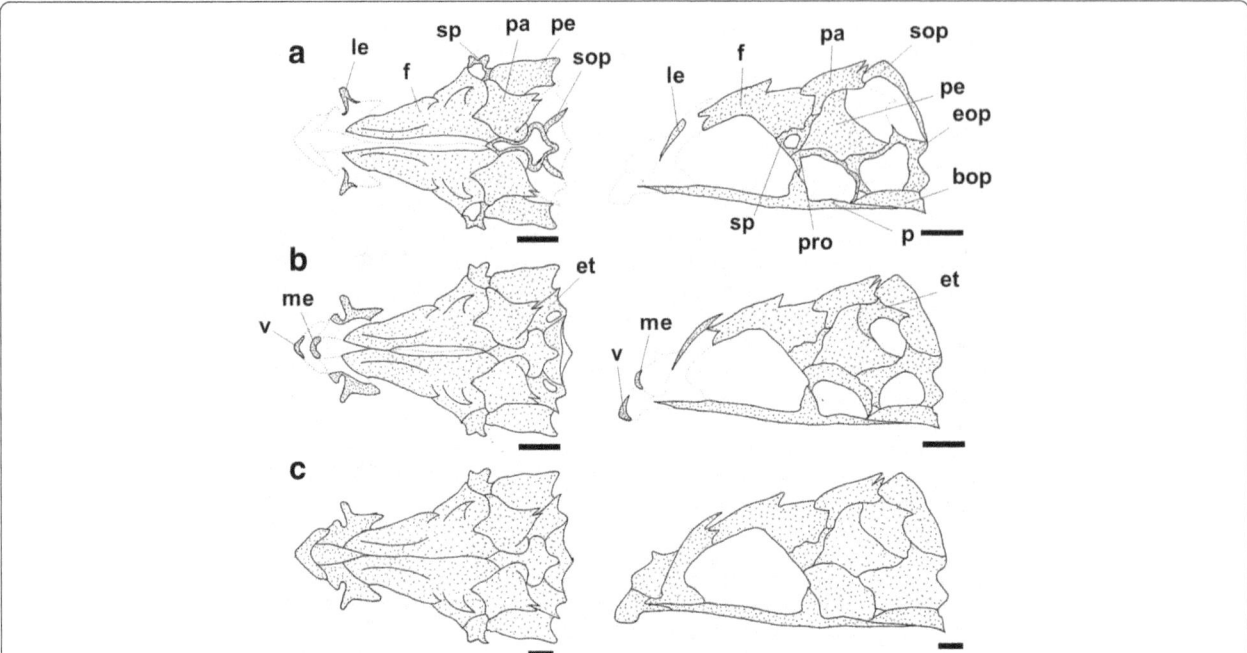

Fig. 2 Developmental sequence of the neurocranium of *Sebastes koreanus*, showing dorsal (*left*) and lateral (*right*) views of postflexion larval to juvenile stages. **a** Postflexion larva; 10.20 mm BL. **b** Postflexion larva; 11.10 mm BL. **c** Juvenile; 18.60 mm BL. *Dotted lines* show the outlines of skeletal structures in the adult. *Open areas* show nonskeletal structures. *Solid lines* show the boundaries of ossified areas. *Dotted areas* show ossified elements. bop, basioccipital; eop, exoccipital; et, epiotic; f, frontal; le, lateral ethmoid; me, medial ethmoid; p, parasphenoid; pa, parietal; pe, pterotic; pro, prootic; sop, supraoccipital; sp, sphenotic; v, vomer. Bars 0.5 mm

(Fig. 7c). The development of the neural spines in the vertebrae occurred more rapidly than did the vertebral centra. Two to three ossified parapophyses appeared on the trunk centra at this stage. At 10.20 mm BL, the anterior centra, neural spines, and hemal spines were almost fully ossified, completely surrounding the notochord, whereas the posterior vertebrae continued to ossify consecutively towards the caudal complex (Fig. 7d). The urostyle had also started to ossify for the first time at this stage, along its anterior margin (Fig. 7d). At 11.10 mm BL, despite the progressive ossification of consecutive vertebrae, a few posterior vertebrae were still only present as cartilaginous structures (Fig. 7e). In the caudal skeleton, the urostyle had fully ossified at this stage, but no additional ossification was visible in the caudal skeleton (Fig. 7e). In the juvenile stage (18.60 mm BL), all the vertebral centra had completely surrounded the notochord, and the adjacent neural spines and hemal spines were also ossified (Fig. 7f). In addition, the hypurals, equals, parahypurals, parapophyses, and uroneural in the caudal skeleton were fully ossified in the juvenile (Fig. 7f). At the juvenile stage, the first and second hypurals and the third and fourth hypurals had also fused to form, together with the fifth hypural, three hypural segments (hy 1 + 2, hy 3 + 4, and hy 5).

Discussion

This study is the first to examine and describe in detail the sequence of osteological development in *S. koreanus* collected in the wild and to provide data with which to infer the phylogenetic relationships of species within the suborder Scorpaenoidei. In *S. koreanus*, ossification of the skeletal elements is first observed in the neurocranium, jaw bones, palatine, opercular, hyoid arch, and pectoral girdle of the preflexion larva with a length of 6.27 mm BL [6.45 mm total length (TL)], and then the only one juvenile (18.60 mm BL) has fully completed the skeletal development of all elements (Table 1). In a previous study of early skeletal development in the genus *Sebastes*, ossification was first observed in *S. inermis* complex at 7 days (7.0 mm mean TL) (Kim et al. 1993), in *S. schlegelii* at 6–8 days (6.85 mm TL) (Kim and Han 1991), and in *S. oblongus* at 3 days after release (8.0 mm TL) (Byun et al. 2012). In a study of early skeletal development in *S. macdonaldi* from southern California, ossification was first observed in the smallest larva (6.11 mm BL) (Moser 1972). Like this, the ossification of wild-captured *S. koreanus* larvae was first observed in larvae smaller than reared larvae of *S. inermis*, *S. schlegelii*, and *S. oblongus* (but not in *S. macdonaldi*). These differences in the size at the onset of ossification are

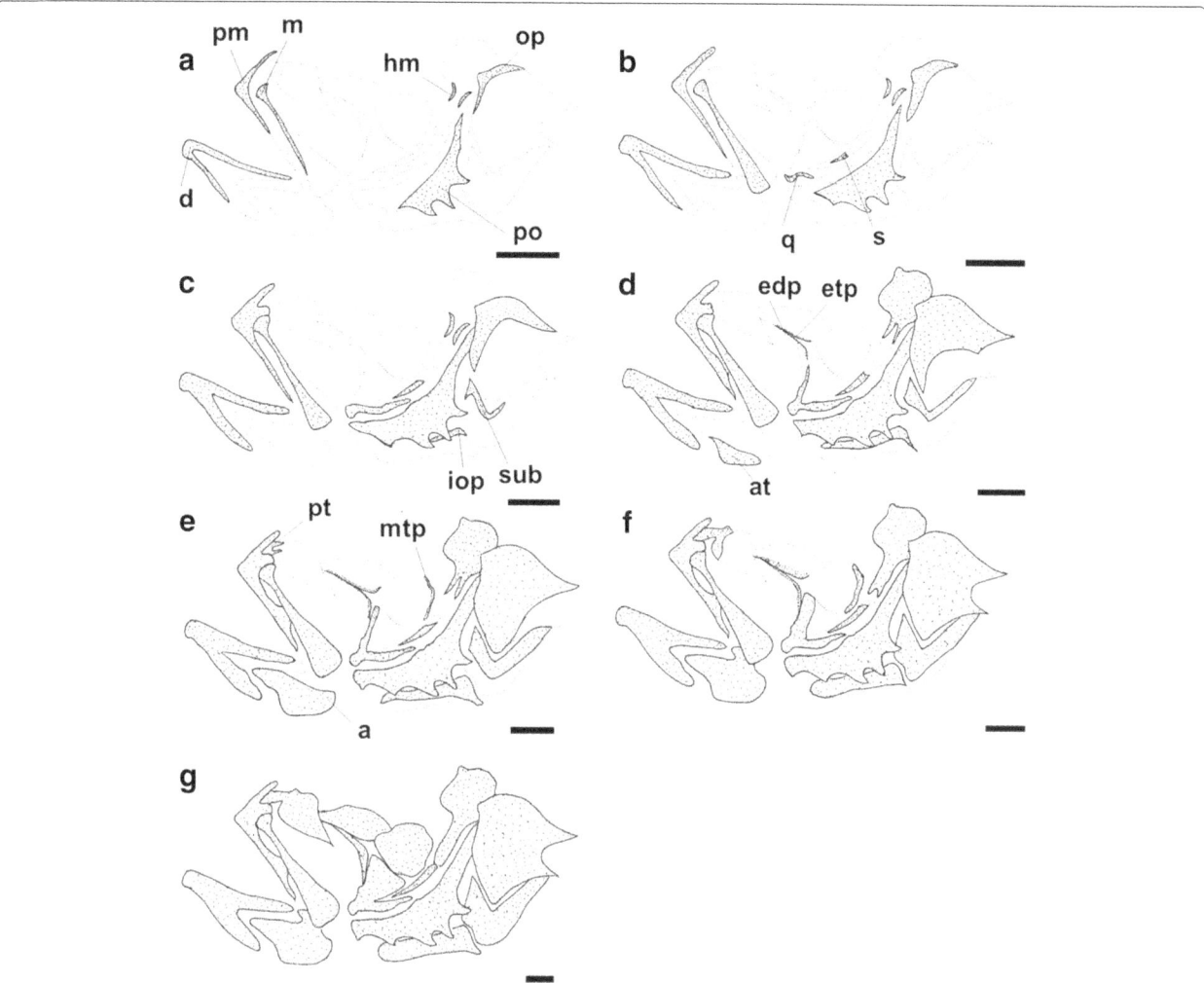

Fig. 3 Developmental sequences of the jaw bone, palate, and opercular series of *Sebastes koreanus* in preflexion larval to juvenile stages (lateral views). **a** Preflexion larva; 6.27 mm BL. **b** Flexion larva; 7.17 mm BL. **c** Flexion larva; 8.17 mm BL. **d** Postflexion larva; 9.06 mm BL. **e** Postflexion larva; 10.20 mm BL. **f** Postflexion larva; 11.10 mm BL. **g** Juvenile; 18.60 mm BL. *Dotted lines* show the outlines of skeletal structures in the adult. *Open areas* show nonskeletal structures. *Solid lines* show the boundaries of ossified areas. *Dotted areas* show ossified elements. a, angular; at, articular; d, dentary; edp, endopterygoid; etp, ectopterygoid; hm, hyomandibular; iop, interopercle; m, maxilliary; mtp, metapterygoid; op, opercle; po, preopercle; pm, premaxilliary; pt, palatine; q, quadrate; s, symplectic; sub, subopercle. Bars 0.5 mm

probably related to the size at which the larvae are released from the adult, which is smaller than 6.11 mm BL in *S. koreanus* (this study), 6.12 mm TL in *S. inermis* (Kim et al. 1993), 5.52 mm TL in *S. schlegelii* (Kim and Han 1991), 7.2 mm TL in *S. oblongus* (Byun et al. 2012), and 4.5 mm BL in *S. macdonaldi* (Moser 1972). These differences may also be affected by external environmental factors, such as temperature and salinity (Fuiman 2002; Ložys 2004; Löffler et al. 2008; Ott et al. 2012), which may cause corresponding osteological differences (Matsuoka 1987; Wimberger 1993; Koumoundouros et al. 1997a) and meristic variations (Fowler 1970; Lau and Shafland 1982) between reared larvae and wild-captured larvae (Boglione et al. 2001). Therefore, despite the similar size of the released larvae of

S. koreanus, *S. inermis*, and *S. schlegelii*, the first ossification size may also differ from each other as larvae of *S. koreanus* were collected in the wild while the other species were cultured in captivity. In addition, ossification was first observed in the reared larvae of *Sebastiscus marmoratus* and *Sebastiscus tertius* at 3.35 and 4.4 mm TL, respectively, even smaller than the larvae of species of *Sebastes* (Kim et al. 1997; Han et al. 2001). These results are also probably related to the size at parturition; at parturition, size is smaller than the larvae of *Sebastes* (Kim et al. 1997; Han et al. 2001).

In most cases, early skeletal development occurs first in elements that are necessary for feeding and respiration and therefore affect the survival of young larvae (Vandewalle et al. 1997; Wagemans and Vandewalle

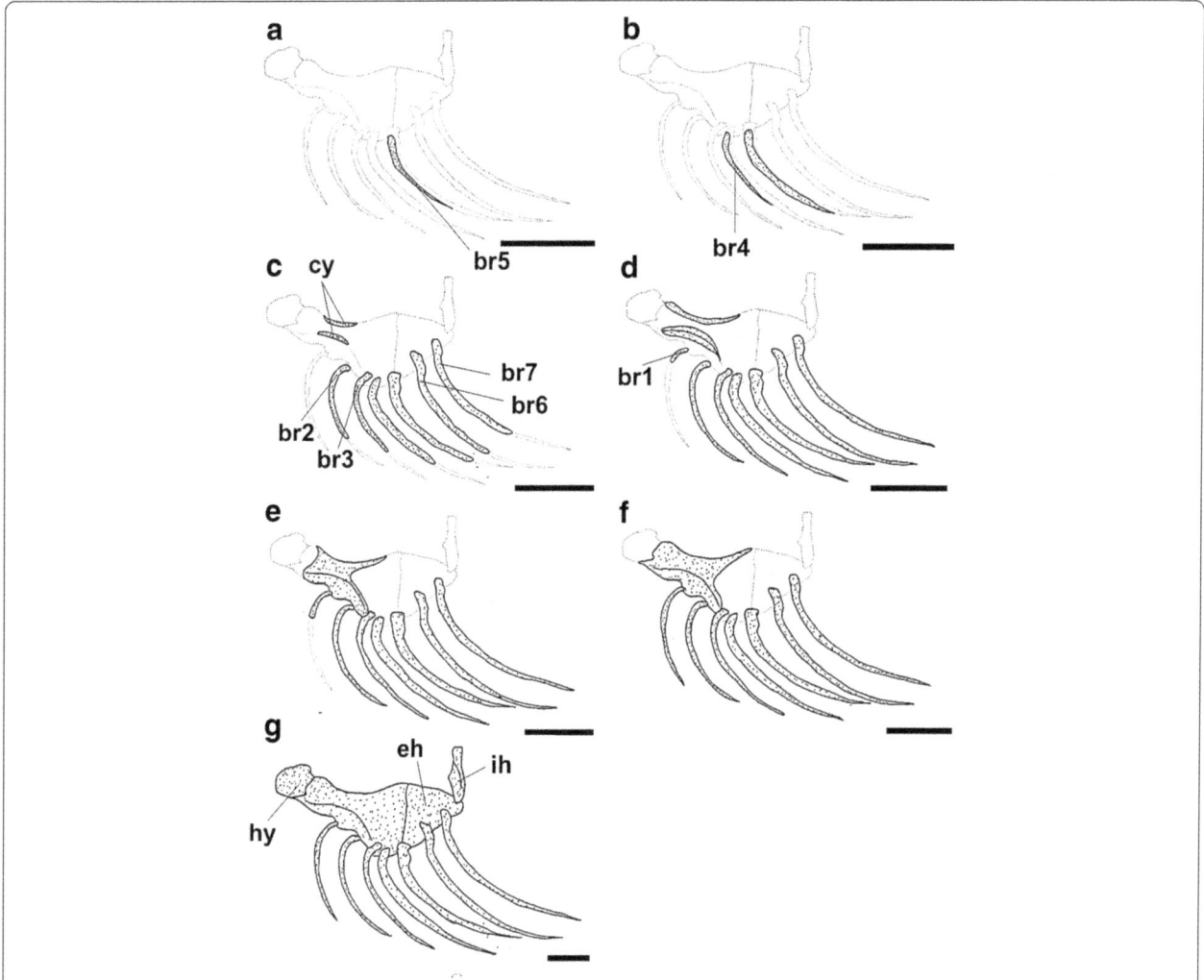

Fig. 4 Developmental sequence of the hyoid arch of *Sebastes koreanus* in preflexion larval to juvenile stages (lateral views). **a** Preflexion larva; 6.27 mm BL. **b** Flexion larva; 7.17 mm BL. **c** Flexion larva; 8.17 mm BL. **d** Postflexion larva; 9.06 mm BL. **e** Postflexion larva; 10.20 mm BL. **f** Postflexion larva; 11.10 mm BL. **g** Juvenile; 18.60 mm BL. *Dotted lines* show the outlines of skeletal structures in the adult. *Open areas* show nonskeletal structures. *Solid lines* show the boundaries of ossified areas. *Dotted areas* show ossified elements. br, branchiostegal ray; cy, ceratohyal; eh, epihyal; ih, interhyal; hy, hypohyal. Bars 0.5 mm

1999). For example, the total resorption of the vitellus is essential for the transition from endogenous to exogenous feeding, because the efficiency of suction feeding increases with increasing prey size and the ossification of the related skeletal elements (Gluckmann et al. 1999). In *S. koreanus*, the skeletal elements that first start to ossify (at 6.27 mm BL) are the premaxillary, maxillary, dentary, preopercle, opercle, hyomandibular, and the fifth branchiostegal ray (Figs. 3 and 4), and the order of ossification is initially defined by the importance of the skeletal elements to feeding, swimming, and respiration. The cleithrum in the pectoral girdle ossifies in the same developmental stage (Fig. 5), and the early ossification of the clavicle produces an attachment site for the sternohyoideus muscle, which is important for swimming in subsequent growth stages

(Wagemans and Vandewalle 1999; Koumoundouros et al. 2001a; Cloutier et al. 2011). Similar patterns of early skeletal development have been observed in other species (e.g., *S. inermis*, *S. schlegelii*, *S. oblongus*, *Sebastiscus marmoratus*, and *Sebastiscus tertius*). However, the timing of the ossification of the hyomandibular is highly variable (Kim and Han 1991; Kim et al. 1993; Kim et al. 1997; Byun et al. 2012). In most teleostei, the parasphenoid is the first element to ossify, except in some species in which the parasphenoid ossifies simultaneously with the frontals (*Pagrus major*; Matsuoka 1987) or the basioccipital (*Heterobranchus longifilis*; Vandewalle et al. 1997), or is ossified after the ossification of the frontals (*Scophthalmus maximus*; Wagemans et al. 1998). In *S. koreanus*, the first-ossified elements in the neurocranium (at 6.27 mm BL) are the

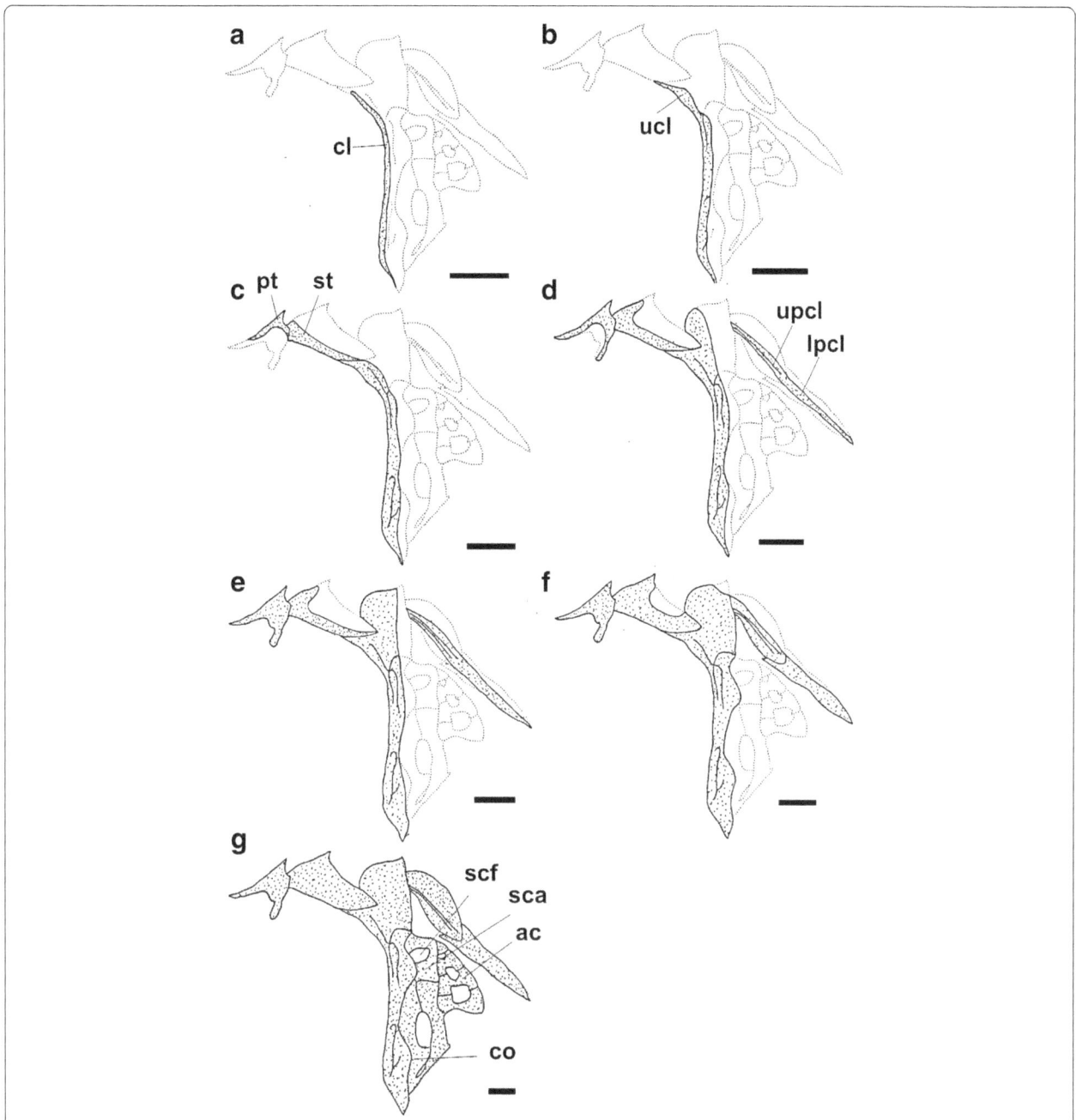

Fig. 5 Development of the pectoral girdle of *Sebastes koreanus* in preflexion larval to juvenile stages (lateral views). **a** Preflexion larva; 6.27 mm BL. **b** Flexion larva; 7.17 mm BL. **c** Flexion larva; 8.17 mm BL. **d** Postflexion larva; 9.06 mm BL. **e** Postflexion larva; 10.20 mm BL. **f** Postflexion larva; 11.10 mm BL. **g** Juvenile; 18.60 mm BL. *Dotted lines* show the outlines of skeletal structures in the adult. *Open areas* show nonskeletal structures. *Solid lines* show the boundaries of ossified areas. *Dotted areas* show ossified elements. ac, actinost; cl, clavicle; co, coracoids; lpcl, lower postclavicle; pt, posttemporal; sca, scapula; scf, scapula foramen; st, supratemporal; ucl, upper clavicle; upcl, upper post clavicle. Bars 0.5 mm

parietal, frontal, and pterotic (Fig. 1). Subsequently, the parasphenoid and basioccipital begin to ossify at 8.17 mm BL (Fig. 1); these elements may help to reinforce the cranial floor to prevent damage to the neurocranium during feeding (Vandewalle et al. 1992) and to promote the balance needed for swimming (Weisel 1967). Therefore, the ossification of the

parasphenoid and basioccipital during early skeletal development is important because they significantly affect feeding and swimming behavior, as do the jaw bones and clavicle, respectively.

The order of ossification of the neurocranial elements appears similar in different *Sebastes* species, but variations exist, particularly in the timing of the ossification

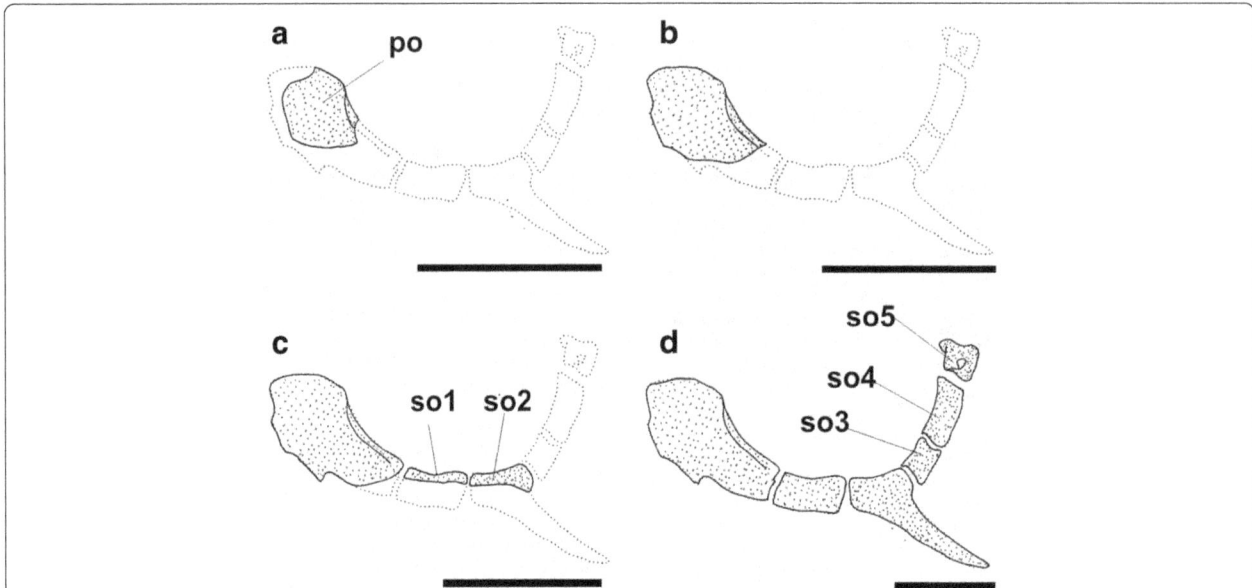

Fig. 6 Development of the infraorbital bone of *Sebastes koreanus* in postflexion larval to juvenile stages (lateral views). **a** Postflexion larva; 9.06 mm BL. **b** Postflexion larva; 10.20 mm BL. **c** Postflexion larva; 11.10 mm BL. **d** Juvenile; 18.60 mm BL. *Dotted lines* show the outlines of skeletal structures in the adult. *Open areas* show nonskeletal structures. *Solid lines* show the boundaries of ossified areas. *Dotted areas* show ossified elements. po, preorbital bone; so, suborbital. Bars 0.5 mm

of the parasphenoid. In many species of *Sebastes* and *Sebastiscus*, such as *S. macdonaldi*, *S. inermis*, *S. schlegelii*, *Sebastiscus marmoratus*, and *Sebastiscus tertius*, the parasphenoid is the first element to ossify (Moser 1972; Kim and Han 1991; Kim et al. 1993; Kim et al. 1997; Han et al. 2001), whereas in *S. koreanus*, the parasphenoid begins to ossify simultaneously with the basioccipital and exoccipital, just after the ossification of the parietal, frontal, and pterotic (present study), or in *S. oblongus*, the parasphenoid begins to ossify simultaneously with the supraoccipital, just after the ossification of the parietal and frontal (Byun et al. 2012). The pterotic and parietal also begin to ossify relatively early in some species, including *S. koreanus*, *S. macdonaldi*, *Sebastiscus marmoratus*, and *Sebastiscus tertius*, but no clear differences between the species of *Sebastes* and *Sebastiscus* are apparent (Moser 1972; Kim et al. 1997; Han et al. 2001). In *S. koreanus*, the early ossification of the hyoid arch appears on the ceratohyal and branchiostegal rays, but there is no additional ossification of elements between 8.17 and 11.10 mm BL (Fig. 4). In contrast, the ossification of the hyoid arch is clearly different in many species of *Sebastes* and *Sebastiscus* from that observed in *S. koreanus* and begins to occur at the same time as the ossification of the ceratohyal and epihyal (in *S. inermis*, *S. schlegelii*, *Sebastiscus marmoratus*, and *Sebastiscus tertius*), or the epihyal begins to ossify just after the ossification of the ceratohyal (as in *S. oblongus*) (Kim and

Han 1991; Kim et al. 1993; Kim et al. 1997; Han et al. 2001; Byun et al. 2012).

Ossification of the pectoral girdle also shows a high degree of variability between different species of *Sebastes* and *Sebastiscus*. In *S. koreanus*, the ossification of the pectoral girdle begins with the clavicle, followed by the upper clavicle and soon thereafter by the supratemporal and posttemporal (Fig. 5). In contrast, in *S. oblongus*, the ossification of the clavicle first begins 3 days after release, and the ossification of the upper clavicle and posttemporal begin at 20 days, soon after the initial ossification of the supratemporal (Byun et al. 2012). In *S. inermis*, the first ossification on the clavicle begins 7 days after release, followed by the ossification of the postclavicle at 45 days and the upper clavicle at 50 days. The supratemporal begins to ossify at 65–69 days (Kim et al. 1993). In *Sebastiscus marmoratus*, the ossification of the clavicle first appears 5 days after release, and the ossification of the upper clavicle, posttemporal, scapula, and coracoid begin at 28 days (Kim et al. 1997). Therefore, it is difficult to determine a common ossification pattern for the pectoral girdle because of the observed variability between species. Also, *S. koreanus* ossified faster than other species of *Sebastes* and *Sebastiscus*, because this species might be to adapt to the harsh environment of the Yellow Sea such as strong current. According to Ishida (1994), adults of *Sebastes*, *Sebastiscus*, and *Hozukius* (suborder Scorpaenoidei) share the derived

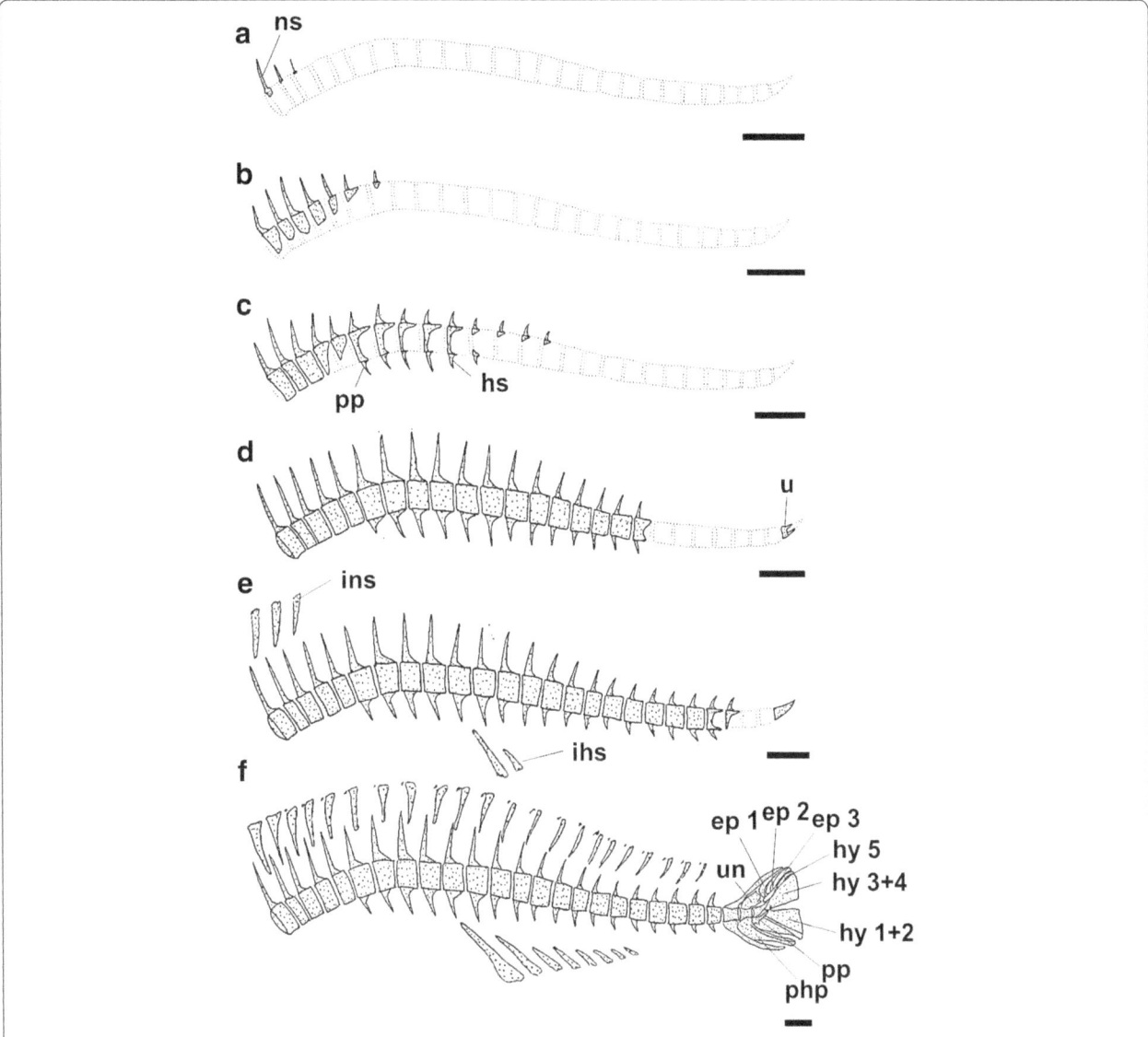

Fig. 7 Development of the vertebrae and caudal skeleton of *Sebastes koreanus* in flexion larval to juvenile stages (lateral views). **a** Flexion larva; 7.17 mm BL. **b** Flexion larva; 8.17 mm BL. **c** Postflexion larva; 9.06 mm BL. **d** Postflexion larva; 10.20 mm BL. **e** Postflexion larva; 11.10 mm BL. **f** Juvenile; 18.60 mm BL. *Dotted lines* show the outlines of skeletal structures in the adult. *Open areas* show nonskeletal structures. *Solid lines* show the boundaries of ossified areas. *Dotted areas* show ossified elements. ep, epural bone; hs, hemal spine; hy, hypural bone; ihs, interhemal spine; ins, interneural spine; php, parhypural; pp, parapophysis; ns, neural spine; u, urostyle bone; un, uroneural. Bars 0.5 mm

characteristic of a fusion of the scapula and uppermost radial in the pectoral girdle. However, although the fusion of the scapula and uppermost radial was observed in adults of *S. koreanus*, fusion was not observed in the *S. koreanus* juvenile (18.60 mm BL) (Fig. 5g). In some species of *Sebastes* (e.g., *S. oblongus*, Byun et al. 2012; *S. schlegelii*, Kim and Han 1991; Omori et al. 1996; *S. macdonaldi*, Moser 1972), fusion between the scapula and uppermost radial is not observed during skeletal development. Thus, it appears that the scapula and uppermost radial fuse slowly after (or starting in) the juvenile stage. Actually, the scapula and uppermost radial were closely adjoined

along a thin boundary line in the *S. koreanus* juvenile observed in this study, presumably just prior to fusion. Since some ontogenetic studies have resolved taxonomic uncertainty that cannot be defined by adult morphology (Hubbs and Kampa 1946; White et al. 1983; Parin 1996), further ontogenetic study is needed to understand interrelationship between *S. koreanus* and *Sebastes* spp.

With respect to locomotion, the swimming of larvae immediately after their release from the adult is possible only through the antagonistic interactions of the notochord and trunk muscles (Ott et al. 2012). With growth, the notochord is gradually replaced by vertebrae, and the ossified vertebrae contribute stronger attachment sites for the powerful dorsalis trunci muscles, which are primarily responsible for swimming (Rojo 1991). In *S. koreanus*, after the ossification of the neural spine at 7.17 mm BL, the ossification of the vertebral centra mainly proceeds from the abdominal to the caudal vertebrae, and the urostyle is fully ossified just before the ossification of the caudal vertebrae is complete (Fig. 7). This pattern is similar to that observed in other species of *Sebastes* and *Sebastiscus*, except in *S. schlegelii* (Kim and Han 1991; Omori et al. 1996), e.g., in *S. oblongus* (Byun et al. 2012), *S. inermis* (Kim et al. 1993), *Sebastiscus marmoratus* (Kim et al. 1997), and *Sebastiscus tertius* (Han et al. 2001). Furthermore, in adults, the caudal skeleton in *Sebastes* and *Sebastiscus* species is formed by three hypurals (hy 1 + 2, hy 3 + 4, and hy 5), because the first and second hypurals and the third and fourth hypurals are fused (Ishida 1994). A similar trend is observed in the skeletal development of *Sebastes* and *Sebastiscus* species, including *S. koreanus*, *S. inermis*, *S. schlegelii*, *S. koreanus*, *S. macdonaldi*, *Sebastiscus marmoratus*, and *Sebastiscus tertius* (Fig. 7f) (Kim et al. 1993; Omori et al. 1996; Kim et al. 1997; Han et al. 2001; Byun et al. 2012). Therefore, the ontogenetic characteristics reflect the taxonomic characteristics of the adults well. In particular, the hypural cartilages fuse before ossification, unlike the fusion of the scapula and the uppermost radial (Omori et al. 1996).

Park et al. (2015) provided a brief overview of the external and osteological development of *S. koreanus* based on the artificial breeding of hatched larvae, using a gravid adult collected from the Korea Strait. However, compared with larvae and juvenile of *S. koreanus* collected from the wild in the Yellow Sea (present study; Yu et al. 2015), there are several differences, such as the pigmentation patterns and sequence of osteological development. We established three hypotheses regarding their differences. The first hypothesis is that it could be caused by the different population, as the present study, and Park et al. (2015) used different sampling sites (Yellow Sea vs. Korea Strait). In this respect, there is a possibility of existence of different population. Similarly, Kim et al. (2010) confirmed that the two populations of *Ammodytes personatus* larvae showed morphological differences in morphometric characters and pigmentation. The second hypothesis is that it could be caused by the difference in sea water temperature. The larvae and juvenile of *S. koreanus* (present study) were collected at the average sea water temperature 12.0–15.1 °C from Taean Peninsula (May and June), whereas larvae and juveniles were reared at the sea water temperature 13.5–15.5 °C. This slight water temperature difference (ca. 1.5 °C)

could possibly cause the ontogenetic differences (Löffler et al. 2008; Ott et al. 2012). Finally, the third hypothesis is that it could be caused by differences of food source between reared larvae and wild-captured larvae. Particularly, calcium deficiency at the food source induces a delay in the ontogeny of skeletal development without affecting final bone mineralization (Fontagné et al. 2009). In addition, the different fixatives, formalin or alcohol, may have reduced or removed some pigmentation. Therefore, further research with microsatellite DNA and comparison of rearing is required to confirm the observed difference between reared larvae and wild-captured larvae.

Conclusions

In summary, although the sequences and periods of osteological development in *Sebastes* and *Sebastiscus* species show some variation, the early ossification of the skeleton proceeds in a sequence that prioritizes the elements required for feeding, swimming, and respiration. In this way, larvae are equipped with functional capacities that enhance the probability of their survival at this stage of their life cycle. Also, the ossification of parasphenoid and epihyal in the neurocranium appeared relatively later than congeneric species, this features indicated that *S. koreanus* is a unique species, which has been evolved in distinctive marine environment of the Yellow Sea.

Acknowledgements
This research was supported by the Marine Fish Resources Bank of Korea (MFRBK) under the Ministry of Oceans and Fisheries, Korea.

Authors' contributions
HJY performed the experiments and wrote the manuscript. JKK suggested all aspects of study design and commented on the earlier drafts of the manuscript. Both authors read and approved the final manuscript.

Competing interests
The authors declare that they have no competing interests.

References
Boglione C, Gagliardi F, Scardi M, Cataudella S. Skeletal descriptors and quality assessment in larvae and post-larvae of wild-caught and hatchery-reared gilthead sea bream (*Sparus aurata* L. 1758). Aquaculture. 2001;192:1–22.

Byun SG, Kang CB, Myoung JG, Cha BS, Han KH, Jung CG. Early osteological development of the larvae and juveniles in *Sebastes oblongus* (Pisces: Scorpaenidae). Korean J Ichthyol. 2012;24:67–76.

Choi Y, Yang AF. Intertidal fishes from the Shandong Peninsula, China. Korean J Ichthyol. 2008;20:54–60.

Cloutier R, Lambrey de Souza J, Browman HI, Skiftesvik AB. Early ontogeny of the Atlantic halibut *Hippoglossus hippoglossus* head. J Fish Biol. 2011;78:1035–53.

Darias MJ, Lan Chow Wing O, Cahu C, Zambonino-Infante JL, Mazurais D. Double staining protocol for developing European sea bass (*Dicentrarchus labrax*) larvae. J Appl Ichthyol. 2010;26:280–5.

Faustino M, Power DM. Development of the pectoral, pelvic, dorsal and anal fins in cultured sea bream. J Fish Biol. 1999;54:1094–110.

Fontagné S, Silva N, Bazin D, Ramos A, Aguirre P, Surget A, et al. Effects of dietary phosphorus and calcium level on growth and skeletal development in rainbow trout (Oncorhynchus mykiss) fry. Aquaculture. 2009;297:141–50.

Fowler JA. Control of vertebral number in teleosts-an embryological problem. Q Rev Biol. 1970;45:148–67.

Fuiman LA. Special considerations of fish eggs and larvae. In: Fuiman LA, Werner RG, editors. Fishery science: the unique contributions of early life history stages. Oxford, UK: Blackwell Science; 2002. p. 1–32.

Gluckmann I, Huriaux F, Focant F, Vandewalle P. Postembryonic development of the cephalic skeleton in Dicentrarchus labrax (Pisces, Perciformes, Serranidae). Bull Mar Sci. 1999;65:11–36.

Han KH, Lim SK, Kim KS, Kim CW, Yoo DJ. Osteological development of the larvae and juveniles of Sebasticus tertius (Barsukov et Chen) in Korea. Korean J Ichthyol. 2001;13:63–8.

Hubbs CL, Kampa EM. The early stages (egg, prolarva and juvenile) and the classification of the California flyingfish. Copeia. 1946;1946:188–218.

Ishida M. Phylogeny of the suborder Scorpaenoidei (Pisces: Scorpaeniformes). Bull Nansei Nat Fish Res Inst. 1994;27:1–11.

Kang CB, Myoung JG, Kim YU, Kim HC. Early osteological development and squamation in the spotted sea bass Lateolabrax maculates (Pisces: Lateolabracidae). Kor J Fish Aquat Sci. 2012;45:271–82.

Kim IS, Choi Y, Lee CL, Lee YJ, Kim BJ, Kim JH. Illustrated book of Korean fishes. Seoul, KR: Kyo-Hak Publishing Co; 2005.

Kim IS, Lee WO. A new species of the genus Sebastes (Pisces; Scorpaenidae) from the Yellow Sea, Korea. Korean J Zool. 1994;37:409–15.

Kim JK, Ryu JH, Kim S, Lee DW, Choi KH, Oh TY, et al. An identification guide for fish eggs, larvae and juveniles of Korea. Busan, KR: Hanguel Graphics Publishing Co; 2011.

Kim JK, Watson W, Hyde J, Nancy L, Kim JY, Kim S, et al. Molecular identification of Ammodytes (Ammodytidae, pisces) larvae, with ontogenetic evidence on separating populations. Genes Genomics. 2010;32:437–45.

Kim YU, Han KH. The early life history of rockfish, Sebastes schlegeli. Korean J Ichthyol. 1991;3:67–83.

Kim YU, Han KH, Byun SK. The early life history of the rockfish, Sebastes inermis. 2. Morphological and skeletal development of larvae and juveniles. Bull Korean Fish Soc. 1993;26:465–76.

Kim YU, Han KH, Kang CB, Kim JK, Byun SK. The early life history of the rockfish, Sebastiscus marmoratus. 2. Morphological and skeletal development of larvae and juveniles. Korean J Ichthyol. 1997;9:186–94.

Koumoundouros G, Divanach P, Kentouri M. Development of the skull in Dentex dentex (Osteichthyes: Sparidae). Mar Biol. 2000;136:175–84.

Koumoundouros G, Divanach P, Kentouri M. Osteological development of Dentex dentex (Osteichthyes: Sparidae): dorsal, anal, paired fins and squamation. Mar Biol. 2001a;138:399–406.

Koumoundouros G, Gagliardi F, Divanach P, Boglione C, Cataudella S, Kentouri M. Normal and abnormal osteological development of caudal fin in Sparus aurata L. fry. Aquaculture. 1997a;149:215–26.

Koumoundouros G, Oran G, Divanach P, Stefanakis S, Kentouri M. The opercular complex deformity in intensive gilthead sea bream (Sparus aurata L.) larviculture. Moment of apparition and description. Aquaculture. 1997b;156:165–77.

Koumoundouros G, Sfakianakis D, Maingot E, Divanach P, Kentouri M. Osteological development of the vertebral column and of the fins in Diplodus sargus (Teleostei: Perciformes: Sparidae). Mar Biol. 2001b;139:853–62.

Lau SR, Shafland PL. Larval development of snook, Centropomus undecimalis (Pisces: Centropomidae). Copeia. 1982;1982:618–27.

Leis JM, Carson-Ewart BM. The larvae of indo-pacific coastal fishes: an identification guide to marine fish larvae. Leiden, Netherlands: Brill; 2000.

Lima ARA, Barletta M, Dantas DV, Ramos JAA, Costa MF. Early development of marine catfishes (Ariidae): from mouth brooding to the release of juveniles in nursery habitats. J Fish Biol. 2013;82:1990–2014.

Liu CH. Early osteological development of the yellowtail Seriola dumerili (Pisces: Carangidae). Zool Stud. 2001;40:289–98.

Löffler J, Ott A, Ahnelt H, Keckeis H. Early development of the skull of Sander lucioperca (L.) (Teleostei: Percidae) relating to growth and mortality. J Fish Biol. 2008;72:233–58.

Ložys L. The growth of pikeperch (Sander lucioperca L.) and perch (Perca fluviatilis L.) under different water temperature and salinity conditions in the Curonian Lagoon and Lithuanian coastal waters of the Baltic Sea. Hydrobiology. 2004;514:105–13.

Matsuoka M. Development of skeletal tissue and skeletal muscle in the red sea bream, Pagrus major. Bull Seikai Reg Fish Res Lab. 1987;65:1–102.

Moser HG. Development and geographic distribution of the rockfish, Sebastes macdonaldi (Eigenmann and Beeson, 1893), Family Scorpaenidae, off Southern California and Baja California. US Nat Mar Fish Serv Fish Bull. 1972;70:941–58.

Nakabo T, Kai Y. Sebastidae. In: Nakabo T, editor. Fishes of Japan with pictorial keys to the species. 3rd ed. Tokyo, JP: Tokai Univ Press; 2013. p. 668–81.

Omori M, Sugawara Y, Honda H. Morphogenesis in hatchery-reared larvae of the black rockfish, Sebastes schlegeli, and its relationship to the development of swimming and feeding functions. Ichthyol Res. 1996;43:267–82.

Ott A, Löffler J, Ahnelt H, Keckeis H. Early development of the postcranial skeleton of the pikeperch Sander lucioperca (Teleostei: Percidae) relating to developmental stages and growth. J Morphol. 2012;273:894–908.

Parin NV. On the species composition of flying fishes (Exocoetidae) in the west-central part of tropical Pacific. J Ichthyol. 1996;36:357–64.

Park JM, Cho JK, Han H, Han KH. Morphological and skeletal development and larvae and juvenile of Sebastes koreanus (Pisces: Scorpaenidae). Korean J Ichthyol. 2015;27:1–9.

Rojo AL. Dictionary of evolutionary fish osteology. Boca Raton, FL: CRC Press; 1991.

Russell FS. The eggs and planktonic stages of British marine fishes. London, U.K.: Academic Press Inc; 1976.

Vandewalle P, Focant B, Huriaux F, Chardon M. Early development of the cephalic skeleton of Barbus barbus (Teleostei: Cyprinidae). J Fish Biol. 1992;41:43–62.

Vandewalle P, Gluckmann I, Baras E, Huriaux F, Focant B. Postembryonic development of the cephalic region in Heterobranchus longifilis. J Fish Biol. 1997;50:227–53.

Voskoboinikova OS, Kudryavtseva OY. Development of bony skeleton in the ontogeny of lumpfish Cyclopterus lumpus (Cyclopteridae, Scorpaeniformes). J Ichthyol. 2014;54:301–10.

Wagemans F, Focant B, Vandewalle P. Early development of the cephalic skeleton in the turbot. J Fish Biol. 1998;52:166–204.

Wagemans F, Vandewalle P. Development of the cartilaginous skull in Solea solea: trends in Pleuronectiforms. Ann Sci Nat. 1999;1:39–52.

Weisel GF. Early ossification in the skeleton of sucker (Catastomus macrocheilus) and the guppy (Poecilia reticulata). J Morphol. 1967;121:1–18.

White BN, Lavenberg RJ, McGowen GE. Atheriniformes: development and relationships. Spec Publ Soc Ichthyol Herpetol. 1983;1:355–62.

Wimberger PH. Effects of vitamin C deficiency on body shape and skull osteology in Geophagus brasiliensis: implications for interpretations of morphological plasticity. Copeia. 1993;2:343–51.

Yu HJ, Im YJ, Jo HS, Lee SJ, Kim JK. Morphological development of eggs, larvae, and juvenile of Sebastes koreanus (Scorpaeniformes: Scorpaenidae) from the Yellow Sea. Ichthyol Res. 2015;62:439–49.

25

Bioaccumulation, alterations of metallothionein, and antioxidant enzymes in the mullet *Mugil cephalus* exposed to hexavalent chromium

Eun Young Min[1], Tae Young Ahn[2] and Ju-Chan Kang[2*]

Abstract

A laboratory experiment was conducted to determine hexavalent chromium (Cr^{6+}) accumulation in the mullet and investigate Cr^{6+} toxicity using a panel of biomarkers including metallothioneins (MTs), glutathione (GSH), glutathione S-transferase (GST), and superoxide dismutases (SODs) for 4 weeks. Cr^{6+} bioaccumulation in all tissues, except muscle, was consistently time- and dose-dependent. The accumulation of Cr^{6+} for 4-week exposures was in the following order: kidney ≈ liver > intestine ≈ gill > spleen > muscle. Compared with the control, Cr^{6+} bioaccumulation was increased in ≥200 µg L^{-1} groups ($P < 0.05$). An independent relation was observed between accumulation factors (AFs) and exposure concentration. But AFs increased with exposure time. In the liver and gill, GST and SOD differed from the control at a high Cr^{6+} concentration at 2 and 4 weeks ($P < 0.05$). This study indicated that the gills were as sensitive as the liver to Cr^{6+} toxicity. However, the latter appeared to influence largely on the organism's adaptive response to Cr^{6+}, since Cr^{6+} may elevate GSH and MT levels by enhancing the hepatic uptake of metal in the mullet.

Keywords: Hexavalent chromium, Mullet, Bioaccumulation, Metallothionein, Antioxidant enzymes

Background

Heavy metals are considered major anthropogenic contaminants in marine environments, as they pose a serious threat to marine organisms due to their toxicity, persistence, and bioaccumulation tendencies (DeForest et al. 2007). Chromium is an essential microelement in insulin-related functions, but it is also a toxic element. A typical feature of chromium in the environment is that it does not disappear but merely changes its form and valence. Demirak et al. (2006) reported that the gills are the major site for Cr accumulation due to their close relationship with the external environment. Hexavalent chromium (Cr^{6+}) induces histological alterations in fish, including hyperplasia of the gill epithelium and fusion of the secondary gill lamella. However, the liver plays a major role in the detoxification of metals through the induction of metal-binding proteins such as metallothionein (MT) (Linde et al. 2005). MT, a metal-binding protein, is a low molecular weight cysteine-rich protein that not only plays an important role in the transport and storage of essential metals but also provides protection against the toxic effect of metals (Lange et al. 2002). In aquatic species, exposure to metal increases the MT level (Lange et al. 2002). The MT induction in fish is known to be high in tissues directly involved in metal uptake, storage, and excretion, such as the gill, liver, kidney, intestine, and muscle (Amiard et al. 2006). The toxic effects of Cr are widely believed to be associated with the stimulation of free radical processes as well as the formation of highly reactive intermediates of Cr^{6+} reduction (Lushchak et al. 2008). Reactive oxygen species (ROS) which are generated during Cr^{6+} degradation exert oxidative stress in cells (Wang et al. 2004). As a result, the biological system induces antioxidants such as superoxide dismutase (SOD), catalase, and glutathione (GSH)-related enzymes to combat the increased levels of ROS. Antioxidant enzymes

* Correspondence: jckang@pknu.ac.kr
[2]Department of Aquatic Life Medicine, Pukyong National University, Busan 608-737, South Korea
Full list of author information is available at the end of the article

serve as excellent biomarkers to study oxidative stress in aquatic organisms. Roberts and Oris (2004) investigated a series of biomarkers in rainbow trout, *Oncorhynchus mykiss*, in response to Cr toxicity. Arillo and Melodia (1998) also pointed out that Cr^{6+} induces alterations in the oxidative function of mitochondria in trout. Furthermore, the intracellular fate of both essential and non-essential metal ions strongly depends on thiol-containing molecules, particularly GSH and MT (Schlenk and Rice 1998). Therefore, the objectives of this study were (a) to investigate Cr^{6+} accumulation in tissues of the mullet *Mugil cephalus*, after waterborne exposure and (b) to investigate the effects of exposure to Cr^{6+} on MT and GSH levels including glutathione *S*-transferase (GST) and SOD activities in the liver and gills to elucidate the cause and effect relationship between metals and antioxidant responses.

Methods

Experimental animals and waterborne exposure test

Mullet *M. cephalus* (weight, 48.25 ± 8.36 g; length, 17.23 ± 0.85 cm) were collected from an aquafarm in Ha-Dong, South Korea, and were acclimatized to laboratory conditions for 4 weeks before experimentation. The water quality parameters measured for the bioassay were as follows: pH, 8.10 ± 0.2; salinity, 33.50 ± 0.6 ‰; DO, 7.14 ± 0.3 mg L^{-1}; and chromium, ≤ 0.1 μg L^{-1}. All experiments were conducted in a seawater temperature of 20 ± 0.5 °C under a 12-h light/12-h dark cycle. Experimental fish were exposed to waterborne treatments of 0, 25, 50, 100, 200, and 400 μg L^{-1} Cr^{6+} concentrations for 4 weeks. Potassium dichromate (Sigma-Aldrich, Inc., USA) was used as the test compound, and the culture water was renewed every 2 days. Fish were anesthetized using benzocaine, and the liver, gill, kidney, spleen, and muscle were sampled and kept at −80 °C until analysis after exposures of 2 and 4 weeks.

Analysis of Cr^{6+} accumulation in tissues

Each tissue was lyophilized to constant weight and then digested using the wet digestion method. Lyophilized organs were dissolved in HNO_3 and re-dried by heating to 120 °C. Final samples were dissolved in 2 % HNO_3 and used for analysis after filtration (Advantec MFS, Inc., 0.2-μm filter). The concentration of total chromium in tissues was measured using an ICP-MS (Elan, PerkinElmer, Inc., USA) equipped with an automatic sampler using argon gas. The accumulation factor (AF) is measured by the following formula: $AF = [Me]_{fw,exp} - [Me]_{fw,control}/[Me]_{water}$, where $[Me]_{fw,exp}$, $[Me]_{fw,control}$, and $[Me]_{water}$ are the metal concentrations in the experimental group, control group, and water, respectively, measured in micrograms per gram (Holwerda 1991).

Analysis of MT and antioxidant enzyme activity

The liver and gill tissues used to determine enzyme activity were rinsed in 0.1 M KCl (pH 7.4) and homogenized (099CK4424, Glas-Col, Germany) in 0.1 M PBS (pH 7.4). The homogenate was centrifuged at $10,000 \times g$ for 60 min (+4 °C), and the supernatant was used for GST and SOD assays. To determine MT and GSH, tissues were homogenized in 20 mM Tris buffer (pH 7.8) containing 0.25 M sucrose. The homogenate was centrifuged at $10,000 \times g$ for 30 min (+4 °C), and the supernatant was used for assays. The supernatant was stored at −75 °C prior to analysis. Protein concentrations were determined using the method of Bradford (1976), with bovine serum albumin as a standard. MT was measured at 412 nm following the method of Viarengo and Nott (1993) using 0.25 M NaCl, 1 N HCl containing 4 mM EDTA, and 0.2 M sodium phosphate solution containing 0.43 mM 5,5-dithiobis-2-nitrobenzoic acid (DTNB). GSH was measured at 412 nm following the method of Srivastava and Beutler (1970) using a precipitation solution containing metaphosphoric acid, disodium ethylene-diamine tetraacetic acid (Na_2EDTA), 0.3 M Na_2HPO_4 solution, and 0.5 mM DTNB. The GSH level was evaluated and determined using the reduced glutathione standard curve. The SOD activity was measured at 450 nm using an SOD Assay Kit-WST (Dojindo Co., Japan) to determine the 50 % inhibitor rate of the reduction of 2-(4-Iodophenyl)-3-(4-nitrophenyl)-5-(2,4-disulfo-phenyl)-2H-tetrazolium, monosodium salt (WST-1). The SOD activity at the 50 % inhibitor rate was expressed as units per milligram protein. GST activity was measured using 0.2 M potassium phosphate (pH 6.5), 10 mM GSH, and 20 μL of 10 mM CDNB. This enzyme assay was according to the methods of Habig et al. (1974) with minor modifications (Anosike et al. 1991). GST activity was determined by an absorbance increase at 340 nm after 5 min and was expressed as nanomoles per minute per milligram protein. For statistical analysis, a one-way analysis of variance (ANOVA) was used followed by Duncan's multiple range tests using SPSS statistical software (SPSS Inc., Chicago, IL, USA). Differences were considered statistically significant when $P < 0.05$.

Results and discussion

Cr^{6+} accumulation in the organs

Cr^{6+} accumulation in the organs of the mullet depending on the exposure time and exposure dose are shown in Fig. 1. As evidenced, the Cr^{6+} accumulation resulted in a net increase of the total Cr^{6+} content in all organs except the muscle with respect to the control. At 400 μg L^{-1}, the liver, kidneys, intestine, and gill showed similar Cr^{6+} contents; however, these four organs statistically differed from the control. The accumulation patterns of Cr^{6+} after 4-week exposure occurred in the following order: kidney

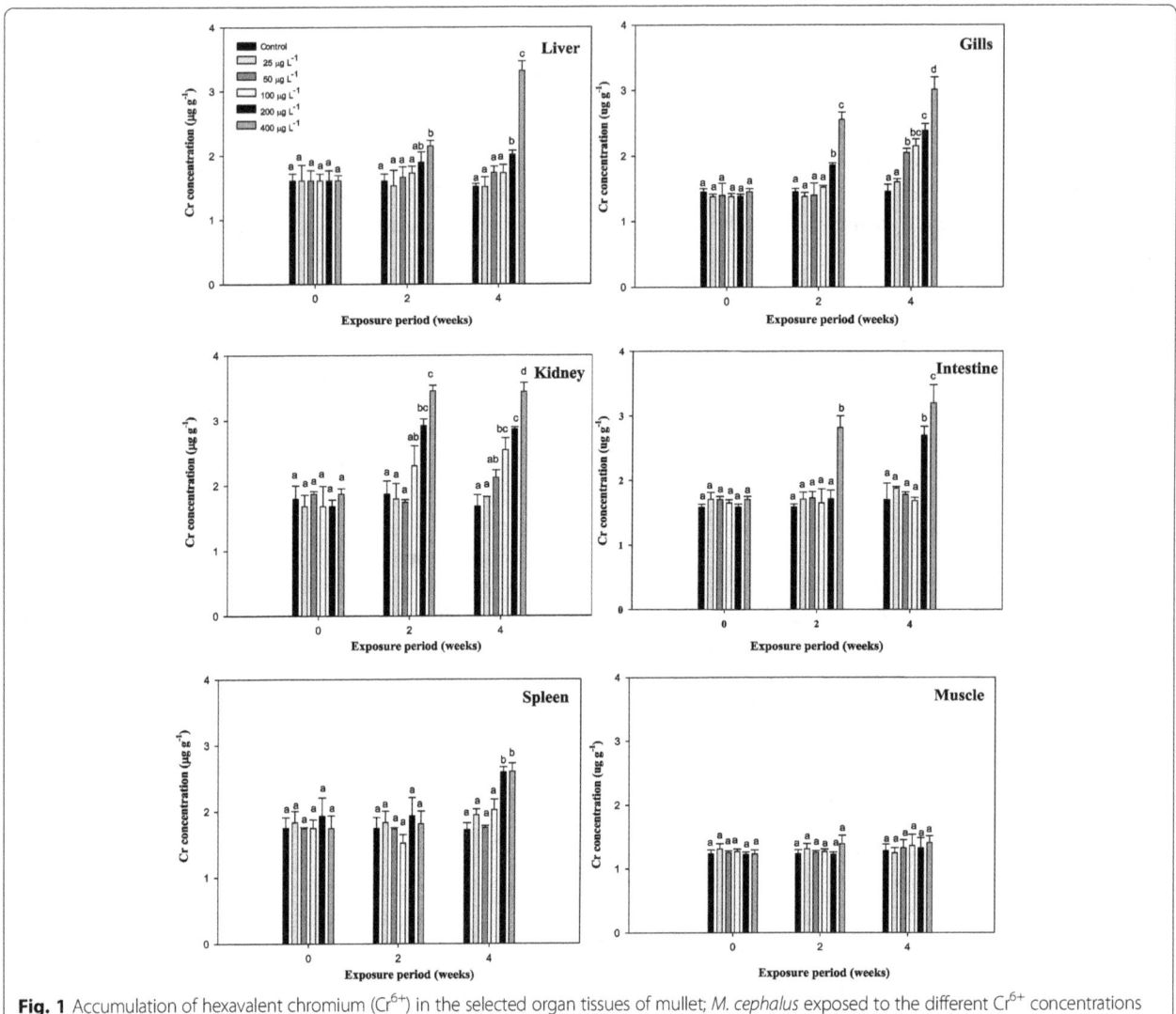

Fig. 1 Accumulation of hexavalent chromium (Cr^{6+}) in the selected organ tissues of mullet; *M. cephalus* exposed to the different Cr^{6+} concentrations

≈ liver > intestine ≈ gill > spleen > muscle. Similar patterns of metal accumulation have also been demonstrated for other aquatic animals (Kim et al. 2004). Metal accumulation in the organs of fish is dependent upon the exposure time and exposure dose as well as other factors, such as temperature, age, interaction with other metals, water chemistry, and metabolic activity of the fish (Campana et al. 2003). Here, Cr^{6+} accumulation in the mullet kidney was higher than that in the liver except at 400 μg L^{-1}, indicating that Cr^{6+} accumulation in the kidney was more effective than that in the liver of the mullet. Similar results have been reported in the freshwater fish *Cirrhinus mrigala*, exposed to chromium (Palaniappan and Karthikeyan 2009); in catfish *Ictalurus punctatus*, exposed to mercuric chloride (Kendall 1977); and in carp (Cinier et al. 1999) and eel (Yang and Chen 1999) exposed to cadmium. Kraal et al. (1995) observed that the accumulation of metal in the kidney was higher than that in the liver during chronic cadmium exposure.

The main location of metal accumulation varies strongly across fish species. In addition, the maintenance of high accumulation in the intestine and kidney has often been observed, as these tissues comprise the principal route of excretion for most toxicants. In the present study, gill tissue contained a substantial amount of Cr^{6+} during the experimental period (Fig. 1). Demirak et al. (2006) and Malik et al. (2010) have indicated that the gills are highly Cr-accumulating organs in fish due to their close relationship with the external environment. On the other hand, the concentrations of Cr^{6+} were lower in the muscles compared to the other organs examined in this study. This result is particularly important because the muscles contribute the greatest mass of the flesh that is consumed as food. Table 1 presents the AFs for various organs after Cr^{6+} exposure in the mullet. The AFs increased with exposure period and were inversely related to the exposure concentration in the organs of the mullet. The AFs were calculated for two major purposes: first, to measure

Table 1 Accumulation factor (AF) over time in liver, kidney, spleen, intestine, gill, and muscle tissues of mullet, *M. cephalus* (mean ± S.E.), exposed to the different concentrations

Tissue	Hexavalent chromium (µg L^{-1})				
	25	50	100	200	400
2 weeks					
Liver	3.56 ± 0.99	2.88 ± 0.49	2.09 ± .0.54	1.885 ± 0.32	1.58 ± 0.25
Kidney	2.52 ± 0.94	3.80 ± 0.12	6.20 ± 0.15	6.22 ± 0.02	4.43 ± 0.25
Spleen	0.08 ± 0.01	0.26 ± 0.05	0.01 ± 0.06	1.04 ± 0.05	0.98 ± 0.05
Intestine	0.24 ± 0.43	0.44 ± 0.30	0.31 ± 0.01	0.45 ± 0.02	2.79 ± 0.54
Gill	0.04 ± 0.01	0.08 ± 0.05	0.69 ± 0.01	2.03 ± 0.07	2.74 ± 0.34
Muscle	0.31 ± 0.03	0.44 ± 0.08	0.39 ± 0.19	0.05 ± 0.00	0.39 ± 0.04
4 weeks					
Liver	3.66 ± 0.01	4.30 ± 0.29	2.11 ± 0.64	2.47 ± 0.13	2.49 ± 0.47
Kidney	2.57 ± 0.02	8.78 ± 0.34	8.68 ± 0.93	6.92 ± 0.07	4.44 ± 0.41
Spleen	0.88 ± 0.03	5.58 ± 0.01	4.78 ± 0.81	4.35 ± 0.15	2.20 ± 0.38
Intestine	0.72 ± 0.09	1.64 ± 0.12	1.69 ± 0.51	4.98 ± 0.28	3.74 ± 0.85
Gill	0.58 ± 0.19	11.84 ± 0.20	6.96 ± 0.51	4.64 ± 0.21	3.90 ± 0.56
Muscle	0.06 ± 0.03	1.88 ± 0.03	1.31 ± 0.08	0.49 ± 0.03	0.44 ± 0.03

how much Cr^{6+} is accumulated with respect to aqueous exposure concentration, and second, to determine the finite limit in the ability of fish to accumulate metals (Sorensen 1991). Similar patterns have also been observed in carp (Cinier et al. 1999), eel (Yang and Chen 1999), and olive flounder after cadmium exposure (Kim et al. 2004).

Metallothionein level and antioxidant enzyme activity

The results of metallothionein (MT) in the mullet are presented in Fig. 2. The liver MT levels were significantly increased compared with the control with exposure period. However, the gill MT level did not show the significance from the control (Fig. 2). We observed that Cr^{6+} concentrations were the highest in the liver of the mullet, which was consistent with the findings of Çogun et al. (2006), who demonstrated that the liver is the main target for heavy metals. The high value of bioaccumulation observed in the liver reflects the affinity of the metal to these tissues. Because the liver is a major producer of metal-binding proteins, MT can be closely related to heavy metal exposure. In accordance with metal residues, Fig. 2 indicates that Cr^{6+} induces hepatic MT in the mullet. Hepatic MT synthesis is induced by cytokines and stress hormones and in fish by bivalent metals, including cadmium, zinc, copper, lead, and mercury. In mullet, MT is predominantly a Cu-binding protein (Linde et al. 2005), which is similar to other conditions of hepatic Cu overload (Klein et al. 1997). In this study, the level of gill MT was not as high as that in the liver and did not significantly differ from the controls in the mullet (Fig. 2). In the rainbow trout *O. mykiss*, hepatic

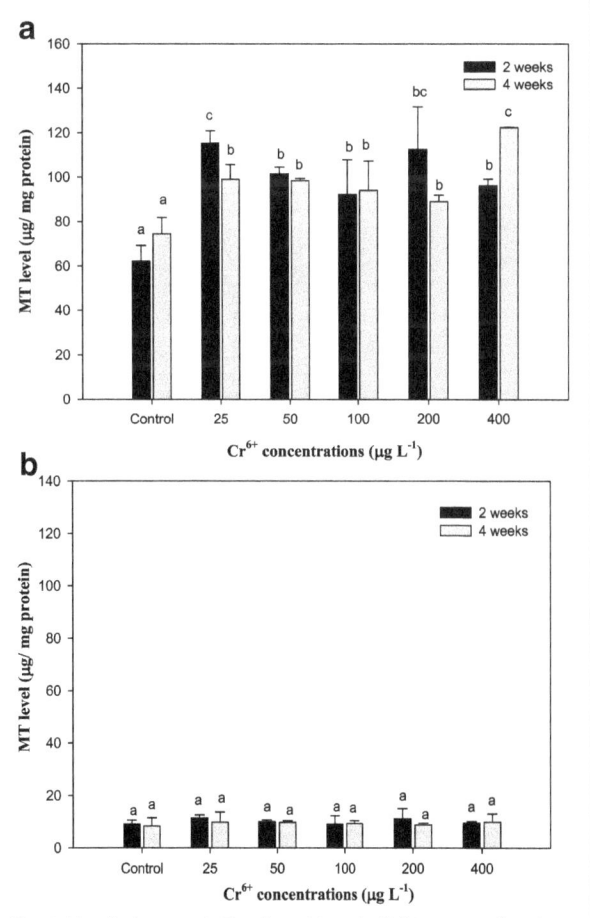

Fig. 2 Metallothionein (MT) in liver (**a**) and gill (**b**) tissues of mullet, *M. cephalus*, exposed to various Cr^{6+} concentrations for a 4-week time period

and gill MT mRNA levels significantly increased over 7 days of Cr^{6+} exposure (Roberts and Oris 2004). These authors reported that although gill MT expression was not as high in the liver, expression occurred much earlier in the experiment (Roberts and Oris 2004). The gill, which lies in direct contact with the water column, would have greater interaction with Cr ions; however, in this study, the hepatic tissues exhibited a much greater capacity for production of MT than the gill. Furthermore, Roberts and Oris (2004) suggested that a greater amount of Cr^{3+} likely reaches the liver than the gill, thus inducing the liver to produce greater amounts of MT. MT has also been recognized as being involved in cellular antioxidant functions (Sato and Bremner 1993).

The capacity for Cr^{6+} conversion to Cr^{3+} in biological systems may be the mechanism of detoxification of Cr^{6+} (Lushchak et al. 2008). The results of reduced glutathione (GSH) in the mullet are reported in Fig. 3. As seen, the GSH level in the liver with Cr^{6+} exposure significantly increased. GSH is considered a first line of cellular defense against metals by chelating and detoxifying the metals, scavenging oxyradicals, and participating in

detoxification reactions catalyzed by glutathione peroxidase (Thomas and Wofford 1984; Potter and Tran 1993; Sies 1999). Sengupta et al. (1990) showed that the acute oral administration of Cr to rats led to an increase of lipid peroxide and decreases in GSH, GST, and SOD of intestinal epithelial cells, whereas a chronic one led to increases in lipid peroxide, SOD, and GPx activity and to a decrease in GST activity. Exposure to heavy metals caused a time- and dose-dependent increase of GSH in various fish species, including mullet (Thomas and Wofford 1984; Lange et al. 2002; Zirong and Shijun 2007; Atli and Canli 2008). Atli and Canli (2008) concluded that the induction of GSH is probably due to the primary defense system for protecting the fish from oxidative stress. Similar results in this study indicated that exposure of the mullet to Cr^{6+} led to a significant increase in hepatic GSH levels, and these changes were associated with the exposure period and concentrations of Cr^{6+} (Fig. 3). However, we observed no changes in GSH in gill tissue. Kubrak et al. (2010) found that the gills of the goldfish *Carassius auratus* treated with Cr^{6+} did not result in any changes in GSH after 96 h and only the liver of goldfish *C. auratus* exhibited an increase in GSH. Consequently, we noted that GSH, including MT, in the mullet gill was not affected by Cr^{6+} treatment. It reflects the mullet liver has more responsibility for the detoxification of toxic effects of Cr^{6+} than the gills.

Glutathione *S*-transferase (GST) activities in the mullet liver and gills exposed to Cr^{6+} for 4 weeks were significantly reduced compared with those in the control as shown in Fig. 4. Similar results have been reported in previous studies addressing the effect of Cr on GST in goldfish *C. auratus* (Lushchak et al. 2009a, b; Kubrak et al. 2010). Elia et al. (2000) reported that high mercury concentrations induced a reduction of GST that was likely responsible for the increased hepatic GSH levels in catfish *Ictalurus melas*. GST is a well-known phase-II enzyme of the metabolism of detoxification, and it conjugates GSH to certain xenobiotic compounds or to their metabolites (Kim and Kang 2015). GST and GSH are important in protecting organisms from oxidative stress, and the fluctuation of GSH in organisms exposed to metals appears to be generally accompanied by variation in GST activity (Paris-Palacios et al. 2000).

Superoxide dismutase (SOD) activity in the mullet exposed to Cr^{6+} for 4 weeks is presented in Fig. 5. Significant changes were observed in the liver in only 400 µg L^{-1} at 2 weeks and 100–400 µg L^{-1} at 4 weeks compared with the control. In the gills, SOD activity was significantly decreased in 400 µg L^{-1} at 4 weeks. Wang et al. (2004) demonstrated that SOD completely abolished ROS generation induced by Cr^{6+}. Roberts and Oris (2004) also suggested that the generation of ROS by the reduction of Cr^{6+} to Cr^{3+} occurred to a greater extent in the gill than the liver

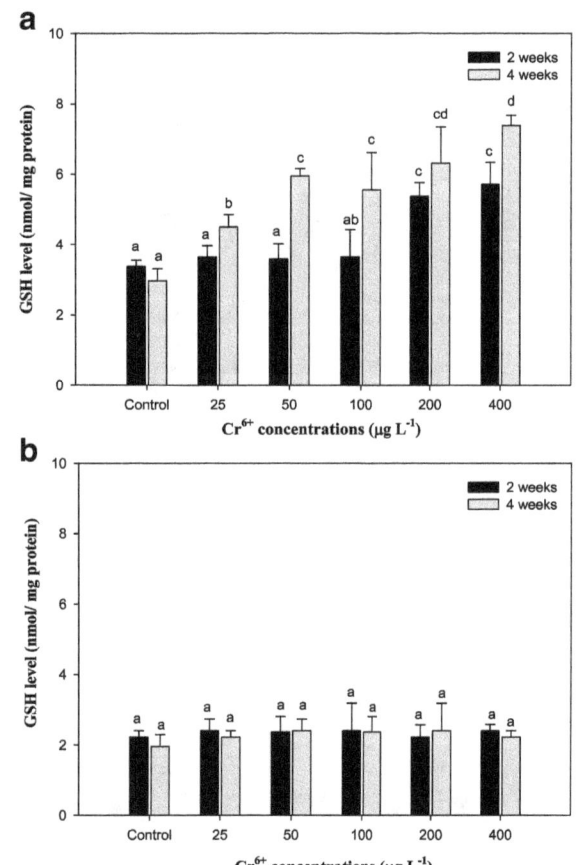

Fig. 3 Reduced glutathione (GSH) level in liver (**a**) and gill (**b**) tissues of mullet, *M. cephalus*, exposed to various Cr^{6+} concentrations for a 4-week time period

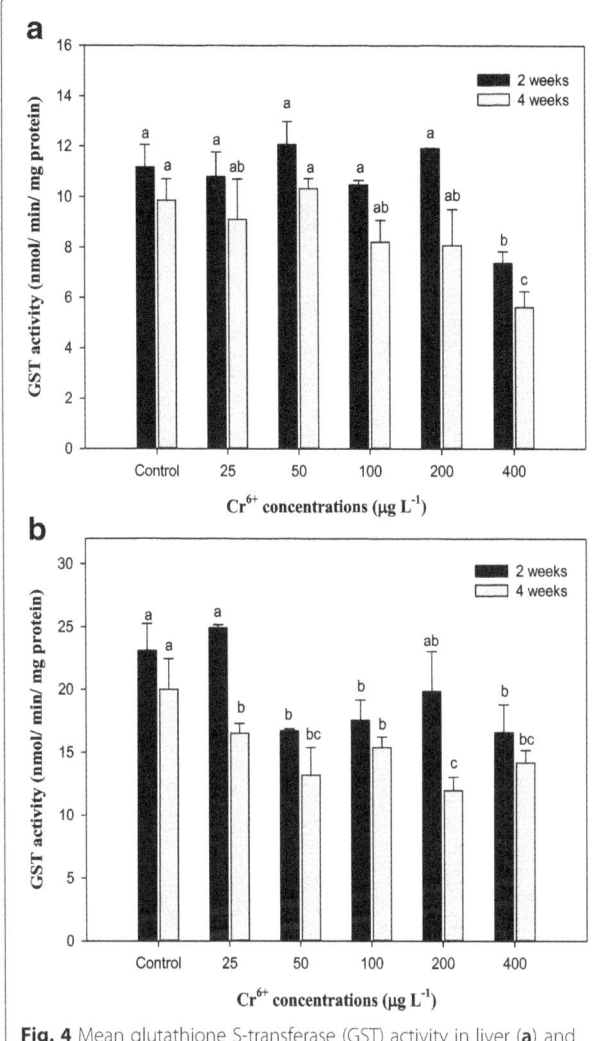

Fig. 4 Mean glutathione S-transferase (GST) activity in liver (**a**) and gill (**b**) tissues of mullet, *M. cephalus*, exposed to various Cr^{6+} concentrations for a 4-week time period

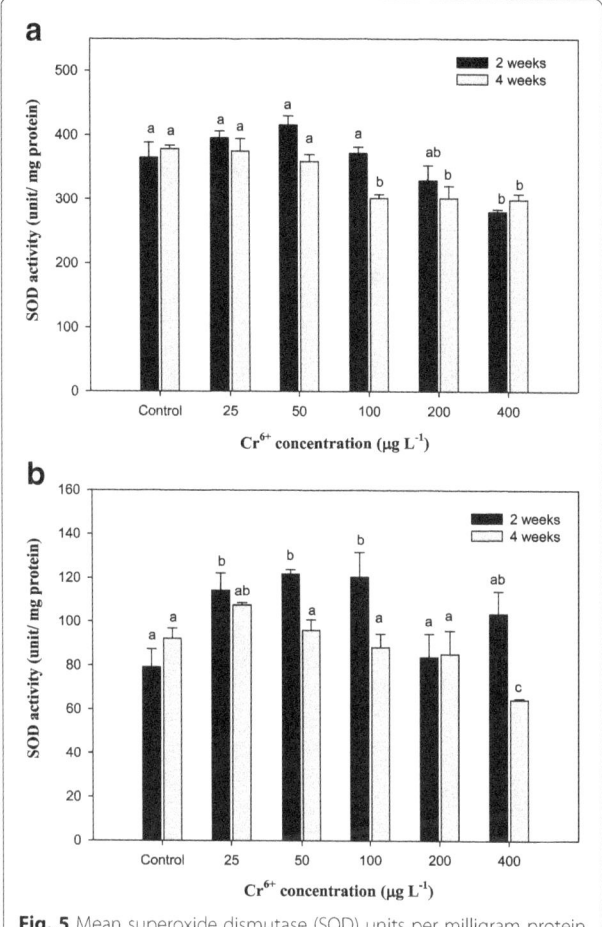

Fig. 5 Mean superoxide dismutase (SOD) units per milligram protein in liver (**a**) and gill (**b**) tissues of mullet, *M. cephalus*, exposed to various Cr^{6+} concentrations for a 4-week time period

and that the gill was more sensitive than the liver to Cr toxicity. In this study, gill SOD activity in the mullet significantly increased at concentrations of 25–100 µg L^{-1}; however, liver SOD activity was significantly reduced at 100–400 µg L^{-1} (Fig. 5). Kubrak et al. (2010) reported that a 48-h exposure to Cr^{6+} reduced SOD activity in the brain of goldfish *C. auratus* and had no effect on liver SOD. Roberts and Oris (2004) demonstrated that SOD activity in rainbow trout, *O. mykiss*, significantly increased in the liver but did not change in the gill after Cr^{6+} exposure.

Conclusions

In conclusion, results from these studies indicate that a high concentration of Cr-mediated oxidative stress could inactivate SOD, although Cr induces a tissue-specific antioxidant response. These results indicated that significant modulation of the activities of these biomarkers occurred in response to Cr^{6+} toxicity. This study highlighted that Cr^{6+} treatment at a level ≥200 µg L^{-1} may affect bioaccumulation and that the AF increased with exposure time in the mullet. The responses of bioaccumulation, MT induction, and GSH to Cr^{6+} exposure were all very likely to be interconnected. Furthermore, this study demonstrated that Cr^{6+} exposure decreased the markers of oxidative stress, SOD and GST, in the mullet liver and gill.

Acknowledgements

This work was supported by a Research Grant of Pukyong National University (year 2015).

Authors' contributions

EYM carried out the experiments with TYA and wrote this article. TYA carried out the experiments. J-CK helped make the design of this experiment and helped write this article. All authors read and approved the final manuscript.

Competing interests

The authors declare that they have no competing interests.

Author details

[1]Institute of Fisheries Sciences, Pukyong National University, Busan, South Korea. [2]Department of Aquatic Life Medicine, Pukyong National University, Busan 608-737, South Korea.

References

Amiard JC, Amiard-Triquet C, Barka S, Pellerin J, Rainbow PS. Metallothioneins in aquatic invertebrates: their role in metal detoxification and their use as biomarkers. Aquat Toxicol. 2006;76(2):160–202.

Anosike EO, Uwakwe AA, Monanu MO, Ekeke GI. Studies on human erythrocytes glutathione-S transferase form HbAA, HbAS and hbSS subject. Biochem Biomed Acta. 1991;50:1051–5.

Arillo A, Melodia F. Effect of hexavalent chromium on trout mitochondria. Toxicol Lett. 1998;44:71–6.

Atli G, Canli M. Response of metallothionein and reduced glutathione in a freshwater fish Oreochromis niloticus following metal exposures. Environ Toxicol Pharmacol. 2008;25:33–8.

Bradford MM. A rapid and sensitive method for the quantitation of microgram quantities of protein utilizing the principle of protein-dye binding. Anal Biochem. 1976;72:248–54.

Campana O, Saraquete C, Blasco J. Effect of lead on ALA-D activity, metallothionein levels, and lipid peroxidation in blood, kidney, and liver of the toadfish Halobatrachus didactylus. Ecotoxicol Environ Safe. 2003;55:116–25.

Cinier CC, Petit-Ramel M, Faure R, Garin D, Bouvet Y. Kinetics of cadmium accumulation and elimination in carp Cyprinus carpio tissues. Comp Biochem Physiol. 1999;122C:345–52.

Çogun HY, Yuzereroglu TA, Firat O, Gok G, Kargin F. Metal concentrations in fish species form the Northeast Mediterranean Sea. Environ Monit Assess. 2006;121:431–8.

Demirak A, Yilmaz F, Tuna AL, Ozdemir N. Heavy metals in water, sediment and tissues of Leuciscus cephalus from a stream in southwestern Turkey. Chemosphere. 2006;63:1451–8.

DeForest DK, Brix KV, Adams WJ. Assessing metal bioaccumulation in aquatic environments: The inverse relationship between bioaccumulation factors, trophic transfer factors and exposure concentration. Aquat Toxicol. 2007;84: 236–246.

Elia AC, Dörr AJM, Mantilacci L, Taticchi MI, Galarini R. Effects of mercury on glutathione and glutathione-dependent enzymes in catfish (Ictalurus melas R.). In: Markert B, Friese K, editors. Trace elements—their distribution and effects in the environment: trace metals in the environment. Vol. 4. Amsterdam, Netherlands: Elsevier Science; 2000. p. 411–21.

Habig WH, Pabst MJ, Jakoby WB. Glutathione S-transferase: the first enzymatic step in mercapturic acid formation. J Biol Chem. 1974;249(22):7130–9.

Holwerda DA. Cadmium kinetics in freshwater clams. V. Cadmium-copper interaction in metal accumulation by Anodonta cyngnea and characterization of metal binding protein. Arch Environ Contam Toxicol. 1991;21:432–7.

Kendall MW. Acute effects of methyl mercury toxicity in channel catfish (Ictalurus punctatus) liver. Bull Environ Contam Toxicol. 1977;18(2):143–51.

Kim JH, Kang JC. Oxidative stress, neurotoxicity, and non-specific immune responses in juvenile red sea bream, Pagrus major, exposed to differnet waterborne selenium concentrations. Chemosphere. 2015;135:46–52.

Kim SG, Jee JH, Kang JC. Cadmium accumulation and elimination in tissues of juvenile olive flounder, Paralichthys olivaceus after sub-chronic cadmium exposure. Environ Poll. 2004;127:117–23.

Klein D, Michaelsen S, Sato S, Luz A, Stampfl A, Summer KH. Binding of Cu to metallothionein in tissues of the LEC rat with inherited abnormal copper accumulation. Arch Toxicol. 1997;71:340–3.

Kraal MH, Kraak MHS, De Groot CJ, Davids C. Uptake and tissue distribution of dietary and aqueous cadmium by carp (Cyprinus carpio). Ecotoxicol Environ Saf. 1995;31:179–83.

Kubrak OI, Lushchak OV, Lushchak JV, Torous IM, Storey JM, Storey KB, et al. Chromium effects on free radical processes in goldfish tissues: comparison of Cr(III) and Cr(VI) exposure on oxidative stress markers, glutathione status and antioxidant enzymes. Comp Biochem Physiol. 2010;152(C):360–70.

Lange A, Ausseil O, Segener H. Alterations of tissue glutathione levels and metallothionein mRNA rainbow trout during single and combined exposure to cadmium and zinc. Comp Biochem Physiol. 2002;131C:231–43.

Linde AR, Klein D, Summer KH. Phenomenon of hepatic overload of copper in Mugil cephalus: role of metallothionein and patterns of copper cellular distribution. Basic Clinical Pharma Toxicol. 2005;97:230–5.

Lushchak OV, Kubrak OI, Nykorak MZ, Storey KB, Lushchak VI. The effect of potassium dichromate on free radical processes in goldfish: possible protective role of glutathione. Aquat Toxicol. 2008;87:108–14.

Lushchak OV, Kubrak OI, Lozinsky OV, Storey JM, Storey KB, Lushchak VI. Chromium (III) induces oxidative stress in goldfish liver and kidney. Aquat Toxicol. 2009;92:45–52.

Lushchak OV, Kubrak OI, Torous IM, Nazarchuk TY, Storey KB, Lushchak VI. Trivalent chromium induces oxidative stress in goldfish brain. Chemosphere. 2009;75:56–62.

Malik N, Biswas AK, Qureshi TA, Borana K, Virha R. Bioaccumulation of heavy metals in fish tissues of a freshwater lake of Bhopal. Environ Monit Assess. 2010;160:267–76.

Palaniappan PR, Karthikeyan S. Bioaccumulation and depuration of chromium in the selected organs and whole body tissues of freshwater fish Cirrhinus mrigala individually and in binary solutions with nickel. J Environ Sci. 2009;21:229–36.

Paris-Palacios S, Biagianti-Risbourg S, Vernt G. Biochemical and (ultra) structural hepatic perturbations of Brachydanio reio (Teleostei, Cyprinidae) exposed to two sublethal concentrations of copper sulfate. Aquat Toxicol. 2000;50:109–24.

Potter DW, Tran TB. Apparent rates of glutathione turnover in rat tissues. Toxicol Appl Pharmacol. 1993;120:186–92.

Roberts AP, Oris JT. Multiple biomarker response in rainbow trout during exposure to hexavalent chromium. Comp Biochem Physiol. 2004;138C:221–8.

Sato M, Bremner I. Oxygen free radicals and metallothionein. Free Radic Biol Med. 1993;14:325–37.

Schlenk D, Rice CD. Effect of zinc and cadmium treatment on hydrogen peroxide-induced mortality and expression of glutathione and metallothionein in teleost hepatoma cell line. Aquat Toxicol. 1998;43:121–9.

Sengupta T, Chattopadhyay D, Ghosh N, Das M, Chatterjee GC. Effect of chromium administration on glutathione cycle of rat intestinal epithelial cells. Indian J Exp Biol. 1990;28:1132–5.

Sies H. Glutathione and its role in cellular functions. Free Radic Biol Med. 1999;27:916–21.

Sorensen EM. Cadmim. In: Metal poisoning in fish. Boca Raton: CRC Press; 1991. p. 175–234.

Srivastava SK, Beutler E. Glutathione metabolism of the erythrocyte. Biochem J. 1970;119:353–7.

Thomas P, Wofford HW. Effects of metals and organic compounds on hepatic glutathione, cysteine, and acid-soluble thiol levels in mullet (Mugil cephalus L.). Toxicol Appl Pharmacol. 1984;76(1):172–82.

Viarengo A, Nott JA. Mechanisms of heavy metal cation homeostasis in marine invertebrates. Comp Biochem Physiol. 1993;104C:355–72.

Wang S, Leonard SS, Ye J, Gao N, Wang N, Shi X. Role of reactive oxygen species and Cr(VI) in Ras-mediated signal transduction. Mol Cell Biochem. 2004;255:119–27.

Yang HN, Chen HC. Uptake and elimination of cadmium by Japanese eel, Anguilla japonica, at various temperatures. Bull Environ Contam Toxicol. 1999;56:670–6.

Zirong X, Shijun B. Effects of waterborne Cd exposure on glutathione metabolism in Nile tilapia (Oreochromis niloticus) liver. Ecotoxicol Environ Safe. 2007;67:89–94.

Evaluation of reference genes for RT-qPCR study in abalone *Haliotis discus hannai* during heavy metal overload stress

Sang Yoon Lee[1] and Yoon Kwon Nam[1,2]*

Abstract

Background: The evaluation of suitable reference genes as normalization controls is a prerequisite requirement for launching quantitative reverse transcription-PCR (RT-qPCR)-based expression study. In order to select the stable reference genes in abalone *Haliotis discus hannai* tissues (gill and hepatopancreas) under heavy metal exposure conditions (Cu, Zn, and Cd), 12 potential candidate housekeeping genes were subjected to expression stability based on the comprehensive ranking while integrating four different statistical algorithms (geNorm, NormFinder, BestKeeper, and ΔCT method).

Results: Expression stability in the gill subset was determined as *RPL7* > *RPL8* > *ACTB* > *RPL3* > *PPIB* > *RPL7A* > *EF1A* > *RPL4* > *GAPDH* > *RPL5* > *UBE2* > *B-TU*. On the other hand, the ranking in the subset for hepatopancreas was *RPL7* > *RPL3* > *RPL8* > *ACTB* > *RPL4* > *EF1A* > *RPL5* > *RPL7A* > *B-TU* > *UBE2* > *PPIB* > *GAPDH*. The pairwise variation assessed by the geNorm program indicates that two reference genes could be sufficient for accurate normalization in both gill and hepatopancreas subsets. Overall, both gill and hepatopancreas subsets recommended ribosomal protein genes (particularly *RPL7*) as stable references, whereas traditional housekeepers such as β-tubulin (*B-TU*) and glyceraldehyde-3-phosphate dehydrogenase (*GAPDH*) genes were ranked as unstable genes. The validation of reference gene selection was confirmed with the quantitative assay of *MT* transcripts.

Conclusions: The present analysis showed the importance of validating reference genes with multiple algorithmic approaches to select genes that are truly stable. Our results indicate that expression stability of a given reference gene could not always have consensus across tissue types. The data from this study could be a good guide for the future design of RT-qPCR studies with respect to metal regulation/detoxification and other related physiologies in this abalone species.

Keywords: Reference gene, RT-qPCR, Gene expression study, Abalone *Haliotis discus hannai*, Heavy metal overload

Background

Quantitative reverse transcription-PCR (RT-qPCR) has long been an established tool for the variety of gene expression assays owing to its cost effectiveness, sensitivity, reproducibility, and simplicity. This methodology has been commonly considered as the most reliable technique particularly for the quantitative validation of differentially expressed genes during a designed experimental treatment (Radonić et al. 2004; Udvardi et al.

2008). Several factors affecting RT-qPCR results include variables associated with qualitative and quantitative differences of input RNA templates as well as efficiency and specificity of reactions. For this reason, in most cases, the amplifications of internal controls (normalization controls) in parallel with the gene(s) of interest have been unavoidable to overcome the noise generated by the variables above (Kubista et al. 2006; Bustin et al. 2009).

Classically known housekeeping genes (HKGs) such as cytoskeletal β-actin (*ACTB*), glyceraldehyde-3-phosphate dehydrogenase (*GAPDH*), and 18S rRNA genes have been commonly used as internal standard in RT-qPCR based on the belief that these HKGs do not change their expression levels irrespective of experimental stimulatory

* Correspondence: yoonknam@pknu.ac.kr

[1]Department of Marine Bio-Materials & Aquaculture, Pukyong National University, Busan 48513, South Korea

[2]Center of Marine-Integrated Biomedical Technology (BK21 Plus Team), Pukyong National University, Busan 48513, South Korea

treatments. However, a considerable number of recent studies have also claimed that these HKGs may not always be stable depending on aquatic species; they could be modulated by certain stimulatory treatments with variable expression levels under different physiological conditions (Li & Shen 2013; Taylor et al. 2013; Yuan et al. 2014). Hence, careful validation of internal controls is now highly necessary prior to expression studies for target genes of interest. Different statistical software tools have become available with this purpose, and the potential utility of computational algorithms to select the best candidate reference gene has been broadly agreed upon in various gene expression studies (Vandesompele et al. 2002; Pfaffl et al. 2004; Silver et al. 2006).

Pacific abalone *Haliotis discus hannai* is one of the most commercially important mollusk species in Korean aquaculture domain. Because many abalone farms in Korea have used the sea-cage system in the coastal areas, waterborne toxicants existing in these areas have become serious concerns, affecting various physiologies of growing abalones (Wan et al. 2011; Park & Kim 2013). In particular, heavy metals are one of the critical pollutants in coastal areas, caused by various anthropogenic and/or industrial activities. Due to their high bio-accumulation capability, negative effects of heavy metals are often thought to be widespread at multilevels of the aquatic food chains (Türkmen et al. 2008; Silva-Aciares et al. 2011; Song et al. 2014). Aquatic invertebrates especially including the benthic mollusks having basically sedentary characteristics have been recognized as a useful platform not only to study the effects of metal toxicants on the physiology of marine animals but also to develop an effective early-warning bioindicator of metal pollution in the coastal environments. Heavy metals are also known as potent pro-oxidant effectors, consequently leading the potential deregulation of various cellular events in the exposed organisms (Amiard et al. 2006; Kim et al. 2007). Hence, comprehensive understanding on the relevant gene expression patterns in response to metal exposures would be a fundamental basis to gain a better insight into the protective mechanism of the exposed animals against metal toxicants (Kim et al. 2007; Wang & Rainbow 2005). Undoubtedly, the selection of suitable reference genes should be a prerequisite requirement for launching an expression analysis project regarding heavy metal exposures. However, despite its importance, the evaluation of stable reference genes from this abalone species in relation with heavy metal exposures has been little studied. Based on this need, the objective of this study was to validate different HKGs associated with heavy metal exposures as the initial step in environmental toxicogenomics of this abalone species.

Methods

Abalone specimens

Abalone specimens used in the present study were the stock that had been maintained in the Experimental Fish Culture Station of Pukyong National University (PKNU). During the pre-acclimation period (2 weeks), abalones have been maintained in rectangular ($1 \times 3 \times 0.5$ m = $W \times L \times D$) tanks equipped with protein skimmers Water temperature and dissolved oxygen throughout this period were adjusted to be 20 ± 1 °C and 8 ± 1 ppm, respectively. Abalones were fed with commercial diet for abalones three times a day on an ad libitum basis, and daily water exchange rate was 50 % using 1-μm-filtered seawater.

Experimental heavy metal exposures and tissue sampling

Three sets of heavy metal exposure treatments were performed. First, juvenile abalones (average body weight = 42.5 ± 3.4 g and shell length = 7.2 ± 0.5 cm) were exposed to cadmium (Cd), copper (Cu), or zinc (Zn). Heavy metal chemicals (analytical grade) were used (Sigma-Aldrich, St. Louis, MO, USA). Six individuals were assigned to one of three 50-L tanks containing 40 L seawater supplemented with each heavy metal. The dose strength of each heavy metal (Cd, Cu, or Zn) was 0.5 ppm as a nominal concentration. The non-exposed control group was also made identically except for the heavy metal. Two replicate tanks were prepared for each treatment group including the non-exposed control. Temperature (in-tank temperature) was adjusted to be ranged at 20 ± 1 °C with room air conditioner, and dissolved oxygen was maintained to be 8 ± 1 ppm with an air-diffuser-connected aerator. Exposure duration was 24 h. After 24 h, three individuals were randomly chosen from each replicate tank (i.e., six individuals per treatment) and tissues (gill and hepatopancreas) were surgically removed from each individual abalone. Tissues were immediately frozen on dry ice and stored at −80 °C until used for gene expression assay. Second, abalones (same sized as above) were exposed to different dose strengths (0, 0.1, 0.25, and 0.5 ppm) of Cu for 48 h. The preparation of two replicate tanks each containing six individuals and other tank conditions were the same with those used for the first exposure experiment. Three individuals were sampled from each replicate tank ($n = 6$ per treatment) in order to obtain the two tissue types (gill and hepatopancreas) from each individual. Third, abalones were exposed to Cd (0 or 0.25 ppm) for different durations of 24, 48, and 72 h. Twelve abalones were assigned into either one of four tanks (two replicate tanks for 0 and 0.25 ppm). Again, all other tank conditions were identically prepared with those mentioned above. During exposure treatments, no feed was provided, and the metal was renewed every 24 h with a daily water exchange rate of 20 %. At each time point, three individuals were randomly chosen from each replicate tank (six for Cd-exposed and six for non-exposed groups). Tissue sampling was carried out as described above. During the three exposure experiments, no mortality was observed.

Selection of candidate internal control genes and primer design

Based on searches against both a local transcriptome next-generation sequencing (NGS) database constructed with abalone tissues (unpublished data) and the public NCBI GenBank database (http://www.ncbi.nlm.nih.gov), 12 candidate HKGs were selected. They included cytoskeletal β-actin (*ACTB*), β-tubulin (*B-TU*), elongation factor-1 alpha (*EF1A*), glyceraldehyde-3-phosphate dehydrogenase (*GAPDH*), peptidyl-prolyl cis-trans isomerase B (PPIB), ubiquitin-conjugating enzyme E2 (*UBE2*), and six kinds of ribosomal proteins (*RPL3, RPL4, RPL5, RPL7, RPL7A,* and *RPL8*). Partial nucleotide sequences of the genes selected from the NGS database were confirmed by RT-PCR cloning followed by direct sequencing. Correct annotation for each gene from the NGS database was confirmed by BLAST homology search against NCBI GenBank. For each gene, multiple primer pairs were designed considering primer length (20–22 bp), G+C contents (40–60 %), melting temperature (55–65 °C), and amplicon size (100–200 bp). The PCR efficiency (E) for each gene was estimated using the known formula [$E = 10^{(-1/k)} - 1$, with $k = $ slop] (Kubista et al. 2006), based on the standard curve prepared with 4-log serial dilution points of abalone complementary DNA (cDNA) samples. LightCycler 480 Real-Time PCR System (Roche Life Science, Mannheim, Germany) was used for all the qPCR experiments in this study. Based on a series of preliminary evaluations, the optimal pair of primers was selected for each gene for further use in expression stability analysis. The summarized information on the genes and oligonucleotide primers used in the present study is provided in Table 1. The amplification of a single specific product using the best primer pair optimized for each gene was confirmed with end-point RT-PCR (Fig. 1).

Table 1 Summary on housekeeping reference genes and qPCR assay conditions used in this study

Gene	Description	Accession no.	Primer sequence (5'–3')	Amplicon size (bp)	PCR (E)[a]	R^2 value[b]
ACTB	Cytoskeletal β-actin	AY380809.1	FW: GGTATTGTTCTGGACTCTGG	162	1.968	0.996
			RV: GGTGGTGGTGAATGAGTAAC			
B-TU	β-tubulin	EF103431.1	FW: ACATTCACTAGGTGGGGGTA	161	1.976	0.998
			RV: GTACTGACAATGTGGCGTTG			
EF1A	Elongation factor-1 alpha	JX002677.1	FW: GCTCTCTGGAAGTTTGAGAC	165	1.974	0.994
			RV: CTCCTTCGAGATACCAGCTT			
GAPDH	Glyceraldehyde-3-phosphate dehydrogenase	ABO26632.1	FW: ACCGCTACACAGAAGACAGT	177	1.912	0.995
			RV: TACATCAGGTACTGGGACAC			
PPIB	Peptidyl-prolyl cis-trans isomerase B	KP698942	FW: CGAGAAAGCAGGACGAATTG	171	1.977	0.993
			RV: AAGTCCCCTCCTTGGATCAT			
RPL3	Ribosomal protein L3	KP698943	FW: TCATTGCACACACCCAGACT	168	1.911	0.997
			RV: CAATGACCTCATCCTGTTCG			
RPL4	Ribosomal protein L4	KP698944	FW: GCTGCTTCAAGACCGCTTAT	176	2.013	0.992
			RV: TGGCCAGCTTTCTCTGAAAC			
RPL5	Ribosomal protein L5	ABO26701.1	FW: AGATGAGGATGGCAAACCAG	168	1.937	0.996
			RV: TCGCTGCTCTCAGAGTCAAA			
RPL7	Ribosomal protein L7	KP698945	FW: CAAGCTGAACACTCCAAACG	156	1.997	0.994
			RV: TCCACAGCACTGATGTTTCC			
RPL7A	Ribosomal protein L7A	KP698946	FW: GCTGTCGAAAAAGGTTGAGC	165	1.972	0.995
			RV: TGCTTCAAGACAGCGAACTG			
RPL8	Ribosomal protein L8	KP698947	FW: TGGAAACTACGCCACAGTCA	161	1.944	0.995
			RV: GTCCTGCCTTCAACATTGGT			
UBE2	Ubiquitin-conjugating enzyme E2	KP698948	FW: CCAAGCTCTTCTTAGTGCAC	170	1.954	0.997
			RV: CTCCCCACTTCCATCACTTT			

[a]PCR efficiency
[b]Determination coefficient

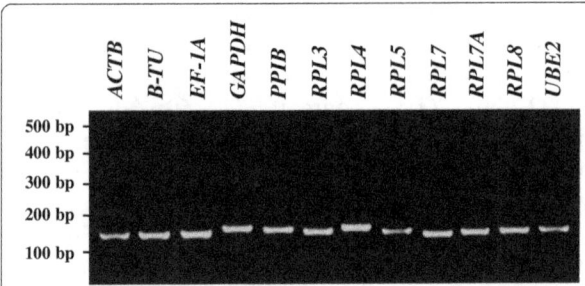

Fig. 1 Representative end-point gel showing the specific amplification of the single PCR band for each of 12 candidate reference genes tested in this study. PCR products were run on 2 % agarose gel and visualized by ethidium-bromide staining. The description for each gene symbol can be referred to in Table 1

RNA purification, cDNA synthesis, and RT-qPCR

Total RNA from either gill or hepatopancreas was extracted using TriPure Reagent (Roche Applied Science) and further purified by using RNeasy Plus Mini Kit (Qiagen, Hilden, Germany) and RNase-Free DNase Set Kit (Qiagen) according to the manufacturers' instructions. The quality and quantity of purified total RNA was checked with Libar S70 spectrophotometer (Biochrom Ltd., Cambridge, UK) at 230, 260, and 280 nm, in which only the RNA samples showing the absorbance values greater than 1.8 for both $A_{260/280}$ and $A_{260/230}$ ratios were considered for use. The integrity of selected RNA sample was validated with 1 % MOPS formaldehyde agarose gel electrophoresis. The absence of genomic DNA contamination in the purified RNA samples was confirmed by no amplification product (the intronic fragment of abalone actin gene) in the PCR reaction without RT reaction (data not shown). An aliquot (2 μg) of total RNA from each sample was reverse transcribed into cDNA using Omniscript RT Kit (Qiagen) according to the manufacturer's recommendation. The RT product (cDNA) was diluted fourfold with DNase-free distilled water and 2 μL was subjected to PCR amplification. Thermal cycling in a reaction volume of 20 μL was performed with Light Cycler 480 System (Roche Applied Science) and SYBR Green I Master (Roche Applied Science) under the following conditions: initial denaturation step (at 95 °C for 10 min) followed by 40 cycles of 20 s at 95 °C, 20 s at 58 °C, and 20 s at 72 °C with an initial denaturation step at 94 °C for 2 min. For each sample, triplicate reactions were performed. The specificity of the amplification product was verified by melting curve analysis with the default setting in the thermal cycler. All the samples were assayed three times for technical replications.

Data subsets and expression stability analysis

Raw quantification cycle (Cq) data obtained were divided into two main datasets on the basis of tissue types

because the transcriptional responses of certain target gene(s) to metal-stimulated treatment should be usually interpreted in a tissue-specific fashion. In the present study, we selected gill and hepatopancreas tissues, the two main organs with respect to the uptake and bioprocess of heavy metal ions in aquatic organisms. In order to evaluate the expression stability of candidate reference genes in either tissue type, the first dataset comprised of gill samples ($n = 84$ including biological replications), while the second dataset consisted of 84 hepatopancreas samples from non-exposed and metal-exposed groups. For each biological sample (i.e., individual cDNA sample) from either tissue types, technical replications of the qPCR assay were made in triplicates to determine the median value of each biological sample, and afterward, the control cycle threshold (Ct) values (i.e., biological replications) within an exposure treatment set were averaged in order to prevent the proportion of data from non-exposed control groups from being too high in the raw datasets.

To evaluate the expression stability of the HKGs tested, the RT-qPCR data was statistically analyzed using different types of software programs, geNorm (qBase-PLUS ver. 3.0) (Vandesompele et al. 2002; Hellemans et al. 2007), NormFinder (ver. 0.953) (Andersen et al. 2004), BestKeeper (Pfaffl et al. 2004), and comparative delta Ct (ΔCT) method (Silver et al. 2006). For geNorm, the lower Cq value (i.e., the maximum expression level) from each gene was used as a calibrator and was arbitrarily set = 1. The geNorm algorithm calculates an expression stability value (M) for each gene and then compares the pairwise variation (V) of this gene with the others. The cutoff levels for M and V values were set to 1.5 and 0.15, respectively. Candidate genes with the lowest M value were considered to be the most stable under tested experimental conditions. The minimum number of reference genes for accurate normalization was also estimated by the V value in the geNorm program. For NormFinder, the same input file format as the geNorm was adopted; however, Cq values were converted into relative quantities after correcting the PCR efficiencies. For BestKeeper, untransformed Ct values and PCR efficiency (E) were used to determine the best suited standards and to combine them into an index by the coefficient of determination and the P value based on the geometric mean of the Cq values. The ΔCT method compares the relative expression of pairs of genes within each sample to confidently identify useful HKGs. Finally, comprehensive ranking to identify the most appropriate control gene was determined using the software RefFinder (http://fulxie.0fees.us/?type=reference&ckattempt=1), in which the four abovementioned software programs were integrated to compare the ranking of the tested candidate reference genes. Based on the rankings from each program, RefFinder assigns an appropriate weight to an individual gene and calculates the geometric mean of their weights for the overall final ranking.

Validation of reference gene selection
Based on comprehensive rankings of the candidate reference genes, the suitability of selected reference genes was validated with the gene expression assay of metallothionein (*MT*) in response to heavy metal exposure treatment. The gill and hepatopancreas cDNA samples both from the Cd-exposed (treated at 0.25 ppm Cd for 24 h) and non-exposed control groups were subjected to qPCR quantification of *MT* transcripts with different normalization genes. A segment of *MT* cDNA (GenBank accession no. KT895222) was amplified by qPCR using a pair of primers (MT-FW: 5′-GGTACCGACTGCAA GTGTAA-3′ and MT-RV: 5′-TCATCGGAAGTCATGT GAGC-3′; amplicon size = 189 bp; E = 1.94) under the amplification conditions as described above for reference genes. The expression level of *MT* transcripts was normalized against the most stable reference gene (ranked to the top; R1), the two most stable reference genes (R1+R2), the second unstable reference gene (R11), or the least stable gene (ranked to the bottom; R12). Expression levels of *MT* transcripts in the gill and hepatopancreas tissues from the Cd-exposed abalones were presented as the fold difference to those from non-exposed control group using the formula $2^{-\Delta\Delta Ct}$ (Schmittgen & Livak 2008) based on the normalization of *MT* transcript levels against the expression levels of its own reference genes (R1, R1+R2, R11, and R12). The effect of the reference gene selection was examined if differential levels of *MT* transcripts could be presented depending on the reference genes selected. Calculated levels of *MT* transcripts in gill and hepatopancreas based on each normalization regime were subjected to one-way ANOVA followed by Duncan's multiple range tests in order to examine statistical significance at $P = 0.05$.

Results
Expression profiles of the candidate reference genes
From technical and biological replicates, the efficiency of RT-qPCR calculated for each candidate reference gene was proven to range from 1.911 (*RPL3*) to 2.013 (*RPL4*) (Table 1). Also, every primer pair used in the present qPCR assays successfully revealed a single peak from the melting curve analysis (data not shown), indicating that the specific amplification for each candidate gene segment could be achieved. Of the 12 candidate reference genes tested, *B-TU* was the least expressed (the highest mean Cq values) gene in the gill subsets, whereas *RPL7* was the most expressed gene in this tissue type. Considering the variation among samples in a given gill tissue,

B-TU, *EF1A*, *GAPDH*, *RPL3*, *RPL5*, and *UBE2* exhibited relatively greater variation in their expression levels than other candidate genes. In the subset of the hepatopancreas also, *B-TU* and *RPL7* showed the least and highest expression levels, respectively, as similarly in the gill, although their absolute mean Cq values were different from those observed in the subset of the gill (Fig. 2).

Expression stability in the subset of abalone gills
Based on the evaluation with four different statistical approaches (geNorm, NormFinder, BestKeeper, and ΔCT method) followed by ranking decision with the RefFinder tool, all the four approaches did not recommend the same reference gene for the most suitable internal control gene (Table 2). According to the geNorm algorithm, the *RPL8* was the most stable candidate gene in the gill, followed by *RPL7* and *ACTB*. On the other hand, geNorm pointed *B-TU* and *UBE2* as the two most unstable HKGs in the gill related with heavy metal exposures. NormFinder also recommended *RPL7*, followed by *RPL8*, for the most stable references while indicated *B-TU* as the most variable gene. According to BestKeeper, *RPL7* and *RPL8* turned out to be the most stable in the gill, whereas *B-TU* and *UBE2* were pointed as the least stable candidates, although all the 12 genes tested showed SD values lower than 1. For the most stable candidate genes, according to the ΔCT method, the *RPL7* gene was the first gene of the rank in the gill, while *B-TU* was the last gene. Collectively, based on the geometric mean-based ranking with RefFinder integrating the four statistical approaches, comprehensive ranking of each candidate gene was provided according to their stability, in which genes with lower ranking were considered to be most stably or invariantly expressed in the gill under heavy metal exposure conditions. As shown in Table 2, the recommended comprehensive ranking of the stability in the gill was determined as *RPL7* > *RPL8* > *ACTB* > *RPL3* > *PPIB* > *RPL7A* > *EF1A* > *RPL4* > *GAPDH* > *RPL5* > *UBE2* > *B-TU*. Overall, the results indicate *RPL7* could be considered as the ideal reference gene in the gill regarding metal exposure experiment in this abalone species, while *UBE2* and *B-TU* would be the worst or unstable references that are not recommended for the use in this case.

Expression stability in the subset of abalone hepatopancreas
The stability ranking of candidate reference genes in the subset for the hepatopancreas was not exactly the same with that ordered in the gill, although the most stable reference gene finally recommended by the RefFinder program was the *RPL7* in both tissue types (Table 3). With the geNorm program, the most stable candidates were *RPL8* and *ACTB* while *B-TU* and *GAPDH* were the

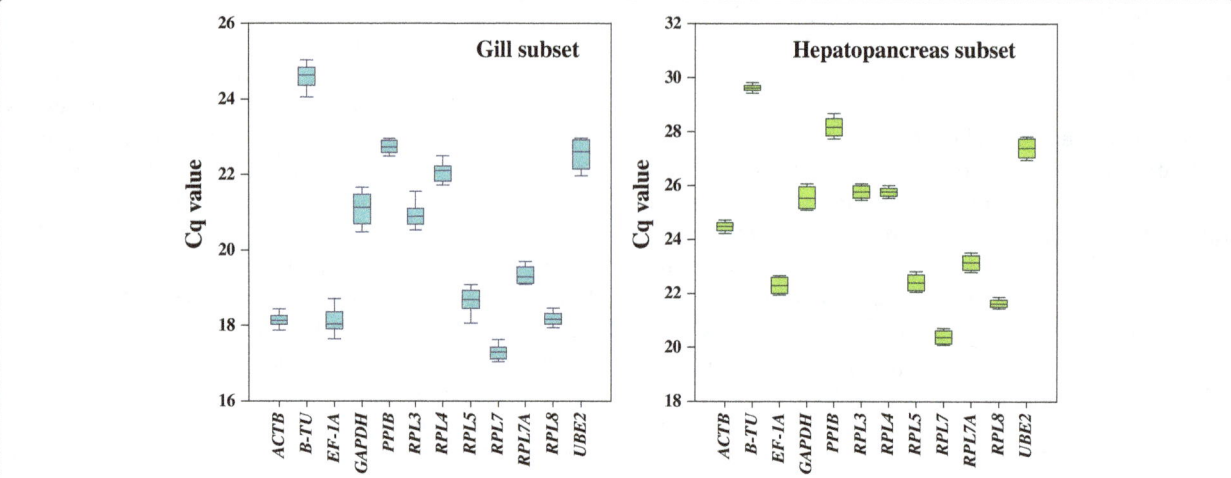

Fig. 2 Expression profiles (Cq values) of the 12 candidate reference genes in the abalone gill and hepatopancreas subsets. In the box plot, the *lower* and *upper boundaries* of each box indicate the 25th and 75th percentiles, respectively. A *line* within the box marks the median value. *Whiskers* above and below the box indicate the 90th and 10th percentiles, respectively

most variable references. NormFinder also pointed *GAPDH* as one of the least stable candidates whereas *RPL7* and *RPL3* were recommended as very stable references. On the other hand, BestKeeper showed slightly different patterns of rankings in the sense that *B-TU*, which was recognized as one of the least stable genes in other algorithms, was recommended as the most stable gene. This highest stability was followed by *RPL8*. However, for the least stable candidate reference genes, the *GAPDH* was called again by the BestKeeper approach. According to the ΔCT method, the most stable candidate was *RPL7* while the most unstable gene was *B-TU*. Taken together, the comprehensive ranking determined by RefFinder for the subset in hepatopancreas

was *RPL7* > *RPL3* > *RPL8* > *ACTB* > *RPL4* > *EF1A* > *RPL5* > *RPL7A* > *B-TU* > *UBE2* > *PPIB* > *GAPDH*.

Pairwise variation analysis using geNorm

With the geNorm software program, the minimum number of reference genes for accurate normalization was determined based on the calculation of the pairwise variation V_n/V_{n+1} between two sequential normalization factors to consider the necessity of adding the next reference gene, in which large variation means that the newly added gene should have a significant effect on the normalization accuracy and hence be added for calculation. Considering the recommended cutoff value of 0.15, the V_2/V_3 value (0.066) observed in the gill criterion was

Table 2 Expression stability ranking of 12 candidate reference genes during heavy metal exposures in the gill of different treatment sets

RefFinder		geNorm		NormFinder		BestKeeper		Delta CT	
Genes	GRV	Genes	SV (M)	Genes	SV	Genes	SD [±CP]	Genes	Ave. SD
RPL7	1.000	RPL8	0.199	RPL7	0.146	RPL7	0.117	RPL7	0.317
RPL8	1.682	RPL7	0.204	RPL8	0.150	RPL8	0.137	RPL8	0.320
ACTB	3.834	ACTB	0.209	RPL3	0.211	ACTB	0.169	RPL3	0.352
RPL3	4.559	RPL7A	0.236	ACTB	0.215	PPIB	0.188	PPIB	0.353
PPIB	4.681	PPIB	0.250	RPL7A	0.218	RPL7A	0.192	RPL7A	0.353
RPL7A	4.729	RPL3	0.266	PPIB	0.219	EF1A	0.199	ACTB	0.355
EF1A	6.735	EF1A	0.280	EF1A	0.227	RPL4	0.203	EF1A	0.360
RPL4	7.7737	RPL4	0.290	RPL4	0.231	RPL3	0.207	RPL4	0.364
GAPDH	9.000	GAPDH	0.304	GAPDH	0.282	GAPDH	0.280	GAPDH	0.397
RPL5	10.488	RPL5	0.329	UBE2	0.376	RPL5	0.316	UBE2	0.465
UBE2	10.719	UBE2	0.352	RPL5	0.408	B-TU	0.321	RPL5	0.479
B-TU	11.742	B-TU	0.392	B-TU	0.539	UBE2	0.341	B-TU	0.593

Table 3 Expression stability ranking of 12 candidate reference genes during heavy metal exposures in the hepatopancreas of different treatment sets

RefFinder		geNorm		NormFinder		BestKeeper		Delta CT	
Genes	GRV	Genes	SV (M)	Genes	SV	Genes	SD [±CP]	Genes	Ave. SD
RPL7	2.340	RPL8	0.059	RPL7	0.023	B-TU	0.100	RPL7	0.103
RPL3	2.991	ACTB	0.060	RPL3	0.047	RPL8	0.127	RPL3	0.113
RPL8	3.364	RPL4	0.061	EF1A	0.055	RPL4	0.129	EF1A	0.115
ACTB	3.600	RPL3	0.074	RPL5	0.058	ACTB	0.133	RPL5	0.117
RPL4	4.409	RPL7	0.079	RPL7A	0.069	RPL3	0.174	RPL7A	0.126
EF1A	4.695	EF1A	0.090	RPL4	0.101	RPL7	0.191	ACTB	0.135
RPL5	5.471	RPL5	0.096	ACTB	0.101	RPL7A	0.202	RPL4	0.136
RPL7A	6.117	RPL7A	0.101	RPL8	0.105	RPL5	0.225	RPL8	0.139
B-TU	6.447	UBE2	0.111	UBE2	0.116	EF1A	0.229	UBE2	0.146
UBE2	9.463	PPIB	0.119	PPIB	0.136	PPIB	0.257	PPIB	0.161
PPIB	10.000	GAPDH	0.127	GAPDH	0.156	UBE2	0.268	GAPDH	0.174
GAPDH	11.242	B-TU	0.139	B-TU	0.185	GAPDH	0.308	B-TU	0.199

much lower than the proposed cutoff value, suggesting that two reference genes are sufficient enough for the precise normalization of gene expression data in the gill samples regarding heavy metal exposures. Pairwise variation in the hepatopancreas subset also indicated that two reference genes could be very sufficient for accurate normalization in relation with the metal exposures in the present study, as evidenced by the V_2/V_3 value of 0.020 (cutoff value = 0.15) (Fig. 3).

Validation of reference genes

Stable or unstable reference genes showed different results in the interpretation of transcriptional responses of the *MT* gene to the Cd exposure (Fig. 4). In the gill, the most stable reference (R1; *RPL7*) and the combination of the two most stable references (R1+R2; *RPL7*+*RPL8* based on the geNorm analysis) resulted in the similar amounts of *MT* transcripts induced during exposure (about 4.4-fold relative to non-exposed control group) ($P > 0.05$). However, when normalized to R11 (*GAPDH*) or R12 (*B-TU*), the calculated amounts of the *MT* transcripts in the gill of Cd-exposed abalones were only less than threefold relative to non-exposed controls ($P < 0.05$). In the hepatopancreas also, differential levels of the *MT* gene expression were visualized depending upon the reference genes selected. Similar with the finding from the gill tissue, the hepatic *MT* expression levels normalized to either R1 (*RPL7*) or R1+R2 (*RPL7*+*RPL3*) were similar with each other (about twofold relative to non-exposed control) ($P > 0.05$). However, the normalization regime using R11 (*PPIB*) or R12 (*GAPDH*) calculated the induced fold of *MT* transcripts in hepatopancreas to be more than fourfold as compared to the basal expression observed in the non-exposed control group ($P < 0.05$).

Discussion

Recently, the necessity to evaluate the stability of reference genes themselves prior to RT-qPCR assay for target genes of interest has been claimed by a number of studies to report that traditional reference transcripts may

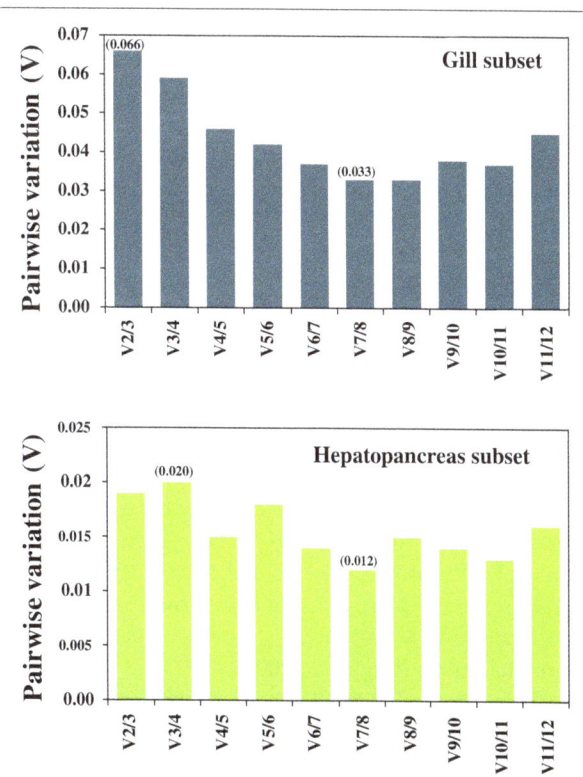

Fig. 3 Pairwise variation (V_n/V_{n+1}) analysis to determine the minimum number of reference genes for accurate normalization in the gill subset or hepatopancreas subset, based on geNorm software. For each subset, the largest and smallest V values are indicated in the *parentheses*

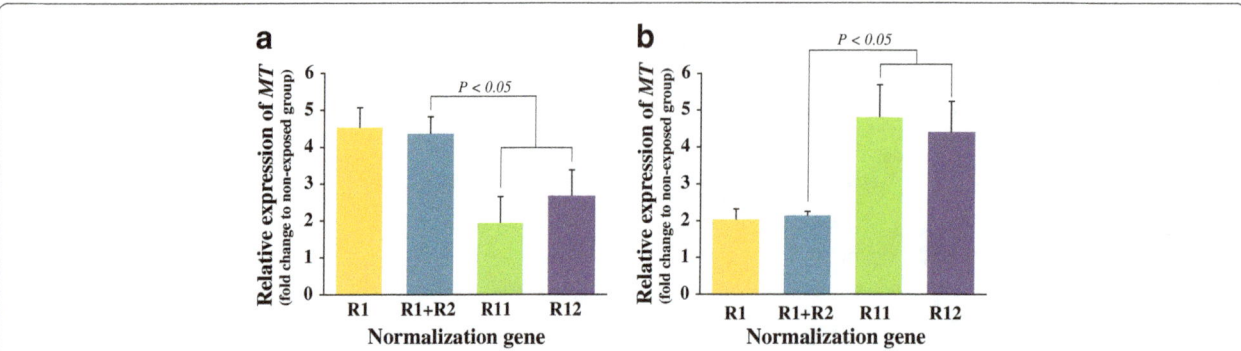

Fig. 4 Validation of reference gene selection with the quantitative RT-qPCR assay of metallothionein (*MT*) transcripts in the gill (**a**) and hepatopancreas (**b**) subsets. *MT* expression levels in the Cd-exposed group were normalized against the most stable reference gene (R1), the two most stable references (R1+R2), the second least stable reference (R11), and the most unstable reference gene (R12). The statistical difference among the expression levels of *MT* transcripts calculated with different normalization genes is indicated based on ANOVA at $P = 0.05$

not always show constitutive expression as has often been assumed. It has already been widely agreed that no given single reference gene can be universally applied to all the experimental conditions and the utility of certain candidate reference genes could significantly vary among biotic and abiotic parameters including species, developmental stages, tissue types, and kinds of experimental treatments (Guénin et al. 2009; De Santis et al. 2011). For this reason, the case-specific choice of best reference gene for RT-qPCR may be unavoidable in many situations. In our case, we aimed to validate suitable reference genes from two subsets of tissue types (gill and hepatopancreas) in abalone regarding heavy metal exposures. Regarding gene expression studies of abalone tissues in response to heavy metal exposures, the gill is one of the most important biological barriers between internal and external environments of aquatic animals (Wang & Rainbow 2005; Hwang et al. 2011), and the hepatopancreas is the main organ responsible for deposits and metabolic detoxification of trace metals in aquatic invertebrates (Rainbow 2007). In this study, we performed two sets of heavy metal exposures (i.e., exposure to different heavy metals, Cu, Zn, Cd, and exposure to different doses of a given metal, Cu), and the raw RT-qPCR data were pooled into two subsets of tissue types (gill and hepatopancreas) in order to evaluate the expression stability of candidate reference genes in a tissue-specific fashion. The exposure conditions similar with those in the present study have already been proven to induce pro-oxidant stresses in abalone species belonging to Haliotidae (Kim et al. 2007).

For the reliable RT-qPCR, the high efficiency and specificity of PCR should be fulfilled. In our case, the calculated PCR efficiency (E) value for each candidate gene tested was at least higher than 90 %. Recommended efficiency values vary between 90 and 110 % (Schmittgen & Livak 2008; Doak & Zair 2012), and high E has been reported to be usually correlated with robust and precise

interpretation of the gene expression (Bustin et al. 2009). The presence of a single peak for melting curve analysis with each PCR reaction also suggests that the specific amplification is carried out in the present conditions. Consequently, our RT-qPCR assay conditions comply well with typically known recommendations.

In the present study, we choose a total of 12 candidate reference genes including traditionally used genes (i.e., *ACTB* and *GAPDH*) but excluded the 18S rRNA, one of the most popular reference genes. The reason of exclusion was that 18S rRNA is not a poly(A+)-tailed RNA; therefore, the random priming for RT reaction is, at least in part, mandatory and the oligo-d(T) priming may not be applicable to this RNA type. In this case, due to its extremely high abundance, signal saturation obtained at very low Ct values might often be problematic unless template cDNA for the samples analyzed with 18S primers were diluted at least a hundredfold relative to all other candidate genes. Further, such a dilution procedure may cause a risk to introduce a random element of variability (Le et al. 2012). Alternatively, the 18S rRNA template could be prepared during an oligo-d(T)-primed RT reaction by adding an 18S rRNA-specific reverse primer at a very low final concentration (also called Co-RT method) (Zhu & Altmann 2005). However, in this case also, potential variability related with differential reverse transcription efficiencies between poly(A+)-tailed mRNA and abundant 18S rRNA should be taken into account. Although 18S rRNA has still been one of the versatile calibrators in many gene expression studies, many other studies also have argued that 18S rRNA should be considered as one of the not-recommended references in organisms belonging to a wide array of taxonomic positions (Taylor et al. 2013; Yuan et al. 2014; Shen et al. 2010). Twelve candidates selected in the present study are assigned to several categories of functional ontology such as cytoskeletal cell shape and mobility (*ACTB* and *B-TU*), energy metabolism (*GAPDH*), translation and protein synthesis (*EF1A* and various *RPLs*), and protein metabolism and degradation (*UBE2*). Such a selection regime might also be of importance for adopting software programs and interpreting outputs, since the algorithms used by each program differ from one another. For example, the geNorm usually requires a gene number ≥7 for a more reliable estimation and has the potential to detect co-regulated genes because its estimate is based on the pairwise comparison of the similarity of all the reference genes concerned (Vandesompele et al. 2002). On the other hand, the NormFinder calculates the variations within groups and between groups and it is less likely to detect co-regulation (Andersen et al. 2004). According to these, for more reliability, enough number of reference genes potentially belonging to different functional categories is recommended, and our selection regime is congruent with this requirement.

From expression stability tests, the two subsets of tissue types (gill and hepatopancreas) for abalones under either metal-exposed or non-exposed conditions shared a similar ranking pattern as judged by RefFinder in the sense that both subsets recommended equally the *RPL7* as the most stable candidate gene while pointing the traditional candidate *GAPDH* as a very unstable (second most unstable and the first most unstable in the gill and hepatopancreas, respectively) reference gene. Our finding on the high stability of *RPL7* reference genes is in accordance with previous reports with aquatic animals including oyster (*Crassostrea gigas*; Du et al. 2013), medaka (*Oryzias latipes*; Zhang & Hu 2007), and halibut (*Hippoglossus hippoglossus*; Øvergård et al. 2010), although the *RPL7* gene has been little recommended in abalone species so far. Meanwhile, the other ribosomal protein gene *RPL5* has been recommended as the most stable gene in different-sized groups of red abalone (*Haliotis rufescens*; López-Landavery et al. 2014) and also disk abalone (*Haliotis discus discus*) broodstock exposed to xenobiotics (Wan et al. 2011). In the present study, *RPL5* showed a clear tendency of tissue dependency in its stability during metal exposures where it was ranked as one of the least stable candidates in the gill; in contrast, it was the third most stable gene in the hepatopancreas. It suggests the importance of validating reference genes according to not only the experimental conditions but also tissue types under evaluation. *GAPDH*, the unstable gene with substantial variations in this study, encodes a known glycolytic enzyme responsible for carbohydrate metabolism. Due to its classical housekeeping role, *GAPDH* mRNA expression has been commonly recognized as an invariant internal standard. However, from the last decades, substantial claims have arisen to report that *GAPDH* is no longer a classical HKG that is involved solely with carbohydrate metabolism. Instead, GAPDH protein has been now recognized

as a multiplayer associated with diverse cellular pathways especially including stress response, apoptosis, and innate immunity (Cho et al. 2008; Lee et al. 2013). Further, from aquatic animals, the upstream regulatory regions (i.e., 5′-flanking regions) of the *GAPDH* gene have been reported to reveal binding motifs targeted by various transcription factors that have been known to be potentially involved in the stress response pathways and/or innate immunity, which is also congruent with the potential variability of *GAPDH* gene expression during experimental challenges or treatments (Lee et al. 2013). As previously known, heavy metals are a potent inducer for cellular oxidative stress often accompanied with inflammations and apoptotic changes in certain cell types; thus, metal exposure has the potential to modulate *GAPDH* genes in aquatic animals, as evidenced by the mud loach (*Misgurnus mizolepis*; Cypriniformes) (Cho et al. 2011).

On the other hand, *ACTB* was recommended as the third and fourth most stable gene in the gill and hepatopancreas, respectively. *ACTB* is one of the most common references used for gene expression experiments also in aquatic animals; however, the utility as an internal control has been reported to be largely controversial among studies. Notably, a recent study with different abalone species (red abalone *H. rufescens*) reported that *ACTB* was the least stable reference in the hepatopancreas (López-Landavery et al. 2014). The unstable feature of the *ACTB* gene was also reported in this abalone species during bacterial challenge (Qiu et al. 2013). The significantly different basal levels of the *ACTB* transcripts across different tissues in aquatic animals might also be, at least in part, supportive of our finding with this abalone species (Cho et al. 2011; Kim et al. 2008). Meanwhile, other two commonly known HKGs, *B-TU* and *UBE2*, did not receive high stability values in either the gill or hepatopancreas subset in this study, which is different from previous observations with other aquatic animals which report fairly good suitability of these HKGs as internal control standards under various experimental conditions (Fernandes et al. 2008; Zheng & Sun 2011). The importance of appropriate selection of the reference gene(s) for quantitative gene expression analysis regarding the exposure of heavy metal to abalone was clearly verified with the validation experiment using the *MT* gene conducted in this study. *MT* protein is a non-enzymatic multiplayer playing essential roles in metal homeostasis and detoxification, and due to its high inducibility at both the mRNA and protein levels, *MT* has been long recognized as a potential biomarker of heavy metal with aquatic organisms (Mao et al. 2012). In the present study, the amounts of heavy metal (Cd)-induced *MT* expression were differentially calculated depending on selected reference genes having different expression stabilities. Our data potentially suggest that the transcriptional response of *MT* could be either underestimated (as evidenced in the gill; see Fig. 4a) or overestimated (in the hepatopancreas; Fig. 4b) if "wrong" references were used, although the direction of transcriptional response (i.e., upregulation or downregulation) was not affected by the kinds of reference genes tested. Collectively, the overall findings from this study confirm not only that the reference genes chosen under a certain experimental treatment conditions may not be universally applicable to other conditions but also that expression stability of a given reference gene could not have consensus across tissue types.

Conclusions

Twelve potential candidate reference genes for RT-qPCR-based expression studies were evaluated in the gill and hepatopancreas tissues of abalone *H. discus hannai* under heavy metal exposure conditions. Based on multiple statistical algorithms, both gill and hepatopancreas subsets recommended ribosomal protein genes (particularly *RPL7*) as stable reference genes while traditionally known HKGs such as *B-TU* and/or *GAPDH* genes as inappropriate references. Our results also highlight the importance of selecting the suitable reference gene(s) for RT-qPCR studies, considering not only the experimental conditions but also tissue types under evaluation. Data from this study could be a good fundamental basis to guide future design of RT-qPCR studies with respect to metal regulation/detoxification and other related physiological aspects in this abalone species.

Abbreviations

ACTB, cytoskeletal β-actin; *B-TU*, β-tubulin; Cq, quantification cycle; Ct, cycle threshold; *EF1A*, elongation factor-1 alpha; *GAPDH*, glyceraldehyde-3-phosphate dehydrogenase; NGS, next-generation sequencing; *PPIB*, peptidyl-prolyl cis-trans isomerase B; *RPL*, ribosomal protein L isoform; RT-qPCR, quantitative reverse transcription-PCR; *UBE2*, ubiquitin-conjugating enzyme E2

Acknowledgements

The authors thank Mr. Min Soo Gwon and Ms. Eun Jeong Kim for their technical assistances on the management and tissue sampling of experimental abalone specimens.

Funding

This study was supported by the grant from the Golden Seed Project (GSP), Ministry of Oceans and Fisheries, Republic of Korea. The GSP has supported funds for the design of study, preparation of experimental abalones, expression analysis, and interpretation of data described in this paper.

Authors' contributions

SYL carried out the biological stimulatory experiments using abalones and all the gene expression studies. YKN designed the study, performed bioinformatics analysis, and drafted the manuscript. All authors read and approved the final manuscript.

Competing interests

The authors declare that they have no competing interests.

References

Amiard JC, Amiard-Triquet C, Barka S, Pellerin J, Rainbow PS. Metallothioneins in aquatic invertebrates: their role in metal detoxification and their use as biomarkers. Aquat Toxicol. 2006;76:160–202.

Andersen CL, Jensen JL, Ørntoft TF. Normalization of real-time quantitative reverse transcription-PCR data: a model-based variance estimation approach to identify genes suited for normalization, applied to bladder and colon cancer data sets. Cancer Res. 2004;64:5245–50.

Bustin SA, Benes V, Garson JA, Hellemans J, Huggett J, Kubista M, et al. The MIQE guidelines: minimum information for publication of quantitative real-time PCR experiments. Clin Chem. 2009;55:611–22.

Cho YS, Lee SY, Kim KH, Nam YK. Differential modulations of two glyceraldehyde 3-phosphate dehydrogenase mRNAs in response to bacterial and viral challenges in a marine teleost *Oplegnathus fasciatus* (Perciformes). Fish Shellfish Immunol. 2008;25:472–6.

Cho YS, Kim DS, Nam YK. Isoform-specific response of two GAPDH paralogs during bacterial challenge and metal exposure in mud loach (*Misgurnus mizolepis*; Cypriniformes) kidney and spleen. J Fish Pathol. 2011;24:269–78.

De Santis C, Smith-Keune C, Jerry DR. Normalizing RT-qPCR data: are we getting the right answers? An appraisal of normalization approaches and internal reference genes from a case study in the finfish *Lates calcarifer*. Mar Biotechnol. 2011;13:170–80.

Doak SH, Zair Z. Real-time reverse-transcription polymerase chain reaction: technical considerations for gene expression analysis. In: Parry JM, Parry EM, editors. Genetic toxicology: principles and methods, methods in molecular biology. New York: Springer; 2012. p. 251–70.

Du Y, Zhang L, Xu F, Huang B, Zhang G, Li L. Validation of housekeeping genes as internal controls for studying gene expression during Pacific oyster (*Crassostrea gigas*) development by quantitative real-time PCR. Fish Shellfish Immunol. 2013;34:939–45.

Fernandes JMO, Mommens M, Hagen O, Babiak I, Solberg C. Selection of suitable reference genes for real-time PCR studies of Atlantic halibut development. Comp Biochem Physiol B. 2008;150:23–32.

Guénin S, Mauriat M, Pelloux J, Van Wuytswinkel O, Bellini C, Gutierrez L. Normalization of qRT-PCR data: the necessity of adopting a systematic, experimental conditions-specific, validation of references. J Exp Bot. 2009;60:487–93.

Hellemans J, Mortier G, De Paepe A, Speleman F, Vandesompele J. qBase relative quantification framework and software for management and automated analysis of real-time quantitative PCR data. Genome Biol. 2007;8:R19.

Hwang PP, Lee TH, Lin LY. Ion regulation in fish gills: recent progress in the cellular and molecular mechanisms. Am J Physiol Regul Integr Comp Physiol. 2011;301:R28–47.

Kim KY, Lee SY, Cho YS, Bang IC, Kim KH, Kim DS, et al. Molecular characterization and mRNA expression during metal exposure and thermal stress of copper/zinc- and manganese-superoxide dismutases in disk abalone, *Haliotis discus*. Fish Shellfish Immunol. 2007;23:1043–59.

Kim KY, Lee SY, Cho YS, Bang IC, Kim DS, Nam YK. Characterization and phylogeny of two β-cytoskeletal actins from *Hemibarbus mylodon* (Cyprinidae, Cypriniformes), a threatened fish species in Korea. DNA Seq. 2008;19:87–97.

Kubista M, Andrade JM, Bengtsson M, Forootan A, Jonák J, Lind K, et al. The real-time polymerase chain reaction. Mol Asp Med. 2006;27:95–125.

Le DT, Aldrich DL, Valliyodan B, Watanabe Y, Ha CV. Evaluation of candidate reference genes for normalization of quantitative RT-PCR in soybean tissues under various abiotic stress conditions. PLoS One. 2012;7:e46487.

Lee SY, Kim DS, Nam YK. Genomic organization, tissue distribution and developmental expression of glyceraldehyde 3-phosphate dehydrogenase isoforms in mud loach *Misgurnus mizolepis*. Fish Aquat Sci. 2013;16:291–301.

Li R, Shen Y. An old method facing a new challenge: re-visiting housekeeping proteins as internal reference control for neuroscience research. Life Sci. 2013;92:747–51.

López-Landavery EA, Portillo-Lopez A, Gallardo-Escarate C, Rio-Portilla MAD. Selection of reference genes as internal controls for gene expression in tissues of red abalone *Haliotis rufescens* (Mollusca, Vetigastropoda; Swainson, 1822). Gene. 2014;549:258–65.

Mao H, Wang DH, Yang WX. Involvement of metallothionein in the development of aquatic invertebrate. Aquatic Toxicol. 2012;110–111:208–13.

Øvergård AC, Nerland AH, Patel S. Evaluation of potential reference genes for real time RT-PCR studies in Atlantic halibut (*Hippoglossus hippoglossus* L.); during development, in tissues of healthy and NNV-injected fish, and in anterior kidney leucocytes. BMC Mol Biol. 2010;11:36.

Park CJ, Kim SY. Abalone aquaculture in Korea. J Shellfish Res. 2013;32:17–9.

Pfaffl MW, Tichopad A, Prgomet C, Neuvians TP. Determination of stable housekeeping genes, differentially regulated target genes and sample integrity: BestKeeper—Excel-based tool using pair-wise correlations. Biotechnol Lett. 2004;26:509–15.

Qiu R, Sun B, Fang S, Sun L, Liu X. Identification of normalization factors for quantitative real-time RT-PCR analysis of gene expression in Pacific abalone *Haliotis discus hannai*. Chin J Oceanol Limnol. 2013;31:421–30.

Radonić A, Thulke S, Mackay IM, Landt O, Siegert W, Nitsche A. Guideline to reference gene selection for quantitative real-time PCR. Biochem Biophys Res Commun. 2004;313:856–62.

Rainbow PS. Trace metal bioaccumulation: models, metabolic availability and toxicity. Environ Intl. 2007;33:576–82.

Schmittgen TD, Livak KJ. Analyzing real-time PCR data by the comparative CT method. Nat Protoc. 2008;3:1101–08.

Shen GM, Jiang HB, Wang XN, Wang JJ. Evaluation of endogenous references for gene expression profiling in different tissues of the oriental fruit fly *Bactrocera dorsalis* (Diptera: Tephritidae). BMC Mol Biol. 2010;11:76.

Silva-Aciares F, Zapata M, Tournois J, Moraga D, Riquelme C. Identification of genes expressed in juvenile *Haliotis rufescens* in response to different copper concentrations in the north of Chile under controlled conditions. Mar Pollut Bull. 2011;62:2671–80.

Silver N, Best S, Jiang J, Thein SL. Selection of housekeeping genes for gene expression studies in human reticulocytes using real-time PCR. BMC Mol Biol. 2006;7:33.

Song Y, Choi MS, Lee JY, Jang DJ. Regional background concentrations of heavy metals (Cr, Co, Ni, Cu, Zn, Pb) in coastal sediments of the South Sea of Korea. Sci Total Environ. 2014;482–483:80–91.

Taylor DA, Thompson EL, Nair SV, Raftos DA. Differential effects of metal contamination on the transcript expression of immune- and stress-response genes in the Sydney rock oyster, *Saccostrea glomerata*. Environ Pollt. 2013;178:65–71.

Türkmen M, Türkmen A, Tepe Y, Ates A, Gökkus K. Determination of metal contaminations in sea foods from Marmara, Aegean and Mediterranean seas: twelve fish species. Food Chem. 2008;108:794–800.

Udvardi MK, Czechowski T, Scheible WR. Eleven golden rules of quantitative RT-PCR. Plant Cell. 2008;20:1736–7.

Vandesompele J, De Preter K, Pattyn F, Poppe B, Van Roy N, De Paepe A, et al. Accurate normalization of real-time quantitative RT-PCR data by geometric averaging of multiple internal control genes. Genome Biol. 2002;3:1–11. research0034.

Wan Q, Whang I, Choi CY, Lee JS, Lee J. Validation of housekeeping genes as internal controls for studying biomarkers of endocrine-disrupting chemicals in disk abalone by real-time PCR. Comp Biochem Physiol C. 2011;153:259–68.

Wang WX, Rainbow PS. Influence of metal exposure history on trace metal uptake and accumulation by marine invertebrates. Ecotoxicol Environ Saf. 2005;61:145–59.

Yuan M, Lu Y, Zhu X, Wan H, Shakeel M, Zhan S, et al. Selection and evaluation of potential reference genes for gene expression analysis in the brown planthopper, *Nilaparvata lugens* (Hemiptera: Delphacidae) using reverse-transcription quantitative PCR. PLoS One. 2014;9:e86503.

Zhang Z, Hu J. Development and validation of endogenous reference genes for expression profiling of medaka (*Oryzias latipes*) exposed to endocrine disrupting chemicals by quantitative real-time RT-PCR. Toxicol Sci. 2007;95:356–68.

Zheng WJ, Sun L. Evaluation of housekeeping genes as references for quantitative real time RT-PCR analysis of gene expression in Japanese flounder (*Paralichthys olivaceus*). Fish Shellfish Immunol. 2011;30:638–45.

Zhu L, Altmann SW. mRNA and 18S-RNA coapplication-reverse transcription for quantitative gene expression analysis. Anal Biochem. 2005;345:102–9.

PERMISSIONS

All chapters in this book were first published in FAS, by BioMed Central; hereby published with permission under the Creative Commons Attribution License or equivalent. Every chapter published in this book has been scrutinized by our experts. Their significance has been extensively debated. The topics covered herein carry significant findings which will fuel the growth of the discipline. They may even be implemented as practical applications or may be referred to as a beginning point for another development.

The contributors of this book come from diverse backgrounds, making this book a truly international effort. This book will bring forth new frontiers with its revolutionizing research information and detailed analysis of the nascent developments around the world.

We would like to thank all the contributing authors for lending their expertise to make the book truly unique. They have played a crucial role in the development of this book. Without their invaluable contributions this book wouldn't have been possible. They have made vital efforts to compile up to date information on the varied aspects of this subject to make this book a valuable addition to the collection of many professionals and students.

This book was conceptualized with the vision of imparting up-to-date information and advanced data in this field. To ensure the same, a matchless editorial board was set up. Every individual on the board went through rigorous rounds of assessment to prove their worth. After which they invested a large part of their time researching and compiling the most relevant data for our readers.

The editorial board has been involved in producing this book since its inception. They have spent rigorous hours researching and exploring the diverse topics which have resulted in the successful publishing of this book. They have passed on their knowledge of decades through this book. To expedite this challenging task, the publisher supported the team at every step. A small team of assistant editors was also appointed to further simplify the editing procedure and attain best results for the readers.

Apart from the editorial board, the designing team has also invested a significant amount of their time in understanding the subject and creating the most relevant covers. They scrutinized every image to scout for the most suitable representation of the subject and create an appropriate cover for the book.

The publishing team has been an ardent support to the editorial, designing and production team. Their endless efforts to recruit the best for this project, has resulted in the accomplishment of this book. They are a veteran in the field of academics and their pool of knowledge is as vast as their experience in printing. Their expertise and guidance has proved useful at every step. Their uncompromising quality standards have made this book an exceptional effort. Their encouragement from time to time has been an inspiration for everyone.

The publisher and the editorial board hope that this book will prove to be a valuable piece of knowledge for researchers, students, practitioners and scholars across the globe.

LIST OF CONTRIBUTORS

Soo Ji Woo, Hee Young Jang and Joon Ki Chung
Department of Aquatic Life Medicine, Pukyong National University, Busan 608-737, South Korea

Hyung Ho Lee
Department of Biotechnology, Pukyong National University, Busan 608-737, South Korea

Kang-Woong Kim, Kyoung-Duck Kim and Hyon Sob Han
Aquafeed Research Center, National Institute of Fisheries Science (NIFS), Pohang 791-923, Republic of Korea

Mohammad Moniruzzaman, Hyeonho Yun, Seunghan Lee and Sungchul C. Bai
Department of Marine Bio-materials and Aquaculture/Feeds and Foods Nutrition Research Center, Pukyong National University, Busan 608-737, Republic of Korea

Jeong-Ho Kim and Woo-Hwa Nam
Department of Marine Bioscience, Gangneung-Wonju National University, Gangneung, Gangwon 210-702, South Korea

Chan-Hyeok Jeon
East Coast Life Science Institute, Gangneung-Wonju National University, Gangneung 210-702, South Korea

Md Mostafizur Rahman and Sang-Min Lee
Department of Marine Biotechnology, Gangneung-Wonju National University, Gangneung 25457, South Korea

Hyon-Sob Han, Kang-Woong Kim, Kyoung-Duck Kim and Bong-Joo Lee
Aquafeed Research Center, National Institute of Fisheries Science, Pohang 37517, South Korea

Sang-Bo Kim, Na Young Yoon, Kil-Bo Shim and Chi-Won Lim
Food Safety and Processing Research Division, National Institute of Fisheries Science, Busan 46083, South Korea

Heyong Jin Roh, Ahran Kim, Gyoung Sik Kang and Do-Hyung Kim
Department of Aquatic life Medicine, College of Fisheries Science, Pukyong National University, Busan, South Korea

Bom-Bi Ha, Eun-Hee Jo and Seon-Bong Kim
Department of Food Science and Technology/Institute of Food Science, Pukyong National University, 45, Yongso-ro, Nam-gu, Busan 48513, South Korea

Suengmok Cho
Research Group of Innovative Functional Foods, Korea Food Research Institute, Seongnam 13539, South Korea

Seunghan Lee, Mohammad Moniruzzaman, Jinho Bae, Minji Seong, Yu-jin Song and Sungchul C. Bai
Department of Marine Bio-materials and Aquaculture/Feeds and Foods Nutrition Research Center (FFNRC), Pukyong National University, Busan 608-737, Republic of Korea

Bakshish Dosanjh
EWOS Canada Ltd., 7721-132nd Street Surrey, Vancouver, British Columbia, Canada

Jong Kyu Lee
Department of Microbiology, College of Natural Sciences, Pukyong National University, Busan 48513, Republic of Korea

Hak Jun Kim
Department of Chemistry, College of Natural Sciences, Pukyong National University, Busan 48513, Republic of Korea

Jung Nyun Kim and Yang Jae Im
West Sea Fisheries Research Institute, National Institute of Fisheries Science, Incheon 22383, South Korea

Mi Hyang Kim
Korea Inter-University Institute of Ocean Science, Pukyong National University, Busan 48513, South Korea

Jung Hwa Choi
Southeast Fisheries Research Institute, National Institute of Fisheries Science, Tongyeong 53085, South Korea

Jung Nyun Kim and Yang Jae Im
West Sea Fisheries Research Institute, National Institute of Fisheries Science, Incheon 22383, South Korea

Jung Hwa Choi
Southeast Sea Fisheries Research Institute, National Institute of Fisheries Science, Tongyeong 53085, South Korea

Hyun-su Jo
Department of Marine Science and Production, Kunsan National University, Kunsan 54150, South Korea

Young Ho Koh, Hyung Woo Lee and Myung Sook Kim
Department of Biology, Jeju National University, Jeju 63243, South Korea

Kyeong Hun Kim
Interdisciplinary Program of Biomedical Engineering, Pukyong National University, Busan 48513, Republic of Korea

Byoung Kwon Kim, Sung Kyun Kim and War War Phoo
Department of Marine Bio-materials and Aquaculture, Pukyong National University, Busan 48513, Republic of Korea

B. A. Venmathi Maran
Fisheries Science and Technology Centre, Pukyong National University, Goseong-gun, Gyeongsangnam-do 52957, Republic of Korea

Chang-Hoon Kim
Interdisciplinary Program of Biomedical Engineering, Pukyong National University, Busan 48513, Republic of Korea
Department of Marine Bio-materials and Aquaculture, Pukyong National University, Busan 48513, Republic of Korea
Fisheries Science and Technology Centre, Pukyong National University, Goseong-gun, Gyeongsangnam-do 52957, Republic of Korea

Joo-Min Kim
Sajodongaone, 873, Deokpyeong Ro, Sunseong-myeon, Dangjin-Si, Chungcheongnam Do 31759, South Korea

G. H. T. Malintha, G. L. B. E. Gunathilaka, Chorong Lee, Min-Gi Kim and Kyeong-Jun Lee
Department of Marine Life Sciences, Jeju National University, Jeju 63243, South Korea

Bong-Joo Lee
Aquafeed Research Center, National Institute of Fisheries Science, Pohang 37517, South Korea

Jeong-Dae Kim
College of Animal Life Science, Kangwon National University, Chuncheon 24341, South Korea

Jae Hyun Im
Research Cooperation Division, National Institute of Fisheries Science (NIFS), Busan 46083, South Korea

Hyun Woo Gil, In-Seok Park, Cheol Young Choi and Tae Ho Lee
Division of Marine Bioscience, College of Ocean Science and Technology, Korea Maritime and Ocean University, 727 Taejong-ro, Yeong do-gu, Busan 49112, South Korea

Kwang Yeol Yoo
Food, Agriculture, Forestry and Fisheries Examination Division, Korean Intellectual Property Office, Daejeon 35208, South Korea

Chi Hong Kim and Bong Seok Kim
Inland Fisheries Research Institute, National Institute of Fisheries Science (NIFS), Cheongpyeung 12453, South Korea

Mi-Jin Yim, Grace Choi, Jeong Min Lee and Dae-Sung Lee
Department of Applied Research, National Marine Biodiversity Institute of Korea, Seocheon 33662, South Korea

Soon-Yeong Cho
Department of Food Processing and Distribution, Gangneung-Wonju National University, Gangneung 25457, South Korea

Yi Kyung Kim
Department of Marine Biotechnology, Gangneung-Wonju National University, 7 Jukheon-gil, Gangneung, Gangwon-do 25457, South Korea

Md Anisuzzaman, Jeong U-Cheol, Jin Feng, Choi Jong-Kuk, Kabery Kamrunnahar and Kang Seok-Joong
Department of Seafood and Aquaculture Science, Gyeongsang National University, Tongyeong 53064, Republic of Korea

Lee Da-In and Yu Hak Sun
Department of Parasitology, School of Medicine, Pusan National University, Yangsan-si, Gyeongsangnam-do 626-870, Republic of Korea

Pil Joon Kang and Ki Wan Nam
Department of Marine Biology, Pukyong National University, Busan 48513, South Korea

Hong Seab Kim and Jung-Ho Son
Noah Biotech Inc., Cheonan 31035, South Korea

Ki-Hwa Chung
Department of Animal Resources and Technology, Gyeongnam National University of Science and Technology, Jinju 52725, South Korea

Agano J. Makori, Paul O. Abuom, Raphael Kapiyo and Douglas N. Anyona
School of Environment and Earth Science, Maseno University, Kenya

Gabriel O. Dida
School of Public Health, Maseno University, Maseno, Kenya

Sang Yoon Lee and Yoon Kwon Nam
Department of Marine Bio-Materials and Aquaculture, Pukyong National University, Busan 48513, South Korea

Gwang-Yeol Yoo
Chungcheongnam-do Fisheries Research Institute, Boryeong 33508, South Korea

Jeong-Yeol Lee
Department of Aquaculture and Aquatic Science, Kunsan National University, Gunsan 54150, South Korea

Hyo Jae Yu and Jin-Koo Kim
Department of Marine Biology, Pukyong National University, 45, Yongso-roNam-gu, Busan 608-737, South Korea

Eun Young Min
Institute of Fisheries Sciences, Pukyong National University, Busan, South Korea

Tae Young Ahn and Ju-Chan Kang
Department of Aquatic Life Medicine, Pukyong National University, Busan 608-737, South Korea

Yoon Kwon Nam
Department of Marine Bio-Materials & Aquaculture, Pukyong National University, Busan 48513, South Korea
Center of Marine-Integrated Biomedical Technology (BK21 Plus Team), Pukyong National University, Busan 48513, South Korea

Index

A

Ahnfeltiopsis Concinna, 136-140

Algae, 81, 85-86, 90, 93, 104, 128-129, 131-135, 140

Alginate Hydrogel, 46

Analog Food, 46

Angiotensin I-converting Enzyme, 33-34, 38

Anisakis Pegreffii, 17, 23

Antioxidant, 8-9, 33-38, 113, 117-118, 134, 136, 140, 188-189, 191-194

Aquaporins, 119, 127

Axiidae, 78, 80

B

Bacterial Fish Pathogen, 39, 45

Bacterial Infection, 119

Balssaxius Habereri, 78-79

Bangiales, 81-85

Bioaccumulation, 188, 191, 193-194, 196, 205

Blue Light, 39-41, 43-45

C

Calcium Alginate, 46, 49, 52

Coi, 18-20, 23, 136-137, 139-140

Cutlass Fish, 23

D

Decapsulated Artemia, 86-93

E

Earthen Fish Ponds, 148-149, 152

Economic Marine Alga, 136

Enzymatic Hydrolysate, 33, 37

Extruded Pellet, 25, 53-55, 57-58, 86-87, 89, 102, 175

F

Feed, 11-13, 15-16, 18, 25, 27-28, 30-32, 34, 38, 45, 53-58, 86-102, 105, 121, 128-130, 132-135, 149-150, 153-154, 156-157, 169-176, 196

Feeding Rate, 86-87, 94, 134

Fish Disease, 1, 39

Fish Feeds, 102, 149

Fluorescent Elastomer, 103-107, 110-112

G

Galatheidae, 66-67, 76-77, 80

Gene Expression Study, 195

Growth, 1, 8-9, 11-16, 23, 25, 30-32, 38-39, 41, 43-45, 53-55, 57-59, 61, 86-89, 91-104, 110-112, 118, 120, 128-135, 141, 146-157, 167-175, 182, 185, 187

Growth Rate, 11, 13, 15, 39, 53-55, 57-58, 86-87, 89, 91-93, 96-101, 128, 130, 132, 134, 148, 151, 153-156, 169-170, 172-173, 175

Gynogenesis, 141-145

H

Heavy Metal Overload, 195

Hematology, 16, 53, 101

Hexavalent Chromium, 188, 190-191, 194

Hyperlipidemia, 113, 115-116, 118

Hysterothylacium Sp., 17, 20, 22

I

Innate Immunity, 5, 40, 58, 95-96, 100, 158, 168, 203

Interleukin (IL)-10, 128-129, 133

J

Juvenile, 2, 9, 11-16, 26, 28-32, 53-58, 86-102, 104, 110-112, 128-129, 133-135, 157, 169-170, 174-187, 194, 196, 205

K

Korean Fish, 176, 187

L

Largehead Hairtail, 17-19, 21-23

Larvae, 17-20, 22-24, 40, 87, 92-93, 158, 160, 165-166, 168, 170, 176-177, 179-181, 185-187

Leap-2 Isoforms, 158-163, 165-167

Light-emitting Diode, 39-40, 45

Lipopolysaccharide, 1-2, 8-9, 119, 121, 126

M

Marine Medaka, 103-106, 108-111, 124, 127

Marphysa Sanguinea, 86, 91, 93-94

Metallothionein, 188, 191, 194, 199, 202, 205

Misgurnus Mizolepis, 119-120, 123, 127, 158-159, 167-168, 203-205

Miuraea Migitae, 81, 83-84

Moist Pellet, 53-55, 57-58, 97

Molecular Analysis, 18, 22, 136

Morphology, 23, 58, 61-64, 81, 83-84, 105, 136-137, 139, 143, 185

Mrna Expression, 1, 3, 95-96, 100, 125, 127, 131, 164, 166, 203-204

Mud Loach Misgurnus Mizolepis, 119-120, 123, 158, 205

Munididae, 66, 70, 76

N

Nile Tilapia, 12, 16, 26, 31, 57, 111, 148-150, 156, 174, 194

Northern Shrimp, 33-34

Nucleolar Organizing Regions, 141-142, 146

O

Oplegnathus Fasciatus, 11-12, 14, 16, 98, 101, 168, 175, 204

Optimum Protein Level, 11

Oryzias Dancena, 103, 105-111, 121, 127

Osteological Development, 176-177, 180, 186-187

P

Photoinactivation, 39

Physico-chemical Parameters, 148, 150-152, 155-156

Polychaete, 86, 94

Polyinosinic: Polycytidylic Acid, 9, 119, 121

Q

Q10, 169-170, 172-174

R

Rbcl, 81-85, 136-137, 139-140

Readability, 103-104, 107

Reference Gene, 4, 97, 125, 131, 195-196, 199-200, 202-205

Response Surface Methodology, 46-47

Rt-qpcr, 160, 164-166, 195, 198-199, 201-202, 204

S

Saccharina Japonica, 113

Sea Cucumber (apostichopus Japonicus), 128, 135

Sea Tangle, 113-118

Sebastes Koreanus, 176, 179, 181-184, 187

Silver Staining, 141-143, 145-147

Squat Lobster, 66

Stocking Density, 149-150, 155, 157, 169-175

Stress Response, 1, 8, 103-104, 110, 163, 174, 203

Subtropical Fish, 59-60, 62

T

Takifugu Obscurus, 169, 175

Taurine, 95-102

Thermal Hysteresis, 59

Trichiurus Japonicus, 17

Triploid, 141-147

Type Iv Antifreeze Protein, 59-60, 65